Biochemistry and Molecular Biology

USHEFHEALTH

William H. Elliott

Department of Biochemistry, University of Adelaide

Daphne C. Elliott

School of Biological Sciences, Flinders University of South Australia

Biochemistry and Molecular Biology

Oxford   New York   Melbourne

OXFORD UNIVERSITY PRESS

Oxford University Press, Great Clarendon Street, Oxford OX2 6DP

Oxford New York
Athens Auckland Bangkok Bogota Bombay
Buenos Aires Calcutta Cape Town Dar es Salaam
Delhi Florence Hong Kong Istanbul Karachi
Kuala Lumpur Madras Madrid Melbourne
Mexico City Nairobi Paris Singapore
Taipei Tokyo Toronto Warsaw
and associated companies in
Berlin Ibadan

Oxford is a trade mark of Oxford University Press

Published in the United States
by Oxford University Press Inc., New York

© W. H. Elliott and D. C. Elliott, 1997

First published 1997
Reprinted (with corrections) 1997

A catalogue record for this book is available from the British Library

Library of Congress Cataloging-in-Publication Data
Elliott, William H.
Biochemistry and molecular biology /
William H. Elliott, Daphne C. Elliott.
1. Biochemistry.   2. Molecular biology.   I. Elliott, Daphne C.
II. Title   QP514.2.E36   1997   574.8'8—dc20   96-9545 CIP

ISBN 0 19 857794 X (Hbk)
ISBN 0 19 857793 1 (Pbk)

Printed in Hong Kong

Preface

This book is aimed at students in science and medicine with some background knowledge of chemistry who are taking their first substantial university course in biochemistry and molecular biology.

The molecular processes of life present us with a seemingly never-ending succession of chemical mechanisms of almost incredible fascination but the subject can be a daunting one for many students. To a considerable degree this is due to the vast amount of detail that is now available and is often presented. This can effectively obscure the fascination of the subject and may sometimes result in students falling back on to rote learning.

Much of the detail is not essential for the understanding of the basic principles that have dictated the molecular processes of the cell that have been adopted to permit life to exist. Nor, for that matter, is much of the detail outside of their specialist fields of interest usually carried in the heads of professional biochemists.

We have designed this book to be 'student-friendly'—to be understood and enjoyed by students, in a self-learning capacity if necessary, but without treating the subject superficially. It is necessary for students, in the face of the avalanche of information available, to know what is expected of them and, indeed, in introductory courses an important role of lectures can be to define which parts of the comprehensive texts are to be learned. The problem of whether complex structures and other material should be learnt in detail can be a worrying one to students new to the subject. To help in this point we have given suggestions as to which parts of the material should be looked at but not memorized, though no doubt this may be modified by teachers of the subject to suit the particular requirements of individual courses. The method we have adopted permits the presentation of complex material which students should see without raising the frequent, and understandable, 'what do we have to know?' question with which most teachers of the subject will be familiar.

The chapters are arranged to flow on, one to another, in an easily understood classification of topics based on biological function. In arranging the sequence of material, we have as far as possible avoided giving information before it is relevant to a biological function. This saves the reader having to cope with a mass of structures and other information prematurely.

We have, in the metabolic sections, largely confined ourselves to animals (photosynthesis excepted) on the grounds that the same systems in prokaryotes and other organisms, however intrinsically interesting and important, are often variations in detail. Where the differences are of major importance, such as in the genome and gene expression areas, we have dealt with both prokaryotes and eukaryotes. The same is true for those areas where much of the information has been derived from prokaryotes.

We have attempted to include more explanation of concepts than is customary and also to give biological context to the molecular systems being discussed. We feel that the latter is valuable in enabling students to appreciate the purpose of the biochemical systems. Metabolic pathways for the transport, storage, and release of energy supplies are more interesting and acceptable to students when viewed against the physiological need for such processes; the same is true for the details of the immune system. We have also preferred to present topics as biological entities wherever possible. Thus blood clotting, superoxide dismutases, the cytochrome P450 system, etc., are collected into a chapter on life-saving protective systems rather than scattered around as isolated examples of specific protein functions. Similarly, hemoglobin is dealt with in a chapter on gas transport in blood rather than as an example in the protein section.

We have adopted the policy of dealing with metabolism in the first half of the book. This permits a progressive increase in complexity and provides groundwork for the subsequent molecular biology sections, which may facilitate learning by students. The extensive use of cross-referencing will, we hope, make it possible for students to return to individual sections of the book for revision purposes without difficulty.

Our hope is that the use of this book will put students in a position from which they can read more widely and deeply on aspects of the subject without too much difficulty, whether their main interest lies in the biological, chemical, or medical aspects of the subject. Just as importantly we hope that it will make them eager to do so. Biochemistry and molecular biology has exploded with new information during the past years and the pace increases. This situation has resulted in a wealth of journals specializing in short, exceptionally clear, and readable reviews on almost every topic, and regular scientific journals sometimes provide similar 'mini reviews'. These would form an excellent extension of this book for students wishing to go more deeply into particular aspects of the subject. To this end we have provided with each chapter (other than preliminary overview ones) lists of references, mainly to such reviews, classified under different topics within the chapter and with a brief description of the contents of each. The majority of the references are likely to be readily available in most libraries. The lists are more extensive in the molecular biology chapters, reflecting the current intense activity in these areas. Many of the articles have intriguing, informative, and stimulating titles. We suggest that even just browsing through the lists themselves would expand the reader's view of the subject and convey some of its excitement.

We also hope that the availability of a basic text with perhaps more explanation than is customary might to some degree permit teachers to concentrate more on areas of the subject that they judge to be particularly relevant to their individual courses.

Adelaide
April 1996

W.H.E.
D.C.E.

Acknowledgements

We are greatly indebted to the following colleagues, who have reviewed sections or given valuable advice:

Dr M. Abbey, Division of Human Nutrition, CSIRO, Adelaide, South Australia

Dr Jan A. Berden, E. C. Slater Institute, University of Amsterdam, The Netherlands

Dr G. W. Booker, Department of Biochemistry, University of Adelaide, South Australia

Professor M. C. Clark, Department of Biochemistry, University of Tasmania

Dr J. C. Coates, Department of Chemistry, University of Adelaide, South Australia

Professor M. J. C. Crabbe, University of Reading, UK

Professor Dr Karel van Dam, University of Amsterdam, The Netherlands

Dr Alan Daws, School of Biological Sciences, University of East Anglia, UK

Dr J. B. Egan, Department of Biochemistry, University of Adelaide, South Australia

Dr Ian W. Flynn, University of Edinburgh, UK

Dr C. Hahn, Department of Biochemistry, University of Adelaide, South Australia

Dr David Hawcroft, De Montfort University, Leicester, UK

Dr Simon van Heyningen, University of Edinburgh, UK

Dr E. M. J. Jaspars, The Leiden Institue of Chemistry, University of Leiden, The Netherlands

Professor John R. Jefferson, Dept of Chemistry, Luther College, Decorah, Iowa, USA

Professor I. Kotlarski, Department of Microbiology and Immunology, University of Adelaide, South Australia

Professor I. de la Lande, Department of Physiology, University of Adelaide, South Australia

Professor S. Lincoln, Department of Chemistry, University of Adelaide, South Australia

Dr B. K. May, Department of Biochemistry, University of Adelaide, South Australia

Dr G. Mayrhofer, Department of Microbiology and Immunology, University of Adelaide, South Australia

Professor J. F. Morrison, John Curtin School of Medical Research, Australian National University, Canberra, Australian Capital Territory

Professor D. L. Nelson, Department of Biochemistry, University of Wisconsin, Madison, USA

Professor John M. Palmer, Department of Biology, Imperial College of Science, Technology and Medicine, London, UK

Professor P. Rathjen, Department of Biochemistry, University of Adelaide, South Australia

Dr A. Robins, Bresatec Limited, University of Adelaide, South Australia

Dr Arthur C. Robinson, Department of Biological Sciences, Napier University, UK

Professor G. E. Rogers, Department of Biochemistry, University of Adelaide, South Australia

Mrs R. Rogers, Department of Biochemistry, University of Adelaide, South Australia

Professor D. Rowley, Queen Elizabeth Hospital, Adelaide, South Australia

Professor R. Saint, Department of Genetics, University of Adelaide, South Australia

Dr T. Sadlon, Department of Biochemistry, University of Adelaide, South Australia

Dr G. Scroop, Department of Physiology, University of Adelaide, South Australia

Professor R. H. Symons, Waite Agricultural Research Institute, University of Adelaide, South Australia

Professor D. Thomas, Department of Paediatrics, University of Adelaide, South Australia

Prof.dr. P. Van de Putte, Department of Molecular Genetics, Leiden University, The Netherlands

Professor J. Veale, Department of Physiology, University of Adelaide, South Australia

Dr E. W. de Vrind de Jong, The Leiden Institute of Chemistry, University of Leiden, The Netherlands

Dr J. C. Wallace, Department of Biochemistry, University of Adelaide, South Australia

Professor John M. Walker, Division of Biosciences, University of Hertfordshire, UK

Dr P. Wigley, Department of Biochemistry, University of Adelaide, South Australia

Dr R. J. H. Williams, Cardiff Institute of Higher Education, UK

Dr A. F. Wilkes, Ludwig Institute for Cancer Research, Royal Melbourne Hospital, Melbourne, Victoria

Dr J. Wiskich, Department of Botany, University of Adelaide, South Australia

Our especial thanks are due to Dr L. A. Burgoyne, Biological Sciences, Flinders University of South Australia, who patiently read the entire typescript in draft form, and made many valuable comments. We are also grateful to Chris Matthews who introduced us to drafting figures using computer graphics and to David Kerry for carrying out computer literature searches.

We wish to thank the staff at Oxford University Press with whom all of our dealings during the development, editing, and production of this book have been a pleasure.

Finally, but not least, we wish to thank Mrs Ros Murrell who prepared the typescript of this book. Her skill and cheerful patience in coping with many difficult drafts are deeply appreciated.

Brief contents

Contents

List of abbreviations

A   adenine
AA   aminoacyl group
ACAT   acyl-CoA:cholesterol acyltransferase
ACP   acyl carrier protein
ACTH   adrenocorticotropic hormone
ADH   antidiuretic hormone (also called vasopressin)
AIDS   acquired immunodeficiency syndrome
ALA   5-aminolevulinic acid
ALA-S   aminolevulinate synthase
AP   apurine or apyrimidine
APC   antigen-presenting cell
apo   apolipoprotein (apoB, apoE, etc.)
AURE   AU-rich responsive element
AZT   azidothymidine

bp   base pair
BPG   2:3-bisphosphoglycerate
1:3-BPG   1:3-bisphosphoglycerate
BSE   bovine spongiform encephalopathy ('mad cow disease')

C   cytosine
c-   'cytoplasmic', denotes protooncogene (c-*ras*, c-*myc*, etc.)
CAM   calmodulin
CAP   catabolite gene activator protein
cDNA   complementary DNA
CETP   cholesterol ester transfer protein
CoA   coenzyme A (A = acyl)
CoQ   ubiquinone (see also Q, UQ)
CDP   cytidine diphosphate
CRE   cAMP-responsive elements
CREB   CRE-binding protein
CSF   colony-stimulating factors

d-   deoxy (dNMP, dNTP, etc.)
DAG   diacylglycerol
dd-   dideoxy
DHAP   dihydroxyacetone phosphate
DNA   deoxyribonucleic acid
DNase   deoxyribonuclease
dopa   dihydroxyphenylalanine

E'_0 redox potential value
EF   elongation factor
EGF   epidermal growth factor
eIF   eukaryote initiation factor
ER   endoplasmic reticulum

F Faraday constant (96.5 kJ V^{-1} mol^{-1})
FAD   flavin adenine dinucleotide
$FADH_2$ reduced form of FAD
Fc   crystallizable fragment (of immunoglobulin)
Fd   ferredoxin
FFA   free fatty acid
FH_2 dihydrofolate
FH_4 tetrahydrofolate
fMet   formylmethionine
FMN   flavin mononucleotide
FSH   follicle-stimulating hormone

G   'gap' phase of cell cycle
G   guanine
G free energy (Gibbs)
$G^{0'}$ standard free energy (Gibbs)
G-1-P   glucose-1-phosphate
G-6-P   glucose-6-phosphate
GABA γ-aminobutyric acid
Gal-1-P   galactose-1-phosphate
GAS γ-interferon-activated sequence
GRB   growth receptor-binding protein
GSH   reduced glutathione
GSSG   oxidized glutathione

H enthalpy
Hb   hemoglobin
HbO_2 oxyhemoglobin
HDL   high-density lipoprotein
HGPRT   hypoxanthine-guanine phosphoribosyltransferase
HIV   human immunodeficiency virus
HMG-CoA   3-hydroxy-3-methylglutaryl-CoA
Hsp   heat shock protein
HTH   helix–turn–helix (DNA recognition motif)

I	inosine	PG	prostaglandin (PGA, PGE, PGF, PGE_2, etc.)
IDL	intermediate-density lipoprotein	3-PGA	3-phosphoglycerate
IF	initiation factor	PDGF	platelet-derived growth factor
Ig	immunoglobulin (IgG, IgG1, IgA, etc.)	PI	phosphatidylinositol
IGF	insulin-like growth factor (IGFI, IGFII)	PIP_2	phosphatidylinositol-4,5-bisphosphate
IP_3	inositol triphosphate	PK	protein kinase (PKA, PKC, etc.)
IRE	iron-responsive element	PK	pyruvate kinase
IS	insertion sequence	pK_a	the pH at which there is 50% dissociation of an acid
ISRE	interferon-stimulated response elements		
		PKU	phenylketonuria
JAK	type of tyrosine kinase (*Janus kinase*)	PLC	phospholipase C
		PLP	pyridoxal-5′-phosphate
K_a	acid dissociation constant	Pol	DNA polymerase (Pol I, Pol II, Pol III in *E. coli*; Pol α, Pol β, etc., in eukaryotes)
K_{eq}	equilibrium constant of a reaction		
K'_{eq}	equilibrium constant at pH 7.0	PP_i	inorganic pyrophosphate
K_m	Michaelis constant: the substrate concentration at which a Michaelis–Menten enzyme works at half-maximal velocity	PrP	prion protein
		PRPP	5-phosphoribosyl-1-pyrophosphate
		PS	phosphatidylserine
kb	kilobase	PS	photosystem (PS I, PS II)
LCAT	lecithin:cholesterol acyltransferase	Q	ubiquinone (see also CoQ, UQ) in mitochondria; plastoquinone in plants
LCK	light chain kinase		
LDL	low-density lipoprotein		
LH	leutinizing hormone	R	gas constant (8.315 J mol^{-1} K^{-1})
LTR	long terminal repeat	RFLP	restriction fragment length polymorphism
		RNA	ribonucleic acid
MHC	major histocompatibility complex	RNase	ribonuclease
mRNA	messenger RNA	R-5-P	ribose-5-phosphate
MTOC	microtubule organizing centre	rRNA	ribosomal RNA
		Rubisco	ribulose-1:5-bisphosphate carboxylase
N	unspecified base in a nucleotide		
NAD^+	nicotinamide adenine dinucleotide (oxidized form)	S	Svedberg unit (measurement of the size of large structures)
NADH	reduced form of NAD	S	'synthesis' phase of cell cycle
$NADP^+$	nicotinamide adenine dinucleotide phosphate (oxidized form)	S	entropy
		SAM	*S*-adenosylmethionine
NADPH	reduced form of NADP	SH2	domain of GRB (Src homology region 2)
NCAM	nerve cell adhesion molecule	sn	stereospecific numbering
		snRNA	small nuclear RNA
Ⓟ	high-energy phosphoryl group	SOS	'son of sevenless' protein
P450	cytochrome P450	SOS	response in *E. coli* to adverse conditions
P_i	inorganic phosphate	SR	sarcoplasmic reticulum
PAF	platelet-activating factor	SRP	signal recognition particle
PBG	porphobilinogen	SSB	single-strand binding protein
PC	phosphatidylcholine (also called lecithin)	STAT	signal transducer and activator of transcription (protein)
Pc	plastocyanin		
PCR	polymerase chain reaction		
PDGF	platelet-derived growth factor	T	thymine
PDH	pyruvate dehydrogenase	T_3	triiodothyronine
PE	phosphatidylethanolamine (also called cephalin)	T_4	thyroxine
PEP	phosphoenolpyruvate	TAF	TBP-associated factor
PEP-CK	PEP carboxykinase	TAG	triacylglycerol
PFK	phosphofructokinase (PFK_1, PFK_2)	TBP	TATA binding protein

TCA	tricarboxylic acid	UQ	ubiquinone (see also CoQ, Q)
TCR	T cell receptor	UTR	untranslated region (of mRNA)
TFIID	transcriptional factor D for polymerase II	UV	ultraviolet (light)
TPA	tissue plasminogen activator		
TPP	thiamin pyrophosphate	v-	'virus', denotes oncogene (v-*ras*, v-*myc*, etc.)
tRNA	transfer RNA	v	velocity of reaction
tRNA^{Phe}	tRNA specific for phenylalanine (by analogy, tRNA^{Leu}, tRNA^{Met}, etc.)	v_{max}	maximum velocity of reaction
		VLDL	very low-density lipoprotein
tRNA$_f^{Met}$	tRNA for bacterial translation initiation		
tRNA$_i^{Met}$	tRNA for eukaryole translation initiation	X-5-P	xylulose-5-phosphate
TSH	thyroid-stimulating hormone		
UDPG	uridine diphosphoglucose	YAC	yeast artificial chromosome

Part 1

Introduction to the chemical reactions of the cell

Chapter summary

Previous page: Electron micrograph of a
mitochondrion in the process of division.
Mitochondria are self-replicating organelles
within cells with their own DNA, but still
dependent on the rest of the cell for the
synthesis of most of their proteins. Mitochondria
are the main site of oxidative generation of
chemical energy used by the cell.

Photograph: Don Fawcett/Science Photo Library

Chapter 1

Chemistry, energy, and metabolism

Life is a chemical process involving thousands of different reactions occurring in an organized manner. These are called **metabolic reactions** and, collectively, **metabolism**.

In studying biochemistry you will therefore be confronted with a fairly large number of chemical equations describing the essentials of cell chemistry (though not with all of the thousands of reactions). The chemical strategies perfected by millions of years of evolution are elegant and fascinating but to appreciate these, or for that matter to easily understand and enjoy biochemistry, you first need to appreciate the problems that faced the establishment of life.

The consideration underlying everything in life is **energy**. A printed chemical equation on its own lacks a vital piece of information and that is the energy change involved. It may not be immediately obvious what is meant by this, for chemical energy change in reactions occurring in solution is not something of which we are necessarily conscious. Energy considerations determine whether a reaction is possible on a significant scale and whether the reverse reaction can occur to a significant degree.

To give an example of how this aspect applies directly to living processes, during vigorous exercise, glycogen (a polymer of glucose) in muscles is converted to lactic acid by a series of chemical reactions. During the ensuing rest period, lactic acid is converted to glycogen but not by the reverse of exactly the same chemical reactions by which it was formed. With a knowledge of simple energy considerations it is possible to see why the cell does what it does. Without such a knowledge, the chemistry can seem pointlessly complicated.

The level of knowledge of thermodynamics needed to understand biochemistry, in general, is little more than common sense. While there are specialist areas of the subject where a sophisticated knowledge of thermodynamics is needed, it is probably true that the majority of biochemists do not often use a thermodynamic equation, but nevertheless understand the simple principles involved. The aim of this section is to give you this simple understanding.

What determines whether a chemical reaction is possible?

As already implied, energy considerations permit certain chemical reactions to occur while others are not allowed. Note that in biochemistry we are concerned with reactions occurring to a *significant* extent. In principle, energy considerations do not completely prevent a reaction occurring (unless reactants and products are precisely at equilibrium concentrations; see below), but if the reaction ceases after a minute amount of conversion has taken place, then, for the practical purposes of life, it has not occurred.

First, what do we mean by energy change in a chemical system (the latter being an assemblage of molecules in which chemical reactions can take place, such as occurs inside a cell)? This is not self-evident as is the case with, say, gravitational energy change in a weight falling. A chemical system involves a huge number of individual molecules each of which contains a certain amount of energy dependent on its structure. This energy can be described as the heat content or **enthalpy** of the molecule. When a molecule is converted to a different structure in a chemical reaction, its energy content may change; the change in the enthalpy is written as ΔH. The ΔH may be negative (heat is lost from molecules and released, so raising the temperature of the surroundings) or positive (heat is taken up from the surroundings, which, correspondingly, cool).

At first sight it may seem surprising that reactions with a positive ΔH can occur since it might seem analogous to a weight raising itself from the floor. This is the point at which physical analogies such as weights falling become inadequate as models for chemical reactions, In the latter, a negative ΔH favours the reaction and a positive ΔH has the opposite effect. But, in a chemical reaction, ΔH is not the final arbiter as is gravitational energy with a weight system, for entropy change, or ΔS, also has a say in the matter.

Entropy can be defined as the degree of randomness of a system. In a chemical system, this can take three forms: first, a molecule is not usually rigid or fixed—it can vibrate, twist around bonds, and rotate. The greater the freedom to do these things—really, to indulge in any molecular movement—the greater the randomness or entropy. Secondly, in a chemical system, vast numbers of individual molecules are involved, which may be randomly scattered or be in some sort of orderly arrangement as, for example, occurs to a great degree in living cells. Thirdly, the number of individual molecules or ions may change as a result of a chemical change. The greater the number of individual molecules, the greater the randomness; the greater the randomness, the greater the entropy. Thus a chemical reaction may change the entropy of the system. Increasing entropy lowers the energy level of the system; decreased entropy (increased order) increases the energy level.

Both ΔH and ΔS have a say in determining whether a chemical reaction may occur. A negative ΔH and a positive ΔS both reinforce the 'yes' decision; a positive ΔH and negative ΔS both reinforce the 'no' decision. A negative ΔH and negative ΔS disagree on the decision, as do a positive ΔH and a positive ΔS, and whether the outcome is yes or no depends on which is quantitatively the larger.

The driving force of increasing entropy is often illustrated by the example of the melting of a block of ice in warm water. Heat is taken up as the ice melts (the ΔH is positive) but the scattering of the organized molecules of the ice crystal, as it dissolves, increases the entropy so that the process proceeds. It must be admitted, however, that it is not necessarily easy to visualize how entropy drives a chemical reaction. A usual statement is that, if energies are similar, the disordered state occurs more readily than the ordered state. If you find it difficult to conceptualize entropy as a driving force, rather than have a mental block it would be best to accept the concept for the moment; it will become more familiar later.

The situation we have described between ΔS and ΔH is not convenient—we have two terms of variable sizes that may reinforce or oppose each other in determining whether a reaction is possible. Moreover, in biological systems it is difficult or impossible to measure the ΔS term directly. The situation was greatly ameliorated by the **Gibbs free energy** concept that combined the two terms in a single one. The change in free energy (ΔG after Gibbs) is given by the famous equation

$$\Delta G = \Delta H - T\Delta S$$

where T is the absolute temperature.

The term 'free' in free energy means free in the sense of being *available* to do useful work, not free as in something for nothing. ΔG represents the *maximum* amount of energy available from a reaction to do useful work. It is somewhat like available cash being free for you to make purchases with.

Useful work includes muscle contraction, chemical synthesis in the cell, and osmotic and electric work. ΔG values are expressed in terms of calories or joules per mole (1 calorie = 4.19 joules); the former term is used in some books, but the latter is now the official version. Since the values are large, the terms kilocalories or kilojoules per mole are used. The latter is usually abbreviated to $kJ\,mol^{-1}$. The ΔG of a reaction is the all-important thermodynamic term; in its application to chemical reactions, the rule is as follows.

A chemical reaction can occur only if the ΔG is negative—that is, that at the *prevailing conditions* the products have less free energy than the reactants have. That's all.

Reversible and irreversible reactions and ΔG values

You may have learned in chemistry that, strictly speaking, *all* chemical reactions are reversible. This may puzzle you because it might imply that the ΔG value must be negative in both directions—remember that a reaction cannot occur unless the ΔG value is negative. The answer to this apparent paradox is that the ΔG of a reaction is not a fixed constant but varies with the reactant and product concentrations. The relationship is given later in this chapter. Thus in the reaction A↔B, if A is at a high concentration and B at a low one, the ΔG may be negative in the direction A→B and, of course, positive from B→A. Reverse the concentrations and the ΔG can be negative for the reverse direction. A reaction will proceed to the point at which A and B are in concentrations at which the ΔG is zero in both directions and then no further net reaction can occur. This is the **chemical equilibrium point**.

If the ΔG of a biochemical reaction A ↔ B is small, significant reversibility may be possible in the cell because changes in reactant concentrations in the cell may be sufficient to reverse the sign of the ΔG of the reaction. If it is large, for all practical purposes the reaction is irreversible. In cellular reactions there is relatively little scope for concentration change in the **metabolites** (as reactants and products are called)—the concentrations are always relatively low: 10^{-3}–10^{-4} would be typical of many. The net result is that reactions with large negative ΔG values are irreversible because concentration changes are insufficient to reverse the sign of those values. (We later explain why, in certain reactions, this may appear to be contradicted; see page 110.)

As a general guide, hydrolytic reactions in the cell—reactions in which a bond is split by water—are irreversible, in the sense that synthesis of substances in the cell does not occur by reverse of the hydrolytic reactions. As you proceed through the chapters on metabolism you will become familiar with which reactions are reversible and which irreversible.

To summarize, a reaction with a small ΔG is likely to be reversible in the cell, the direction being determined by small changes in metabolite concentrations. A reaction with a large ΔG value will, in cellular terms, proceed in one direction only and, moreover, will proceed to virtual completion because the equilibrium point is so far to the side of the reaction. Put in another way, in the latter case, the ΔG of the reaction does not become zero until virtually all of the reactant(s) have been converted to product(s).

The importance of irreversible reactions in the strategy of metabolism

In the preceding section we have emphasized the effect of the ΔG of a reaction in determining whether that reaction can proceed in the reverse direction. Why is there such emphasis on reversibility? We will now explain this. The major physiological chemical processes of the cell usually involve, not single reactions, but series of reactions organized into metabolic pathways in which the products of the first reaction are the reactants of the next and so on. In the example of the glycogen → lactic acid conversion in muscle mentioned earlier, a dozen successive reactions are involved.

An important general characteristic of metabolic pathways is that they are, as a whole, irreversible. Many of the individual reactions in a pathway may be freely reversible but it virtually always contains one or more reactions that cannot be directly reversed in the cell. Such irreversible reactions act as one-way valves and ensure that from a thermodynamic viewpoint the pathway can proceed to completion. Metabolic pathways are decisive—they go with a bang as it were.

All this is not the same as saying that overall physiological chemical processes are irreversible. Lactic acid *is* converted back to glycogen in the body and, while many steps in the process are the simple reversal of those in the forward process, there are steps that cannot be reversed directly and alternative reactions are necessary. These involve the input of energy, which makes the alternative reactions also irreversible, but in the opposite direction. So the forward pathway (glycogen → lactic acid) is directly irreversible as in the reverse pathway (lactic acid - → glycogen). A typical metabolic situation is

$$A \rightleftharpoons B \;\circlearrowleft\; C \rightleftharpoons D \rightleftharpoons E \;\circlearrowleft\; F \rightleftharpoons \text{Products}.$$

The red arrows represent irreversible reactions with large negative ΔG values. You will want to know how energy is provided to achieve irreversibility in both directions. This will become clear later.

A general biochemical principle emerges. *Whenever the overall chemical process of a metabolic pathway has to be reversed, the reverse pathway is not exactly the same as the forward pathway—some of the reactions are different in the two directions.*

Why is this metabolic strategy used in the cell?

There are basically two reasons. The alternative to the strategy we have outlined is that *all* the reactions of a metabolic pathway are reversible, that is,

$$A \rightleftharpoons B \rightleftharpoons C \rightleftharpoons D \rightleftharpoons E \rightleftharpoons F \rightleftharpoons \text{products}.$$

The major drawback of this arrangement is that the whole process is subject to mass action. If the concentration of A increases (perhaps due to something you ate), the reaction would swing to the right and more products would be formed. If substance A decreased in amount, some of the products would revert to A to maintain the equilibrium. Imagine that the pathway is for the synthesis of molecules such as the DNA of genes or of vital proteins, etc. and you can see how impossible this scenario is. It would be rather like building the walls of a house with the laying of bricks being a reversible process. The walls would rise and fall to maintain a constant equilibrium with the number of bricks lying on the ground. In the cell, if the concentration of A were to rise, more DNA and protein would be synthesized; if it were to fall, the DNA or proteins would break down—a totally unworkable situation for a living cell.

There is a second, important but related reason for having dissimilar reactions in the forward and reverse direction of metabolic pathways. Metabolism must be controlled. As already stated, during violent exercise muscles convert glycogen to lactic acid. During rest, the lactic acid is converted to glycogen and the forward conversion switched off. To independently control the two directions (that is, to switch one on and the other off) there must be separate reactions to control; otherwise it would be possible only to switch both directions on and off together. Thus, the irreversible reactions are usually the control points. Metabolic control is a major subject that we will deal with in Chapter 12.

How are ΔG values obtained?

The ΔG value of a reaction, as explained is *not* a fixed constant but the ΔG value of a reaction under specified standard conditions is a fixed constant. The standard conditions are with reactions and products at $1.0\,M$, $25°C$, and $pH\,7.0$, that is, the free energy difference between separated one-molar solutions of reactants and products. The ΔG value, *under these conditions*, is called the **standard free energy change of a reaction**. It is denoted as $\Delta G^{0'}$, the prime being used in biochemical systems to indicate that the pH is 7.0 rather than the value at $pH\,0$, which is used in physical sciences.

The $\Delta G^{0'}$ value may often be calculated from physical chemistry tables. In these, the standard free energies of formation of large numbers of compounds are listed. For many reactions, the $\Delta G^{0'}$ value can be calculated by adding up the free energies of formation of the reactants and (separately) those of the products. The difference is the $\Delta G^{0'}$ value. An alternative is to determine experimentally the equilibrium constant for the reaction and from this the $\Delta G^{0'}$ value is easily calculated.

There is a simple direct relationship between $\Delta G^{0'}$ values and ΔG values at given reactant and product concentrations. If, therefore, we know the relevant metabolite concentrations in the cell, the ΔG value for the reaction in the cell is readily determined (see page 12 for an illustrative calculation). This has been done for a good many biochemical reactions so that such values are often quoted.

There is a snag—determining the actual concentrations of the thousands of metabolites in a cell is a nontrivial matter. They are present at low concentrations and are changing anyway. So, for many biochemical reactions, we do not have this data and therefore do not have the ΔG values. However, it is found that $\Delta G^{0'}$ values usually correlate very well with known cellular happenings so that they are a useful guide in understanding metabolic reactions. Thus, although such values are not directly applicable to cells since metabolite concentrations are never 1.0 M, they are frequently quoted to explain why certain reactions behave as they do. It is a compromise but a very useful one.

Standard free energy values and equilibrium constants

A particularly useful aspect of knowing the $\Delta G^{0'}$ value of a reaction is that, from it, the **equilibrium constant** of that reaction is readily determined. The equilibrium constant K'_{eq} of a reaction represents the ratio at equilibrium of the products to reactants. (The prime indicates that it is the K_{eq} at pH 7.0.) Thus, if one considers the reaction $A + B \leftrightarrow C + D$, the K'_{eq} is calculated from the concentrations of A, B, C, and D present after the reaction has come to equilibrium—that is, when there is no net change in their concentrations,

$$K'_{eq} = \frac{[C][D]}{[A][B]}.$$

The relationship between the value of the $\Delta G^{0'}$ for a reaction and the K'_{eq} for that reaction is

$$\Delta G^{0'} = -RT \ln K'_{eq} = -RT\,2.303 \log_{10} K'_{eq}.$$

R is the gas constant ($8.315\,\text{J mol}^{-1}\,\text{K}^{-1}$) and T is the temperature in degrees Kelvin ($298\,\text{K} = 25°\text{C}$). At $25°\text{C}$, $RT = 2.478\,\text{kJ mol}^{-1}$. Thus, if the K'_{eq} is determined, the $\Delta G^{0'}$ for the reaction can be calculated and vice versa. Table 1.1 shows the relationship between $\Delta G^{0'}$ value and the K'_{eq} value for chemical reactions.

Given that a reaction has a negative ΔG value, what determines whether it actually takes place at a perceptible rate in the cell?

The change in free energy of a chemical reaction must, as described above, be negative for that reaction to occur, but, given that condition, it does not follow automatically that such a reaction actually proceeds at a perceptible rate. Thermodynamic considerations of sugar reacting with oxygen (burning) to form CO_2 and H_2O are highly favourable but sugar is quite stable in air. There is a barrier to the occurrence of chemical reactions, even if they involve large negative free energy changes. If this weren't so, all combustible material on earth would burst into flame or, rather, would have done so long ago. However, sugar

Table 1.1 Relationship of the equilibrium constant (K_{eq}) of a reaction* to the $\Delta G^{0'}$ value of that reaction

Approximate $\Delta G^{0'}$ (kJ mol^{-1})	K'_{eq}
+17.1	0.001
+11.4	0.01
+5.7	0.1
0	1.0
−5.7	10.0
−11.4	100
−17.1	1000

*For a reaction $A + B \leftrightarrow C + D$, the equilibrium constant is the molar concentration of $C \times D$ divided by that of $A \times B$.

$\left(K_{eq} = \frac{[C][D]}{[A][B]} \right)$; K'_{eq} is the K_{eq} at pH 7.0.)

molecules, so stable in the bowl, can be rapidly oxidized when they enter a living cell. The question then, is what causes reactions to occur in the cell that do not occur outside the cell (or do so at an imperceptible rate)?

The answer lies in enzyme catalysis.

The nature of enzyme catalysis

An **enzyme**, in its simplest form, is a **catalyst** that brings about one particular chemical reaction. There are many thousands of biochemical reactions, each catalysed by a separate enzyme. Multifunctional enzymes with several catalytic activities on the same molecule and multienzyme complexes exist for special situations but the general principle remains.

An enzyme is a **protein** molecule. The structure of proteins is dealt with in Chapter 2 but a brief description here will be useful. Proteins are built up of 20 different species of amino acids that are linked together to form long chains. Enzymes are therefore large molecules—a molecular weight of 10 000 corresponds to a small enzyme; they range in size up to hundreds of thousands in molecular weight. One of the reasons why they are so large is that the long chain(s) of which they are constituted must be folded such that there is an 'active site' so shaped that it is a pocket into which the compounds attacked by the enzyme fit with exquisite precision. By 'attack' we mean to cause a chemical reaction in the compounds. The molecules attacked by an enzyme are called its **substrates**.

The overall enzymic process is represented by the **Michaelis–Menten model**,

$$E + S \rightleftharpoons ES \rightleftharpoons EP \rightleftharpoons E + P,$$

where E denotes the enzyme and S the substrate. ES is the enzyme–substrate complex and EP the enzyme–product complex.

What the equation says is that substrates bind reversibly at the active site of the enzyme, reaction occurs, and then products diffuse away into the solution. We shall return to the Michaelis–Menten model later on in the book (page 161) when we deal with control of enzymic catalysis. But meanwhile let us continue with the fundamental nature of the catalysis.

How does an enzyme work?

To answer this question we must look at the nature of chemical reactions. A reaction occurs in two stages. Consider the reaction, $S \rightarrow P$. S must first be converted to the transition state, $S^{\ddagger}$,

$$S \rightarrow S^{\ddagger} \rightarrow P.$$

$S^{\ddagger}$ might be thought of as a 'halfway house' in which the molecule is distorted to an electronic conformation that readily converts to P. (The transition state has an exceedingly brief existence of 10^{-14}–10^{-13} seconds.) Appreciation of the energy considerations involved is needed.

The overall reaction $S \rightarrow P$ must have a negative free energy change; otherwise it couldn't occur. An important thermodynamic principle is that the free energy change of a reaction is determined solely by the free energy difference between the starting and final products—the 'energy pathway' or **energy profile**, the route by which the reaction takes place, does not affect the free energy change of the reaction. Hence it is in no way a contradiction that the transition state ($S^{\ddagger}$) is at a higher free energy level than S. The free energy change for $S \rightarrow S^{\ddagger}$ is positive. It is called the **energy of activation** for that reaction (Fig. 1.1).

This energy hump constitutes a barrier to chemical reactions

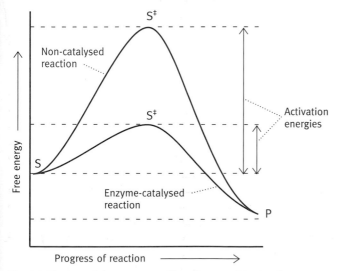

Fig. 1.1 Diagram of the energy profiles of noncatalysed and enzyme-catalysed reactions. The inverse relationship between the rate constant and the activation energy of the reaction is exponential so that the rate of the reaction is extremely sensitive to changes in the activation energy. S, substrate; $S^{\ddagger}$, transition state; P, products.

occurring. If it weren't present and if the energy profile $S \rightarrow P$ were a straight downwards slope then, as stated earlier, everything that could react would do so.

It follows that energy of activation must be supplied to permit the reaction to occur. In a noncatalysed reaction this can be done by collisions between molecules in solution. Provided that colliding molecules are appropriately oriented, reactant molecule(s) can be distorted into the appropriate higher-energy **transition state** that enables the reaction to occur. The rate of formation of the transition state therefore determines the reaction rate. High temperatures, which increase molecular motion and increase collision frequency between molecules, facilitate this. Hence the organic chemist usually employs high temperatures to promote reactions. At physiological temperatures however, around 37°C, most biochemical reactions, uncatalysed, proceed at imperceptible rates.

It is worth reflecting on what a formidable obstacle this problem posed for the development of life, for it leads to an appreciation of what an astonishing phenomenon enzyme catalysis is. It was necessary to develop a means whereby, at low temperatures, at almost neutral pH, in aqueous solutions, and at low reactant concentrations, otherwise stable molecules undergo the rapid chemical conversions needed for life, that is, a means had to be found whereby reactant molecules rapidly form transition states for the required reactions at physiological temperatures.

As stated, each enzyme has one or more combining sites within the active sites, to which the substrates bind. The binding has several effects. First, it positions substrate molecules in the most favourable relative orientations for the reaction to occur. Secondly, the active site is perfectly complementary, not to the substrate in its ground state (as the inactivated substrate molecule is referred to), but to the transition state, which is intermediate between the reactant molecules and products. The transition state binds to the enzyme more tightly than does the substrate in its ground state. The transition state has too ephemeral an existence to measure just how tightly it binds to the enzyme, but transition-state analogues have been synthesized. These are stable molecules similar in structure to the transition state. It has been found that these bind to enzymes with remarkably high affinity—in one case, binding thousands of times more tightly than the substrate.

Another way of looking at this is that formation of the transition state involves a redistribution of electrons. The amino acid groups in the structure of the active centre of the enzyme are of such a nature and are so positioned that they stabilize the electron distribution of the transition state. The fact that the active centre is a less perfect fit to the substrate than it is to the transition state results in the substrate being strained on binding to the active centre and this favours transition state formation. The net effect is to lower the activation energy of the reaction; the energy hump barrier to the reaction is largely removed (Fig. 1.1). Once formed, the transition state rapidly converts to give

(a)

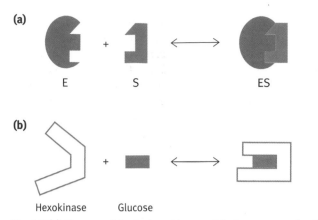

(b)

Hexokinase   Glucose

Fig. 1.2 (a) Diagram of the lock and key model of enzyme mechanism. E, Enzyme; S, substrate. **(b)** Diagram of the induced fit model of the hexokinase mechanism.

products at a rate that is independent of the structure of the $ES^{\ddagger}$ complex. The products bind less tightly to the enzyme and diffuse away. The catalysis rate is very sensitive to changes in the activation energy. A very small reduction in this, equivalent to the small amount of energy released on formation of a single average hydrogen bond (described later), can increase the rate of the reaction by a factor of 10^6. Enzymes increase the rate of chemical reactions, usually by a factor of 10^7–10^{14}. The enzyme, urease, that destroys urea, reduces the energy of activation by $84\,\mathrm{kJ\,mol^{-1}}$ and increases the reaction rate 10^{14} times.

The above description of the relationship between an enzyme and its substrates is simply a **lock and key model** in which the active site is a rigid structure in an unchanging form and the substrate fits to it (Fig. 1.2(a)). A more recent concept, however, is the **induced fit mechanism**, which is based on the view that the enzyme is not a rigid structure analogous to a lock, but rather is a flexible structure capable of changing its conformation slightly in a interactive way when its substrate binds. This induced fit mechanism has been most fully established for the enzyme hexokinase where the postulated conformational change has been proven to occur by X-ray crystallographic studies. The enzyme catalyses the transfer of a phosphoryl group from ATP to glucose. (The nature of ATP will be explained shortly so do not worry about what it is at this stage.) The enzyme has two 'wings' to its structure. In the absence of glucose, these have an 'open' conformation, but on binding of glucose the wings close in a jaw-like movement, which results in the creation of the catalytic site (Fig. 1.2(b)). The conformational change explains why the enzyme is specific for glucose. A structure that binds this molecule would be expected to be capable to some extent of binding other polyhydric molecules or even water. On the basis of the lock and key model this might be expected to result in some activity with other molecules. In the case of water being bound, this might result in hexokinase hydrolysing ATP, an undesirable reaction that does not, in fact, occur. Since the catalytic site forms only on binding of the correct substrate,

glucose, the enzyme phosphorylates only that substrate. Other molecules may bind, but, unless they are capable of bringing about the conformational change, no reaction occurs.

Most molecules are capable of participating in many different chemical reactions, each with its own transition state. In an uncatalysed reaction, promoted by high temperatures, molecules collide unpredictably and different transition states are formed, resulting in a whole variety of side reactions. By contrast, an enzyme catalyses a specific reaction for which there are only a limited number of well-defined products.

The enzyme may have additional roles in facilitating chemical reactions by general acid–base catalysis of a reaction in which a proton is transferred in the transition state. The active sites of an enzyme may be structured so that a proton donor or accepting group (part of the enzyme molecule itself) is precisely positioned such that the reaction is facilitated. Alternatively or additionally, the enzyme may position a metal ion at precisely the correct site to promote metal catalysis of a reaction.

In some cases the enzyme covalently participates as a transitory reactant though without itself being altered at the end of the process. As an example, an important class of enzymes are the serine proteases, so called because they have at their active centres the amino acid serine (page 26), which participates covalently in the catalytic process. They catalyse the hydrolysis of peptide bonds of proteins

$$R{-}CO{-}NH{-}R' + H_2O \rightarrow RCOO^- + R'NH_3^+$$

In the reaction, the acyl group of the peptide bond is first transferred to the —OH of the reactive serine group and then to water

1. $R{-}CO{-}NH{-}R' + ENZ{-}OH + H^+ \rightarrow$
    $ENZ{-}O{-}CO{-}R + NH_3^+{-}R';$
2. $ENZ{-}O{-}CO{-}R + H_2O \rightarrow$
    $ENZ{-}OH + R{-}COO^- + H^+.$

Different serine proteases have specificities towards different peptides but all have the reactive serine as part of their catalytic centres. The mechanism by which these enzymes perform their reactions is understood in very great detail.

The rate at which an enzyme-catalysed reaction occurs is basically an intrinsic property of the enzyme. All enzymes work very fast compared with the noncatalysed rate, the latter usually being imperceptible at cell concentrations of reactants. The increase in the catalysed rate over the noncatalysed rate, as already mentioned, can be as high as 10^{14} or more. The rates of catalysis vary from 4×10^7 molecules of substrate attacked per molecule of enzyme per second down to a few molecules per second. The rate at which an enzyme catalyses a reaction in the cell depends partly on the substrate concentration, a concept to be dealt with later (page 161), but at a given substrate concentration, in the simplest case, the rate of a given reaction in the cell depends on the number of molecules of the responsible enzyme in the cell. In addition, however, many enzymes have control mechanisms built into them that regulate their catalytic rate according to physiological needs—this topic belongs to the important area of metabolic control discussed later in Chapter 12.

The **activity** of an enzyme is influenced by several factors. The stability of the protein is influenced by the state of its ionizable groups and the function of the active centre may also be dependent on this. The ionization of the substrate itself may also be affected. The rate of catalysis is therefore usually dependent on the pH. Enzyme pH activity profiles vary from one to another but the optimum is generally around neutral pH; a typical plot is shown in Fig. 1.3(a). Exceptions occur, such as the case of the digestive enzyme, pepsin, which functions in the acidic stomach contents; its pH optimum is near 2.0. Temperature also affects enzyme activity rates. As the temperature increases, the rate of chemical reactions increases (approximately twofold for each 10°C), but, because of the

inherent instability of most protein molecules (page 24), the enzyme is inactivated at higher temperatures. Thus a typical enzyme optimum temperature plot would be similar to that shown in Fig. 1.3(b), but this has little absolute significance since the optimum will depend on the experimental time period used in measuring the rates; the shorter the time, the less will be the destructive effect of higher temperatures. A few enzymes are stable to high temperatures but, in general, temperatures over 50°C are destructive and some enzymes are more labile still. Many enzymes require a metal ion such as Mg^{2+}, Zn^{2+}, Fe^{2+}, or Cu^{2+} to be present, without which there is no catalytic activity.

The activity of an enzyme can also be affected by inhibitory compounds. One class, called competitive inhibitors, simply mimic the substrate and compete with the latter for the active site. Noncompetitive inhibitors affect the activity by combining with the enzyme in other ways. For example, a heavy metal such as mercury may covalently react with a thiol group essential for catalytic activity. In other cases, an inhibitor may covalently acylate the enzyme in an irreversible manner. Thus aspirin inactivates an enzyme (cyclooxygenase) involved in prostaglandin synthesis (page 146) as follows

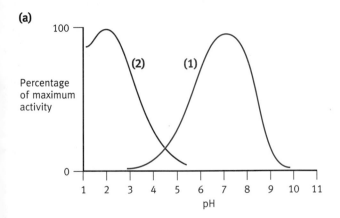

(a)

Percentage of maximum activity

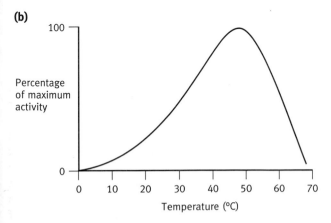

(b)

Percentage of maximum activity

Fig. 1.3 Effect of pH and temperature on enzyme activity. **(a)** Effect of pH on enzyme activity. Curve (1) is typical of the majority of enzymes with maximal activity near physiological pH. Curve (2) represents pepsin, an exceptional case, since the enzyme functions in the acid stomach contents. **(b)** Effect of temperature on a typical enzyme. The precipitous drop at high temperatures is due to enzyme destruction (though a number of heat-stable enzymes are known). Note that, as described in the text, the temperature 'optimum' of an enzyme has little significance.

The acetylated group is a serine residue which is part of the active centre of the enzyme.

How is food breakdown in cells coupled to supply energy-requiring reactions in the cell?

The conversion of simple precursor molecules to larger cellular molecules (such as DNA and proteins) involves large increases in free energy and therefore cannot occur spontaneously. 'Spontaneous' in a thermodynamic context has a special limited meaning, namely *energetically* unassisted. (Enzymes are still needed to catalyse biochemical reactions even if spontaneous in this sense.) The required energetic assistance comes ultimately from food oxidation. Chemical conversions involving positive free energy changes—the synthetic or 'building up' processes—are collectively called **anabolism** or anabolic reactions. (The anabolic steroids of sporting ill-repute promote increase in body mass, hence their name.) The other half of metabolism consists of the 'breaking down' reactions with negative free energy changes—the catabolic reactions, collectively, **catabolism**. Metabolism is comprised of catabolism and anabolism. Catabolism of food liberates free energy, which is used to drive the energy-requiring processes of anabolism. A chemical *conversion* involving a positive free energy change can, as stated, occur only if energetically assisted but the ΔG of the *actual reaction(s)* by which the conversion is accomplished must be negative—as is the case with all reactions. Catabolic reactions are often referred to as exergonic since they give out energy and those requiring energetic assistance are sometimes referred to as endergonic reactions. A distinction needs to be made between a chemical reaction and a chemical conversion or process occurring by a number of separate reactions. Thus, compound A may be converted in the cell to compound X even if X has a much higher standard free energy level than A, and this necessitates energetic assistance. But, the conversion must, by definition, be achieved by chemical reactions, each of which has a negative ΔG value. We will deal with how energetic assistance is given shortly.

We can summarize the overall situation as in Fig. 1.4. Food oxidation releases energy, which is used to drive energetically unfavourable processes. To keep the system going on a global scale, CO_2 and H_2O are reconverted during photosynthesis to food molecules (glucose) using light energy. Glucose is converted by organisms to other food molecules such as fat. Although the assembly of large cellular structures involves a decrease in entropy (unfavourable), the oxidation of food molecules involves a greater entropy increase (favourable). The entropy change of the total system (cell and surroundings) is positive, and so the **second law of thermodynamics** is obeyed.

(This law states that all processes must result in an increase in the total entropy of the universe.)

An analogy can be made of petroleum oxidation driving a car uphill, but, if you placed a bucket of petroleum under the bonnet and set it alight, while energy would be liberated, it wouldn't be useful in driving the car—the negative ΔG of petroleum oxidation must be appropriately coupled to the wheels of the car. Similarly, oxidation of food in the cell without appropriate coupling to the energy-requiring reactions would simply liberate heat and this cannot be used to do chemical or other work in the cell. Instead, the free energy change involved in food oxidation must somehow be coupled to the energy-requiring processes.

How is this done? It cannot be done by heat liberation. Heat at body temperature is waste energy—it cannot, as indicated, be used to perform work. With reference also to this problem, energy release from different foods occurs by a variety of reactions and this energy release has to drive many forms of energy-requiring processes in the cell. These include mechanical (muscular) work, osmotic work, and electrical work, in addition to chemical work. It would be impracticable to *directly* and inflexibly link up the different energy-producing reactions to each of the many energy-requiring processes. Instead, there is a common energetic intermediary that accepts the energy from *all* types of food oxidation and that can deliver it to *all* types of processes involving work and requiring energy. This is a more flexible arrangement. There is an analogy in electrical energy generation. The free energy released by burning coal, oil, wood, or gas may be converted to electrical energy, which can be transmitted to where it is needed and used for a wide variety of applications. (The analogy isn't quite right for electricity generation involves heat as an intermediary which is *not* true of the cell.) In the case of the cell, the energy intermediary has to be chemical in nature. Which leads to the next important question—what is this universal intermediary?

The high-energy phosphate compound

The answer to the above central question is remarkably simple and applies universally throughout life (to be strictly correct there are exceptions, but they are so rare that assertion of a single universal energy intermediary is justified).

The concept can be stated very simply. As a result of catabolic reactions of food molecules, inorganic phosphate ions are converted into phosphoryl groups in molecules of a **high-energy phosphate compound** (defined below). The high-energy phosphate compound is transported to wherever work is to be performed in the cell, where the phosphoryl groups are converted back to inorganic phosphate ions, with the liberation of the free energy that went into the formation of the phosphate compound. It is very important to note, however, that this does not mean that *direct* hydrolysis of the phosphate compound occurs, for this could only liberate the energy as useless heat.

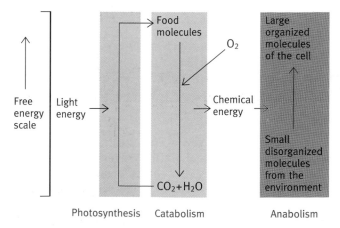

Fig. 1.4 Diagram of the energy cycle in life.

The mechanisms by which the energy is harnessed for work will be described shortly.

What is a 'high-energy phosphate compound'?

If we consider cellular compounds containing a phosphoryl group, they can be divided into two categories—low-energy phosphate compounds, whose hydrolysis to liberate inorganic phosphate (P_i) is associated with negative $\Delta G^{0'}$ values in the range of about 9–20 kJ mol^{-1} and high-energy phosphate compounds with corresponding negative $\Delta G^{0'}$ values larger than about 30 kJ mol^{-1}. The concept of a high-energy phosphate compound used to be described in terms of the 'high-energy phosphate bond', but this usage has been abandoned because a high-energy bond in chemical terms refers to a bond whose breakage requires a large input of energy, the reverse of the intended biochemical concept. For clarity of presentation we will occasionally use the term 'high-energy phosphoryl group' as a shorthand way of referring to a phosphoryl group in a high-energy phosphate compound. (The energy resides in the molecule as a whole and not in the actual phosphoryl group.) With that important qualification, the **high-energy phosphoryl group** can be usefully regarded as the universal energy currency of the living cell. This concept is illustrated in Fig. 1.5 which is a refinement of Fig. 1.4.

What are the structural features of high-energy phosphate compounds?

Phosphoric acid, H_3PO_4, is an oxyacid of phosphorus with three dissociable protons as shown below.

$$
\begin{array}{c}
O \\
\parallel \\
HO-P-OH \\
\mid \\
OH
\end{array}
\quad
\underset{pK_{a_1} = 2.2}{\rightleftharpoons}
\quad
\begin{array}{c}
O \\
\parallel \\
HO-P-O^- \\
\mid \\
OH
\end{array}
+ H^+
$$

$$pK_{a_2} = 7.2$$

$$
\begin{array}{c}
O \\
\parallel \\
HO-P-O^- \\
\mid \\
O^-
\end{array}
+ H^+
$$

$$pK_{a_3} = 12.3$$

$$
\begin{array}{c}
O \\
\parallel \\
{}^-O-P-O^- \\
\mid \\
O^-
\end{array}
+ H^+
$$

At normal cell pH, a mixture of the singly negatively and doubly negatively charged phosphate ions predominates in the solution. In biochemistry, this mixture of phosphate ions is symbolized as

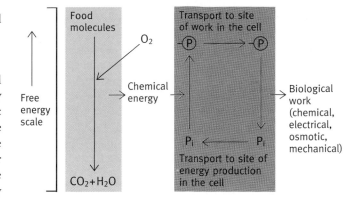

Fig. 1.5 Diagram showing the role of the phosphoryl group in the energy economy of a cell. —Ⓟ represents the phosphoryl group in a molecule whose hydrolysis to liberate P_i is associated with a $\Delta G^{0'}$>30 kJ mol^{-1}.

P_i, the i indicating 'inorganic'; it represents the lowest energy form of phosphate in the cell and can be regarded as the ground state of phosphate when considering its energetics. The degree of ionization of the molecule is a function of the three dissociation constants of phosphoric acid. These are known as **pK_a values**; a pK_a is the pH at which there is 50 per cent dissociation of the group. An explanation of pK_a values and buffers is given in the appendix to this chapter; this should be studied and learned now if you are not already familiar with the subject. At physiological pH (say 7.4) the group with a pK_a of 2.2 will be fully ionized and that with a pK_a of 12.3 will be undissociated. The group with a pK_a value of 7.2 will be partially dissociated. The actual degree of the latter can be calculated from the **Henderson–Hasselbalch equation**,

$$pH = pK_{a_2} + \log \frac{[\text{Salt}]}{[\text{Acid}]};$$

$$7.4 = 7.2 + \log \frac{[\text{HPO}_4^{2-}]}{[\text{HPO}_4^-]}.$$

Therefore,

$$\log \frac{[\text{HPO}_4^{2-}]}{[\text{H}_2\text{PO}_4^-]} = 0.2$$

so that

$$\frac{[\text{HPO}_4^{2-}]}{[\text{HPO}_4^-]} = 1.58$$

Inorganic phosphate, esterified with an alcohol, is a phosphate ester.

$$
H_2O +
\begin{array}{c}
O \\
\parallel \\
R-O-P-O^- \\
\mid \\
O^-
\end{array}
\longrightarrow
ROH +
\begin{array}{c}
O \\
\parallel \\
HO-P-O^- \\
\mid \\
O^-
\end{array}
$$

Phosphate ester   Alcohol   Inorganic phosphate (P_i)

The $\Delta G^{0'}$ for the hydrolysis of this ester is roughly of the order of -12.5 kJ mol^{-1}, resulting in the equilibrium for the hydrolytic reaction being strongly to the side of hydrolysis. In the cell the reverse reaction does not occur.

As well as forming phosphate esters with alcohols, inorganic phosphate can form a phosphoric anhydride called **inorganic pyrophosphate** (*pyro* means fire; pyrophosphate can be made by driving off water from P_i at high temperatures) and in biochemistry it is often written as PP_i. The $\Delta G^{0'}$ for the hydrolysis of this compound is -33.5 kJ mol^{-1}, which is much higher than the value for hydrolysis of the ester phosphate.

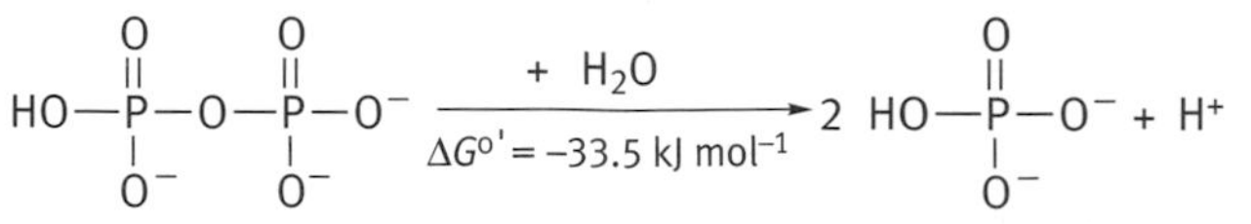

Inorganic pyrophosphate (PP$_i$)   Inorganic phosphate (P$_i$)

The very high free-energy release associated with the hydrolysis of the phosphoric anhydride group is due to several factors. When the pyrophosphoryl group is hydrolysed, several things result.

1 The destabilization of the structure caused by electrostatic repulsion of the two negatively charged phosphoryl groups is relieved. This factor is made clear by the reflection that, in the reverse reaction to synthesize a pyrophosphoryl bond, the electrostatic repulsion of the two ions has to be overcome in bringing them together.

2 The products of the reaction (in this case, two P_i ions) are **resonance stabilized**—they have a greater number of possible resonance structures than has the pyrophosphate structure. This is because the two phosphoryl groups in the latter have to compete for electrons in the bridging oxygen atom; these electrons are essential for the resonating structure. Therefore, in the pyrophosphate molecule one or other group can resonate—they compete in this. The two P_i ions can resonate simultaneously. In an inorganic phosphate ion, all of the P—O bonds are partially double-bonded in character rather than the proton being associated with any one oxygen, resulting in an increase in entropy. In the following diagram δ^- symbolizes a partial negative charge (δ^+ would symbolize a partial positive charge).

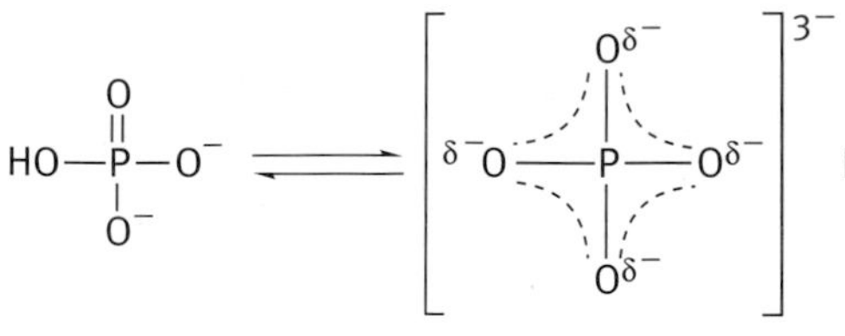

Resonance stabilization of phosphate ion

These factors all favour the hydrolysis reaction, which has an equilibrium constant decisively in favour of P_i formation (equals a large negative $\Delta G^{0'}$ value). These considerations apply not only to inorganic pyrophosphate but to any phosphoric anhydride group. Other 'high-energy' phosphoryl groups will

also be referred to later. Points 1 and 2 above are not applicable to hydrolysis of ester phosphates, which explains why the energy release is so much less.

The phosphoric anhydride structure discussed above is not the only biological high-energy phosphate compound (though it is the predominant one). Three other types are found.

The first is the anhydride between phosphoric acid and a carboxyl group, hydrolysis of which by the reaction shown has a very large negative $\Delta G^{0'}$ value (-49.3 kJ mol^{-1} in a typical case). The large free-energy change is associated with the resonance stabilization of both products, namely, P_i and the carboxy acid (the latter illustrated below).

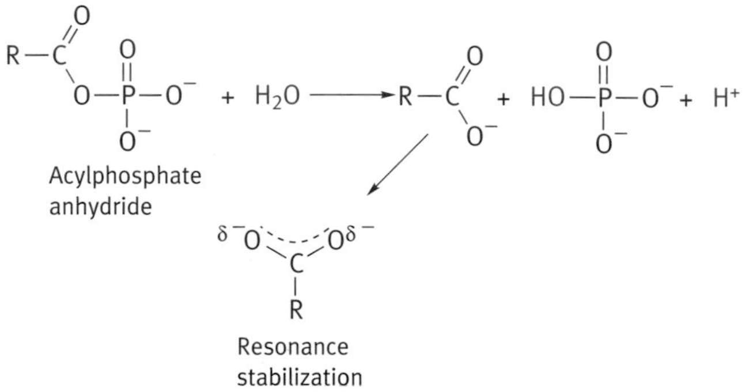

Acylphosphate anhydride

Resonance stabilization

The second structure is that of guanidino-phosphate, whose hydrolysis to produce P_i also has a very large negative $\Delta G^{0'}$ value (-43.0 kJ mol^{-1} in a typical case). Again this is due to resonance stabilization of both products.

Guanidino-phosphate

Resonance stabilization

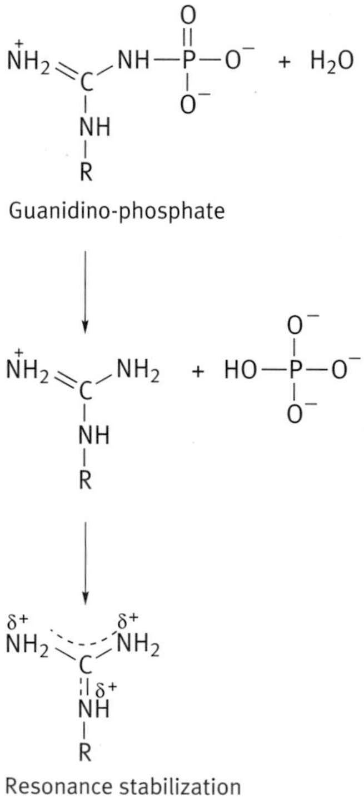

A third type is in a different category—an enol-phosphate structure. This is found in the metabolite, phosphoenolpyruvate.

This looks an unlikely candidate for high-energy status, but, on removal of the phosphate group by hydrolysis, the enol structure so formed spontaneously tautomerizes into keto form of pyruvate, the equilibrium of this being far to the right.

Phosphoenolpyruvate
(Enol phosphate)

Unstable

Stable

The overall conversion of phosphoenolpyruvate to P_i and the keto form of pyruvate has a $\Delta G^{0'}$ value of $-61.9\,\text{kJ mol}^{-1}$.

With that description of the 'high-energy phosphoryl group', we can refine the situation in Fig. 1.4 to that in Fig. 1.5 which, in the overall sense, is a complete description of energy generation and utilization.

As you progress through the book you will meet a good number of phosphate compounds and we will give their individual structures and $\Delta G^{0'}$ values for their hydrolysis as we come to them. However, there is one of absolutely central importance that we will deal with now.

We have so far spoken of the phosphoryl groups present in high-energy phosphate compounds in general terms. Thus, in Fig. 1.5, P_i is shown as being elevated to a 'high-energy phosphoryl group' and then transported around the cell to where it is needed to supply the energy for work. Clearly, the phosphoryl group (—Ⓟ) must be covalently attached to another molecule that acts as its carrier within the cell. Which brings us to the next question.

What transports the —Ⓟ around the cell?

The general carrier is AMP or **adenosine monophosphate**, whose structure is given in Fig. 1.6. AMP carrying a single —Ⓟ group is called ADP or **adenosine diphosphate**. AMP carrying two —Ⓟ groups is ATP or **adenosine triphosphate**. ATP therefore can be written as AMP—Ⓟ—Ⓟ. The two terminal phosphates are attached to AMP by phosphoric anhydride linkages of the same type as already met in PP_i, and the reasons given earlier for the large $\Delta G^{0'}$ for hydrolysis off of these two groups apply equally here. Note that AMP itself is a low-energy ester phosphate and does not itself directly participate in the energy cycle except as carrier for the two —Ⓟ groups. You will see from Fig. 1.6 that the $\Delta G^{0'}$ of hydrolysis of each of the two terminal groups is $-30.5\,\text{kJ mol}^{-1}$; that is, of the high-energy type. When food is oxidized, the energy released, as explained, is

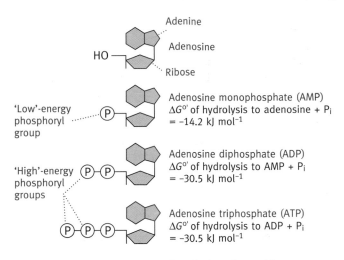

Fig. 1.6 Diagrammatic representation of adenosine and its phosphorylated derivatives. Adenosine is a nucleoside with ribose and adenine as its components. Details of their structures are irrelevant here but are dealt with in later chapters.

coupled to the conversion of ADP (adenosine diphosphate or AMP—Ⓟ) to ATP,

$$ADP + P_i \xrightarrow[+\text{energy}]{} ATP + H_2O.$$

Each cell has only a very small quantity of ATP in it at any one time. The amount would last only a very short time—and it can't get any from the outside; neither ATP, ADP, nor AMP can spontaneously diffuse through the cell membrane because they are highly charged (see Chapter 3). Each cell has to synthesize the entire molecule itself. ATP thus 'turns over' or cycles very rapidly, by which we mean it breaks down to ADP and P_i and is resynthesized to ATP. We can modify the diagram given in Fig. 1.5 to include this fact (Fig 1.7). The poison, cyanide, stops most of the resynthesis of ATP and death results in a very short time because insufficient energy can be supplied for cellular processes once the small amount of pre-existing ATP is broken down.

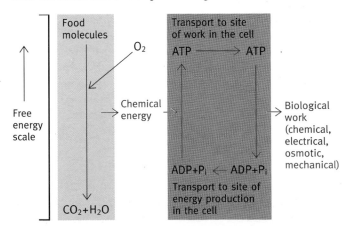

Fig. 1.7 Diagram of the role of ATP in the energy economy of a cell. Note that some types of work involve breakdown of ATP to AMP, but, as described in the text, this does not change the concept given here.

How does ATP perform chemical work?

Suppose the cell needs to synthesize X—Y from the two reactants, X—OH + Y—H, and the $G^{0'}$ change involved in the conversion is $12.5\,\text{kJ mol}^{-1}$. The simple reaction $XOH + YH \rightarrow XY + H_2O$ cannot occur to any significant extent because the $\Delta G^{0'}$ is positive and the equilibrium is far to the left. The solution is to use **coupled reactions** involving ATP breakdown, as shown below. Coupled reactions are two or more reactions in which the product of one becomes the reactant for the next. No physical coupling is necessarily involved—they simply have to be present in the same chemical system. Note that all reactions in the cell involving ATP must be enzymically catalysed—the fact that hydrolysis of each of the two phosphoric anhydride groups is strongly exergonic does not mean that ATP is an unstable or highly reactive molecule. (It can be bought in bottles as a white powder and can be stored indefinitely in the freezer.)

Reaction 1: $XOH + AMP–Ⓟ–Ⓟ \rightarrow X–Ⓟ + AMP–Ⓟ$

Reaction 2: $X–Ⓟ + YH \rightarrow X–Y + P_i$

Sum of 1 + 2: $XOH + YH + AMP–Ⓟ–Ⓟ \rightarrow$
$\quad X–Y + AMP–Ⓟ + P_i$

The overall $\Delta G^{0'}$ for coupled reactions is the arithmetic sum of the $\Delta G^{0'}$ values of the component reactions. The $\Delta G^{0'}$ for the $XOH + YH \rightarrow X–Y + H_2O$ we take to be $12.5\,\text{kJ mol}^{-1}$; the $\Delta G^{0'}$ for the reaction $ATP + H_2O \rightarrow ADP + P_i$ is $-30.5\,\text{kJ mol}^{-1}$. Therefore, the $\Delta G^{0'}$ of the overall process is $-18.0\,\text{kJ mol}^{-1}$, a strongly exergonic process that will proceed essentially to completion—the equilibrium constant will be about 10^3—which means that the reactions can proceed to approximately 99.9% to the side of X—Y formation (see Table 1.1). It is to be noted that the use of ATP here involves phosphoryl group transfer from ATP to one of the reactants and only then is P_i liberated. *Direct* hydrolysis of ATP to ADP and P_i is not occurring. In the cell, the actual ΔG value of the ATP breakdown to ADP and P_i is considerably larger than the $\Delta G^{0'}$ value for this, because ΔG values are affected by reactant concentrations and cellular levels of ATP, ADP, and P_i are, of course, nowhere near $1.0\,\text{M}$ (the concentrations specified in $\Delta G^{0'}$ determinations). If the actual cellular concentrations of these components are known, the ΔG value for the hydrolysis of ATP to ADP + P_i can be calculated from the equation

$$\Delta G = \Delta G^{0'} + RT\,2.303 \log_{10} \frac{[\text{Products}]}{[\text{Reactants}]}$$

where R is the gas constant ($8.315\,\text{J mol}^{-1}\text{K}^{-1}$); and T is the temperature in degrees Kelvin ($298\,\text{K} = 25^\circ\text{C}$).

The concentrations of ATP, ADP, and P_i will vary from cell to cell. ATP 'levels' generally fall in the range 2–8 mM, ADP levels are about a tenth of this, and P_i values are similar to those of ATP. However, for simplicity let us assume a situation in which

all three are at 10^{-3} M. (For reactions in dilute aqueous solution, water has the value of 1.) Substituting these values into the equation

$$\Delta G = \Delta G^{0'} + RT\,2.303 \log_{10} \frac{[\text{ADP}][\text{P}_i]}{[\text{ATP}]},$$

we obtain

$$\Delta G = -30.5\,\text{kJ mol}^{-1} + (831.5 \times 10^{-3})(298)$$
$$\times\, 2.303 \log_{10} \frac{[10^{-3}][10^{-3}]}{[10^{-3}]}$$
$$= -30.5 - 17.1 = -47.6\,\text{kJ mol}^{-1}.$$

The reaction sequence given above is used for many biochemical reactions coupled to ATP breakdown. But, more commonly, for the synthesis of molecules such as nucleic acids and proteins, the cell uses an even more energetically effective trick. Instead of breaking off one —Ⓟ of ATP to P_i it releases two. The negative $\Delta G^{0'}$ value that results from this is so large that the equilibrium is such that the reaction is completely irreversible in the cell. The way this is done is as follows, again taking $XOH + YH \rightarrow X–Y + H_2O$ as the example.

Reaction 1: $XOH + AMP–Ⓟ–Ⓟ \rightarrow X–AMP + PP_i$

Reaction 2: $X–AMP + YH \rightarrow X–Y + AMP$

Reaction 3: $PP_i + H_2O \rightarrow 2\,P_i$

Sum: $XOH + YH + AMP–Ⓟ–Ⓟ + H_2O \rightarrow$
$\quad X–Y + AMP + 2\,P_i$

Assume that the $\Delta G^{0'}$ for the reaction from $XOH + YH \rightarrow X–Y + H_2O$ is $12.5\,\text{kJ mol}^{-1}$: the $\Delta G^{0'}$ for the reaction $ATP + H_2O \rightarrow AMP + PP_i$ is $-32.2\,\text{kJ mol}^{-1}$; that for PP_i hydrolysis is $-33.5\,\text{kJ mol}^{-1}$.

The $\Delta G^{0'}$ for the reaction $ATP + H_2O \rightarrow AMP + 2\,P_i$ is $-65.7\,\text{kJ mol}^{-1}$. The overall process of synthesizing X—Y at the expense of ATP breakdown thus has a negative $\Delta G^{0'}$ of $65.7 - 12.5 = 53.2\,\text{kJ mol}^{-1}$, a very large value indeed.

The mechanism depends on the fact that inorganic pyrophosphate, PP_i, is rapidly broken down in the cell. It explains why cells contain an enzyme, called **inorganic pyrophosphatase** that catalyses the reaction (already shown above)

Inorganic pyrophosphate (PP$_i$)   Inorganic phosphate (P$_i$)

This enzyme, once regarded as an unimportant one, is a driving force in biochemical syntheses and is widely distributed.

How does ATP drive other types of work?

As well as performing chemical work, ATP breakdown powers muscle contraction, the generation of electrical signals, and pumping of ions and other molecules against concentration gradients. The mechanisms are, in principle, the same. Whatever the process, provided ATP breakdown is coupled into the mechanism the resultant free-energy liberation will drive it. Many of these processes will be dealt with in subsequent chapters.

A note on the relationship between AMP, ADP, and ATP

There might seem to be a problem. As explained, many ATP-requiring chemical syntheses in the cell produce AMP + 2P$_i$, not ADP + P$_i$. AMP cannot, itself, be converted to ATP by the foodstuff oxidation system; only ADP is accepted as shown in Fig. 1.6. However, AMP is easily rescued by an enzyme that transfers —Ⓟ from ATP to AMP. The reaction is

AMP + AMP—Ⓟ—Ⓟ ↔ 2AMP—Ⓟ

or, as written more conventionally,

AMP + ATP ↔ 2ADP.

The enzyme is called AMP-kinase. **Kinase** is the term for carrying a phosphoryl group from ATP; AMP-kinase means it transfers the group to AMP (AMP is also called adenylic acid; hence AMP-kinase may also be called adenylate kinase). Note that the —Ⓟ group is directly transferred from one molecule to another without hydrolysis or significant release of energy. You will find as you go through the book that such freely reversible 'shuffling' at the 'high-energy level' occurs frequently. The ADP (AMP—Ⓟ) so produced is now accepted by the food oxidation system and converted to ATP.

To give perspective to the subject, we have not in this account dealt with the mechanisms by which ATP is synthesized from ADP and P$_i$ at the expense of food catabolism. This is a very large topic that forms a substantial part of the chapters on metabolism later in this book. Also, we have dealt with the utilization of ATP energy to perform work only in general terms so far—as you progress through the book you will encounter example after example of this in later chapters, for ATP utilization is involved in virtually all biochemical systems where work is performed.

We come now to a change of topic though it is still concerned with free-energy changes in chemical processes.

Weak bonds and free-energy changes

The chemistry we have been discussing so far in this chapter involves covalent bonds; now we come to a different type of molecular interaction involving **noncovalent bonds**, also referred to as secondary or **weak bonds**. The latter two terms belie their importance in life.

In enzymatically catalysed reactions of the type discussed so far, atoms or groups of atoms are rearranged so that new compounds result and covalent bonds are broken and made.

However, although the covalent bonds may have different atoms attached to them, at the end of the reaction there is the same total number of covalent bonds.

There's a good reason why, in biochemical reactions, there is no net destruction of covalent bonds. When two atoms become joined to form a chemical bond, a certain amount of energy is released,

X + X → X — X + energy.

(X here represents a free *atom*, two of which are becoming joined by a bond *that did not previously exist*. We are *not* talking of ordinary chemical reactions in which there is no increase in the number of covalent bonds.)

To reverse this process, to break the chemical bond, exactly the same amount of energy has to be supplied as was released in its formation,

X — X + energy → X + X.

The formation of a typical covalent bond releases a very large amount of energy and, therefore, its destruction also requires a very large amount of energy. Such bond destruction does not occur in biochemical systems. To illustrate the point, oxygen exists as O$_2$—the $\Delta G^{0'}$ value of its formation from atomic oxygen is so large (about 460 kJ mol^{-1}), and negative, that the equilibrium is such that the number of single atoms in oxygen gas is an infinitely small proportion of the total existing as O$_2$.

In chemical reactions, the breakage of a covalent bond is accompanied by the simultaneous formation of another covalent bond via the transition state so that the net energy change involved is far less than the very large value required to *destroy* a covalent bond. However, noncovalent, weak or secondary bonds between atoms and groups of atoms in solution are continually being formed and broken. They are sufficiently strong to form structures and yet sufficiently weak for those structures to be readily disassembled—they allow very great structural flexibility. This general statement may have little reality to you at this stage but their central role will become clear when we've dealt with membranes, proteins, and DNA. They are weak or secondary, not in importance, but in energy of formation, involving $\Delta G^{0'}$ values in the range of only 4–30 kJ mol^{-1} so that correspondingly small amounts of energy are required to break them.

There are three types of weak bonds, ionic bonds, hydrogen bonds, and **van der Waals** attractions. **Ionic bonds** arise from the electrostatic attraction between two permanently charged groups that are close to each other. A typical example would be that between a negatively charged carboxyl group and a positively charged amino group,

—COO$^-$---H$_3^+$N—.

Next we have **hydrogen bonds**. These also are due to electrostatic attractions between atoms of polar molecules, but the attractions are due not to ionized groups but to a much weaker form of positive and negative charge separation, the electric dipole moment. Consider a water molecule as shown below.

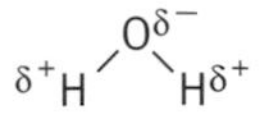

$$\delta^+ H \diagdown \overset{O^{\delta-}}{\diagup} H^{\delta+}$$

The oxygen atom is electronegative—it has a greater attraction for the bond electrons than has a hydrogen atom—it is greedy for electrons and takes more than its half share. The molecule has a partial negative charge on the oxygen atom and corresponding partial positive charges on the hydrogen atoms. This results in weak attractions between the O and H atoms of adjacent water molecules known as hydrogen bonds. Each water molecule is bonded to four other molecules because the oxygen atom can form two hydrogen bonds and the hydrogen atoms one each. A hydrogen atom of any molecule can participate in hydrogen bond formation provided it is attached to an electronegative atom. The main relevant electronegative atoms in biochemistry are nitrogen and oxygen. The bond must involve also an electronegative acceptor atom—also usually nitrogen or oxygen.

The types of hydrogen bonds common in biochemistry are illustrated here.

—O—H---O—

—O—H---N—

—N—H---O—

—N—H---N—

The hydrogen bond is highly directional and of maximal strength when all of the atoms are in a straight line. Van der Waals forces is the collective term used to describe a group of weak interactions between closely positioned atoms; fluctuations in electron density of the atoms cause induced dipoles. They are nonspecific and are the only type of weak bond possible between nonpolar molecules that cannot form hydrogen or ionic bonds. Van der Waals forces can operate between *any* two atoms, which must be positioned close together so that their electron shells are almost touching. The attraction between them is inversely proportional to the sixth power of the distance, so close positioning is essential. If atoms are forced to become too close together, their electron shells overlap and a repulsive force is generated. In short, precise positioning is a prerequisite for van der Waals attraction to arise.

What are the characteristics of the three types of weak bonds? First, while they vary in length, *they are all short-range attractions*; they vary somewhat in their energy of bond formation. The $\Delta G^{0'}$ values are, on average, about $20\,kJ\,mol^{-1}$ for ionic bonds, $12–29\,kJ\,mol^{-1}$ for hydrogen bonds, and about $4–8\,kJ\,mol^{-1}$ for the average **van der Waals** bond. Compare these values with bond energy of covalent bonds—several hundred $kJ\,mol^{-1}$.

The importance of weak bonds is that they promote associations between and within molecules. In this context we must add another phenomenon that also promotes this but is not an actual bonding—it is called **hydrophobic force** or sometimes **hydrophobic attraction**.

A polar molecule such as sugar that can form hydrogen bonds is soluble in water because, although it disrupts hydrogen bonding between water molecules, it can itself form bonds with the latter, making the process energetically feasible. Salts such as NaCl are soluble because the Na^+ and Cl^- ions become surrounded by hydration shells in which the ion–water attractions exceed those between the ions themselves and, when separated, there is a large increase in entropy from the crystalline state. If an attempt is made to dissolve a nonpolar substance such as olive oil in water, the oil molecules get in the way of hydrogen bonding between water molecules. Since hydrogen bonding is highly directional, the water molecules around the oil molecules must arrange themselves so that they can still be fully bonded—that is, with none of their bonding sites aimed at the oil molecules. This more highly ordered arrangement is at a lower entropy level than that of randomly arranged water molecules (that is, a higher energy level) and so the solubilization of the oil is opposed. The nonpolar molecules are forced to associate together so as to present the minimum oil/water interface area. The olive oil forms droplets and then a separate layer which is the minimum free energy state. This repelling effect, pushing together nonpolar molecules by water molecules, is called a hydrophobic force. Hydrophobic groups occur in proteins and DNA and other cellular molecules. Hydrophobic forces on these play a crucial role in the structure of these molecules as will be seen later in this book; it is remarkable how the necessity of 'hiding' hydrophobic groups from water determines so much in living cells.

What causes weak bond formation and breakage?

Formation of weak bonds involves a fall in free energy and the energies of activation involved in their formation are very low. They therefore form spontaneously between appropriate atoms that are sufficiently close, *without the need for enzyme catalysis*. It has already been mentioned that there is a direct relationship between the $\Delta G^{0'}$ of a reaction and its equilibrium constant. Table 1.1 gives the equilibrium constants for given $\Delta G^{0'}$ values. Although those for weak bond formation are small, nonetheless from Table 1.1 you can see that a $\Delta G^{0'}$ as small as $-5.7\,kJ\,mol^{-1}$ gives an equilibrium well to the side of bond formation. When you get up to a $\Delta G^{0'}$ value of around $-11.4\,kJ\,mol^{-1}$, the equilibrium is 99% in favour of bonding. In short, weak bond formation is favoured by energy considerations.

However, such a bonding equilibrium is a dynamic one—individual bonds are continually being broken and reformed. What causes the breakage? All that is required to break a weak bond is to supply the same amount of energy as was released when the bond was formed. This energy comes from the kinetic energy of molecules in motion (thermal motion) when they collide with the atoms forming the weak bonds; thermal motion of molecules at the higher end of the energy distribution spectrum, even at physiological temperatures, has sufficient kinetic energy to disrupt weak bonds.

If weak bonds are so easily broken, what is their advantage in biochemical systems?

Given that, say, a van der Waals attraction has a $\Delta G^{0\prime}$ of formation only slightly above the average thermal energy of molecules in solution and can be made or broken spontaneously, it might be thought that such ephemeral bonds can't be of much importance. To a lesser degree the same could be thought about the other weak bonds.

Paradoxically, it is their very weakness that makes them of central importance to life. The essence of almost everything in life involves biological specificity, which depends on one molecule binding to another with such exquisite precision that only the particular molecules designed by evolution to interact do so. To give a few examples at random, an enzyme is specific for one reaction because its active site will combine only with its own substrate(s). A hormone has specific effects because it combines only with its own receptor protein on cells. A gene is controlled because controlling proteins recognize and bind to a particular piece of DNA. An antibody combines only with its relevant antigen. The list is very long. You'll come across one case after another as you proceed through the subject.

How are these precise molecular interactions achieved? The binding processes in the examples referred to depend on weak bonds and the specificity arises from: (1) the fact that individual bonds are so weak that several must be formed for the attachment to occur; and (2) the weak bonds, being short-range ones, can form only if the relevant atoms are precisely positioned very close to each other. The combination of (1) and (2) means that only molecules shaped to fit their designated partner(s) can bind in this way. The relationship can be so precise that, if a single atom is changed in an enzyme substrate, or in the active centre of an enzyme, the interactions do not occur.

Another important consideration is that biological processes often demand that specific bonding between molecules be transitory. Binding of molecules to an enzyme must not be so tight that the products of the reaction can't diffuse away; gene action and reversible gene control demand that interactions are easily reversible. This is where $\Delta G^{0\prime}$ values for bond energies of the weak bonds are exactly right. They are sufficiently large to give equilibrium constants that favour bond formation but sufficiently small that bonds can be easily broken and the attachments freely reversible. In the case of antibody–antigen interactions, a sufficiently large number of weak bonds results in bonding that is essentially irreversible. In summary, weak bonds

permit precise binding between molecules and give the basis for biological specificity. Small numbers of bonds permit easy reversibility of the binding while large numbers can produce stable inter- and intramolecular interactions. The role of weak bonds in biochemistry will be more apparent when we come to the chapters on membranes, protein structure, gene structure, and gene action.

With that introduction we will deal in the next two chapters with two important topics. These are protein structure and membranes. Both topics are necessary for understanding metabolism which is the major area covered in the first half or so of the book.

Appendix. Buffers and pK_a values

It is very important in biochemistry to understand what buffers and pK_a values are. We have placed this is an appendix to avoid disrupting the text and because many will have dealt with this in their chemistry studies. If not, this material *should be learned now*.

The pH of a living cell is maintained in the range 7.2–7.4. Special situations occur, such as in the stomach where HCl is secreted and in lysosomes into which protons are pumped to maintain an acid pH, but, otherwise, the pH of cells and of circulating fluids is maintained within narrow limits. This is despite the fact that metabolic processes, such a lactic acid and acetoacetic acid formation and CO_2 conversion to H_2CO_3 in the blood, occur on a large scale.

This pH stability is largely due to the buffering effect of weak acids. An acid in this context is defined as a molecule that can release a hydrogen ion.

A carboxylic acid dissociates, liberating a proton

$$R\ COOH \leftrightarrow R\ COO^- + H^+.$$

The proton donor and proton acceptor in the above equation is called a conjugate acid and base pair. Written in a more general form, the equation for acids is

$$HA \leftrightarrow H^+ + A^-.$$

Acids vary in their tendency to dissociate. Stronger acids do so more readily than weaker ones—this is why, say, a 0.1 M solution of formic acid (HCOOH) has a lower pH than a 0.1 M solution of acetic acid (CH_3COOH). The tendency to dissociate is quantitated as a dissociation constant, K_a, for each acid: the larger the value of K_a, the greater the tendency to dissociate and the stronger the acid,

$$K_a = \frac{[H^+][A^-]}{[HA]}.$$

For acetic acid $K_a = 1.74 \times 10^{-5}$. This value is not much used, as such, by biochemists, for there is another way of expressing the strength of the acid by a much more convenient

(a) for the molecule RCOOH with a pK_a of 4.2:

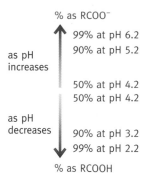

(b) for the molecule RNH_2 with a pK_a of 9.2:

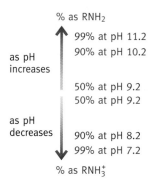

Fig. 1.8 Effect of pH on the ionization of **(a)** —COOH and **(b)** —NH₂ groups. Kindly provided by R. Rogers.

term—the pK_a value. The two are related by the equation

$$pK_a = -\log K_a.$$

Thus, acetic acid has a pK_a of 4.76 and formic acid a pK_a of 3.75. These values represent the pH at which an acid is 50% dissociated. As the pH increases, the acid becomes more dissociated; as it decreases, the reverse. The $HA \leftrightarrow H^+ + A^-$ equilibrium is affected by the H^+ concentration as would be expected (see Fig. 1.8).

An amine base also has a pK_a value because the ionized form can dissociate to liberate a proton and, in this sense, is an acid

$$R\,NH_3^+ \leftrightarrow R\,NH_2 + H^+.$$

In this case, as the pH increases, the dissociation decreases; with both carboxylic acids and amine bases, increase in H^+ concentration causes increased protonation. The difference, of course, is that protonation of a carboxylic acid reduces the amount of ionized form while, with amine bases, the proportion in the ionized form increases (Fig. 1.8).

pK_a values and their relationship to buffers

If you take a 0.1 M solution of acetic acid and gradually add 0.1 M NaOH measuring the pH at each step, the curve shown in

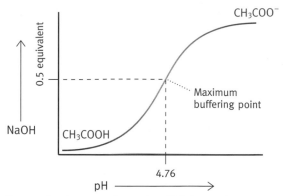

Fig. 1.9 Titration curve for acetic acid (pK_a=4.76).

Fig. 1.9 is obtained.

At the beginning of the titration, the added OH^- neutralizes existing H^+ (the acetic acid is slightly dissociated) and the pH rises rapidly. However, as the pH begins to approach the pK_a value of acetic acid, as more NaOH is added, the acetic acid dissociates into the acetate ion liberating hydrogen ions, which neutralize the hydroxide ions so that the pH change is relatively small. The reaction is

$$CH_3COOH + OH^- \rightarrow CH_3COO^- + H_2O.$$

(Note that sodium acetate is fully dissociated.)

Similarly, from the other side of the pH scale, additions of H^+ cause little change to the pH because of the reaction

$$CH_3COO^- + H^+ \rightarrow CH_3COOH.$$

This is the pH buffering effect of acetic acid. It is maximal at the pK_a so that an equimolar mixture of acetate and acetic acid gives its maximum buffering effect at that pH. As a rule of thumb, in biochemistry, a useful buffer covers a range of 1 pH unit centred at the pK_a. The pH of a solution containing an acid–base conjugate pair can be calculated from the Henderson–Hasselbalch equation

$$pH + pK_a + \log \frac{[\text{Proton acceptor}]}{[\text{Proton donor}]}$$

or, as often expressed,

$$pH = pK_a + \log \frac{[\text{Salt}]}{[\text{Acid}]}.$$

The pH of any mixture of acetic acid and sodium acetate can be calculated from this equation so that the composition of a buffer of any desired pH can be determined.

As an example for a solution containing 0.1 M acetic acid and 0.1 M sodium acetate,

$$pH = 4.76 + \log \frac{[0.1]}{[0.1]}$$
$$= 4.76 + 0 = 4.76.$$

If the mixture contains 0.1 M acetic acide and 0.2 M sodium acetate

$$pH = 4.76 + \log \frac{[0.2]}{[0.1]}$$

$$= 4.76 + \log 2 = 4.76 + 0.30 = 5.06.$$

Suppose that to the buffer containing 0.1 M acetic acid and 0.1 M sodium acetate you added NaOH to a concentration of 0.05 M; the pH of the resultant solution would be

$$4.76 + \log \frac{[0.15]}{[0.05]}$$

(since half of the acetic acid would be converted to sodium acetate)

$$= 4.76 + \log 3 = 4.76 + 0.48 = 5.24.$$

Acetic acid/acetate mixture has no significant buffering power at physiological pHs, but compounds with pK_as close to pH 7 exist in the body, and these are effective buffers. Among the most important are the phosphate ion and its derivatives. Phosphoric acid has three dissociable groups (page 9), the second one ($H_2PO_4 \leftrightarrow HPO_4 + H^+$) has a pK_a value of 6.86 so that phosphate is an excellent buffer in this region.

Another buffering structure is the imidazole nitrogen of the histidine residue in proteins with a pK_a around 6. The dissociation involved is shown below.

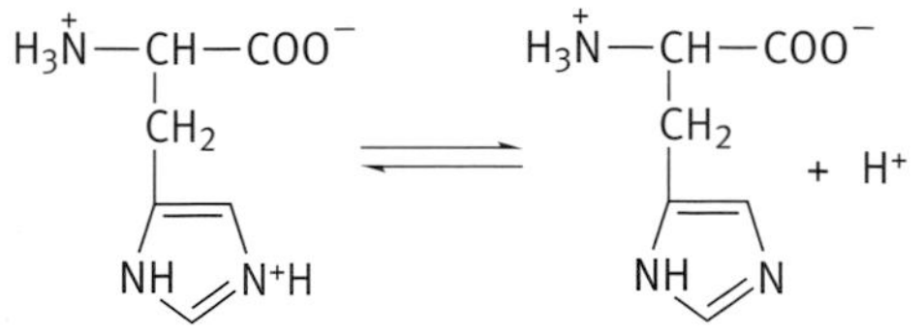

The phenomenon of buffering can be illustrated by a very simple observation. If you take a test tube containing distilled water and adjust the pH to 7.0 (dissolved CO_2 makes it slightly acid) with NaOH and then add a few drops of 0.1 M HCl there will be a precipitous drop in pH. If you take a test tube containing 0.1 M sodium phosphate, also adjusted to pH 7.0, and add the same amount of HCl, the pH will hardly change, due to the buffering action of the phosphate ion.

The buffering action of compounds with pK_a values near 7 thus protects the cell and body fluids against large pH changes.

Further reading

Thermodynamics

Newsholme, E. A. and Leach, A. R. (1983). *Biochemistry for medical sciences*, pp 10–45. Wiley.
Gives an account of the laws of thermodynamics with special emphasis on their application to metabolism.

Nelsestuen, G. L. (1989). A partial remedy for the non-relationship between reversibility of a reaction in the cell and the value of $\Delta G^{0'}$. *Biochemical Education*, 17, 190–2.
Discusses why reactions such as aldolase with a large $\Delta G^{0'}$ value are freely reversible

Enzyme catalysis

Steitz, T. A., Shoham, M., and Bennett, Jr., W. S. (1981). Structural dynamics of yeast hexokinase during catalysis. *Phil. Trans. Roy. Soc. Series B, London.*, 293, 43–52.
Discusses the conformational change of the enzyme on binding of glucose.

Srere, P. A. (1984). Why are enzymes so big? *Trends Biochem. Sci.*, 9, 387–90.

A necessarily speculative, but interesting article.

Koshland Jr., D. E. (1987). Evolution of catalytic function. *Cold Spring Harbor Symp. Quant. Biol.*, LII, 1–7.
A concise clear summary of the basic roles of active sites and of enzyme conformational change in catalysis.

Kraut, J. (1988). How do enzymes work? *Science*, 242, 533–40.
Enzyme catalysis explained by the principle of transition state stabilization. Examples from the zinc protease mechanism and the generation of catalytic antibodies specific for transition state analogues.

The high energy phosphate compound concept

Lipmann, F. (1941). Metabolic generation and utilisation of phosphate bond energy. *Adv. Enzymol.*, 1, 99–162.
The famous review which initiated the concept of ATP driving cellular processes. Very old, but the concepts still apply.

Problems for Chapter 1

1   Suppose that a 70-kg man, moderately active, has a food intake for his energy needs of 10 000 kJ per day. Assume that the free energy available in his diet is utilized to form ATP from ADP and P_i with an efficiency of 50%. In the cell, the ΔG for the conversion of ADP + P_i is approximately 55 kJ mol^{-1}. Calculate the total weight of ATP the man synthesizes per day, in terms of the disodium salt (molecular weight = 551).

2 The $\Delta G^{0\prime}$ for the hydrolysis of ATP to ADP and P_i is 30.5 kJ mol^{-1}. Explain why, in Problem 1 a value of 55 kJ mol^{-1} was suggested as the amount of free energy required to synthesize a mole of ATP from ADP and P_i in the cell.

3 The hydrolysis of ATP to ADP + P_i and that of ADP to AMP + P_i have $\Delta G^{0\prime}$ values of -30.5 kJ mol^{-1}, while the hydrolysis of AMP to adenosine and P_i has a value of -14.2 kJ mol^{-1}. What are the reasons for the large difference?

4 What information can be drawn from a plot of the rate of activity of an enzyme against temperature?

5 Explain

    (a) why benzene will not dissolve in water to any significant extent;

    (b) why a polar molecule such as glucose is soluble in water;

    (c) why NaCl is soluble in water.

6 In the cell, ADP is converted to ATP using the energy derived from food catabolism, but AMP is not utilized by the ATP synthesizing system; many synthetic processes convert ATP to AMP. How is AMP brought back into the system?

7 An enzyme catalyses a reaction that synthesizes the compound XY from XOH and YH, coupled with the breakdown of ATP to AMP and PP_i. The $\Delta G^{0\prime}$ of the reaction $XOH + YH \rightarrow XY + H_2O$ is 10 kJ mol^{-1}. Determine the $\Delta G^{0\prime}$ of the reaction: (a) in the cell; and (b) using a completely pure preparation of the enzyme. Explain your answer. (You are told that the $\Delta G^{0\prime}$ values for ATP hydrolysis to AMP and PP_i and for PP_i hydrolysis are -32.2 and -33.4 kJ mol^{-1}, respectively.)

8 What are the different types of weak bonds of importance in biological systems and their approximate energies. Why is it that their formation does not require enzymic catalysis? If they are so weak, why are they of importance?

9 The pK_a values for phosphoric acid are 2.2, 7.2, and 12.3.

    (a) Write down the predominant ionic forms at pH 0, pH 4, pH 9, and pH 14.

    (b) What would be the pH of an equimolar mixture of NaH_2PO_4 and Na_2HPO_4?

    (c) Suppose that you wanted a buffer fairly close to physiological pH and all you had available were the 20 amino acids found in proteins; which one would be the most suitable?

10 The oxidation of glucose to CO_2 and water has a very large negative $\Delta G^{0\prime}$ value, and yet glucose is quite stable in the presence of oxygen. Why is this so?

11 Explain how an enzyme catalyses reactions.

Chapter summary

Previous page: Structure of a bacterial DNA methylating enzyme (the protein in white) attached to its substrate, double-stranded DNA (DNA backbone sugars in blue and phosphates in purple; stacked base pairs in orange; bases in the recognition sequence in red). The target cytosine residue (red) has 'flipped out' of the B DNA helix to occupy the enzyme catalytic site. The structure in green is the other product of the reaction, *S*-adenosyl homocysteine (see page 203).

Photograph: Drs Xiaodong Cheng and Rich Roberts in Blackburn, G M and Gait, M J, (1996), *Nucleic Acids in Chemistry and Biology,* second edition, Oxford University Press.

Chapter 2

The structure of proteins

It has been said, with considerable justification, that it is improbable that there exists in the entire universe any type of molecule with properties more remarkable than those of proteins. To understand biochemistry it is essential to appreciate the overwhelmingly dominant role of proteins in life. If a specific job is to be done, it is almost always a protein that does it. As stated in the previous chapter, life depends on thousands of different proteins whose structures are fashioned so that individual protein molecules combine, with exquisite precision, with other molecules. Chemical reactions in the cell depend on enzymes combining with substrates and these are often controlled by allosteric effectors combining with specific sites on the proteins. Structures such as muscle depend on protein–protein interactions, gene control depends on protein–DNA combinations, and hormone control depends on hormone interactions with protein receptors. Transport across membranes involves protein–solute interactions, nerve activity involves transmitter substance–protein interactions, and immune protection involves antibody–antigen interactions. All are specific interactions; in detail, the list is almost endless.

This chapter deals with the chemical structure of these giant molecules that carry out such a range of activities.

The primary structure of proteins

Proteins are **polypeptide chains**. A protein molecule may have more than one such chain but for the moment we will refer to single-chain proteins, which are quite common. A polypeptide chain consists of a large number of **amino acids** linked together. Twenty different amino acid structures are used; if we regard these different amino acids as an alphabet of 20 letters, a polypeptide chain is a word usually hundreds of letters in length. The biggest polypeptide chain has about 5000 amino acids, but most have less than 2000. There can, in principle, be an almost infinite number of different 'words' or proteins. Evolution is not limited by the number of different primary structures that could

exist. This is presumably one major reason why proteins were selected for the vast number of different roles that they have. A limited number of different primary structures to select from wouldn't satisfy the needs of evolution for, as you will see, the chance of a random series of amino acids linked together forming a functional protein is small.

An α-amino acid, written in the nonionized form has the structure (a); in aqueous neutral solution it exists as the **zwitterionic** form (b).

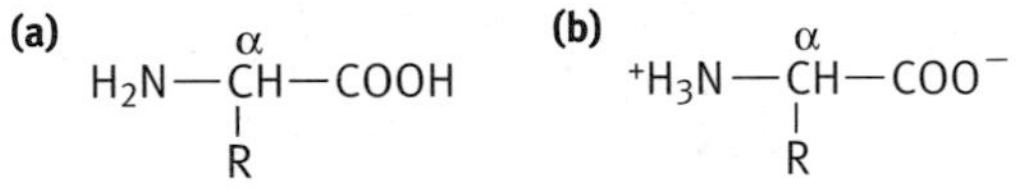

(a)

$$H_2N-\overset{\alpha}{\underset{R}{CH}}-COOH$$

(b)

$$^+H_3N-\overset{\alpha}{\underset{R}{CH}}-COO^-$$

Every amino acid, with the exception of the amino acid proline (see below), has the same $H_2N-CH-COOH$ part—only the R group attached to the α carbon atom varies. All amino acids in proteins are of the L-configuration (except glycine which has no asymmetric carbon atom).

Two amino acids can be linked together by the removal of H_2O (it doesn't happen by such a direct process in the cell) to form a dipeptide, as shown below (structures written in the non-ionized form for clarity). The —CO—NH— bond is the **peptide bond** or link.

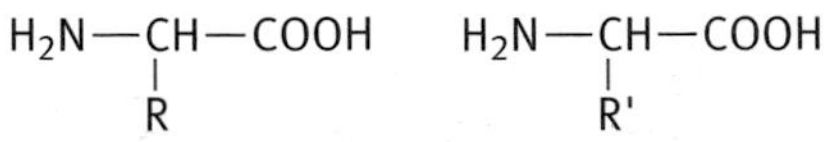

$$H_2N-\underset{R}{CH}-COOH \qquad H_2N-\underset{R'}{CH}-COOH$$

Two amino acids

$$H_2N-\underset{R}{CH}-\overset{O}{\overset{||}{C}}-NH-\underset{R'}{CH}-COOH$$

Peptide bond

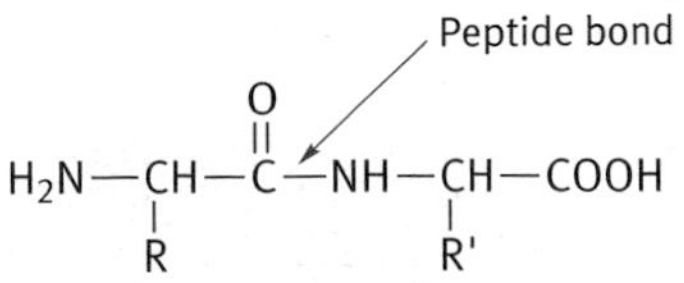

A dipeptide

The dipeptide still has an amino group at one end and a carboxyl group at the other, so that more and more amino acids can be added, giving the polypeptide structure shown.

$$H_2N-CH-\overset{\overset{O}{\|}}{C}-NH-CH-\overset{\overset{O}{\|}}{C}\left[NH-CH-\overset{\overset{O}{\|}}{C}\right]_n NH-CH-COOH$$
$$\qquad | \qquad\qquad | \qquad\qquad\quad | \qquad\qquad\quad |$$
$$\qquad R \qquad\qquad R^2 \qquad\qquad R^n \qquad\qquad R^3$$

A polypeptide

There are no absolute dividing lines in naming peptides of different lengths, but a short chain, say up to 20 amino acids, is called a peptide or oligopeptide (*oligo* = few), a polypeptide has more than about 80 amino acids, while, as stated, a protein may have several such chains. These may be as discrete subunits, which assemble together, by weak bonds, into a multimeric or multisubunit protein, or they may be covalently linked.

A few terms first: the $-NH_3^+$ end of a protein is referred to as the amino terminal or *N*-terminal end and the other as the carboxy terminal or *C*-terminal end. The central chain, without the R groups ($-CH-CO-NH-CH-CO-NH-CH-$, etc.), is called the **polypeptide backbone**. An amino acid in a protein is referred to as an amino acid residue; the R groups are called amino acid side chains, protein side chains, or quite simply side chains.

The order in which the 20 different amino acids are arranged in the polypeptide is the amino acid sequence. Determining the sequence is referred to as protein sequencing, or amino acid sequencing. The sequence is the **primary structure** of a protein. Sequencing is very important in protein chemistry. Fred Sanger in Cambridge was awarded the Nobel Prize in 1958 for determining the amino acid sequence of a protein (insulin) for the first time. Edman developed a machine for doing the job automatically, using different chemistry. Nowadays machines will sequence as little as 0.01 μg of protein. There are also amino acid analysers for determining the overall amino acid composition in proteins; the proteins are first hydrolysed in acid for this purpose

What is a native protein?

The primary structure of a protein gives the impression that proteins resemble extended pieces of string in shape. This is misleading as can be illustrated by reference to egg white, which is largely composed of the protein egg albumin, a water-soluble colourless protein. This can be purified and crystallized as sharp, shiny crystals. The crystallizability depends on the molecules being defined in their compact three-dimensional structure so that they can pack together in the ordered array of a crystal, for crystallization normally demands that all the molecules are identical in shape. The polypeptide chain is, in its natural state in globular proteins, folded up in a complex, specific way into a compact structure. There is no exact definition of a **globular protein** except that it has a small axial ratio in contrast to a **fibrous protein**, which has an elongated shape. The three-

dimensional structure of globular proteins—the precise folding of the polypeptide chain—is usually destroyed by heat. The folded form is called **native** protein, the unfolded form, a randomly arranged polypeptide chain, is called the **denatured** protein, and the conversion from one to the other is known as **denaturation** (see also page 63 for denaturation by acid conditions).

When an egg is fried, the transparent egg white, as it is subjected to mild heat, turns opaque and becomes an insoluble white mass. The discrete folded chains of the native protein molecules unfold with heat and become irreversibly tangled up together to form this (Fig. 2.1). The phenomenon illustrates that the folding of the native protein is held together completely or largely by weak noncovalent bonds (page 13), for covalent bonds are not disrupted by mild heat. The polypeptide chain itself remains intact with this treatment. The biochemical function of proteins, such as enzyme catalysis, is almost always destroyed during the heating process (though a few heat-stable proteins do exist). Heat is lethal to cells because of this.

Having discussed what the primary structure of a protein is, we must now look at the folded three-dimensional arrangement of that polypeptide chain.

What are the basic considerations that determine the three-dimensional structure of a protein?

As already indicated, most proteins are globular, compact

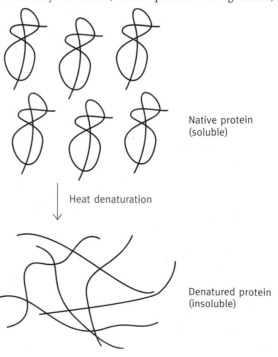

Native protein
(soluble)

Heat denaturation

Denatured protein
(insoluble)

Fig. 2.1 Hypothetical representation of egg albumin denaturation. (The folded configuration is drawn arbitrarily.)

molecules and, except for fibrous proteins, are not elongated chains. From Chapter 1 (page 14) you will realize that, for a molecule to achieve a stable globular shape in an aqueous medium, polar groups must be on the outside in contact with water, while the hydrophobic groups must be, to the maximum possible extent, inside, out of contact with water. Any groups inside the molecule that have the potential to form hydrogen bonds, or ionic bonds, but are not, in fact, involved in appropriate bonding would tend to destabilize the molecule. Thus an ionized group, for example, inside the hydrophobic interior of a protein would tend to destabilize it.

The problems of making a functional globular protein might well appear to verge on the impossible. There is a chain of perhaps hundreds of amino acid residues, all of which must fold up to form a tightly packed solid globular structure with polar side chains on the outside and hydrophobic side chains on the inside, shielded from water. Also on the outside there are pockets so shaped as to fit to, for example, substrate molecules, with exquisite precision, plus, often, sites for **allosteric effectors**—molecules that bind to and control the activities of proteins (described in Chapter 12).

Evolution, in essence, involves the development of new proteins and, with the above considerations in mind, it is not surprising that it is a slow business, being measured timewise in millions of years; the chances of accidentally fulfilling all the requirements for a properly folded protein to perform a specific task are exceedingly small. This helps to explain why, repeatedly, it is found that proteins with different functions are related in structure as if they have evolved one from another. It seems that, once evolution has found a polypeptide structure that folds nicely and performs a function, it may often have been easier to tinker with it and adapt it to other uses rather than start all over to find totally different structures *de novo*.

Before we can deal with the way proteins are folded we must first look at the building blocks from which they are constructed—the 20 different amino acids.

Structures of 20 amino acids

Learning 20 different structures of amino acids may be more attractive if you keep in mind what the different side groups are for. They are the building blocks with which evolution juggles to produce polypeptide chains that will fold up to satisfy the almost impossible requirements listed above. Clearly, there must be a variety of amino acids with different shapes, sizes, and polarity characteristics to permit the evolution of proteins. You can almost imagine the process—a small hydrophobic group needed here to fill a hole adjacent to a large group, a strongly polar group here and weaker one there, and so on. Having 20 different units to play with gives evolution flexibility in designing protein structures.

There are many non-protein amino acids found in different organisms, but only 20 are used for protein synthesis—what Crick has called the 'magic 20'. They are the only ones specified in the genetic code. If the latter statement means nothing to you, don't worry for it is explained in later chapters. We will start with **aliphatic hydrophobic chains** presented in order of increasing size.

$$^+H_3N-CH-COO^-$$
$$|$$
$$H$$

Glycine

Glycine with H as its side group is the smallest amino acid and has no marked hydrophilic or hydrophobic properties. The rest of the aliphatic nonpolar side chains have increasing hydrophobicity.

$$^+H_3N-CH-COO^- \qquad ^+H_3N-CH-COO^-$$
$$|$$
$$CH_3 \qquad\qquad CH$$
$$CH_3 \quad CH_3$$

Alanine   Valine

$$^+H_3N-CH-COO^- \qquad ^+H_3N-CH-COO^-$$
$$|$$
$$CH_2 \qquad\qquad CH$$
$$| \qquad\qquad CH_2 \quad CH_3$$
$$CH \qquad\qquad |$$
$$CH_3 \quad CH_3 \qquad CH_3$$

Leucine   Isoleucine

The latter three are known as the **branched-chain aliphatics**. Methionine is a rather special hydrophobic aliphatic amino acid (special, more because of its role in methyl group metabolism than its role in protein structure, page 202).

$$^+H_3N-CH-COO^-$$
$$|$$
$$CH_2$$
$$|$$
$$CH_2$$
$$|$$
$$S$$
$$|$$
$$CH_3$$

Methionine

Then we come to the really large hydrophobic side chains—**aromatic** ones.

$$^+H_3N-CH-COO^- \quad ^+H_3N-CH-COO^- \quad ^+H_3N-CH-COO^-$$
$$| \qquad\qquad | \qquad\qquad |$$
$$CH_2 \qquad\qquad CH_2 \qquad\qquad CH_2$$

Phenylalanine   Tyrosine   Tryptophan

Tyrosine, with its —OH group, can form a hydrogen bond but its aromatic ring is large and hydrophobic so it is of somewhat mixed classification.

The hydrophilic amino acids

An ionized group is very hydrophilic. Acidic amino acids (with an extra —COOH on the side chain) are negatively charged at the pH of the cell and thus hydrophilic, and basic amino acids (extra —NH$_2$ groups on the side chains) are positively charged and also hydrophilic. Aspartic and glutamic acids are the acidic pair, lysine and arginine the strongly basic pair, and histidine is a weakly basic amino acid. The ring structure of histidine is known as the **imidazole ring**.

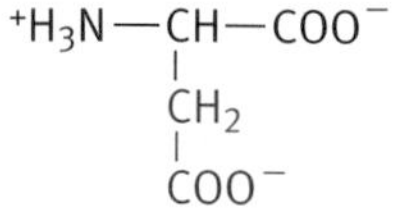

Aspartic acid

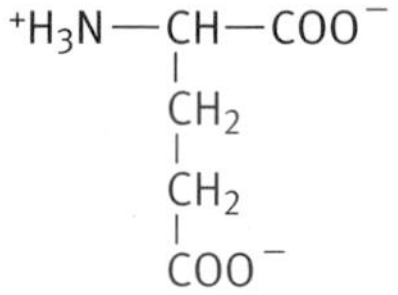

Glutamic acid

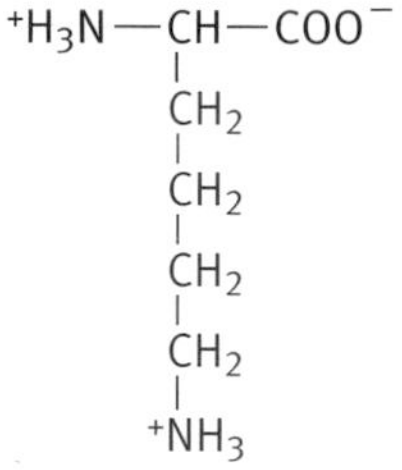

Lysine

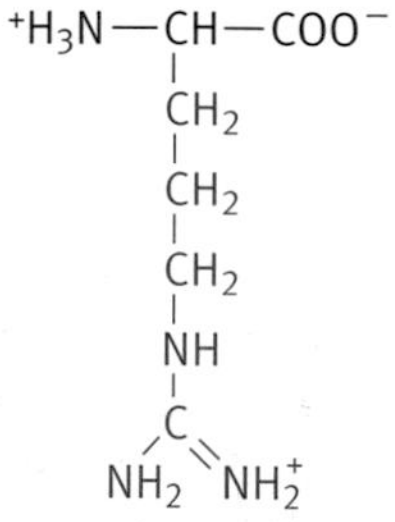

Arginine

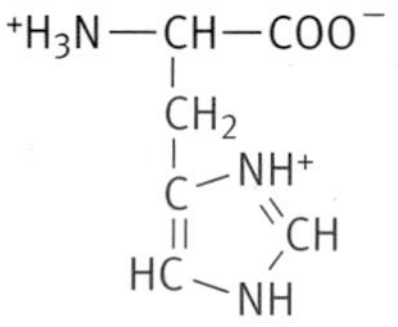

Histidine

The two acidic amino acids, aspartic and glutamic acids, exist also as the amides, asparagine and glutamine, respectively.

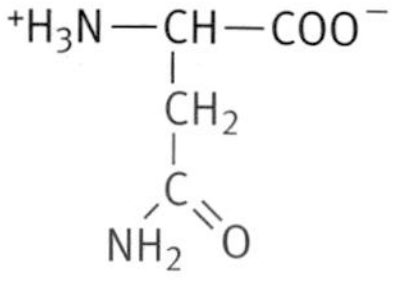

Asparagine

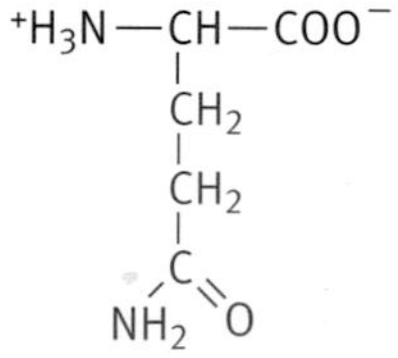

Glutamine

The two hydroxylated amino acids, serine and threonine, are hydrophilic.

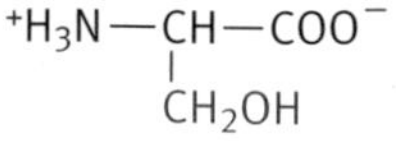

Serine

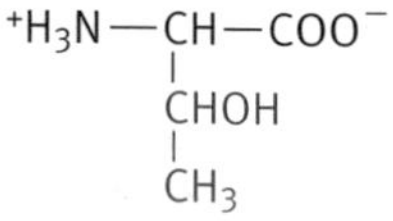

Threonine

Amino acids for special purposes

Cysteine is similar to serine but with —SH instead of —OH; it plays two special roles in protein structure—supplying external —SH groups such as in active centres of enzymes and in forming covalent —S—S— bonds or disulfide bonds internally—but more of this later.

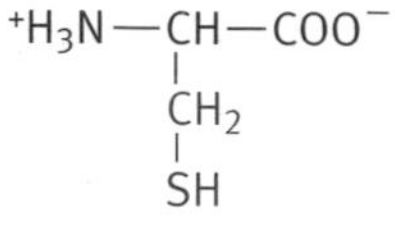

Cysteine

Proline is really the oddity—literally a kinky amino acid in that it can put a kink into the conformation of polypeptide chains (described later). It has an imino (—NH) group rather than an amino group.

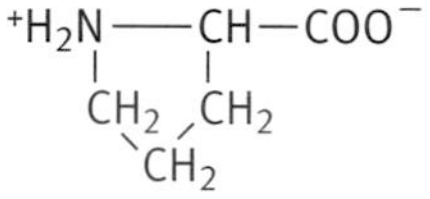

Proline

Ionization of amino acids

As already mentioned, free amino acids in aqueous solution have the zwitterionic structures shown above, in which the α-amino and α-carboxyl groups are ionized. Their pK_a values are around 8–10 and 2, respectively (see the appendix to Chapter 1 if you need to be reminded about such values). However, when incorporated into a polypeptide chain, these dissociable groups are blocked by peptide bond formation, the terminal amino and carboxyl groups excepted. The ionized state of a protein is therefore almost entirely dependent on the amino acid residues of aspartic and glutamic acids, lysine, arginine, and histidine since their ionizable side-chain groups are not blocked by peptide bond formation, as shown for glutamic acid and lysine in the illustration below.

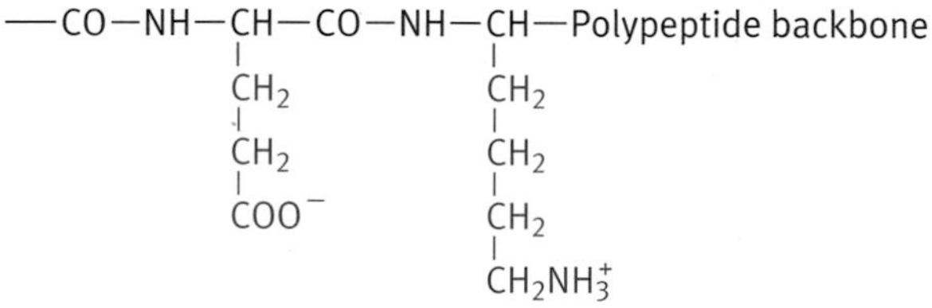

Glutamic acid side chain

Lysine side chain

The side chain —COO$^-$ groups of aspartic and glutamic acids have pK_a values around 4 so that they are virtually dissociated at pH 7.4—these provide the negatively charged acidic side groups of proteins. The basic amino acids, lysine and arginine, with pK_a values for their — NH$_3^+$ side-chain groups of 10.5 and 12.5, respectively, are fully ionized at physiological pH (Fig. 1.8). The third basic amino acid, histidine, has an imidazole ring as the side-chain group whose pK_a, in proteins, is near neutrality (6.04) so that histidine is often found in active sites of enzymes where movement of a proton is involved in the reaction catalysed; the imidazole group can accept or donate a proton at a pH near that existing in the cell.

The distribution of charged amino acid residues in a protein has an important effect on the conformation that a polypeptide chain can adopt. Negative charges close to each other repel each other and the same is true of positive charges. Closely positioned positive and negative charges will attract each other. Phosphorylation of proteins, so commonly used to control enzyme activity (page 165), introduces a strong negative charge that can affect the conformation of the molecule.

(a) Primary

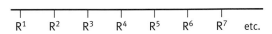

$R^1 \quad R^2 \quad R^3 \quad R^4 \quad R^5 \quad R^6 \quad R^7 \quad$ etc.

This structure will be represented below as a simple line.

(b) Secondary

The polypeptide **backbone** exists in different sections of a protein either as an α helix, a β-pleated sheet, or random coil. (These have yet to be explained.)

α helix   β-pleated sheets   Random coil or loop region

(c) Tertiary

The secondary structures above are folded into the compact globular protein.

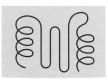

This protein will be represented below as:

(d) Quaternary

Protein molecules known as subunits assemble into a multimeric protein held together by weak forces.

Fig. 2.2 Diagrammatic illustration of what is meant by primary, secondary, tertiary, and quaternary structures of proteins. Note that the structures referred to will be described shortly in the text.

The different levels of protein structure—primary, secondary, tertiary, and quaternary

The sequence of amino acids that are linked together covalently into a polypeptide chain is, as stated, its primary structure (Fig. 2.2(a)). It says nothing about how that polypeptide is arranged in a three-dimensional space.

The polypeptide **backbone** itself is arranged in a particular conformation known as the secondary structure (Fig. 2.2(b)). The secondary structure is yet again folded up to give the tertiary structure (Fig. 2.2(c)). The complete molecule so formed by the primary, secondary, and tertiary structures may be the final functional protein or it may be a protein monomer or subunit so that to form a functional protein it must associate with other protein monomers (which may be the same or different). This association is known as the quaternary structure (Fig. 2.2(d)), the resultant multicomponent molecules being known as polymeric proteins or multisubunit proteins.

With this overview we will now deal in more detail with the different levels of protein structure; the primary structure we have already covered.

Secondary structure of proteins

This is concerned primarily with the arrangement of the polypeptide backbone rather than of the side chains. We have said that in a water-soluble globular protein, as far as possible, polar side chains are arranged (due to tertiary structure) on the outside and the hydrophobic ones on the inside but there is a separate problem with the polypeptide backbone. As it crosses back and forth to fold the whole molecule into a compact shape, it cannot avoid being exposed to the hydrophobic interior of the protein molecule. The problem is that the backbone has polar groups with the capability of hydrogen bonding—in fact, two bonds per amino acid unit, since the C=O and N—H of the peptide bond are capable of hydrogen bond formation. Unless this bonding potentiality is satisfied by actual bond formation, with a resultant liberation of free energy, the structure will be destabilized. The nonpolar side groups of amino acids in the interior of the molecule cannot hydrogen bond to the polypeptide backbone groups. So what can the latter hydrogen bond with? The answer is with groups on the same, or an adjacent, polypeptide backbone.

It should be emphasized again that we are not (yet) talking about side-chain packing, which comes into play in tertiary structure. We are still dealing with the problem of how the hydrogen-bonding requirements of the C=O and N—H groups of the polypeptide backbone are satisfied. There are two main classes of structures that solve the problem—the α **helix** in which the backbone is arranged in a spiral-like coil and the β-**pleated sheet** in which extended polypeptide backbones are side

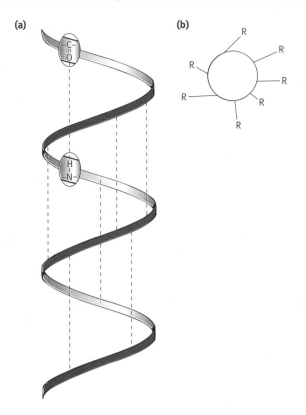

Fig. 2.3 Diagram of the α helix form of a polypeptide chain. **(a)** Illustration of hydrogen bonding between C=O and N—H groups of the polypeptide backbone (side chains not shown). The hydrogen bonds (broken lines) are shown in approximate positions only. **(b)** Diagram of an α helix looking down its axis, with the amino acid side chains projecting from the cylindrical structure (each at a different distance below the plane of the paper). The R groups are not drawn in their exact orientation from the axis. Since there are 3.6 residues per turn, each residue occurs every 100° around a circle (360°/3.6=100°).

by side. These structures are very stable; they can occur at the exterior of proteins with appropriate hydrophilic side chains or in the hydrophobic interior of proteins, with appropriate hydrophobic side chains.

The α helix

In this, the polypeptide backbone is twisted into a right-hand helix, called an α helix, a name used by Pauling because the structure was first recognized in α-keratin (which is described later). For L-amino acids, a right-handed helix is more stable than a left-handed one. You can visualize the direction of twist of the right-handed helix by imagining that you are a right-handed person driving home a screw with a screwdriver. Your fingers then follow the path of the chain. Its shape is given in Fig. 2.3(a).

It is almost natural to assume that in such a helix the number of amino acids constituting one turn would be a whole number,

and this seductive idea bedevilled the first attempts to produce a model of the α helix since the workers tried to build it with an integral number of amino acids per turn. Pauling was the first to discard the assumption and discovered the α helix structure with a pitch of 3.6 amino acid units per turn. This results in the C=O of each peptide bond being aligned to form a hydrogen bond with the peptide bond N—H of the fourth distant amino acid residue. The C=O groups point in the direction of the axis of the helix and are nicely aimed at the N—H groups with which they hydrogen bond, giving maximum bond strength and making the α helix a very stable structure. Thus, every C=O and N—H group of the polypeptide backbone are hydrogen bonded in pairs, forming a stable, cylindrical, rod-like structure (Fig. 2.3(a)). In cross-section an α helix is a virtual solid cylinder with all the R groups projecting outwards (Fig. 2.3(b)). The van der Waals atomic radii are such that there is almost no space along the axis of the α helix, again making it a stable structure. Not all of the polypeptide chain of a globular protein is in the α helix form. The sections that are average 10 residues in length, but the range of lengths varies a great deal from this average in different proteins.

Amino acids vary in their tendency to form α helices. Proline in particular is an α helix breaker or terminator—a proline residue forces a bend in the structure. With the other amino acids, the polypeptide chain can rotate freely around two bonds per residue, as shown below. Note that, although the CO—NH peptide bond is written as an ordinary single bond (about which free rotation might be expected), in fact the peptide bond is a hybrid between two structures: (1) in which the bond between the carbon and oxygen atoms is a pure double bond and (2) in which it is a single bond.

(1)

$$\begin{array}{cc} -C & H \\ \quad \diagdown C-N \diagup \\ O^{} \quad \diagdown C- \end{array}$$

(2)

$$\begin{array}{cc} -C & H \\ \quad \diagdown C=N^+ \diagup \\ O^{-} \quad \diagdown C- \end{array}$$

The actual electron density is between the two giving the carbon–nitrogen bond about 40% double-bonded character. This is sufficient to prevent rotation about it, making the polypeptide chain more rigid. With proline in peptide linkage, there is no H atom on the nitrogen available for hydrogen bonding and the structure of the residue restricts rotation so that it cannot assume the conformation needed to fit into an α helix. Some amino acids are excellent helix formers; a few, like proline, tend to be helix breakers.

$$\begin{array}{c} \quad\quad O \quad\quad\quad\quad O \\ \quad\quad \| \quad\quad\quad\quad \| \\ R-CH-C-N-CH-C-NH- \\ \quad\quad\quad\quad | \quad\quad | \\ \quad\quad\quad\quad CH_2 \;\; CH_2 \\ \quad\quad\quad\quad\quad CH_2 \end{array}$$

Imino nitrogen in
peptide linkage not
able to hydrogen bond

We will return shortly to how runs of amino acids in α helices fit into protein structures but, before that, we must deal with the alternative to the α helix, namely, the β-pleated sheet. Proteins are often made up of mixtures of the two types, which are present in different sections of the polypeptide chain.

The β-pleated sheet

This also forms a stable structure in which the polar groups of the polypeptide backbone are hydrogen bonded to one another, thus enabling the structure to be stable, even in the hydrophobic interior of a globular protein.

The principle is exceedingly simple. The polypeptide chain lies in an extended or β form with the C=O and N—H groups hydrogen bonded to those of a neighbouring chain (which may be formed by the same chain folding back on itself, or may be a separate chain lying alongside; Fig. 2.4). The term 'β form' of a polypeptide chain comes from the fact that it was first recognized in β-keratin, described later. Several chains can thus form a sheet of polypeptide. It is pleated because successive α-carbon atoms of the amino acid residues lie slightly above and below the plane of the β sheet alternately. The adjacent polypeptide chains bonded together can run in the same direction (parallel) or opposite directions (antiparallel). In the latter case, a polypeptide may make tight 'β turns' to fold the chain back on itself.

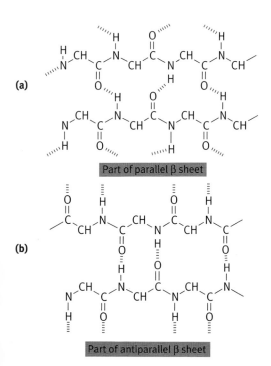

(a)

Part of parallel β sheet

(b)

Part of antiparallel β sheet

Fig. 2.4 (a) Diagram of hydrogen bonding between two polypeptide chains running in the same direction forming a parallel β sheet. The R groups attached to the —CH— groups are above and below the plane of the paper. **(b)** Adjacent polypeptide chains running in opposite directions can also mutually hydrogen bond forming an antiparallel β sheet. This enables a single chain to form a β sheet by folding back on itself.

Random coil or loop regions of the polypeptide chain

The **random coil** is neither random, nor a coil. The term indicates a section of polypeptide in a protein whose conformation is not recognizable as one of the defined structures above. A more common term is a **connecting loop region**. The structure of a random coil in a protein is mainly determined by side-chain interactions and, within a given protein, is fixed rather than varying in a random way as implied by the term. The structure may be in any conformation as determined by the various group interactions in the protein structure despite the term 'coil'. Since the structure of a random coil may not intrinsically satisfy the hydrogen-bonding potentials of the C=O and N—H groups of the backbone, or those of the side groups, such sections are found at the exterior of proteins, in contact with water.

Tertiary structure of proteins

To sum up so far—evolution has 20 'units', the amino acids, to play with in designing proteins. These are assembled into a polypeptide chain, their sequence being the primary structure, and this is folded into secondary structures, namely the α helix, the β-pleated sheet, and the random coil. The former two internally satisfy the hydrogen-bonding potential of the polypeptide backbone groups. A single polypeptide chain in a protein can be arranged as a mixture of these various structures in different parts of the chain and these secondary structures have to be folded up and packed together to form the protein molecule. This arrangement of the various secondary structures into the compact structure of a globular protein is referred to as the **tertiary structure**.

To simplify structural diagrams of proteins, conventions have been adopted to depict secondary structures. An α helix is represented either as a solid cylinder, sometimes with a helix inside it, or alternatively as a helical ribbon (Fig. 2.5(a)). The individual sections of polypeptide chains or β strands that participate in β-pleated sheet formation are represented as broad arrows (Fig. 2.5(b)); an arrow, let it be clear, is not a sheet but a polypeptide in a β-pleated sheet). An extended polypeptide chain in the β configuration has been shown always to twist slightly to the right so that the arrows are usually drawn in protein structures with this twist. Random coils are shown as any sort of line.

How are proteins made up of the three motifs—the α helix, the β-pleated sheet, and the random coil?

In principle, the answer is that any combination of the three can be used to assemble a globular protein provided that the side chains can be packed together in allowed ways, that is, ways appropriate to their various affinities and repulsions, and that the random coil sections have their backbone polar groups satisfactorily bonded, usually on the outside of the molecule. 'Allowed ways' refers to arrangements conforming with the

(a)

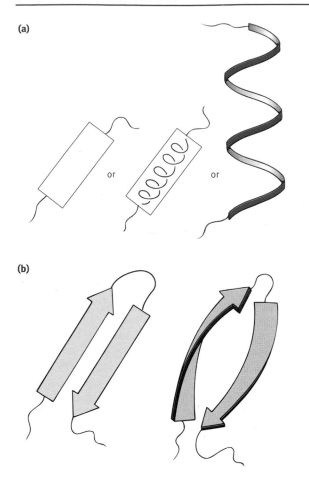

(b)

Fig. 2.5 Symbols used to indicate: **(a)** α helix structures; or **(b)** β-pleated sheets. In **(b)** a pair of polypeptide stretches are shown making an antiparallel sheet. Because of the slight right-handed twist the arrows are represented as on the right. The single lines connecting structures represent random coil or loop sections.

requirements that the structure must have secondary bond formation maximized, and exposure of hydrophobic groups to water or other polar groups minimized. These are formidable provisions, which probably only one in countless numbers of random amino acid sequences can fulfil. Quite large numbers of proteins have had their three-dimensional structure determined by X-ray diffraction studies. The picture is emerging that certain patterns of structure are preferred, so that the number of basic 'recipes' for protein structures may be limited. The three-dimensional structures of proteins are not things a non-specialist biochemist usually memorizes but they are very useful, for example, in discussing biochemical mechanisms. Figure 2.6 gives representations of a few protein structures. Myoglobin (Fig 2.6(a)), described later (page 379), has only α helices connected by loop sections with its heme group inserted into a cleft. The staphylococcal nuclease (Fig. 2.6(b); an enzyme hydrolysing nucleic acid) has a mixture of an antiparallel β sheet and three connected α helices. A common arrangement is the so-called α/β barrel. In this, there is a central core of β-strands arranged like

the staves of a wooden barrel except that they are twisted. The barrel encloses tightly packed hydrophobic side chains. Surrounding this barrel are α helices. The diagram of triosephosphate isomerase (Fig. 2.6(c)) illustrates this. Note that, although the arrows representing the eight β-strands appear to have insufficient contact for the mutual hydrogen bonding involved in forming a β sheet, this is the effect of representing the structure as viewed from the top of the barrel. The enzyme, pyruvate kinase (Fig. 2.6(d)), shows another type of structure with three such sections.

What forces hold the tertiary structure in position?

We have seen that the *secondary* structures of proteins (α helices and β sheets) depend on hydrogen bonding and, as explained, these secondary structures are arranged in specific ways to form the tertiary structure of the protein. Something must hold these sections in their tertiary structure. Hydrogen bonding and ionic interactions, between side chains, influence the folded structure, and the hydrophobic force (page 14) is a major contributor to the drive that folds a protein and results in the hydrophobic side chains being forced into the centre of the molecule, thus minimizing hydrophobic–water interactions and maximizing van der Waals bonding of hydrophobic groups. Most globular proteins depend entirely on the weak forces described above for their structure—but now see below for a qualification of this.

Where do the disulfide or S—S covalent bonds come into protein structure?

It was mentioned earlier that cysteine has special functions in proteins. One is that the thiol group often functions in specific reaction centres on proteins. (The glycolytic enzyme, glyceraldehyde 3-phosphate dehydrogenase, on page 110 is a good example of this.) However, cysteine has another function in proteins. Although the tertiary structure of proteins, as stated, is largely the result of weak forces between the side chains, evolution seems to have 'reasoned', as it were, that this is satisfactory for proteins protected in the sheltered environment of a cell but for other proteins, liberated into the blood (for example, insulin) or the intestine (for example, digestive enzymes), an extra bit of stability would be a good thing in the extracellular hard world. This is achieved by pairs of thiol groups of the cysteine side chains, brought together by polypeptide folding, forming a **covalent disulfide bond**. This provides a very strong interlocking bond—more or less like a steel rivet in the structure. A few of these makes the folded shape much more stable. Insulin has three disulfide or **S—S bridges**, as they are also called. Since covalent bonds are not disrupted by mild heat, proteins with disulfide bonds are often less easily denatured by heat.

To understand how these bonds form, a simple observation might be mentioned. If a water clear solution of the amino acid cysteine is left at neutral pH in an open beaker at room

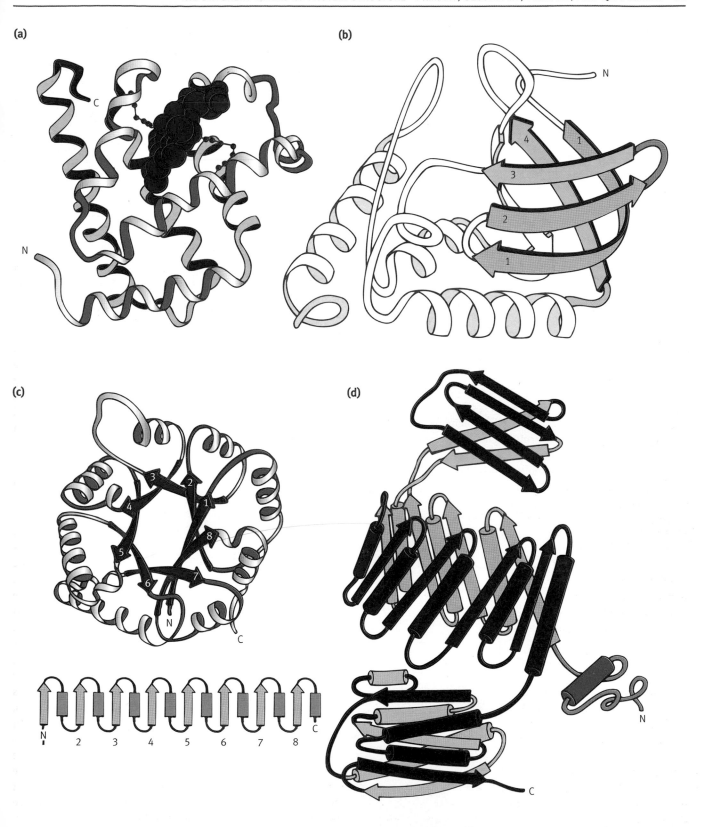

Fig. 2.6 Diagrams of the structures of different proteins. **(a)** Myoglobin; **(b)** staphylococcal nuclease; **(c)** triosephosphate isomerase (the lower diagram shows the arrangement of the numbered β sheets and α helices); **(d)** pyruvate kinase. (See text for description.)

temperature, the surface of the liquid, within hours, becomes covered with a layer of white insoluble crystals of cystine. The reaction is

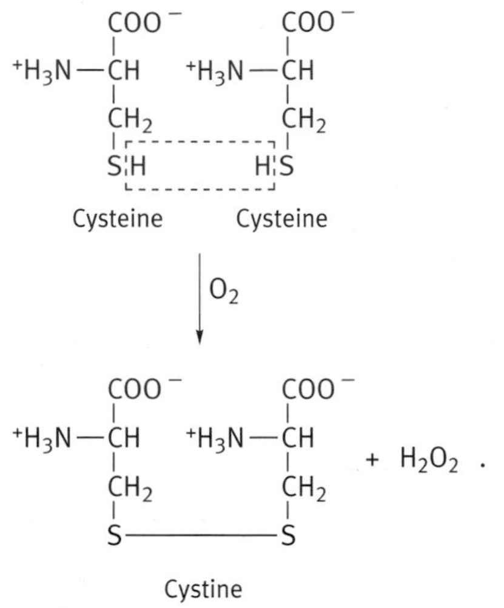

$$
\begin{array}{cc}
\text{COO}^- & \text{COO}^- \\
| & | \\
{}^+\text{H}_3\text{N}-\text{CH} & {}^+\text{H}_3\text{N}-\text{CH} \\
| & | \\
\text{CH}_2 & \text{CH}_2 \\
| & | \\
\text{S}\text{H} & \text{H}\text{S} \\
\end{array}
$$

Cysteine   Cysteine

$$\downarrow \text{O}_2$$

$$
\begin{array}{cc}
\text{COO}^- & \text{COO}^- \\
| & | \\
{}^+\text{H}_3\text{N}-\text{CH} & {}^+\text{H}_3\text{N}-\text{CH} \\
| & | \\
\text{CH}_2 & \text{CH}_2 \\
| & | \\
\text{S}\text{------}\text{S} & \\
\end{array}
\quad + \text{H}_2\text{O}_2 \ .
$$

Cystine

The same formation of —S—S— bonds occurs between appropriate thiols in proteins.

An extreme example of stabilization of a protein by disulfide bridges is the keratin protein of hair. The long polypeptides of this (described below) are interlinked by many disulfide bonds, which are important in determining the configuration of the hair. In permanent waving, these are broken by reduction followed by setting the hair into a new configuration. This is made permanent by the 'neutralizer', which reoxidizes cysteine —SH groups to reform disulfide bonds. However, these bonds are new ones between —SH groups that have been brought together in the curled configuration of the hair.

Quaternary structure of proteins

With the tertiary structure, we now have a protein molecule, and for many proteins this is the end. However, many functional proteins have more than one such protein molecule in them, which remain as discrete molecules but are held together in a single complex by secondary bonds. The molecules have to be structured so as to fit via complementary surface patches so that only the correct subunits complex together. The molecules formed by such a structure are called oligomeric, multimeric, or multisubunit proteins. Allosterically regulated enzymes are mostly of this type (see page 163), and so is hemoglobin (see page 379). This arrangement of subunits into a single functional complex is referred to as the **quaternary structure** of a protein. (Fig. 2.2).

Membrane proteins

We have so far described globular proteins as being water-soluble but many membrane proteins have to have an external section that is water-soluble at each end and a hydrophobic

section in the form of an α helix in the middle (see page 47). Polar groups in contact with the hydrocarbon layer in the centre of a membrane would be in an unstable situation. The structure of membrane proteins will be dealt with in the next chapter.

Conjugated proteins

Many proteins are simply as already described—they need nothing but the folded polypeptide chain(s) and any cofactors such as a metal ion or coenzyme noncovalently attached. However, some also have firmly bound attachments. When the protein is an enzyme and an additional molecule must be firmly attached to form the active enzyme, the attachment is called a **prosthetic group**, the active complex a **holoenzyme**, and the protein alone, the **apoenzyme** (*apo* = detached or separate). The cytochromes (page 121) with heme prosthetic groups and the dehydrogenases with bound flavin adenine dinucleotide (FAD) (page 97) are examples of such structures. Other proteins have carbohydrate attachments and are called **glycoproteins**. Most membrane proteins have oligosaccharides attached on the outside surface (page 48) to the —OH of serine or threonine side chains (*O*-linked) or to an asparagine side chain (*N*-linked). The latter attachment method is illustrated below.

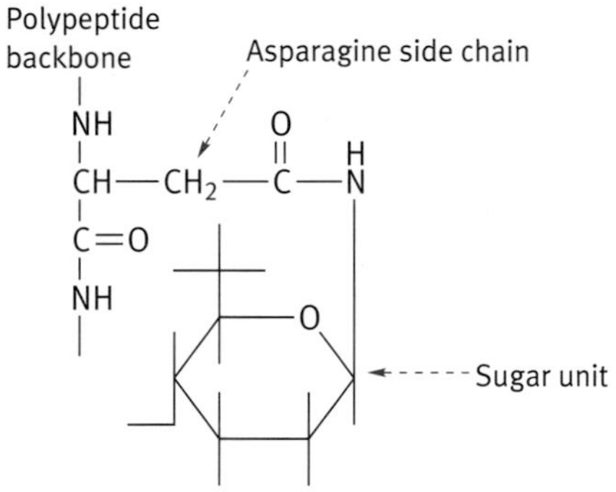

Polypeptide backbone   Asparagine side chain

$$
\begin{array}{c}
| \\
\text{NH} \\
| \quad\quad\quad\quad\quad \text{O} \\
\text{CH}-\text{CH}_2-\overset{\displaystyle \overset{\|}{}}{\text{C}}-\overset{\text{H}}{\underset{}{\text{N}}} \\
| \\
\text{C}=\text{O} \\
| \\
\text{NH} \\
|
\end{array}
$$

Sugar unit

Many secreted proteins such as blood proteins are glycoproteins. The function(s) of the carbohydrate attachments are not always, or perhaps not usually, understood. They may in some cases be related to the stability of the protein to which they are attached. They may also be involved in determining the longevity of the structure of which the glycoprotein is a component. For example, the degradation of carbohydrate attachments to serum proteins and to erythrocyte membrane proteins appears to mark them for uptake and destruction by liver cells—the carbohydrates are acting as indicators of age of the components almost. Conversely, glycosylation of some proteins may protect them from proteolytic attack. In some cases they are involved in recognition, for many combinations of different sugars can exist and thus might constitute unique 'labels'. In the sorting of proteins by the Golgi apparatus they play this role in some cases (discussed in Chapter 22).

In the case of the **mucin glycoproteins** which protect mucous membranes, the *O*-linked carbohydrate attachments cause the

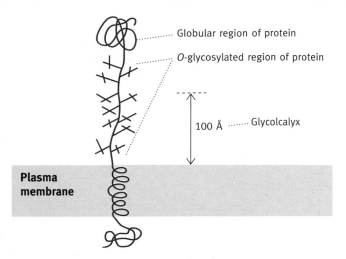

Fig. 2.7 Diagram of the low-density lipoprotein (LDL) receptor showing the *O*-glycosylation region as a stiff extended structure projecting >100 Å above the plasma membrane, with a globular protein domain (the functional receptor) beyond this.

polypeptide to adopt an extended configuration that facilitates their forming the networks found in mucins, even in very dilute solutions. This ability of *O*-linked carbohydrate attachment to cause the polypeptide to adopt an extended form probably has other applications. In **low-density lipoprotein (LDL) receptors** (page 88) the membrane protein is anchored into the membrane while the functional receptor domain is extended well clear of the membrane surface, readily accessible to its binding protein, by a rigid 'stalk' formed by *O*-glycosylation of the polypeptide in this region (Fig. 2.7). A whole family of **proteoglycans** exist in which the carbohydrates are rich in amino sugars, sulfated sugars, and carboxylic acid sugars.

They are believed to have important roles on cell surfaces, but their structures and functions are beyond the scope of this book.

What are protein modules or domains?

If you consider a protein molecule consisting of a single polypeptide chain, its folded structure may be a single compact entity, no part of which could exist on its own, for the folding of the protein would be disrupted.

However, especially in proteins that are larger than about 200 amino acids in length, this may not be the case. In such proteins when the three-dimensional structure is determined it is often seen that there are two or more regions that form compact 'islands' of folded structure usually linked together by unstructured polypeptide. Inspection of the structure leads to the impression that, if these folded sections were to be obtained separately, they would remain neatly folded in their native state. Indeed, in definite cases, they *have* been obtained separately and their integrity of structure confirmed. A definition of a **protein**

module or **domain**, as these sections are called, is that it is a subregion of the polypeptide that possesses the characteristics of a folded globular protein. A domain must be formed from a discrete section of polypeptide chain—it can't involve the folded structure being formed by the chain leaving the domain and then doubling back into it. It must be self-contained. A (somewhat fanciful) analogy is to liken it to individual movements of a symphony; each is self-contained with its own structural characteristics and could reasonably be listened to as a separate piece of music, but nonetheless is part of a single musical entity, linked to the others to form the whole and each with its purpose within the whole. The structure of enzyme pyruvate kinase (Fig. 2.6(d)) clearly illustrates three domains joined together to form a single protein.

Why should domains be of interest?

Protein domains are often associated with different partial activities of a protein that are involved in the function of the protein. The case of mammalian fatty acid synthase with seven catalytic activities needed for fat synthesis (page 140) being accommodated on a single polypeptide chain is an extreme example. In this case, since the same activities reside on separate proteins in bacteria, it is reasonable to speculate that the separate domains or modules of the mammalian enzyme arose from gene fusion. Many enzymes with a single catalytic function must combine with at least two substrates. The nicotinamide adenine dinucleotide (NAD$^+$) dehydrogenases (to be described later; page 96) are a typical case. They must bind NAD$^+$ and also the substrate to be oxidized. Different NAD$^+$ dehydrogenases all bind NAD$^+$, but each binds a different substrate and all catalyse the reaction

$$AH_2 + NAD^+ \leftrightarrow A + NADH + H^+.$$

It is found that enzymes catalysing such reactions have separate domains for binding NAD$^+$ and the substrate (AH$_2$). However, the NAD$^+$-binding domains of the different enzymes examined have a similar basic structure, while the AH$_2$-binding domains are different in the different enzymes. There are at least two possible interpretations of this. One is that it represents convergent evolution—that the NAD$^+$-binding structure is the only one capable of binding NAD$^+$ and therefore it is inevitable that all would be similar. The other is that, once evolution had successfully developed an NAD$^+$-binding module, it was used over and over again. This gives rise to the concept of **domain-shuffling** in evolution.

This envisages that modular construction of enzymes and other proteins permits rapid evolution of new functional proteins. It is somewhat like the modular assembly of electronic instruments. If you have 'boards' (made up of many components) with different broad functions, they can be assembled into different instruments by using various combinations of 'plug-in' boards. Similarly, many different functional proteins might be made by linking together different domains from various pre-existing proteins. Admittedly, it wouldn't do to construct musical symphonies by movement shuffling, to return to our previous analogy, but selection ruthlessly discards the disharmonies and preserves the successes.

All this might seem to be pure speculation based on similarities between proteins with similar functions, which, as mentioned, could have the alternative explanation of convergent evolution. But, as you'll see later (Chapter 21), domains sometimes appear to be coded for by specific separate gene sections, called exons (don't be concerned with what these are just yet). This strongly suggests that 'exon shuffling' and the consequent domain shuffling may well have been an evolutionary mechanism of importance. Developing new proteins by using different combinations of existing domains (with appropriate 'tinkering' for limited modified function) could be a more rapid process than relying on random changing of individual amino acids in entire proteins. We will return to other examples of protein domains when we deal with antibodies (Chapter 25).

The proteins of hair and connective tissues

It seems remarkable that proteins should be the stuff of cells and also of intractable, insoluble, tough materials such as hair, hooves, horns, tendons, skin, and the like.

There are different biological requirements for such proteins. Hair simply has to be a long, tough, totally insoluble fibre. The structural protein is α-keratin. Tendons are the long white structures which link muscles to bones—they have to be completely inelastic and are, per unit mass, stronger than steel. It wouldn't do if, when you contracted a muscle, the connecting link stretched. The structural protein is **collagen**. Skin cells have to be sheet-tough in all directions rather than in that of one axis, as is the case with tendons. A quite different requirement exists for arteries and lungs. Here you need a tough material that is elastic so that it can expand but that always snaps back to its original shape when the pressure is released. You can imagine why the structural protein for this is called **elastin**.

Let us now look at the structures of the proteins of keratin, collagen, and elastin. An important general point is that, in dealing with proteins so far, the weak bond has been of paramount importance in determining the three-dimensional characteristics of the molecule. In the proteins we are now discussing, covalent bonds are of great importance because we are in an area where strength is needed and weak bonds alone are too delicate to link together appropriate structures.

The α-keratins of hair, wool, horn, and hooves

The keratins of structures such as hair and wool are made from intermediate filaments (see page 398). The components of these

are long polypeptide chains with extensive domains of α helical regions containing repeating similar heptapeptides forming a central rod with non-helical domains at each end which vary from one keratin to another. Two of these polypeptides are lined up parallel with their helical portions in register. The central domain is a coiled coil in which the nonpolar amino acid residues of the two polypeptides are in contact and shielded from water. The structure is additionally stabilized by large numbers of disulfide bonds between cysteine residues in the adjacent polypeptides. These coiled-coil dimers further aggregate in an antiparallel fashion to form tetramers analogous to a twisted rope made of four strands. There is also tentative evidence that in intermediate filaments the tetramers may be aggregated octamer structures. Individual α-keratins vary in the detail of their molecular structures.

The structure of collagens

Collagen occurs outside of cells. The protein from which it is assembled is secreted by cells in the form of procollagen which is subjected to a variety of chemical changes catalysed by enzymes, resulting in the mature collagen. **Procollagen** consists of a triple superhelix—three helical polypeptides twisted around each other (see Fig. 2.8). At the ends are extra peptides, which, after secretion of the procollagen, are cleaved to give **tropocollagen** molecules from which collagen is assembled. Each of the polypeptides in the triple superhelix of tropocollagen is an unusual left-handed helix (cf. the common right-handed α helix of globular proteins). About one in three amino acid residues is proline and every third residue is glycine.

After synthesis of the molecule, many of the proline and also lysine residues are hydroxylated to form hydroxyproline and hydroxylysine in the polypeptide. These hydroxylated amino acids are not in the 'magic 20' used to synthesize proteins and are formed after the parent amino acids are in polypeptide form. Hydroxylation of proline in the polypeptide requires ascorbic acid or vitamin C, which keeps an essential Fe^{2+} atom in the enzyme, prolyl hydroxylase, in the reduced form. In deficiency of this vitamin, connective tissue fails to be properly formed, resulting in the painful consequences of scurvy.

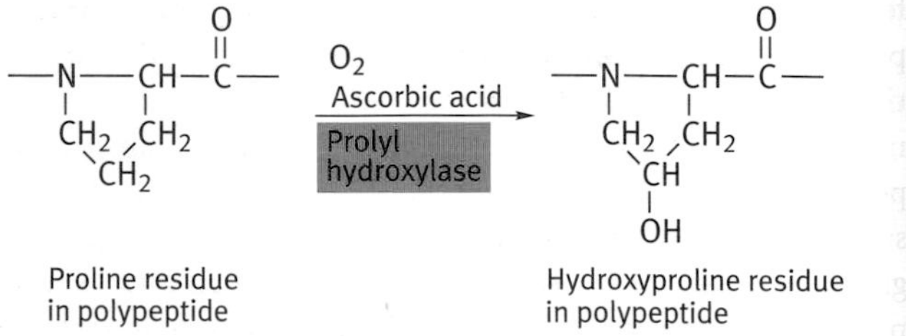

Proline residue in polypeptide → Hydroxyproline residue in polypeptide

(Note that the hydroxylation reaction is more complex than shown here and also involves α-ketoglutarate)

As stated, the peculiar amino acid sequence of collagen results in each polypeptide being a left-handed helix with three residues

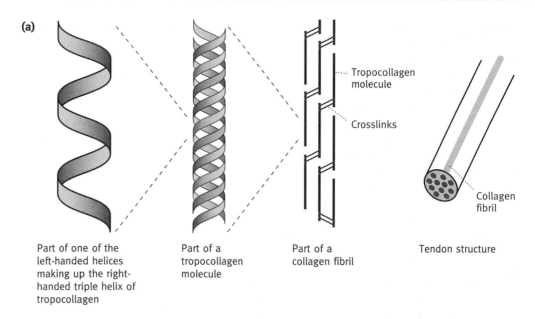

(a)

Part of one of the
left-handed helices
making up the right-
handed triple helix of
tropocollagen

Part of a
tropocollagen
molecule

Part of a
collagen fibril

Tropocollagen
molecule

Crosslinks

Collagen
fibril

Tendon structure

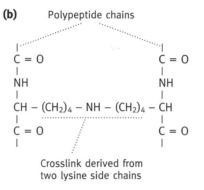

(b) Polypeptide chains

Crosslink derived from
two lysine side chains

Fig. 2.8 (a) Diagram of arrangement of collagen fibrils in collagen fibres. The bonds in red are covalent links formed between lysine residues. They exist also within the triple helix. **(b)** One type of crosslink formed between two adjacent lysine residues. Note that several specific types of collagen exist for individual functions.

per turn, three strands of this helix being twisted together into a triple superhelix. The design of the collagen superhelix illustrates rather beautifully what can be done by the protein 'engineering' that occurs in evolution. The three strands of the triple superhelix are in very close association with one another, forming a very strong structure. The bulky side chains of polypeptide chains would normally prevent such close association. In collagen, the helical structure of each polypeptide is more extended than in an α helix with three amino acid residues per turn. Every third residue is glycine and, in the triple superhelix, the contacts between the chains occur always at glycine residues whose side chain is a hydrogen atom that does not get in the way of close contact. The hydroxylated lysines and prolines form hydrogen bonds between the three chains, thus stabilizing the superhelix.

The tropocollagen molecule so far described has about 1000 amino acid residues. These assemble into collagen fibrils by staggered head-to-tail arrangement as shown in Fig. 2.8(a). The

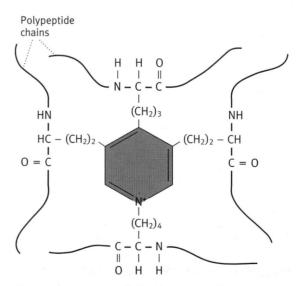

Fig. 2.9 Desmosine crosslink between four polypeptide chains of elastin. The structure is formed by enzyme modification of four lysine residues.

'holes' in this structure are believed to be sites where hydroxyapatite crystals ($Ca_5(PO_3)OH$) are laid down to initiate bone mineralization. The structure described so far would not have the required strength for a tendon. This is achieved by the formation of unusual covalent links between ends of the tropocollagen units; two adjacent lysine side chains are modified to form the link one type of which is shown in Fig 2.8(b). The collagen fibrils so formed aggregate to form tendons by a parallel arrangement. In skin, it is more of a two-dimensional network. Different subclasses of collagen exist dependent on the precise structure of the three polypeptides in the triple superhelix. Variations between the different types of chains include the content of hydroxylysine and hydroxyproline and the degree of glycosylation of these residues. A variety of genetic diseases exist in which processing of the procollagen is defective.

The structure of elastin

The elastin molecule has its own unique structure, different from that of collagen or α-keratin. It contains α helices, which can reversibly stretch. Unusual covalent crosslinks are formed between polypeptide chains, forming an elastic network that can stretch in any direction. The crosslinks are derived from four lysine residues forming a loose network that, together with extendible helices, produces a structure more elastic than rubber. The structure of the four-way crosslinkage (desmosine) is shown in Fig. 2.9.

The mystery of protein folding

How do proteins fold up into their secondary and tertiary configuration?

This chapter deals with protein structure; how the proteins are made is the subject of Chapter 21. However, we would at this point like to draw your attention to the fact that, while the primary structure (amino acid sequence) of a polypeptide chain dictates what protein structure is *possible*, the question of *how* the three-dimensional folding is actually achieved is one of the biggest unresolved biological problems. An 'incorrect' amino acid sequence may prevent the correct structure being achievable, but how does a 'correct' structure become folded, in the correct way? When synthesized in the cell, a typical polypeptide chain of a protein folds up in a couple of minutes or so. If it had to do this by randomly trying every theoretically possible folded structure, it would take countless millions of years to do so. In other words, the solution of trying every conformation until the lowest free energy state achievable is reached, is not feasible. When viewed in this light it looks to be one of the greatest problems to be overcome for life to be established. All of the answers are not yet known, but a good deal of information is available and this is presented later (page 304).

In this chapter we have described the structure of proteins from a general viewpoint. You will now understand what proteins are, some of the structural problems involved, and how these are solved. The best understood examples of how protein structure is related to function are the hemoglobins in oxygen transport and the antibodies in immune protection. However, instead of dealing with these now, it will be more convenient to deal with hemoglobins and antibodies when we deal with their biological functions.

The general plan of the book is to deal with metabolism before the molecular biology of the genome. The prerequisites for understanding metabolism include, as well as these first two chapters, a knowledge of membrane structure, for reasons that will become clear. We will therefore now go on to membranes in the next chapter, following which we will be in a position to deal with the metabolic section of the subject.

Further reading

Branden, C. and Tooze, J. (1991). *Introduction to protein structure*, Garland Publishing.
A comprehensive account of all aspects of protein structure, including the basic aspects and the structure of membrane proteins dealt with in this chapter. In addition it covers specific classes such as DNA-binding proteins, antibodies and receptors relevant to later chapters. Superbly illustrated with protein structures.

Protein size and structure

Chothia, C. (1984). Principles that determine the structure of proteins. *Ann. Rev. Biochem.*, **53**, 537–72.
Excellent for basics of protein structure.

Goodsell, D. S. (1991). Inside a living cell. *Trends Biochem. Sci.*, **16**, 203–6.

An attempt to present a picture of the distribution of molecules on a proper scale in a cell of *E. coli*.

Goodsell, D. S. (1993). Soluble proteins: size, shape and function. *Trends Biochem. Sci.*, **18**, 65–8.
A survey of protein structures—their size and shape. Why are they so big? Why are they oligomeric?'

Protein modules or domains

Doolittle, R. F. (1995). The multiplicity of domains in proteins. *Ann. Rev. Biochem.*, **64**, 287–314.
A discussion of domains, domain shuffling, exons, and introns. (The latter two topics are dealt with in Chapter 21 of this book).

Proteins of hair and connective tissue

Francis, M. J. O. and Duksin, D. (1983). Heritable disorders of

collagen metabolism. *Trends Biochem. Sci.*, **8**, 231–4.
A relatively simple account of this very complex and medically important field.

Hulmes, D. J. S. (1992). The collagen superfamily—diverse structures and assemblies. *Essays in Biochem.*, **27**, 49–67.
An account of the very complex structures found in different collagens, together with medical aspects.

Problems for Chapter 2

1 What is the primary structure of a protein?

2 What is meant by denaturation of a protein?

3 Write down the structure and name of an amino acid with each of the following side chains:
   (a) H;
   (b) aliphatic hydrophobic;
   (c) aromatic hydrophobic;
   (d) acidic;
   (e) basic.

4 Give the approximate pK_a values of:
   (a) acidic amino acid side chains;
   (b) basic amino acid side chains;
   (c) the histidine side chain.

5 Which amino acids are the major determinants of the charge of a polypeptide chain containing all 20 amino acids?

6 What are the four levels of protein structure?

7 Give an account of the secondary structures of proteins, together with their apparent purpose.

8 Explain how proteins, which form the delicate molecules of enzymes, etc., can also form tendons with tremendous tensile strength.

9 What is the peculiar structural feature of elastin that gives it its elastic properties?

Chapter 3

The cell membrane— a structure depending only on weak forces

In Chapter 1 we discussed in general the importance of three types of weak, noncovalent bonds as well as the hydrophobic force resulting from water molecules rejecting nonpolar molecules that interfere with hydrogen bonding. Formation of weak bonds involves a decrease in free energy so that only when formation of such bonds is maximized do you have the preferred most stable structure. This forms a link to the topic now to be dealt with—the cell membrane and its functions.

The plan of the next section of the book, on metabolism, is to start with food and then look at how it is digested, absorbed, and carried around the body, and its fate in the various tissues. After this, the metabolic pathways by which the process occur will be dealt with.

However, before food can enter the bloodstream it must pass through intestinal cell membranes. It is therefore necessary to deal with the biochemistry of membranes before that of food utilization. Also, the constituents of cell membranes are involved in the transport of fats around the body and it will help to first understand what these are.

Why are cell membranes needed?

The obvious answer is to hold together the contents of the cell. It is not certain how life originated, though knowledge has reached the point where possible mechanisms can be formulated. But, whatever the mechanism, one of the prerequisites must, at some stage, have been to contain the primitive self-replicating molecular system. If it weren't contained, presumably the molecular constituents would have been dispersed and the process aborted.

This requirement seems to pose a difficult problem, for a cell membrane is a complex structure. How could a primitive self-replicating system produce a containment membrane? The early containment structure may have been different from modern ones, but it may be no coincidence that, given the right type of molecule, structures resembling the permeability barrier of cells can form with ease—in fact, by agitation of the appropriate molecules in water. If the primitive replicating system happened to have available the right type of substance, agitation could have enclosed minute drops of solution containing the replicative molecular system inside structures similar to the basic structure of the membrane of a modern living cell. In other words the primitive cell membrane may have been self-assembling. The modern cell, of course, synthesizes its membranes, as described later.

What is the type of substance with such properties? The material extractable by organic solvents from egg yolk is one. The egg yolk extract contains polar lipids. A polar lipid (Fig. 3.1) is called an **amphipathic molecule** (*amphi* = both kind of) because it has a polar part (the so-called **head group**) and a nonpolar hydrophobic part (usually called a **hydrophobic tail**). The amphipathic molecules that might have formed the (speculative) primitive membrane need not have been the

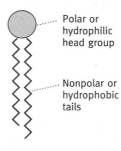

Polar or hydrophilic head group

Nonpolar or hydrophobic tails

Fig. 3.1 Diagram of an amphipathic molecule of the type found in cell membranes.

(a)

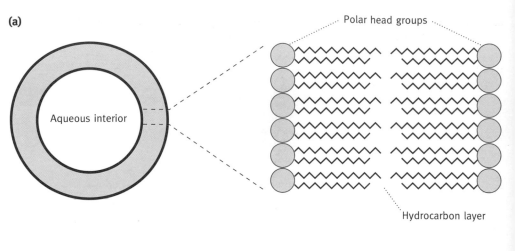

(b)

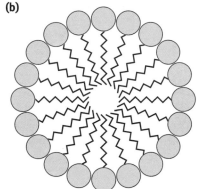

Fig. 3.2 (a) Diagram of a synthetic liposome made of a lipid bilayer structure. **(b)** Cross-section of a micelle. The possession of two hydrophobic tails on the amphipathic component molecule favours the lipid bilayer structure over the micelle structure.

modern polar lipids for, as long as their amphipathic properties were similar, the same forces would operate on them with the same result.

In the presence of water, such molecules are forced into structures in which the hydrophilic head groups have maximum contact with water, to maximize weak bond formation. Conversely, the hydrophobic tails are forced into minimum contact with water because such contact forces water into a higher energy arrangement (page 14) and therefore is resisted. Also, the contact between hydrophobic tails maximizes van der Waals interactions, again minimizing the free energy of the structure and therefore maximizing stability.

When polar lipids are agitated in an aqueous medium, one of the structures formed is called a **liposome**—a hollow spherical **lipid bilayer**. A lipid bilayer has two layers of polar lipids with their hydrophobic tails pointing inwards and their hydrophilic heads outwards in contact with water and each other as shown in Fig. 3.2(a).

If a synthetic liposome is sectioned and stained with a heavy metal that attaches to the polar heads, the lipid bilayer appears in the electron microscope as a pair of dark 'railway lines' due to absorption of electrons by the metal stain. A living cell membrane treated in the same way has an identical appearance.

All biological membranes have, as their basic structure, the lipid bilayer.

Another type of structure taken up by amphipathic molecules in water is the solid cylindrical **micelle** (Fig. 3.2(b)), but the dual hydrophobic tails on membrane polar lipids favour the bilayer structure, rather than the micelle structure.

The lipid bilayer

What are the polar lipid constituents of cell membranes?

As a general comment, there are a number of membrane polar lipids whose structures on paper look quite different from one another, but in space-filling models they all have the same basic shape as the molecule shown in Fig. 3.1 with a polar head and two hydrophobic tails. We will now deal with the structures of the various polar lipids, but the nonpolar lipid structure will be discussed also for comparison.

A lipid is a fat. A neutral fat is derived from **fatty acids**; neutral fats never occur in membranes but it will be helpful to discuss them first. A fatty acid has the structure R COOH where R is a long hydrocarbon chain. The most common lengths are C_{16} and C_{18}. They can be called hexadecanoic and octadecanoic acids, respectively, or by their common names, palmitic and stearic acids. They can also be represented as 16:0 and 18:0 which indicate the number of carbon atoms and the number of unsaturated bonds.

Stearic acid (C_{18}) has the structure

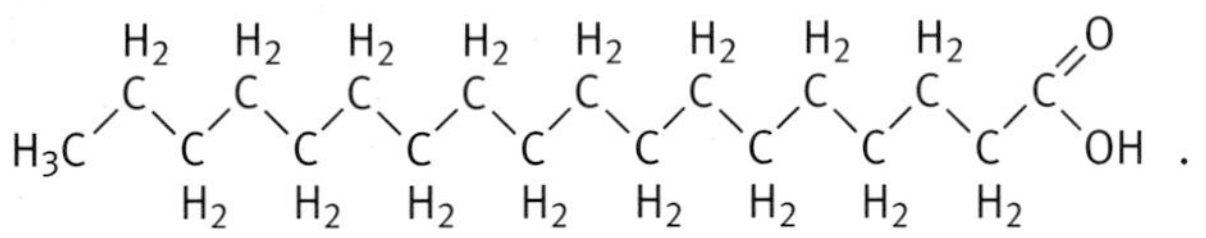

For convenience, fatty acids are often represented simply as

or $CH_3(CH_2)_{16}COOH$ (where the aim is simply to indicate a long hydrocarbon chain we usually don't bother to count the number of zigzags). The corresponding C_{16} compound is palmitic acid. At neutral pH, as say, the sodium salt, a fatty acid is a soap—what you wash yourself with.

Humans eat large quantities of fats which provide a substantial proportion of our food, but soaps are not a suitable part of the diet. They don't taste nice and are detergents. Instead we eat mainly **neutral fat** in which three molecules of fatty acids are esterified to the three hydroxyl groups of glycerol as shown. A carboxylic acid attached in ester linkage is an acyl group, and, since there are three acyl groups attached to one glycerol molecule, a neutral fat is called a **triacylglycerol** or TAG for short; the term **triglyceride** is also used.

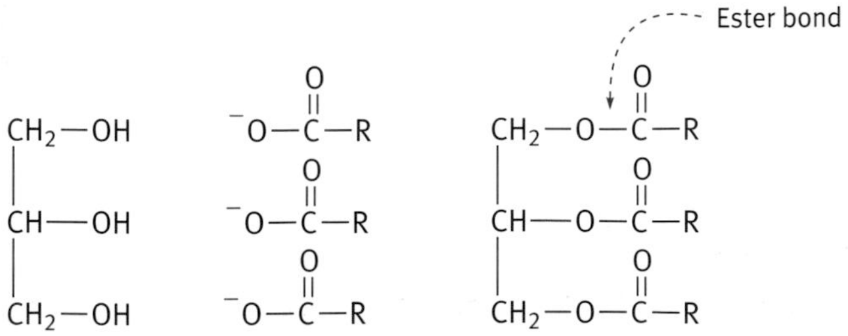

Glycerol | Three molecules of fatty acid | One molecule of neutral fat (or triacylglycerol (TAG) or triglyceride)

If neutral fat is boiled with NaOH or KOH, the ester bonds are hydrolysed, forming soaps and glycerol, which is how soap is made. Neutral fat is suitable as a food; it is not a detergent. As stated, *neutral fat does not occur in membranes*. We have discussed it to help in understanding what a polar lipid is by contrast. A TAG molecule has no polar groups and therefore forms oily droplets or particles in water—it could never produce

a lipid bilayer structure.

Membrane lipids containing glycerol are the most abundant form and are known as **glycerophospholipids**. They are based on glycerol-3-phosphate.

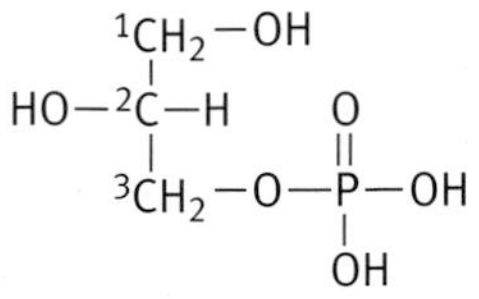

sn-Glycerol-3-phosphate

The central carbon atom of this compound is asymmetric; in natural glycerol-3-phosphate the secondary hydroxyl is represented on the Fischer projection to the left and the carbon atoms numbered from the top. The prefix sn, for stereospecific numbering, refers to the nomenclature system used here.

If two fatty acids are esterified to the primary and secondary hydroxyl groups of glycerol-3-phosphate, the product is **phosphatidic acid**.

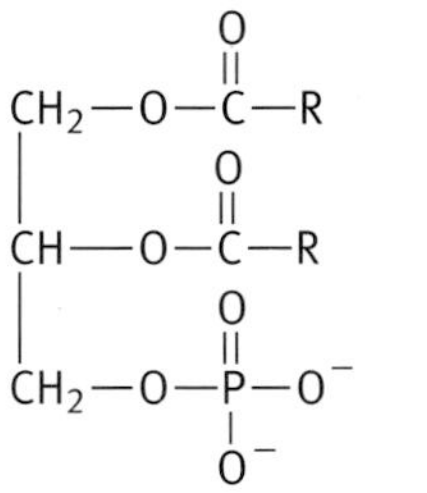

Phosphatidic acid

Fix the name, phosphatidic acid, in your mind for it makes the nomenclature of the rest simple. We can attach other polar molecules to the phosphoryl group. If phosphatidic acid is attached to something else (for example, 'X') it becomes a phosphatidyl-group (for example, phosphatidyl-X) with the structure

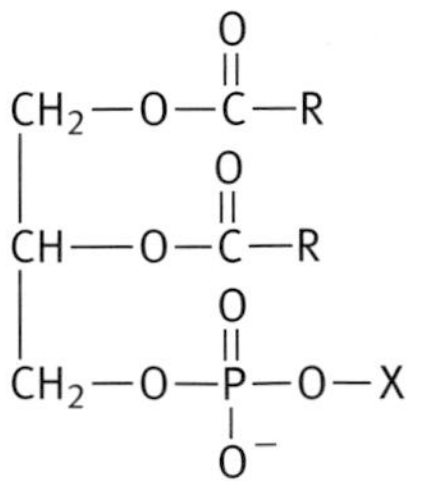

If X is also a highly polar molecule, we now have an extremely polar 'head group'.

The structure above gives a misleading impression of the shape of the molecule. A space-filling model would be somewhat like this:

Diagram of a glycerophospholipid

As emphasized in the introductory statement, it is the amphipathic membrane lipid shown in diagrammatic form in Fig. 3.1 that assembles into a lipid bilayer—a highly polar head group and two hydrophobic tails.

What are the polar groups attached to the phosphatidic acid?

In principle, only one species of polar lipid molecule is needed to form a lipid bilayer. However, living cells use a variety of structures to form their membranes, some quite complex—but all structures used have the same overall shape and amphipathic properties. We therefore have several different polar groups attached to the phosphatidic acid.

The first polar substituent is **ethanolamine** ($HOCH_2CH_2NH_3^+$) giving the phospholipid, **phosphatidyletha-nolamine** (PE for short) or the trivial name **cephalin**. Its structure is

If the nitrogen of ethanolamine is trimethylated, it is **choline**

and the phospholipid derived from it is **phosphatidylcholine** (PC) or **lecithin**. Its structure is the same as that given for PE above, apart from the three methyl groups on the nitrogen atom.

If the ethanolamine is carboxylated we have a **serine** substituent giving **phosphatidylserine** or PS (serine is $HOCH_2CHNH_3^+COO^-$). The attachment is like this.

Another quite different polar substituent is the hexahydric alcohol, **inositol**,

giving **phosphatidylinositol**, usually abbreviated to PI.

Many different ways have been used to ring the changes on glycerol-based polar lipids to form components of lipid bilayers. Here's yet another one—called **cardiolipin** or **diphosphatidyl-glycerol** in which two phosphatidic acids are linked by a third glycerol unit.

Central glycerol unit linking two phosphatidic acid molecules

The resultant structure still has the amphipathic shape to fit into lipid bilayers. It occurs in the inner mitochrondrial membrane and in bacterial cell membranes.

All of the above polar lipids are based on glycerol. However, the process of evolution has, with fascinating ingenuity, produced molecules of almost identical overall shape derived from a different structure called **sphingosine**.

The basic molecule isn't all that different from glycerol; the middle (C—2) hydroxyl group of glycerol is replaced by an —NH$_2$ group, while a hydrogen on carbon atom 1 is replaced by a C$_{15}$ hydrocarbon group—more or less a permanently 'built-in' hydrocarbon tail. One tail isn't enough but another can be added by attaching a real fatty acid to the central —NH$_2$ group by means of a —CO—NH-linkage.

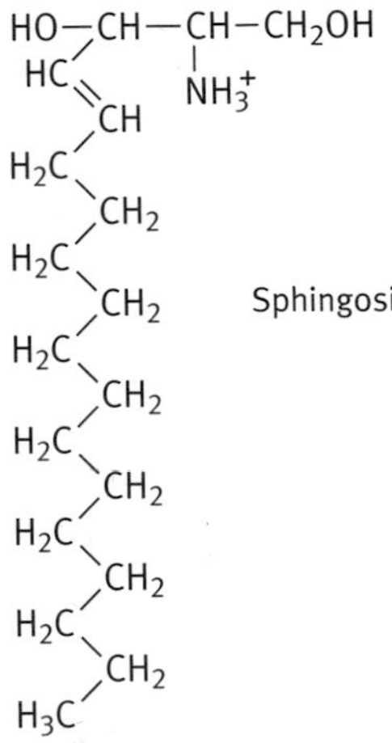

Sphingosine

This gives a molecule, called a **ceramide**, that is very similar in shape to a diacylglycerol.

Diagram of a ceramide

If now we add a phosphorylcholine group, as in lecithin, the ... sphingomyelin

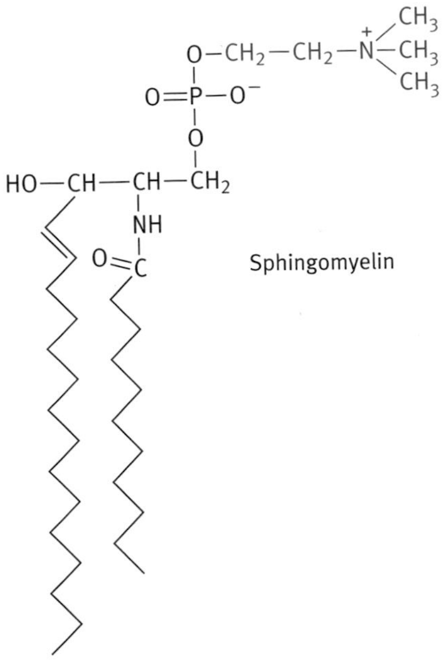

Sphingomyelin

which is similar in shape to lecithin. It is prevalent in the myelin sheath of nerve axons.

Evolution has been described as a tinkerer—it goes on modifying things and, if the change is beneficial, it is preserved. It has tinkered with the sphingosine-based polar molecules by adding different polar groups to the free —OH of the ceramide molecule. **Sugars** are highly polar and are used for this purpose.

Using a single sugar as the polar group we get a **cerebroside**, important in brain cell membranes as the name implies. Cerebrosides contain either **glucose** or **galactose**. The latter is a stereoisomer of glucose in which the C$_4$ hydroxyl is inverted.

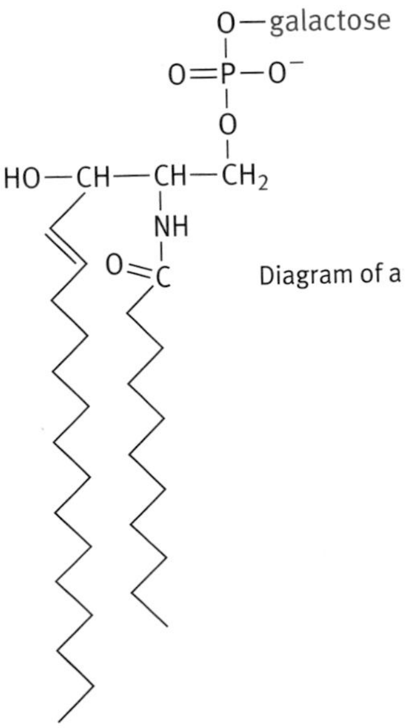

Diagram of a cerebroside

There is a wide variety of sugars in nature. Some are **amino sugars** such as glucosamine (whose structure is given below) or derivatives of them.

Combinations of the different sugars (**oligosaccharides**) can produce a variety of molecules in which a small number of different sugars are linked together in a branched oligosaccharide structure. (*Oligo* saccharide means a small polymer of sugars—perhaps 3–20 sugars rather than the hundreds or thousands found in polysaccharides.) Such an oligosaccharide is highly polar. When one of these is attached to a ceramide we have a **ganglioside**. This gives the molecule a very large highly polar head.

Diagram of a ganglioside

Not many biochemists would memorize the carbohydrate structures in these unless they had a special interest in them. However, you should be familiar with structures such as *N*-acetyl glucosamine and **sialic acid**.

N-Acetylglucosamine

Sialic acid
(also called *N*-acetylneuraminic acid

...ds are constituents of gangliosides and of ...ne proteins (see page 32).

Sialic acid is particularly interesting since it is involved in infection of cells by the influenza virus (page 318). Gangliosides, based on sphingosine, are involved in determination of the human blood groups O, A, and B. The carbohydrate portions are antigenic and differ in the terminal sugars of the oligosaccharide head groups of the membrane lipids.

A note on membrane lipid nomenclature

The names of individual polar lipids have been given above, but alternative collective terms are sometimes used. The terms 'membrane lipids' and 'polar lipids' cover any lipid found in cell membranes. 'Phospholipid' covers any lipid containing phosphorus. Those based on glycerol can be called glycerophospholipids as opposed to sphingomyelin which is a single specific sphingosine-based phospholipid. The ceramide-based membrane lipids with carbohydrate polar groups and lacking any phosphoryl group are called 'glycolipids' or 'glycosphingolipids' to indicate that they are based on sphingosine. A **plasmalogen** is a glycerophospholipid in which one of the hydrophobic tails is linked to glycerol by an ether bond (structure not given). (Cholesterol, another membrane component described below, is not classified as a lipid so none of the above terms include it.)

Why are there so many different types of membrane lipids?

The answer to this question is not known. The different membrane lipids confer different properties on the membrane surface. Lecithin (PC) has a positive charge, PS has a negative charge, and cerebrosides no charge. Different cells have quite different membrane lipid compositions. Cerebrosides and gangliosides are common in brain cell membranes, while the cell membranes of the myelin sheath of nerve axons are rich in glycosphingolipids. As well as different cells having different lipid compositions, the outer and inner halves of the bilayer of the one membrane may be different from one another and different membranes within the cell have different compositions. For example, glycolipids are always on the outer side of the bilayer so that their sugars point outwards from the cell into the external aqueous environment. This asymmetry is preserved by the fact that transverse movement of lipids known colloquially as 'flip-flop' (from one side of the membrane to another) is severely restricted; such movement would involve pushing the polar heads through the central hydrocarbon layer to get to the other side. This is energetically unfavourable and so the asymmetry is preserved. Proteins catalysing energy-dependent membrane flip-flop are known and may be involved in the creation and maintenance of asymmetry.

The very limited flip-flop movement of membrane lipids contrasts with their potential for rapid lateral movement within the plane of the bilayer. The assembly of lipids into the bilayer structure does not involve covalent bonds and, in general, they move around freely—a molecule can cover the length of the cell very rapidly.

In the case of one particular membrane lipid, phosph... inositol, we know of a very special role...

chemical signals that have profound effects on control of cellular events; this is dealt with later (page 359) but, in general, we have little or no idea why membranes contain so many different polar lipid components or why the composition varies from cell to cell and between inner and outer halves of a given bilayer. Some membrane lipids are of special interest. For example, there is a clinical interest in gangliosides, because of children's diseases called glycosphingolipidoses (Tay–Sachs disease and Gaucher's disease). In these, the glycosphingolipids are not broken down properly and residues of them accumulate causing severe brain disorders. However, while these diseases show that turnover of such molecules normally occurs, they do not indicate their specific function. The essential message is that cells use a variety of different polar lipids in forming lipid bilayers, but little is known of the reason for this variation.

What are fatty acid components of membrane lipids?

The fatty acyl 'tail' components of a membrane lipid such as lecithin are not fixed. In length they may range from C_{14} to C_{24} (almost always an even number, for reasons explained in Chapter 9) but C_{14}–C_{18} are the commonest. The two fatty acyl residues in a single phospholipid molecule may be the same or different, but usually in a glycerophospholipid the fatty acid attached to the carbon atom one hydroxyl group is **saturated** and that to the carbon atom two hydroxyl group, **unsaturated**. The degree of saturation of fatty acid tails is of great importance because the central hydrocarbon core of the bilayer must be fluid rather than solid and unsaturated hydrocarbon tails lower the solidification point of bilayer. A saturated fatty acid tail is straight, while a *cis*-double bond introduces a kink into it as illustrated below. Natural unsaturated fatty acids are almost always in the *cis*-configuration.

COOH   COOH   COOH

| Saturated fatty acid | Unsaturated fatty acid with *cis* double bond | Unsaturated fatty acid with *trans* double bond |

The saturated chains comfortably pack together and interact, while the kinked ones cannot do so. The physical effect of unsaturation can be seen by comparing hard mutton fat with olive oil. Both are triglycerides but the olive oil is rich in unsaturated fatty acyl tails. (But note again that such neutral fats never occur in membranes.)

The importance of maintaining membrane fluidity is illustrated by the fact that bacteria adjust the solidification point of their bilayers according to growth temperature. In special situations in animals, such as during hibernation, the degree of saturation of cell membrane components may be altered to cope with lower body temperature.

A variety of fatty acids, sometimes with more than one double bond, are present in membrane lipids. The usual ones are C_{18} and C_{16} with one double bond in the middle (oleic and palmitoleic acids, respectively), linoleic (C_{18}) with two double bonds, and arachidonic acid (C_{20}) with four double bonds. The nomenclature of such acids is dealt with in more detail in Chapter 10 (page 143).

What is cholesterol doing in membranes?

From its structure, as conventionally drawn (Fig. 3.3(a)), **cholesterol** seems an improbable membrane constituent. However, the actual conformation of the ring system is more like that shown in Fig. 3.3(b) (it is not suggested that you memorize this conformation). The molecule is elongated, the steroid nucleus being rigid and the hydrocarbon chain flexible. It is an amphipathic molecule, the —OH group being weakly polar.

Cholesterol in membranes acts as a 'fluidity buffer'. Inserted between the membrane lipids (with the —OH level with the polar heads) it prevents close packing of the hydrocarbon chains and thereby lowers the melting point. The observed effect is that cholesterol 'blurs' the melting point of a lipid bilayer. Without cholesterol, the transition from solid to liquid is sharper than when cholesterol is present. It is reminiscent of impurities in a

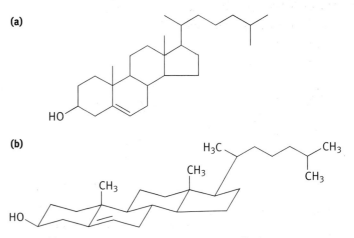

Fig. 3.3 (a) Structure of cholesterol as conventionally drawn. **(b)** Structure of cholesterol drawn to give a better indication of its actual conformation.

crystal giving a diffuse melting point instead of the sharp one of a pure substance. A red blood cell membrane may be about 25% cholesterol. On the other hand, bacterial membranes have no cholesterol and animal cell mitochondria have very little. Plants contain other sterols, known as **phytosterols**.

The self-sealing character of lipid bilayers

The lipid bilayer is effectively a two-dimensional fluid. If you had a small enough probe, you could poke it through without difficulty and the lipids would seal around it. The bilayer gives cells flexibility and a self-sealing potential, the latter being essential when a cell divides (Fig 3.4(a)).

This versatility is also essential in **endocytosis**—the process by which cells can engulf structures much too large to pass through the bilayer. The cell engulfs the object by forming an invagination, which is nipped off to form a vacuole containing that object inside the cell (Fig. 3.4(b)). The internalized object in the vacuole can be processed in ways that will be described later.

The reverse can also happen. If the cell needs to release a substance such as a protein, it is synthesized inside the cell and enclosed within a membrane-bounded vesicle that migrates to

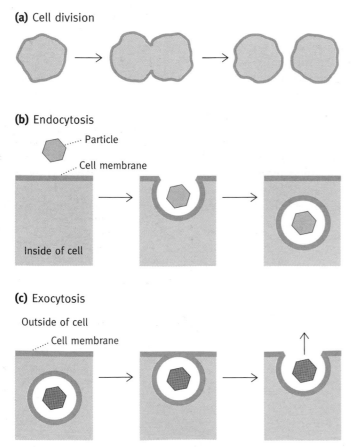

(a) Cell division

(b) Endocytosis

Particle

Cell membrane

Inside of cell

(c) Exocytosis

Outside of cell

Cell membrane

Fig. 3.4 (a) Diagram of cell division. **(b)** Diagram of endocytosis. **(c)** Diagram of exocytosis for the release of, for example, a digestive enzyme from cells.

the plasma membrane, fuses with it, and releases its contents to the outside. The process is called **exocytosis**. A good example of this is the secretion of digestive enzymes into the intestine by pancreatic cells (Fig. 3.4(c)).

Permeability characteristics of the lipid bilayer

The ability of a substance to diffuse through a lipid bilayer is largely related to its fat solubility. Fat-soluble molecules penetrate easily. Strongly polar molecules such as ions traverse the bilayer very slowly. Large, more weakly polar molecules such as glucose penetrate very slowly, while smaller ones such as ethanol or glycerol diffuse across more readily. Ionized groups of polar molecules and inorganic ions are surrounded by a shell of water molecules, which must be stripped off for the solute to pass through the lipid bilayer hydrocarbon centre, and this is energetically unfavourable. The lipid bilayer is therefore almost impermeable to such molecules and to ions, but a slow leak inevitably occurs.

The majority (but not all) of molecules of biochemical interest are polar in nature and cannot pass through the lipid bilayer at rates commensurate with cellular needs. This means that, while the bilayer structure is ideal for holding in the contents of a cell, special arrangements must be made to permit rapid movement of molecules across the membrane as needed.

Surprisingly, water molecules, despite being polar, pass through the lipid bilayer with sufficient ease for the needs of the cell. Presumably, this is due to the small size of the molecule and its lack of charge, but a degree of uncertainty exists as to how it traverses the bilayer with such apparent ease. Cells involved in water transport, such as those of renal tubules and secretory epithelia, have a transmembrane protein, **aquaporin**, which allows free movement of water. The lipid bilayer also allows gases in solution, such as oxygen, to readily diffuse through it.

To sum up so far, cells must be bounded by a barrier that prevents leakage of cell constituents, which are mainly polar. This is achieved by the lipid bilayer structure dependent on weak noncovalent forces acting on a variety of amphipathic or polar lipids. The two halves of the lipid bilayer have different polar lipid compositions, the difference established at the time of synthesis being preserved by the fact that flip-flop movement of lipids across the bilayer is an energetically unfavourable process. Lateral movement within the bilayer halves occurs readily. Different cells have different lipid compositions in their membranes. It is not understood why so many different lipids are used though one component, phosphatidylinositol, has recently emerged as a key compound in cellular control mechanisms. The hydrocarbon layer of the lipid bilayer must be in a liquid, 'non-crystalline' form. This is achieved in animal cells by a proportion of the hydrocarbon tails of the lipids containing *cis*-double bonds that confer a kink in the hydrocarbon chains thus preventing close packing and lowering the solidification temperature. Branched chains achieve the same result in bacteria. Cholesterol also plays a role in maintaining the correct fluidity of the hydrocarbon layer. The lipid bilayer is effectively impermeable to most polar substances, except weakly polar ones of small molecular size, a condition obviously essential for retaining cellular constituents against leakage. The lipid bilayer allows for fusion of membranes and for nipping off of vacuoles—it is self-sealing. These characteristics are essential for cell division, endocytosis, and exocytosis.

Where do we go from here? The lipid bilayer alone isn't enough for a cell membrane. One vital need is to allow essential traffic across the bilayer and the traffic has to be specific to selected molecules. Membrane proteins are responsible for this.

Membrane proteins and membrane design

A beautiful aspect of cell membrane structure is that it provides a totally flexible system that can evolve. Different cells need different membrane functions and hence different cells have quite different protein molecules in their membranes. If a new functional protein is evolved to act in a membrane, it can be inserted into the membrane to function there. (We explain in Chapter 22 how this is achieved.) Proteins can be inserted into the bilayer provided that they have the requisite properties—the bilayer self-seals around the proteins. To emphasize this, functional proteins, experimentally isolated from membranes, have been incorporated into synthetic phospholipid liposomes where they function exactly as in a cell membrane. It is the ultimate in flexible design. The

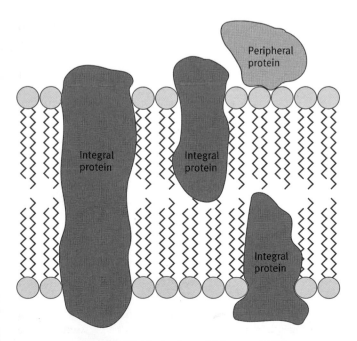

Fig. 3.5 Diagram of lipid fluid mosaic model for membrane structure.

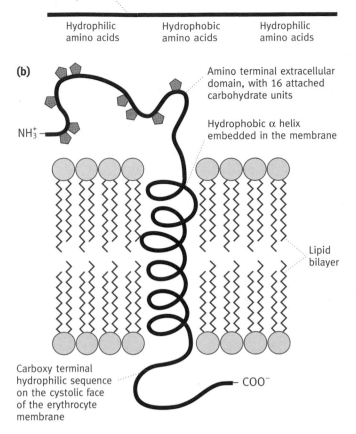

(a) Polypeptide chain of protein

Hydrophilic amino acids   Hydrophobic amino acids   Hydrophilic amino acids

(b)

NH_3^+

Amino terminal extracellular domain, with 16 attached carbohydrate units

Hydrophobic α helix embedded in the membrane

Lipid bilayer

Carboxy terminal hydrophilic sequence on the cystolic face of the erythrocyte membrane

COO^-

Fig. 3.6 (a) Diagram showing the structural plan of an integral membrane protein. **(b)** Diagram of glycophorin—a protein of the erythrocyte membrane. The α helix, containing about 19 hydrophobic amino acid residues, is approximately 30 Å in length, which is sufficient to span the nonpolar interior of the lipid bilayer.

membrane structure is called a lipid fluid mosaic one (Fig. 3.5). In this, 'icebergs' of protein are floating in a two-dimensional sea of lipid. Such proteins are called **integral proteins** (labelled I) because they are integrated into the membrane. Other proteins are called **peripheral proteins** (labelled P) because they associate with the periphery of the membrane by weak forces but are not incorporated into the bilayer.

What holds integral proteins in the lipid bilayer?

Integral membrane proteins are so structured that they are held in position in the lipid bilayer by the weak forces with which you are already familiar.

As stated in the previous chapter, proteins are, in primary structure, long strings of amino acids covalently joined together by peptide bonds. If all the side chains are hydrophobic, the polypeptide will be extremely hydrophobic and so on. Integral membrane proteins have ends made of hydrophilic amino acids and middle sections made of hydrophobic ones as shown diagrammatically in Fig. 3.6(a).

In Fig. 3.6(b) is a diagram of **glycophorin**, a major component of the erythrocyte membrane. On the external surface of the membrane is a large amino terminal section of polypeptide, rich in hydrophilic amino acids; attached to serine, threonine, and asparagine residues are carbohydrates (see page 32), making this part of the protein extremely hydrophilic. On the inside face of the bilayer is a shorter carboxy terminal section devoid of carbohydrate attachments but containing hydrophilic amino acids. (Carbohydrate attachments to membrane proteins are always external.) Connecting the two external sections and spanning the lipid bilayer is a stretch of 19 amino acid residues which, in the form of an α helix, is sufficient to just span the hydrocarbon central layer of the membrane which is about 30 Å wide. In this section of the chain, isoleucine, leucine, valine, methionine, and phenylalanine predominate; all are strongly hydrophobic and good α-helix formers. There are no strongly hydrophilic residues.

The essential feature of an integral membrane protein is that the hydrophobic section corresponds in length with the hydrocarbon middle zone of the membrane lipid bilayer. The parts of the protein projecting from the membrane that are in contact with water, or the polar head groups of membrane lipids, have hydrophilic residues as illustrated in Fig. 3.6(b). Such a protein, with sections of different polarity is called an **amphipathic protein**, indicating that the molecule has *both* polar parts and hydrophobic parts.

In some proteins the polypeptide chain loops back and crosses the bilayer several times—for this, alternating hydrophilic and hydrophilic sections are required, each of the latter being about 19 amino acids long and forming α helices that nicely span the bilayer. One of the best studied examples is **bacteriorhodopsin**, a protein present in the membrane of the purple bacterium, *Halobactium halobium*, found in salty ponds. The protein criss-crosses the membrane seven times, forming a cluster of seven α helices spanning the membrane and connected by hydrophilic loops (Fig. 3.7). The cluster has a light-absorbing pigment at its centre to capture light energy, which drives the pumping of protons from the cell to the outside. The energy of the proton gradient so formed is used to drive the synthesis of ATP by a mechanism to be described in Chapter 7.

The arrangement of membrane protein described maximizes hydrophilic bonding and hydrophobic bonding. If the protein tended to move out of the membrane, the hydrophobic groups of the protein would come into contact with water and hydrophilic groups into contact with the hydrocarbon layer. Both are energetically resisted and the protein is fixed in the transmembrane sense. Unless otherwise restrained, it could move laterally in the bilayer. The mechanism that puts the proteins into the membrane with the correct inside–outside orientation is described in Chapter 22 (page 309).

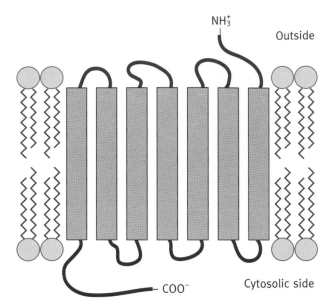

Fig. 3.7 Topological representation of bacteriorhodopsin, with seven α helices spaning the lipid bilayer. The α helices are not actually arranged linearly, as shown here for convenience, but are clustered compactly together.

Anchoring of peripheral membrane proteins to membranes

A peripheral water-soluble membrane protein may be associated with a membrane by hydrogen bonding and ionic attractions. However, an alternative method exists in the case of certain proteins. In these, the proteins have attached to them a fatty acid whose hydrocarbon chain is inserted into the lipid bilayer thus anchoring the protein to the membrane (Fig. 3.8). As is so often the case, evolutionary tinkering has produced variations on the one theme. *N*-myristoylation of the protein is one in which **myristic acid** (C_{14}) is linked to an *N*-terminal glycine of the protein,

$$R-CO-NH-CH_2-CO-polypeptide.$$

In addition, C_{14}, C_{16}, and C_{18} fatty acids may anchor proteins in a similar fashion, but attached to serine or threonine groups by ester linkage or to cysteine by thioester linkage. More complex acids are also found in ether linkage for the same purpose. A much more elaborate anchoring method is for phosphatidylinositol (PI) to be linked via an oligosaccharide to a protein; the PI structure itself is inserted into the bilayer exactly as a normal membrane phospholipid.

Glycoproteins or membrane proteins with sugars attached on the exterior surface

We have already seen that glycolipids occur in cell membranes (page 43). In these, single sugars or small branched complexes of

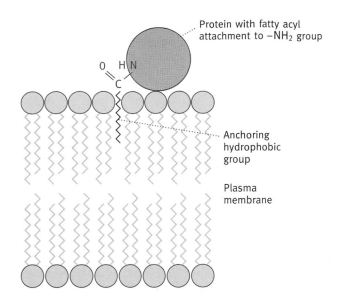

Protein with fatty acyl attachment to $-NH_2$ group

Anchoring hydrophobic group

Plasma membrane

Fig. 3.8 Diagram of a fatty acid anchoring molecule joined by CO—NH linkage to the amino group of a protein. Alternative linkages such as ester and thiol ester can occur with appropriate amino side groups of the protein.

sugars (oligosaccharides) form the polar groups of membrane lipids—the cerebrosides and gangliosides are on the external surface of the cell. Many membrane proteins also have attached to them branched oligosaccharides, again on the external surface. The carbohydrates are covalently attached to the side chains of asparagine or serine (page 32).

The sugars that make up the oligosaccharides include fucose, galactose, mannose, *N*-acetylgalactosamine, *N*-acetylglucos-amine, and sialic acid. The last two you have already met (page 44). Fucose and mannose are isomers of glucose.

In glycophorin of the red blood cell, the glycoprotein described earlier, half the weight of the protein molecule is carbohydrate. The exact function of these carbohydrate attachments is an area of uncertainty but may, in some cases, have recognition roles on cell surfaces (see also page 32).

To summarize the topic of membrane proteins. Since membranes have many functions there are many different membrane proteins. Integral proteins are embedded in the membrane; peripheral proteins are superficially attached (by secondary bonds). Integral proteins have an amphipathic structure with a hydrophobic 'middle section' in contact with the nonpolar centre of the bilayer and hydrophilic ends in contact with the aqueous phase on either side of the membrane. Membrane proteins are often **glycosylated** (have sugars or sugar-complexes covalently attached) on the exterior surface of the cell. An evolutionary advantage of a membrane with functions determined by its inserted proteins is the great flexibility of the system. Any number of different proteins can be put into the lipid bilayer.

Functions of membranes

Membranes have a wide variety of functions; for the moment we'll deal with the plasma membrane surrounding the cell. We'll come to the internal membranes later. The main functions of such membranes, apart from retaining cell contents, are;

(1) transport of substances in and out of the cell;

(2) signal transduction (this will be explained);

(3) maintaining the shape of the cell;

(4) cell–cell interactions.

Let us now look at these more closely.

Transport of substances in and out of the cell

As mentioned, many, or most, of the substances that must be taken up by the cell cannot diffuse across the lipid bilayer at anywhere near the rate required. (There are exceptions to this statement, for a small number of membranes have quite large pores in them that indiscriminately permit the passage of molecules smaller than about 600 molecular weight. This is true of the outer mitochondrial membrane and the outer chloroplast membrane.) Polar structures such as sugars, amino acids, inorganic ions, etc., require specific proteins to allow passage of such substances through the membrane. Simple holes in the membrane will not do—they would allow nonspecific leakage in and out and would be lethal. Each transport system has to handle only specific molecules so there are many different transport systems and therefore many different **transport proteins**.

Passive transport or facilitated diffusion

We can separate transport systems into active and passive types. In the latter, a protein permits the movement of a substance across the membrane so that the movement will be down whatever concentration gradient exists across the membrane, no energy being involved in the actual transport process. This is **facilitated diffusion**. A good example is the anion transport protein in red blood cells, which lets HCO_3^- and Cl^- ions pass through the membrane in either direction (Fig. 3.9). The purpose of this is discussed in Chapter 27 (page 382). Another important example of facilitated diffusion is the glucose transporter possessed by many animal cells. It allows glucose to diffuse passively across the membrane. When we come to discuss the utilization of blood glucose by cells this becomes very important (page 171).

An extremely important class of passive transport systems are the **gated pores** or **channels**. These are transport systems for specific ions that open and close on receipt of a signal. The latter may be a change in the charge across a membrane (referred to as depolarization; this can wait until we deal with nerve conduction in Chapter 26). Such a channel is called a **voltage-gated pore**. Others are opened by acetylcholine or other

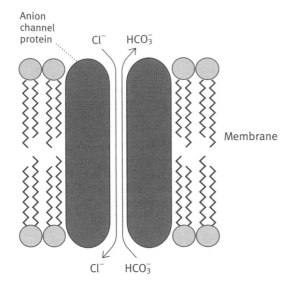

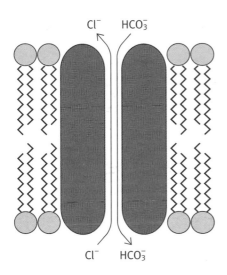

Fig. 3.9 Diagram of anion channel of red blood cells. Cl^- and HCO_3^- may move in either directions, according to concentration gradients across the membrane. The counterflow of the ions prevents any electrical potential developing across the membrane.

chemical signals (**ligand-gated pores**). The most important gated pores are those for Ca^{2+}, Na^+, and K^+ (each specific for one ion). The role of these will be described in Chapter 26.

Active transport

This is transport that requires performance of work which, from Chapter 1, you will know involves (with exceptions in bacteria, see page 12) **ATP hydrolysis**. It requires work because the movement of substances occurs against a concentration gradient.

The amount of energy required to transport a solute into a cell, against a gradient, can be calculated from the equation for chemical reactions given in Chapter 1,

$$\Delta G = \Delta G^{0'} + RT\ 2.303 \log_{10} \frac{[\text{Products}]}{[\text{Reactants}]}.$$

For transport of a solute whose structure does not change, $\Delta G^{0'}$ is zero and the equation becomes

$$\Delta G = RT\ 2.303 \log_{10} \frac{[C_2]}{[C_1]}$$

where C_1 is the concentration outside and C_2 the concentration inside, taken in this example to be in the ratio of 10/1 so that

$$\Delta G = (8.315\ \text{J mol}^{-1}\ \text{K}^{-1})(298\ \text{K})2.303 \log_{10} \frac{10}{1} = 5706\ \text{J mol}^{-1}.$$

Thus the transport of 1 mol under these conditions (at 25°C) requires 5706 J. A cell such as a nerve cell that is very active in Na^+ and K^+ pumping uses a large proportion of its total ATP production in this activity. Note that the above calculation applies to the transport of an uncharged solute. With a charged solute, generation of an electrical potential requires an additional term to correct for this.

A good example of active transport is the **Na^+/K^+ pump** present in animal cells. Almost all animal cells have a high internal K^+ concentration (140 mM) and a low Na^+ concentration (12 mM) as compared with those of the blood (4 mM and 145 mM, respectively)—it maintains a membrane potential of 50–70 mV (positive outside, negative inside). This is energetically expensive—it consumes perhaps a third of the total energy usage in a resting animal. The interior of the cell requires high K^+ and low Na^+ levels to function properly. The ion gradients are necessary for electrical conduction in excitable membranes and for driving solute transport across membranes.

Mechanism of the Na^+/K^+ pump

The Na^+/K^+ pump is also called the **Na^+/K^+ ATPase** because ATP is hydrolysed to ADP+P_i as Na^+ is pumped out and K^+ in. The protein is a complex of four polypeptides or subunits. There are two identical so-called α subunits and two β subunits. Some proteins have the property of slightly changing their shape when specific ligands bind to them. (A ligand is any molecule which binds to a specific site.) This may be due to a change in the conformation of a simple protein, or to a relative change in position of subunits of a protein complex (see hemoglobin on page 380). The covalent attachment of a phosphoryl group to the Na^+/K^+ ATPase by transference from ATP causes a **conformational (shape) change**. Such a change can alter the ability of a protein to bind a given ligand (if it has been designed to do so). Thus, in one form, the protein of the pump is believed to bind Na^+ but not K^+ and in another form, K^+ but not Na^+. With that preamble let's look at the pump mechanism that has been postulated.

In the model shown in Fig. 3.10, the Na^+/K^+ ATPase protein in effect exists in two conformations. Form (a) is open to the interior of the cell and binds Na^+ but not K^+. When Na^+ is bound, ATP phosphorylates the protein, yielding ADP + phosphorylated protein. In this form it is open to the outside but no longer binds Na^+ (which diffuses away) but does bind K^+ (which therefore attaches). This gives form (b) (Fig. 3.10). The $-\textcircled{P}$ group in the presence of bound K^+ is now hydrolysed from the protein giving P_i and the protein reverts to form (a) to which the K^+ no longer binds. The latter therefore enters the cell and more Na^+ attaches to the pump, triggering a new round of phosphorylation. The net result is that hydrolysis of ATP pumps Na^+ out and K^+ in (Fig. 3.10(c)). The ratio is 3 Na^+ out and 2 K^+ in. The overall equation is

$$3Na^+(in) + 2K^+(out) + ATP + H_2O \rightarrow$$
$$3Na^+(out) + 2K^+(in) + ADP + P_i + H^+.$$

There is an interesting medical aspect. **Cardiac glycosides** are a group of compounds found in the digitalis species of plants (foxgloves). They are steroids or glycoside derivatives of these: steroids resemble cholesterol in structure; glycosides are molecules with sugars attached. Such compounds inhibit the Na^+/K^+ ATPase by preventing the removal of the phosphoryl group from the transport protein. This 'freezes' the pump in one form (Fig. 3.10(b)) and stops the ion transport. The cardiac glycosides have long been used clinically as treatment for congestive heart failure. The compounds are lethal in sufficient amounts, but, in appropriate doses, the partial inhibition of the Na^+/K^+ ATPase increases the Na^+ concentration inside heart muscle cells and thus lowers the Na^+ gradient from outside to inside. This has the effect of raising the cytoplasmic Ca^{2+} level because there is another system that transports Na^+ into the cell and Ca^{2+} out of the cell. The ejection of Ca^{2+} is driven by the Na^+ gradient (see below); if the latter is lowered by cardiac glycosides then ejection of Ca^{2+} is reduced and the internal level of Ca^{2+} rises. The raised Ca^{2+} level stimulates heart muscle contraction; the role of Ca^{2+} in contraction is dealt with in Chapter 28 (page 392). Ouabain, the African arrow tip poison, has similar effects.

Cotransport of molecules across membranes (symport)

The Na^+/K^+ ATPase, described above, produces a steep Na^+ gradient across the cell membrane. Any gradient has potential energy and, given a chance, the accumulated exterior Na^+ will flow back into the cell. What happens in Na^+ cotransport is that Na^+ is allowed to flow into the cell and something else hitches a thermodynamic ride on the back of this flow. Such a transport system is called a **symport** (it simultaneously transports two components) or **cotransport system**. The glucose symport protein system permits movement of Na^+ and glucose across the membrane *when both are present* but not when only one is present. It can transport glucose against a concentration gradient at the expense of the Na^+ gradient. The Na^+ entering the cell is pumped out by the Na^+/K^+ ATPase to restore the Na^+

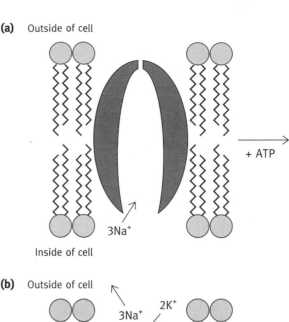

(a) Outside of cell

3Na⁺

Inside of cell

+ ATP

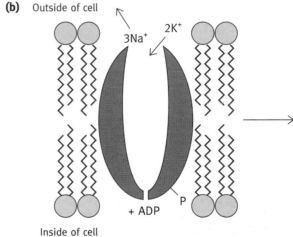

(b) Outside of cell

3Na⁺   2K⁺

Inside of cell

+ ADP   P

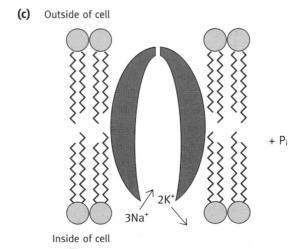

(c) Outside of cell

Inside of cell

2K⁺   3Na⁺

+ P_i

Fig. 3.10 Diagrammatic model of a possible mechanism for the Na^+/K^+ pump. **(a)** Na^+/K^+ ATPase in conformation (a); **(b)** Na^+/K^+ ATPase in conformation (b); **(c)** Na^+/K^+ ATPase after $-\textcircled{P}$ is hydrolysed off the protein, returning the pump to conformation (a).

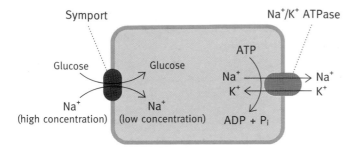

Fig. 3.11 The sodium/glucose cotransport system—a symport.

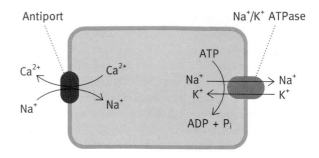

Fig. 3.12 The sodium/calcium cotransport system—an antiport.

gradient so that it is ATP hydrolysis that indirectly supplies the energy for glucose uptake. Absorption from the gut of glucose and amino acids occurs by this mechanism (Fig. 3.11), separate symport proteins being required for the different substances transported.

Antiport transporters

As mentioned above, in connection with cardiac glycoside action on the heart, the cotransport of Na^+ can be used in another way—to pump out Ca^{2+}. This is an **antiport system** (Fig. 3.12). Na^+ is transported into the cell only if, at the same time, Ca^{2+} is transported out. Again, the energy in the Na^+ gradient is the driving force and this is established by the Na^+/K^+ ATPase system. When we come to muscle contraction you will find a different type of Ca^{2+} pump that directly uses ATP (page 393).

Uniport transporters

In this, a transport protein transfers a single specific molecule across the bilayer. The Ca^{2+}/ATPase of sarcoplasmic reticulum just referred to is such a case. You will later meet ATP-driven proton pumps that cause acidification of lysosomal vesicles.

To summarize transport across plasma membranes, most substances cross membranes by courtesy of proteins designed to transport specific solutes and therefore there are many different transporting systems. Transport can be passive, the systems simply allowing passage either way depending on relative concentrations inside and outside, the anion channel of erythrocytes and the glucose transporter of animal cells being important examples. There are many situations where solutes must be transported against a concentration gradient. Animal cells possess a Na^+/K^+ ATPase that pumps Na^+ out and K^+ in to maintain steep gradients. This is inhibited by the cardiac glycosides that are used clinically. The Na^+ gradient outside a cell can be used to drive uptake of, for example, glucose or amino acids, by a cotransport mechanism of the symport type. Such systems are involved in absorption from the gut. The Na^+ gradient can also be used in antiport systems where Na^+ goes in and Ca^{2+} goes out, again with the drive being supplied by the Na^+ gradient, which is continuously maintained by the Na^+/K^+

ATPase. A different Ca^{2+}/ATPase is used in muscle contraction and will be discussed later.

Now for the next membrane function.

Signal transduction

The dictionary defines transduction as 'carrying' or 'leading across'. The cells of an animal must work cooperatively according to physiological needs and instructions must therefore be sent to cells. Life is a chemical process and signals are sent in chemical form. Hormones liberated by glands and neurotransmitter substances released by nerve cell endings are examples.

One class of chemical signals, such as steroid hormones, are lipid-soluble, and enter the cell directly by diffusion through the lipid bilayer (because of their nonpolar nature). The second type are water-soluble and do not directly enter the cell but combine with specific receptor proteins on the external membrane surface and in doing so deliver an 'instruction' to the cell. The combination of the chemical signal with the exterior of the membrane receptor results in molecular events inside the cell. This is the **signal transduction** role of the cell membrane. It is a very important area of biochemistry and is the subject of Chapters 12 and 26 (see pages 169 and 349).

Role of the cell membrane in maintaining the shape of the cell and in cell mobility

Eukaryote cells have an internal scaffolding that maintains the shape of the cell and is involved in amoeboid motility. It is called the **cytoskeleton** (the subject of Chapter 29). This is not a rigid inflexible structure like the bony skeleton of an animal but rather is composed of protein microfilaments that pervade the cytoplasm and, at various places, are attached to proteins of the cell membrane. Such membrane proteins are not then free to move laterally in the lipid bilayer as are other proteins.

An extreme example of the attachment of the membrane proteins to the cytoskeleton is in the red blood cell which has special 'cell shape' proteins. The cell is a biconcave disc, which has the advantage of presenting a large surface area for gaseous exchange, but it is always on the move and therefore subject to shearing forces as it squeezes through capillaries, demanding a robust but flexible cell membrane. Underneath the membrane is a dense scaffolding of protein fibres, called **spectrin**. The name spectrin comes from the fact that you can release the red cell contents but retain the empty membrane with its cytoskeleton, producing a red cell ghost (or spectre). The spectrin is linked by a protein appropriately called **ankyrin** to the anion channel protein described earlier (page 50). The latter is a large protein that projects into the cytoplasm providing a cytoskeleton attachment point (Fig. 3.13). Spectrin is also linked to glycophorin by other specific linking proteins.

Patients are known in whom the cytoskeleton of red blood cells is deficient because of faulty spectrin or ankyrin. The cells are abnormally shaped and tend to be destroyed by the spleen. The diseases are called hereditary spherocytosis or hereditary elliptocytosis.

Cell–cell interactions—tight junctions and gap junctions and cellular adhesive proteins

In an epithelial tissue, such as the lining of the intestine, the products of food digestion are selectively taken up by the cells and then transported from the cell by appropriate systems in the membrane of the opposite side of the cell facing the blood

vessels. (More of this in the next chapter.) It would not do if molecules from the gut leaked between the cells. Special bands of membrane proteins encircle the cells and these bind to corresponding proteins of the neighbouring cells, creating 'tight' junctions—that is, non-leaky cell–cell contacts. The tight junctions are also believed to prevent lateral diffusion of membrane proteins. The epithelial cell surface facing the intestinal lumen requires transport proteins different from those needed in the opposite cell surface facing blood capillaries. Tight junctions, by preventing lateral diffusion of membrane proteins across them, preserve the compartmentation of membrane functions in different parts of the cell membrane. On the other hand it can be desirable for adjacent cells to exchange molecules so as to coordinate the chemical activities throughout a tissue. This is achieved by special proteins forming a tunnel or hole between cells—known as **gap junctions**. Gap junctions will not allow large molecules such as proteins to pass, but molecules as large as ATP can do so. The coordination of heart cell contractions depends on gap junctions.

Another function of membrane proteins is to promote adhesion between cells of a tissue to form that tissue. If different types of embryo cells are mixed together they will reassociate with cells of their own type—kidney cells to kidney cells and so on. This is a function of tissue-specific cell–cell

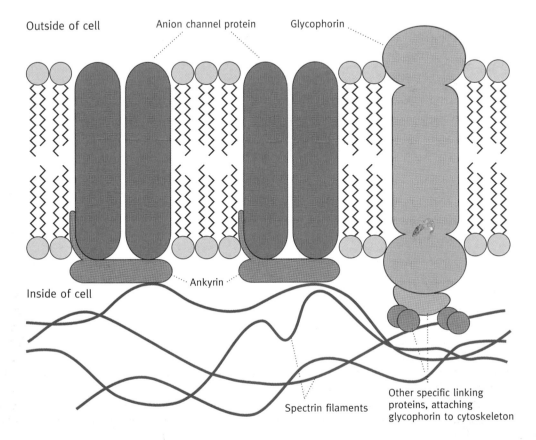

Fig. 3.13 Diagram to illustrate attachment of anion channel protein and glycophorin to the cytoskeleton.

Outside of cell

Anion channel protein

Glycophorin

Inside of cell

Ankyrin

Spectrin filaments

Other specific linking proteins, attaching glycophorin to cytoskeleton adhesion proteins called **cadherins**. Another family of proteins, the **NCAMS** (for *Nerve Cell Adhesion Molecules*), are important in nervous tissue formation.

An overview of cellular membranes

So far in this chapter we have dealt with the external or plasma membrane of the cell, but inside most eukaryote cells there is a variety of membranous structures. All have the same lipid bilayer structure but of differing composition, and have their own retinue of proteins for their various functions. The internal membranous structures separate the cell into separate compartments. The main components of an animal cell are illustrated in Fig. 3.14.

The functions of the various organelles will be described in later chapters, but a brief summary of their salient roles will be given here.

1 The nucleus contains the DNA and is the site of DNA and RNA synthesis.
2 The endoplasmic reticulum (ER) is a reticulum whose lumen may occupy half the total volume of the cell. It is the site of synthesis of proteins destined for secretion or transport to specific sites in the cell. Part is studded with small bodies called ribosomes that synthesize protein; this part is called the rough ER and that without ribosomes, the smooth ER.
3 The Golgi apparatus is a set of flattened membranous bags which sort out proteins synthesized by the ER and send them to their appropriate destinations. Both the ER and Golgi are completely closed structures; material is dispatched from them by vesicles which bud off from the structures.
4 Mitochondria are the site of oxidative metabolism, which generates most of the ATP; the inner mitochondrial membrane is the actual site of ATP synthesis.

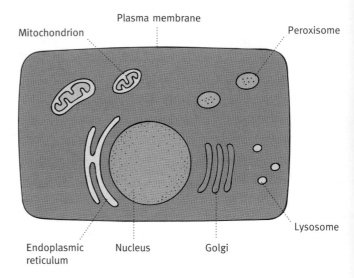

Fig. 3.14 Diagrammatic representation of membranous structures in a typical animal cell.

5 Lysosomes contain hydrolytic enzymes which destroy unwanted material.
6 Peroxisomes metabolize certain molecules by oxidative reactions.

Different cells have different amounts of the various membrane organelles in keeping with their roles. The smooth ER shown in the EM of sectioned testis (Fig. 3.15) is associated with the enzymes of steroid synthesis. Cells active in secreting proteins (such as digestive enzymes or antibodies) are rich in ER. Cells with large ATP needs, such as those of muscle, are rich in mitochondria. Mature erythrocytes have no internal membranes.

Further reading

Membrane lipids and movement

Higgins, C. F. (1994). Flip-flop: the transmembrane translocation of lipids. *Cell*, **79**, 393–5.
Raises the general question of how lipids get to where they should be and discusses particularly how asymmetry of the lipid bilayer is achieved.

Rooney, S. A., Young, S. L., and Mendelson, C. R. (1994). Molecular and cellular processing of lung surfactant. *FASEB J.*, **8**, 957–67.
Reviews a more physiological role of phospholipid—that of lining the alveoli.

Membrane proteins

Macdonald, C. (1985). Gap junctions and cell-cell communication. *Essays in Biochem.*, **21**, 86–118.
A comprehensive review.

Neher, E. and Sakmann, B. (1992). The patch clamp technique. *Sci. Amer.*, **266**(3), 44–51.
Enables studies to be made on ligand-gated and voltage-gated ion channels.

Von Heijne, G. (1994). Membrane proteins: from sequence to structure. *Ann. Rev. Biophys. Biomolec. Structure*, **23**, 167–92.
A review tracing the steps from sequence to structure of integral membrane proteins.

Von Heijne, G. (1995). Membrane protein assembly: rules of the game. *BioEssays*, **17**, 25–30.
Interesting discussion of how proteins can be 'stitched' into membranes with multiple transmembrane α helices.

(a) Steroid secreting cell

Macrophage

Plasma membrane

Smooth ER

Lysosome

Clathrin coated pit

Nucleus

Rough ER

Nuclear pore

Heterochromatin

Euchromatin

Nuclear envelope

Mitochondrion

(b)

Vesicles arriving at, or budded from, Golgi cisternae

Nucleus

Nuclear envelope

Golgi membranes

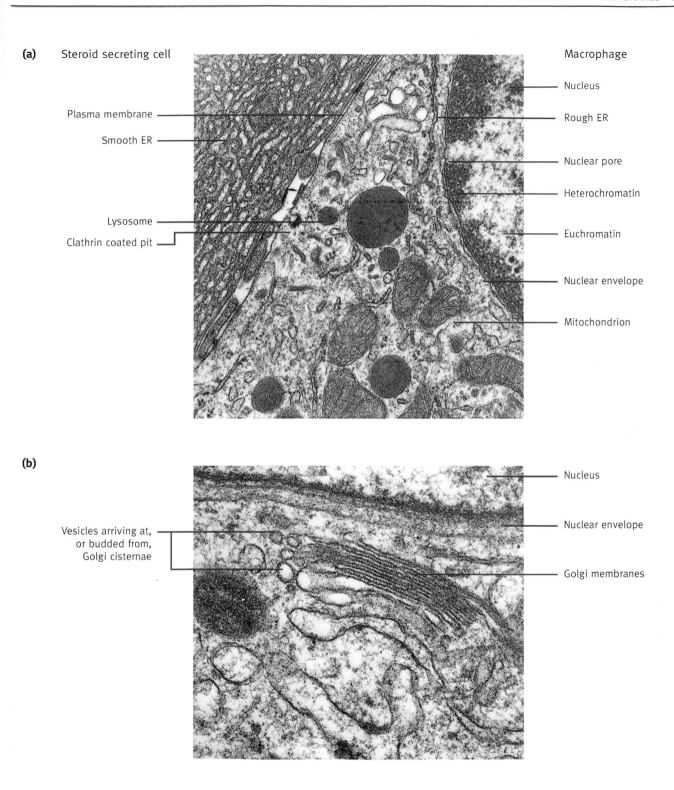

Fig. 3.15 Electron micrograph of part of sectioned testis **(a)** showing a steroid secreting cell (upper left) packed with smooth ER, alongside a macrophage (right); **(b)** a higher magnification EM in which a prominent Golgi apparatus is evident. The terms in this figure not yet explained will be referred to later in the text. Photographs kindly provided by Professor W. G. Breed, Department of Anatomy, University of Adelaide.

Problems for Chapter 3

1 What structural characteristics of amphipathic molecules favours liposome (lipid bilayer) formation, rather than micelle formation?

2 Are triacylglycerol (TAG) molecules constituents of membranes?

3 Can protein molecules traverse the lipid bilayer?

4 What is the role of cholesterol in eukaryote membranes?

5 Give the structure of phosphatidic acid.

6 Name three glycerophospholipids based on phosphatidic acid. Name the substituent in each case attached to the latter.

7 Write down the structure of sphingomyelin.

8 How does a cerebroside and a ganglioside differ from sphingomyelin?

9 Give the structure of sialic acid (*N*-acetylneuraminic acid).

10 What is the significance of *cis*-unsaturated fatty acyl tails in membrane phospholipids?

11 Why don't polar molecules readily pass through membranes?

12 What is meant by facilitated diffusion through a membrane? Give an example.

13 Explain how the amount of energy required to transport a solute into a cell can be calculated. Assuming that the concentration of a solute outside and that inside to be in the ratio of 10/1, calculate the energy required to transport 1 mol into the cell.

14 By means of notes and diagrams, explain how cells maintain a high $K_{inside}/K_{outside}$ ratio and a low $Na_{inside}/Na_{outside}$ ratio.

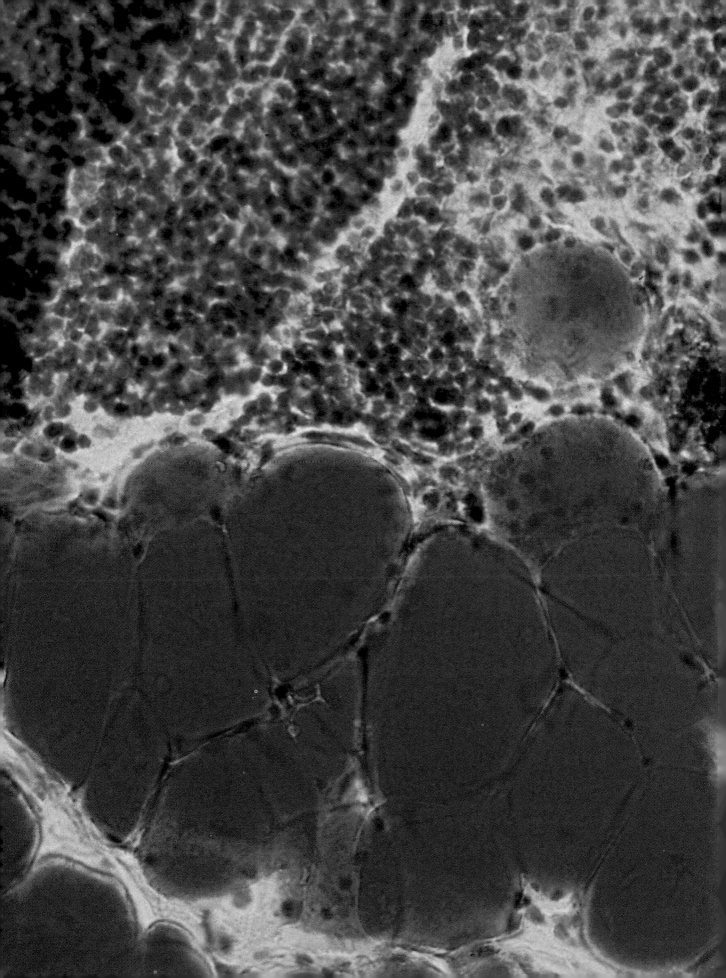

Part 3

Metabolism

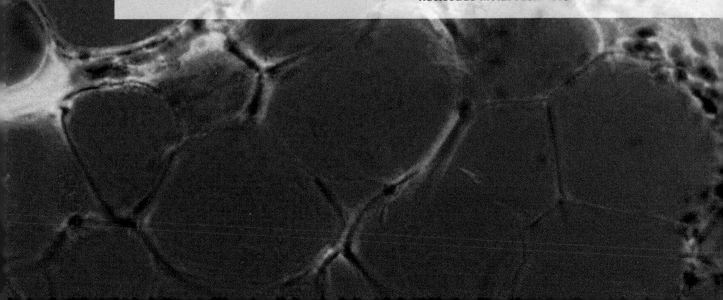

Previous page: Light micrograph of human tissue showing white fat cells (adipose cells). Such cells are dedicated to accumulating fat as the major energy store of the body. Stored fat accumulates as lipid droplets, which fuse to form a single large droplet, reducing the cytoplasm to a thin rim around the periphery.

Photograph: Manfred Kage/Science Photo Library

Chapter 4

. . . .

Digestion and absorption of food

First, where does this topic fit in the overall arrangement of the book? In Chapter 1 we looked at what is necessary for chemical reactions to occur and at the chemistry of weak bonds. Next, in Chapter 2, the structure of protein molecules laid the basis for the key reactions and processes on which life depends. Chapter 3 dealt with the cell membranes because this information is essential for understanding digestion and absorption. We now progress to the major area of metabolism. Food digestion and absorption, is a logical place at which to start, for that's where metabolism begins. At a gross level, the subject may seem mundane; at the molecular level it is as elegant and interesting as are the topics with a more exciting image.

In this chapter we will discuss what food is in chemical terms, how it is prepared for absorption into the bloodstream, and how it reaches the bloodstream. What happens to the food once it gets there will be taken up in the next chapter.

Chemistry of foodstuffs

There are three main classes of food—proteins, carbohydrates, and fats.

- **Proteins** are, as described in Chapter 2, large polymers of 20 species of amino acids linked together to form polypeptide chains.
- **Carbohydrates** are the sugars and their derivatives; the name comes from their empirical formula with carbon atoms and the elements of water in the ratio of 1:1 (CH_2O). Although simple monosaccharide sugars such as glucose occur in food, most of the carbohydrate is in the form of disaccharides such as sucrose or polysaccharides such as starch.
- **Fats** in the diet are mainly in the form of neutral fat or triacylglycerols (page 41). Butter, olive oil, and the visible fat of meat are typical examples. Polar lipids are also present.

Digestion and absorption

With the exception of simple free sugars (monosaccharides) such as glucose, all of the foodstuffs mentioned above must be digested by hydrolysis into their constituent parts in the intestine. To be absorbed, substances must enter the epithelial cells lining the gut and this involves crossing the cell membranes. Triacylglycerols cannot do this, nor can proteins, and amongst the carbohydrates only monosaccharides can be absorbed. Thus, digestion consists mainly of the conversions

Proteins → amino acids;

Carbohydrates → sugar monomers (monosaccharides);

Neutral fats → fatty acids and monoacylglycerol.

Anatomy of the digestive tract

The following regions are involved.

- In the **mouth** food is masticated and lubricated for swallowing; limited starch digestion occurs.
- The **stomach** contains HCl which 'sterilizes' food and denatures proteins; partial digestion of proteins occurs.
- The **small intestine** is the major site of digestion and absorption of all classes of food. It is lined by fine finger-like processes, the villi which are covered by epithelial cells—known as brush border cells because the microvilli of the epithelial cells resemble bristles on a brush (Fig. 4.1). The microvilli are on the external membrane of the epithelial cells of the villi and such an arrangement, the combination of the villi and the microvilli, gives the enormous surface area needed for absorption.
- The **large intestine** is mainly involved in the removal of water.

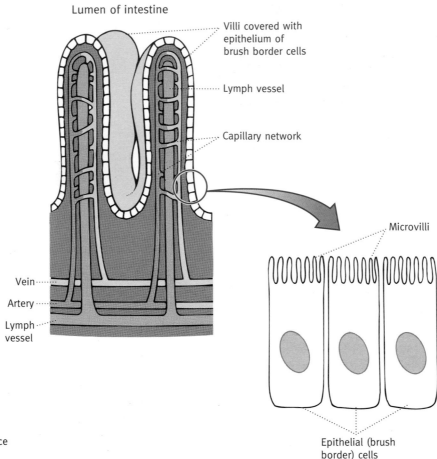

Fig. 4.1 Diagram of the surface of the small intestine.

What are the energy considerations in digestion and absorption?

So far as digestion goes, there are no thermodynamic problems. Hydrolytic reactions such as hydrolysis of proteins to amino acids, of di- and polysaccharides to monosaccharides, and of fats to fatty acids and monoacylglycerols are all exergonic processes—they have negative ΔG values sufficient to push the equilibrium entirely to the side of hydrolysis. Hydrolytic reactions (splitting or lysis by water) in biochemistry are invariably of this type. Absorption is a different matter since it is often an active process in which molecules are absorbed against a concentration gradient and energy is needed.

A major problem in digestion—why doesn't the body digest itself?

Food is, chemically, little different from the tissues of the animal that eats it. A fearsome array of enzymes is produced by the

digestive system to completely digest the food into its smaller components, and those enzymes have to be produced inside living cells, which, if exposed to their action, would be destroyed. There are two major types of defence against this.

Zymogen or proenzyme production

This refers to the production of enzymes as inactive **proenzymes** or **zymogens** which are activated only when they reach the intestine. Thus, glands producing digestive enzymes secrete most of them as inactive proteins and are never themselves exposed to the destructive processes. In the case of enzymes that pose no threat (amylase, the enzyme that hydrolyses starch, is one), proenzymes are not involved. The question also arises as to how cells selectively secrete digestive enzymes, but this is best left until we deal with protein synthesis (Chapter 22) for the mechanism is complex and not directly related to digestion. We will deal with the mechanisms of zymogen activation when we come to particular enzymes.

Protection of intestinal epithelial cells by mucus

The cells lining the intestinal tract are protected from the action of activated digestive enzymes, the main line of defence being the layer of mucus that covers the entire epithelial lining of the gut. The essential components of mucus are the **mucins**. These are extremely large glycoprotein molecules—proteins with large amounts of carbohydrates attached to their polypeptide chains in the form of oligosaccharides that contain a mixture of sugars you've already met in membrane glycoproteins (pages 48 and 32)—glucosamine, fucose, sialic acid, etc. The mucins form a network of fibres, interacting by noncovalent bonds and resulting in a gel containing more than 90% water that protects intestinal cells. The carbohydrates may protect the mucin proteins themselves from digestion. The mucins are synthesized and secreted by special goblet cells in the epithelial lining of the gut. The amount of mucin secreted is controlled.

Digestion of proteins

In a normal or 'native' protein, the polypeptide chain is folded up, the shape being largely determined by weak bond interactions. In this compact folded form, many of the peptide bonds are hidden away inside the molecule where they are not accessible to hydrolytic enzymes. An important step is to **denature** the native proteins. This is done by acid in the stomach where, due to HCl secretion, the pH is about 2.0. This disrupts many weak bonds and the polypeptide folding is disrupted, making the polypeptide chain very susceptible to **proteolysis**, as the hydrolytic process is called. The HCl also kills microorganisms.

HCl production in the stomach

The stomach epithelial lining contains **parietal** or **oxyntic cells** that secrete acid. In essence, the process consists of ejecting H^+ against a concentration gradient—this is similar to the ejection of Na^+ from cells against a concentration gradient (page 50) which is achieved by a Na^+/K^+ ATPase in the membrane. In a similar manner, the acid-secreting cells have a H^+/K^+ ATPase. Using energy from ATP hydrolysis, they eject H^+ and import K^+, the latter recycling back to the exterior of the cell. The process is shown in Fig. 4.2. Where do the protons come from? CO_2 enters the cell from the blood. An enzyme, **carbonic anhydrase**, converts this to carbonic acid, which dissociates as shown.

$$CO_2 + H_2O \longleftrightarrow H_2CO_3 \longleftrightarrow H^+ + HCO_3^-$$
Carbonic anhydrase

The bicarbonate ions exchange via an anion transport protein with Cl^- in the blood (Fig. 4.2), the latter moving to the stomach lumen to form HCl. We have already met this type of exchange in the anion channel of the red blood cell (page 50).

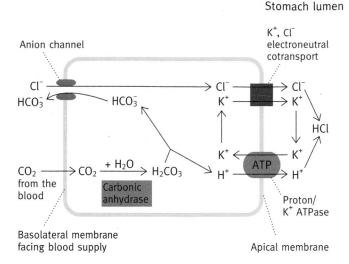

Fig. 4.2 Mechanism of gastric HCl secretion.

Pepsin, the proteolytic enzyme of the stomach

A note on nomenclature: enzymes have names usually ending in 'ase', but many digestive enzymes were amongst the earliest to be discovered and were given names such as pepsin, trypsin, chymotrypsin. To indicate an inactive precursor, the suffix 'ogen' is used; for example, pepsinogen is capable of generating, or giving rise to, **pepsin**.

Cells of the stomach epithelium secrete pepsinogen. The secretion is stimulated by the hormone gastrin released by stomach cells into the blood in response to food entering the stomach. Pepsinogen is pepsin with an extra stretch of 44 amino acids added to the polypeptide chain. This additional segment covers, and blocks, the active site of the enzyme. When pepsinogen meets HCl in the stomach, a slight change in conformation exposes the catalytic site which self-cleaves—it cuts off the extra peptide from itself to form active pepsin. As soon as a small amount of active pepsin is produced in this way, it rapidly converts the rest of the secreted pepsinogen to pepsin. The activation is therefore autocatalytic.

Pepsin is unusual for an enzyme in that it works optimally at acid pH (Fig. 1.3(a)), most enzymes requiring a pH near neutrality. It hydrolyses peptide bonds of proteins within the molecule producing a mixture of peptides. It is therefore an **endoenzyme** or **endopeptidase**; that is, it does not attack terminal peptide bonds at the end of molecules, but only those within the molecule. The derivation is Greek, not English; *endo* = within.

Proteolytic enzymes (proteases) are usually relatively specific in their action—they hydrolyse only peptide bonds adjacent to certain amino acids. Pepsin has this characteristic so that only partial digestion of proteins can occur in the stomach. It hydrolyses only those peptide bonds in which one of the aromatic amino acids, tyrosine, phenylalanine, or tryptophan (page 25), supplies the NH of the peptide bond.

Infants take most of their protein in the form of milk. Another stomach enzyme, rennin, acts on the casein of milk, causing clotting, so that it doesn't pass too rapidly through the stomach and escape gastric digestion.

Completion of protein digestion in the small intestine

The chyme, as the partially digested stomach contents are called, enters the duodenum at the start of the small intestine. The acid stimulates the duodenum to release hormones (secretin and cholecystokinin) into the blood, which stimulate the pancreas to release pancreatic juice. This is alkaline and (together with bile juice) neutralize the HCl giving a slightly alkaline pH suitable for pancreatic enzyme action and terminating pepsin activity.

A battery of pancreatic proteases is produced from clusters of cells in the pancreas and secreted into the intestine as pancreatic juice via the main pancreatic duct, which is formed by the joining together of smaller ducts draining the cell clusters. There are three endopeptidases—**trypsin**, **chymotrypsin**, and **elastase**—all entering the intestine in the form of the inactive proenzymes, trypsinogen, chymotrypsinogen, and proelastase, respectively. An exopeptidase, **carboxypeptidase**, secreted as an inactive proenzyme, chops off terminal amino acids from the carboxy terminal end of peptides.

$$H_2N-CH-CO-NH-CH-CO-NH-----NH-CH-COOH$$

with the groups R, R', R'' on the respective CH carbons.

Amino terminal end
or *N*-terminal end

Carboxy terminal end
or *C*-terminal end

Activation of the pancreatic proenzymes

As with pepsinogen activation, pancreatic proenzymes are activated by proteolytic cleavage of the proenzymes. The activation process is autocatalytic and is triggered off by an enzyme produced by the cells of the small intestine, an active enzyme (not a proenzyme in this case) called **enteropeptidase**, which hydrolyses a single peptide bond of trypsinogen thus activating it to trypsin. The initial amount of active trypsin so produced now activates all proenzymes (including trypsinogen itself) so that all are rapidly activated in an autocatalytic cascade (Fig. 4.3). The activation details differ for different enzymes. This is an elegant mechanism for ensuring that the fearsome array of *active* enzymes are present only in the intestinal lumen. Their premature activation in the pancreas or even the pancreatic duct is deleterious—if it occurs, the disease pancreatitis ensues. Blockage of the pancreatic duct, or damage to the gland, can trigger this disease. After synthesis, the proenzymes are stored in the cells producing them in membrane-bound secretory vesicles. On hormonal or neurological stimulation these fuse with the cell membrane and release

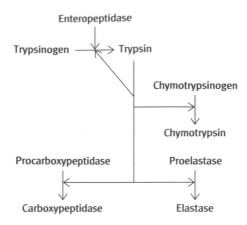

Fig. 4.3 Activation of the pancreatic proteolytic enzymes. Proenzymes (zymogens) are shown in red; activated proteolytic enzymes are shown in green.

their contents by exocytosis (this is described in detail later; page 366). Presumably as a safeguard against accidental leakage of the vesicles before secretion, the cells contain a trypsin-inhibitor protein capable of inactivating any trypsin that might accidentally come in contact with the cytoplasm. The inhibitor protein fits the trypsin active site so perfectly that a sufficient number of weak bonds are formed to make the combination of the two almost completely irreversible. The mechanism by which enzymes to be secreted are produced and enveloped in the secretory vesicles is more suitably dealt with later (Fig. 22.14).

An additional enzyme, called **aminopeptidase** (an **exopeptidase**), attached to the outside (lumen side) of intestinal cells hydrolyses off terminal amino acids from the amino ends of peptides to be digested. Thus the three endopeptidases (trypsin, chymotrypsin, and elastase), each with a different preference for the peptide bonds they attack, chop away in the middle of polypeptides, while carboxypeptidase and aminopeptidase chop away at each end, and proteins are completely converted to free amino acids in the lumen of the intestine.

Absorption of amino acids into the bloodstream

Amino acids are transported from the intestine across the cell membrane of epithelial cells (the brush border cells; Fig. 4.1) and into the cell. The brush border cells actively concentrate amino acids inside themselves from where they diffuse into the blood capillaries inside the villi. Because there are so many different amino acids, there are several different transport pumps in the membrane. The cotransport mechanism has already been described in Chapter 3 (page 51) in which the Na^+ gradient is used to drive the uptake of some of the amino acids. Figure 3.11 shows glucose being transported by this mechanism but it applies equally well to amino acids using a different transport protein.

We will take up the subject of what happens to the amino acids in the blood in the next chapter.

Digestion of carbohydrates

The main carbohydrates in the diet are starch and other polysaccharides, and the disaccharides sucrose and lactose (the latter from milk). Free glucose and fructose also occur. As stated, only free sugars or monosaccharides are absorbed from the intestine so the aim of digestion is to hydrolyse the dietary components to such monosaccharides.

Sugars are linked together by **glycosidic bonds** to form glycosides. This is an important bond from several viewpoints and needs to be described. Glucose has the following structure.

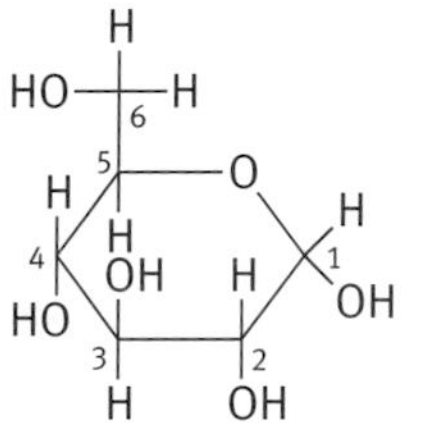

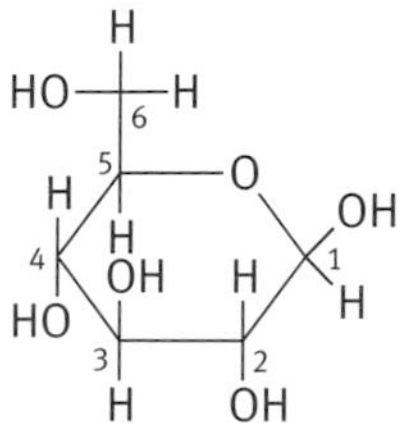

In α-D-**glucose** the —OH on carbon atom 1 points below the plane of the ring, and in β-D-**glucose** upwards. In free sugars, the two are in equilibrium in solution by a **mutarotation** (via an open-chain structure). Suppose we have two glucose molecules.

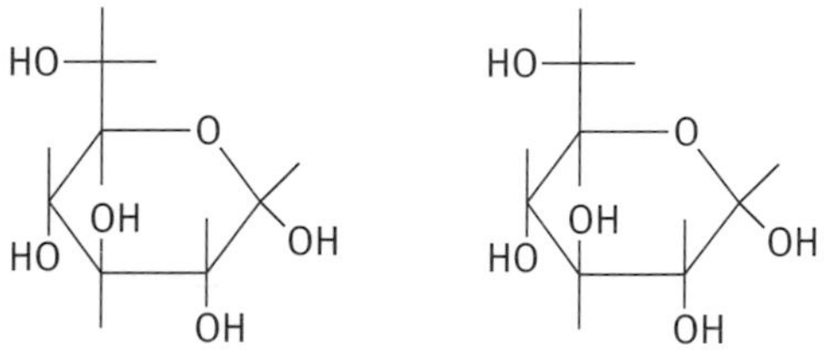

They can (by appropriate chemistry) be joined together by a glycosidic bond.

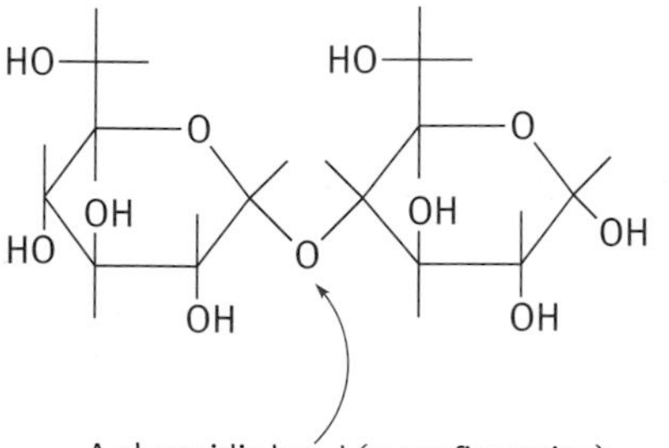

The glycosidic bond fixes the configuration on carbon atom 1 of the first sugar into one form—it no longer mutarotates. In the example shown, the glycosidic bond is between carbon atoms 1 and 4 of the two units and is in α-configuration. The compound is therefore glucose-$\alpha(1 \rightarrow 4)$-glucose. It is a disaccharide plentiful in malted barley and has the trivial name **maltose**. As will be described shortly, disaccharides with β-glycosidic bonds also exist.

Digestion of starch

From the structure of maltose above it can be seen that glucose units can be joined together indefinitely to form a huge polysaccharide molecule. The **amylose** component of starch, a molecule hundreds of glucose units (glucosyl units) long, is precisely this. If we represent an α-glucosyl unit as 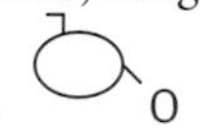 then amylose is

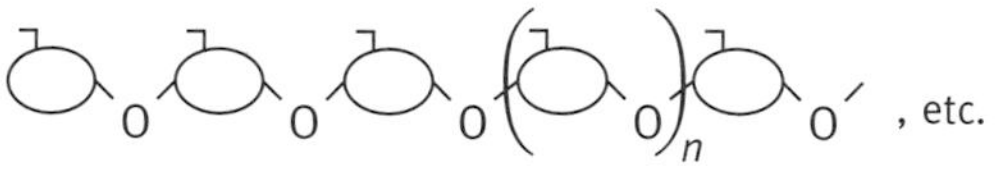

where n is a large number.

The second component of starch is **amylopectin**. This molecule also is a huge polymer of glucose but instead of it being one long chain it consists of many short chains (each about 30 glucosyl units in length) crosslinked together. The glucosyl units of the chains are $1 \rightarrow 4$-α-glycosidic bonds but the crosslinking between chains is by $1 \rightarrow 6$-α-glycosidic bonds, as shown below.

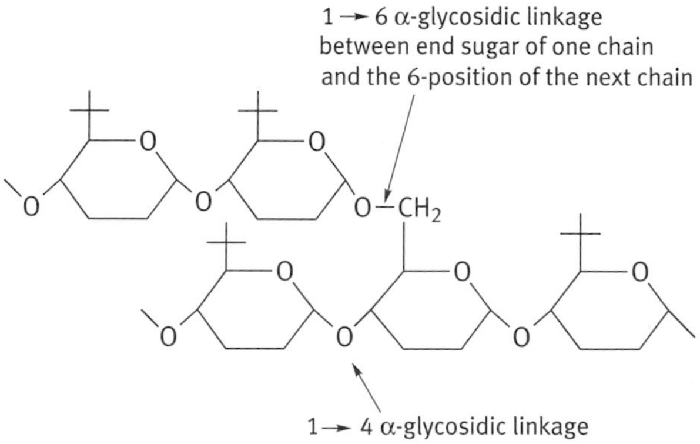

Amylopectin is therefore like this.
Many chains are linked together in this way.

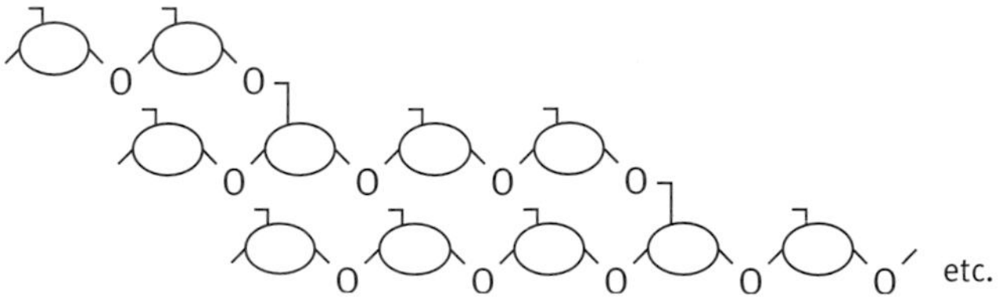

Digestion of starch is done by α-**amylase**, which is present in the saliva and pancreatic juice. It is an **endoenzyme**, hydrolysing glycosidic bonds anywhere inside the molecule except that, in amylopectin, it can't attack a bond near the $1 \rightarrow 6$ linkage. The amylase thus trims off the molecule but leaves a core containing the $1 \rightarrow 6$ bonds plus the nearby glucosyl units. This limit dextrin, as it is called (the limit of hydrolysis), is hydrolysed by an intestinal enzyme (an amylo α-$1 \rightarrow 6$ glucosidase) that hydrolyses the $1 \rightarrow 6$ linkages. The 2 and 3 unit sugars produced by amylase digestion is attacked by intestinal α-**glucosidase** (**maltase**) which produces free glucose. Salivary amylase has only a brief time to attack starch while food is in the mouth, for stomach acid destroys the enzyme. Most of the starch digestion therefore occurs in the intestine. The digestion of the other two main dietary carbohydrates, lactose and sucrose, also occurs in the small intestine.

Lactose is galactose-β $(1 \rightarrow 4)$ glucose, the principal sugar in milk. Galactose has the carbon 4-OH of glucose inverted.

The angled bond shown is normally used to simplify the presentation of the structure. An intestinal enzyme attached to the external membrane of the epithelial cells, **lactase**, hydrolyses lactose to the free sugars. Because lactose is a β-galactoside, lactase is also known as β-**galactosidase**. Many adults, particularly Asians, lose the ability to produce lactase and develop milk intolerance as they leave childhood. They are unable to hydrolyse the sugar and, since the disaccharide is not absorbed, it passes to the large intestine where it is fermented by bacteria giving rise to severe discomfort, water intake, and diarrhoea. The other major disaccharide of food, sucrose, is digested by the intestinal enzyme **sucrase**, liberating glucose and fructose which are absorbed.

Absorption of glucose into epithelial cells occurs by the same Na^+ cotransport mechanism described earlier (page 51) as shown in Fig. 4.4. The Na^+ is continually pumped to the outside by the Na^+/K^+ ATPase so this maintains the Na^+ concentration

difference and drives the cotransport. The glucose has to exit the cells on the side opposite that of the lumen so as to enter the blood capillaries for transport to the liver by the hepatic portal vein. The glucose transporter for this movement is the facilitated diffusion type (page 49); the sugar moves down the cell/blood concentration gradient created by active uptake from the intestine (Fig. 4.4). Fructose is absorbed from the gut by a Na^+-independent passive transport system.

Amino acids and sugars and other non-lipid molecules, absorbed from the intestine, are collected by the portal blood system which delivers digestion products directly to the liver. This arrangement means that these digestion products pass through the liver before being released into general circulation—an advantageous arrangement since the liver is responsible for removing most toxic foreign compounds entering the body via the intestine, and for processing some of the absorbed nutrients.

Digestion and absorption of fat

In Chapter 3 (page 41) we described what fat is—neutral fat, or triacylglycerol (TAG). Since there are no polar groups, liquid fat in water forms droplets with minimum contact between the lipid and water. The TAG molecules cannot be absorbed as such.

The main digestion of fat occurs in the small intestine by the action of the pancreatic enzyme, **lipase**, which mainly attacks *primary* ester bonds; the middle ester bond is a secondary ester bond and is not significantly attacked.

Simple as this digestion reaction is, there are problems. Fat is physically unwieldy in an aqueous medium. Lipase can attack only at the oil/water interface and the area of this in a simple fat/water mixture is insufficient for the required rate of digestion. The available surface area is increased by emulsifying the fat. The monoacylglycerol and free fatty acids, produced by lipase action together with bile salts (see below), help to disperse the oily

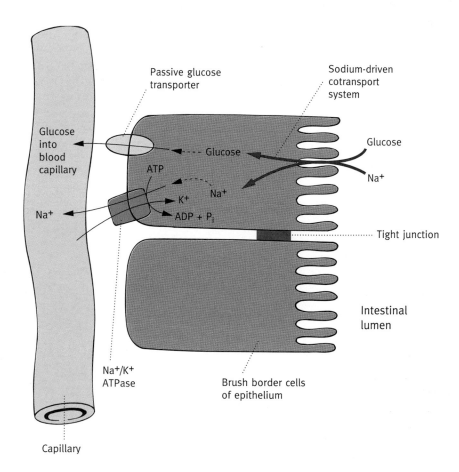

Fig. 4.4 Absorption of gluclose from the intestinal lumen by cotransport with sodium.

liquid into droplets (**emulsification**) and phospholipids in the food also help this process. The products of lipase action are relatively insoluble in water but they must be moved from the emulsion to the cells lining the intestine to be absorbed. This movement is facilitated by bile acids or bile salts as they are interchangeably called.

Bile acids are produced in the liver and stored in the gall bladder until discharged into the duodenum. They are produced from cholesterol, whose structure they resemble. The main change is that the hydrophobic side chain of the cholesterol molecule is converted to a carboxyl group, and OH groups may be added.

Cholesterol

Cholic acid

The main bile acid is cholic acid; others varying in the number and position of hydroxyl groups exist. The cholic acid mostly has attached to it either glycine or the sulfonic acid, taurine. Glycine is $NH_3^+CH_2COO^-$. Taurine is $NH_3^+CH_2CH_2SO_3^-$ (it does *not* occur in proteins). If cholic acid is represented as $RCOO^-$ then glycocholic acid is $RCONHCH_2COO^-$ and taurocholic acid $RCONHCH_2SO_3^-$. These conjugated acids have lower pK_a values (approximately 3.7 and 1.5, respectively) than the parent cholic acid (approximately 5.0). Possibly the reason for the conjugation is to ensure full ionization in the intestinal contents.

The monoacylglycerols and fatty acids, in the presence of bile salts, form **mixed micelles**. These are disc-like particles in which bile salts are arranged around the edge of the disc surrounding a more hydrophobic core containing digestion products of fat, together with cholesterol and phospholipids. The particles are much smaller than emulsion droplets giving a clear solution and carry higher concentrations of lipid digestion products than is possible in simple solution. In this form, the lipid digestion products diffuse to enter the epithelial cells; possibly the micelle breaks down at the cell surface and the lipids diffuse in. Bile salts are also partially reabsorbed and transported back to the liver.

What happens to the fatty acids and monoacylglycerol absorbed into the intestinal brush border cells?

Resynthesis of neutral fat in the absorptive cell
The products of digestion of fat absorbed into the intestinal cell are resynthesized into fat, the fatty acids being re-esterified producing triacylglycerols.

$$
\begin{array}{c}
CH_2OH \\
| \\
CH-O-\overset{\overset{\displaystyle O}{\|}}{C}-R \\
| \\
CH_2OH
\end{array}
\quad
\xrightarrow[\text{ATP energy}]{\text{Fatty acids}}
\quad
\begin{array}{c}
CH_2-O-\overset{\overset{\displaystyle O}{\|}}{C}-R \\
| \\
CH-O-\overset{\overset{\displaystyle O}{\|}}{C}-R \\
| \\
CH_2-O-\overset{\overset{\displaystyle O}{\|}}{C}-R
\end{array}
$$

The mechanism of this resynthesis is not given here because it would divert too much from the topic in hand—however, in Chapter 10 (page 143) on fat metabolism, the matter is dealt with. The neutral fat and cholesterol must now be transported from the intestinal cell to the tissues of the body. Neutral fat cannot diffuse out of the cell through membranes and, in any case, it would be insoluble in the blood—it cannot simply be ejected into the circulation.

The solution is that neutral fat and cholesterol are arranged inside the epithelial cells into fine particles, called **chylomicrons**. They are enclosed in membrane vesicles and released from the cell by exocytosis (see page 46), since they are too large to traverse the membrane in any other way.

What are chylomicrons?

Chylomicrons (Fig. 4.5) are spherical particles, in the middle of which are hydrophobic molecules—neutral fat and cholesterol esters. Cholesterol has a hydrophilic —OH group and a hydrophobic part; a **cholesterol ester** has a fatty acid attached to that —OH group.

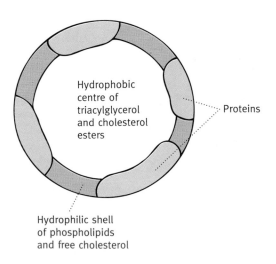

Cholesterol

Fatty acid
ATP energy

$R-\overset{\overset{\displaystyle O}{\|}}{C}-O$

Cholesterol ester

In making a cholesterol ester from cholesterol, the polar —OH group, which would interfere with packing cholesterol into the hydrophobic centre of chylomicrons, is eliminated. It illustrates the importance that the polarity–hydrophobicity characteristics of biological molecules play in life. Hydrophobic particles of

Hydrophobic
centre of
triacylglycerol
and cholesterol
esters

Proteins

Hydrophilic shell
of phospholipids
and free cholesterol

Fig. 4.5 Diagram of the cross-section of a chylomicron.

neutral fat and cholesterol ester would coalesce into an insoluble mass unless they were stabilized. For this, there is a 'shell' containing phospholipids, some free cholesterol which is weakly amphipathic, and, very importantly, some special proteins.

There are several different proteins involved, a major one being **apolipoprotein B (apoB)**, necessary for chylomicron synthesis. ApoB is a glycoprotein (page 32), the carbohydrate attachment providing a highly polar group. The prefix *apo* means 'detached' or 'separate'. Thus an apolipoprotein is a protein normally in lipoproteins but now is detached or separate from the lipoprotein. The whole chylomicron structure is called a lipoprotein. The hydrophilic shell stabilizes the chylomicron so that it remains as a suspended particle in the lymph chyle (see below) and blood. The overall composition of chylomicrons is about 90% or more neutral fat, giving them a low density.

The chylomicrons are not released directly into the blood (unlike absorbed amino acids and sugars) but into the lymph vessels, called lacteals, that drain the villi. The suspension of chylomicrons in lymph fluid is called chyle. A few words on lymph might be useful here. As blood circulates through the capillaries, a clear lymph fluid containing protein, electrolytes, and other solutes filters out to bathe cells in an interstitial fluid. All tissues have a fine network of lymph capillaries closed at the fine ends into which lymph drains. The lymph capillaries join into lymphatic ducts and discharge the lymph back into the

major veins of the neck via the thoracic duct. It is not actively pumped but movements of the body propel it along. After a fatty meal, chylomicrons, entering the bloodstream via lymph, give the blood a milky appearance. They circulate in the blood and their contents are used by tissues, the subject of the next chapter.

We want to temporarily end the story here—we have the food components, amino acids, sugars, and fat, in the blood where they are now utilized by cells. It is convenient to deal with the next phase in the next chapter. Meanwhile, we would like to finish the topic of digestion with a few notes.

Digestion of other components of food

There are components in food other than the ones dealt with so far and digestive enzymes exist to hydrolyse them into their constituent parts. Thus phospholipids in food are hydrolysed by phospholipases; nucleic acids (which you'll meet later) are hydrolysed by ribonuclease (RNase) and deoxyribonuclease (DNase) also in the pancreatic juice. Plant material contains carbohydrate molecules such as cellulose, for which animals do not produce digestive enzymes (herbivores have microorganisms in the rumen which do produce them). In man, these components constitute the fibre of the diet.

Further reading

Digestion and absorption of fats

Riddihough, G. (1993). Picture of an enzyme at work. *Nature*, 362, 793.

A news and views summary of the fascinating way pancreatic lipase is activated in the digestion of fats.

Problems for Chapter 4

1 Which digestive enzymes are produced in an inactive zymogen form?

2 Why should this be? Why do you think amylase is not produced as an inactive zymogen? What prevents activated enzymes attacking intestinal cells?

3 Explain how the inactive proteases are activated in the intestine.

4 If pancreatic proenzymes are prematurely activated due to physical and chemical damage or pancreatic duct blockage, what is the result?

5 Why is digestion of foods necessary?

6 Epithelial cells of the gut take up glucose, and many amino acids, against a concentration gradient, and therefore energy must be supplied to the process. Nevertheless, the transport system actually carrying those components into the cells does not use ATP or other high-energy phosphate compounds. Explain the answer to this apparent paradoxical situation.

7 Many people, especially Asians, suffer severe intestinal distress on consuming milk or milk products. Why is this so?

8 Write down the structure of a neutral fat. What is the alternative name for this? Indicate which groups in the molecule are primary esters.

9 Fat in water forms large globules with little fat/water interface for lipase to attack. Explain how the digestive system copes with the physical intractability of lipid.

10 Amino acids and sugars absorbed into the intestinal cells move into the portal bloodstream and are carried via the liver and thence to the rest of the body. What happens to the absorbed digestion products of fat?

Chapter 5

Preliminary outline of fuel distribution and utilization by the different tissues of the body

In the previous chapter we dealt with food digestion and absorption from the intestine and left the various products of digestion just having reached the blood—fat and cholesterol into the general circulation via the lymph system as chylomicrons, and everything else headed for the liver in the portal vein and thence into the general circulation. With this chapter we start on the subject of metabolism; if you have already heard of metabolic pathways such as glycolysis, the citric acid cycle, etc. we suggest you put them out of your mind for the time being.

We will, first, deal with the traffic of food molecules between the blood and tissues and between different tissues, and with how this is regulated to satisfy different physiological needs—in essence, the logistics of fuel movements in the body. In this chapter we do not give any details of the metabolites involved—the chapter is designed to give a broad picture of the situations. The details of metabolic pathways covered in subsequent chapters will make more sense through this arrangement.

Purposes of metabolism

The purposes of metabolism can be summed up in these terms.

- The major role of metabolism is to oxidize food to provide energy in the form of ATP.
- Food molecules are converted to new cellular material and essential components.
- Waste products are processed to facilitate their excretion in the urine.
- Some specialized cells in human babies oxidize food to generate heat. This is exceptional since heat is, in general, a byproduct of metabolism.

Storage of food in the body

Animals take in food at periodic intervals; they do not eat continuously. The body is, in effect, continually exposed to cyclical feast and famine conditions. The periods between meals can for humans vary from short intervals during the day to longer periods during sleep and very long periods during starvation. The biochemical machinery in the body copes with these situations.

After a meal, the blood becomes loaded with food absorbed from the intestine. To state the obvious, this food isn't held in the blood until it is all used up by metabolism, but rather it is rapidly cleared from the blood by uptake into the tissues so that blood levels quickly return to normal. After a fatty meal, the lipemia (presence of milky blood plasma) is cleared within a few hours. Similarly, blood glucose may increase after a meal from its normal 90 mg dl^{-1} to 140 mg dl^{-1} in an hour (5 and 7.8 mM, respectively) but it reverts back to normal, in a non-diabetic person, in 2 hours. The tissues are not using all of this food at once; most of it is stored.

How are the different foods stored in cells?

Glucose storage as glycogen

It would not be practicable for cells to store glucose as the free sugar—the osmotic pressure of the sugar at high concentrations would be too high. The osmotic pressure of a solution is proportional to the number of particles of solute in the solution. If, therefore, large numbers of glucose molecules are joined together to form a single macromolecule the osmotic pressure, exerted by the store of glucose residues, is accordingly reduced. The resultant large polymerized molecule may fall out of solution as a granule. In animals, glucose is polymerized into a highly branched so-called 'animal starch', or **glycogen**, with the same chemical bondings as in amylopectin (page 65), but more highly branched. Its synthesis requires energy. When needed, glycogen is broken down again into glucose. A very important fact is that *glycogen storage in animals is extremely limited*. In man, the glycogen reserves of liver, which are used to supply glucose to the blood for utilization by other tissues, are exhausted after about 24 hours of starvation. This fact has a profound effect on the biochemistry of an animal as will become apparent later in the chapter.

In all of the above, we have talked about glucose. What about other sugars such as galactose and fructose, present in lactose and sucrose respectively? Glucose is the sugar of central metabolic importance; other sugars are converted to glucose (or glycogen), or else to compounds on the main glucose-metabolizing pathways.

Storage of fat in the body

Fat is stored by cells as neutral fat or triacylglycerols (page 41). The bulk of it is stored in adipose tissue or fat cells, which are distributed in many areas of the body and which specialize in fat storage. A loaded fat cell can, under the microscope, look like droplets of oil surrounded by a thin layer of cytoplasm and membrane as shown in Fig. 5.1.

Unlike glycogen storage, that of fat is essentially unlimited and large reserves of energy can be stored in the body as neutral fat in the adipose cells. In a typical human the reserves of fat, in terms of stored energy, are about 50 times those of the glycogen reserves.

Since this limited glycogen storage causes the metabolic problems described later, it may be asked why the body stores so

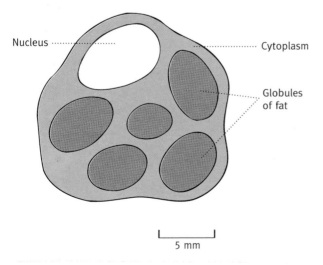

Fig. 5.1 Diagram of a fat cell.

Nucleus

Cytoplasm

Globules of fat

5 mm

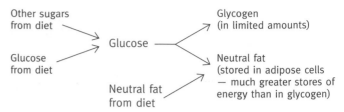

Other sugars from diet

Glucose from diet

Glucose

Neutral fat from diet

Glycogen (in limited amounts)

Neutral fat (stored in adipose cells — much greater stores of energy than in glycogen)

Fig. 5.2 Diagram of storage of sugars and fat from the diet in the post-absorptive period.

much fat, and so little glucose, when the latter is essential to life in animals? Fat is more highly reduced (less oxidized) than is carbohydrate and hence contains more energy per unit mass, particularly since glycogen in the cell is hydrated while fat is not. This means that the latter occupies much less volume per unit of stored potential energy than does glycogen. If the energy equivalent of fat stored in our bodies was in the form of glycogen we'd need to be much larger than our existing sizes.

Since glucose storage is so limited and sugar intake in the diet potentially almost unlimited, it follows that glucose in excess of that used for glycogen synthesis must be stored in another way—in fact as fat. The position then is as shown in Fig. 5.2.

An important fact is that, while glucose is readily converted to fat, in the human body there is no net conversion of fatty acids to glucose. (The glycerol moiety of neutral fat *can* be converted to glucose but this represents a small fraction of the energy available in the three long-chain fatty acids.)

Are amino acids stored by the body?

Digestion of the third main class of food, protein, results in amino acids being absorbed and taken into the blood, but there is no dedicated storage form of amino acids in animals if we exclude egg and milk proteins which don't negate the general concept. Plant seeds have storage proteins whose only function appears to be storage of amino acids in a convenient form to supply the developing embryo. In animals then, dietary amino acids in the blood are taken up by tissues as needed for the synthesis of cellular proteins, neurotransmitters, etc., and then those in excess of immediate needs are destroyed, the amino group ($-NH_2$) being removed and, in mammals, converted into urea and excreted in the urine. Urea is highly water-soluble, neutral, and nontoxic, even in relatively high concentration. This leaves the carbon–hydrogen 'skeletons' of the amino acids, which contain chemical energy. They are converted to glycogen or fat, depending on the particular amino acid, or they can be oxidized (Fig. 5.3), according to the physiological needs at the time.

In one sense, there is a storage of amino acids in the body—in the proteins of all cells. Muscle proteins are quantitatively of greatest importance. However, they are all functional proteins mainly involved in contraction and are not dedicated to storage. When amino acids must be made available to the body in general, these proteins are broken down and the result is muscle

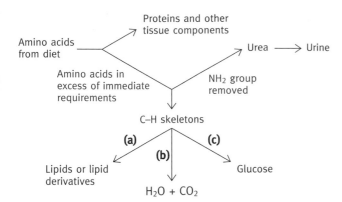

Fig. 5.3 Summary of the fate of amino acids in the diet. Which of the metabolic routes **(a)**, **(b)**, or **(c)** is followed depends on the particular amino acid, the physiological state, and biochemical control mechanisms.

wasting. You will shortly see in what situation this becomes essential.

Characteristics of different tissues in terms of energy metabolism

The different tissues in the body have special biochemical characteristics. However, in our present context of looking at overall food traffic in the body, several organs are of overriding importance. These are the liver, the skeletal muscles, the brain, the adipose cells, and the red blood cells. (We are excluding regulatory organs such as the pancreas and adrenal glands from the list though their roles are crucial.)

1.   The **liver** has a central role in maintaining the blood glucose level; it is the glucostat of the body. When blood glucose is high, such as after a meal, the liver takes it up and stores it as glycogen. When blood glucose is low, it breaks down the glycogen and releases glucose into the blood.

In starvation, after about 24 hours, the reserves of glycogen in the liver are exhausted by this activity; the concentration of glucose in the blood would decline to lethal levels if nothing were done about it. During starvation, there will, for a prolonged period, be reserves of fat that are released into the blood by the fat cells of adipose tissue. This will keep the muscles and other tissues happy, but not the brain and red blood cells, which must have glucose for they cannot use fatty acids. The liver saves the situation by converting amino acids to glucose, a process called **gluconeogenesis**. The main sources of amino acids in this situation are muscle proteins, which are broken down to provide them. This means destruction of functional muscle proteins, causing muscle wasting, but this is preferable to death resulting from low blood glucose.

The liver also plays an important role in fat metabolism. In starvation, the fat cells release fatty acids into the blood that can be directly used by most tissues—brain and red blood cells

excepted, as stated. However, in such conditions, where massive fat utilization is occurring, the liver converts some of the fatty acids to small compounds called ketone bodies and releases them into the blood for use by other tissues (red blood cells again excepted). Of special importance, the brain can adapt to utilize ketone bodies which, during starvation, can supply about half of that organ's energy needs. The rest must come from glucose. For the moment, ketone bodies might be regarded as fatty acids partially metabolized by the liver. The liver is also the major site of fatty acid synthesis, which is exported to adipose cells and other tissues (as TAG).

So, energy-wise, the liver stores glucose as glycogen and releases glucose when needed, as long as there are glycogen reserves left. In starvation it produces glucose from amino acids and ketone bodies from fat, which are turned out into the blood for use by other tissues (Fig. 5.4). It synthesizes fatty acids from glucose and exports them. It preferentially oxidizes fatty acids for energy supplies. The liver has many other functions but for now we are concentrating on energy supplies in the body.

2.   The **brain** must, as already stated, have a continuous supply of glucose in the blood. It has no significant fuel reserves. If the blood becomes hypoglycemic (low glucose level), the rate of glucose entry into the brain is reduced since it involves passive transport and neuron function is impaired. Convulsions and coma result. However, as already emphasized, although the brain *must* have glucose, in starvation it can adapt to use ketone

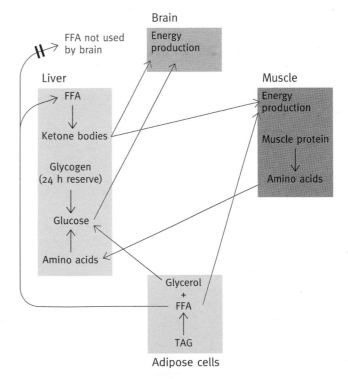

Fig. 5.4 Simplified diagram of main fuel movements in the body during starvation. TAG, Triacylglycerols; FFA, free fatty acids.

bodies produced by the liver from fat for perhaps half its energy needs. This helps to economize on scarce glucose.

3. **Skeletal muscles** require large-scale energy production for muscle contraction. They can utilize most types of energy sources. They take up glucose from the blood and store it as glycogen but *in contrast to the liver, they do not release any glucose back into the blood.* They can oxidize fatty acids, ketone bodies, and amino acids as well as glucose. In the presence of a high level of free fatty acids and ketone bodies in the blood, these are preferentially oxidized. In starvation, muscles sacrifice their protein to supply the liver with amino acids for glucose synthesis.

4. The role of the **adipose cells** can be stated simply. After a meal, they take up fatty acids, which they store as neutral fat, and take up glucose and convert it also to stored neutral fat. As soon as the blood glucose retreats as in mild fasting (or in starvation) they release reserves of fat into the blood as fatty acids for the rest of the body to use. Adipose cells and liver are the major sites of triacylglycerol formation, the fatty acids coming mainly from the diet and from synthesis in the liver. Glucose is needed to provide the glycerol moiety.

5. **Red blood cells** can utilize only glucose, producing lactate. They are terminally differentiated cells (that is, fully formed and don't divide) without mitochondria and cannot oxidize food-stuffs as can other cells. But their Na$^+$/K$^+$ ATPase pumps must run and there are other energy-requiring processes.

Overall control of the logistics of food distribution in the body by hormones

With that description of the major organs involved in food logistics in the body, let us now see in overview how food distribution is directed—how the various organs are coordinated in their activities to cope with different physiological situations. These situations are as follows.

- Excess food available—the postprandial condition after a good meal.
- Mild fasting—this refers to the condition between meals—for example, before breakfast after the night's fasting.
- Starvation—fasting that goes on for longer than about 1 or 2 days.
- Abnormal conditions—for example, in diabetes, glucose cannot be adequately utilized so that, irrespective of its nutritional state, the body is biochemically starved of carbohydrate.
- Emergency responses—this relates to conditions, for example, where violent muscular action is needed to avoid a threat.

The overall picture of food traffic in the body in the various nutritional situations is summarized below.

Post-absorptive phase

All food components are at a high level in blood. Glucose is taken up by the liver, muscle, and adipose cells and used to replenish glycogen stores. Beyond this, excess glucose is taken up by the adipose tissue and by the liver and other tissues and converted into neutral fat. The liver does not retain this in large amounts as do the adipose cells but transfers it to other tissues. A 'fatty liver' is pathological.

Amino acids are taken up by all tissues and used to synthesize protein and other cell components. Fat is taken from chylomicrons and absorbed by tissues including adipose tissue, which stores it, and by the lactating mammary glands, which utilize it in milk production. While active storage of glycogen and fat is going on, the breakdown of these components in the cells is stopped.

The main 'signal' for all of this storage work to take place is the pancreatic hormone, **insulin**, whose release from the pancreas into the blood is signalled by high blood sugar levels. The high insulin level is the signal to the tissues that times are good and that food should be stored both as glycogen and fat. Correlated with this is the *low* level of **glucagon**, another pancreatic hormone, whose release is *inhibited* by high blood sugar levels. Glucagon has the opposite effect to insulin. It signals that the tissues are short of fuel and that storage organs should release some into the blood. Note that the brain is not affected by insulin; it goes on using glucose from the blood all the time.

Fasting condition

The insulin-stimulated storage activity in the post-absorptive period lowers the blood glucose and amino acids to normal levels and clears the chylomicrons. With time, the blood glucose level begins to fall and, with it, insulin secretion. The hormone is rapidly destroyed after release, so its blood level rapidly falls when secretion stops. In concert with this, the pancreas releases glucagon, whose level rises in response to low blood glucose. This stimulates the liver to break down glycogen and release glucose into the blood. It also stimulates adipose tissue to hydrolyse neutral fat and release free fatty acids into the blood. Muscles and liver, and other tissues, use the fatty acids to provide energy and this conserves glucose for the brain. Glycogen synthesis and fat synthesis from glucose cease due to the low insulin/high glucagon ratio.

Prolonged fasting or starvation

After about 24 hours the liver has no glycogen left to provide blood sugar and no other tissue, other than the quantitatively unimportant kidney, is capable of releasing glucose. The insulin level is low and that of glucagon high. In this situation the adipose cells pour out free fatty acids into the blood, for use by muscle and other tissues; the fat reserves in adipose cells are sufficient perhaps for weeks. Muscles break down their proteins to amino acids, which the liver uses to synthesize glucose, a process activated by glucagon. As starvation proceeds, the high glucagon causes the adipose tissue to turn out more and more free fatty acids into the blood and the liver metabolizes some of this into **ketone bodies** which it turns out into the blood. Ketone bodies can be regarded in general terms as 'predigested' or partly metabolized fatty acids. There are two components, acetoacetate and β-hydroxybutyrate.

$CH_3COCH_2COO^-$ Acetoacetate

$CH_3CHOHCH_2COO^-$ β-Hydroxybutyrate

The name 'ketone bodies' is a misnomer for they are not 'bodies', nor is β-hydroxybutyrate a ketone but the term is an old one. Once regarded as occurring only in pathological situations, production of ketone bodies is recognized to be an important normal process.

The situation in diabetes

If a patient lacks insulin, even in the presence of abundant glucose, the sugar is not utilized by muscles. The uptake of glucose by the brain and red blood cells is not dependent on the presence of insulin and so can survive in the diabetic state. Thus, the diabetic state resembles that of extreme starvation resulting in the adipose cells turning out large amounts of fatty acids into

the blood because the insulin/glucagon ratio is obviously very low. The liver, in this situation, produces ketone bodies in large amounts. Acetoacetate, a component of ketone bodies, spontaneously decarboxylates to acetone, which is not metabolized, giving the characteristic sickly sweet smell of the breath of untreated juvenile-onset diabetics.

$$CH_3COCH_2COO^- \rightarrow CH_3COCH_3 + CO_2$$
acetoacetate   acetone

The emergency situation—fight or flight

When an animal such as man is presented with a dangerous situation, the adrenal glands release **epinephrine** (**adrenalin**) into the blood in response to a neurological signal from the brain. Adipose tissue is innervated and can be stimulated by epinephrine and norepinephrine released from nerve endings. Epinephrine, as it were, presses the biochemical panic button. It stimulates the liver to pour out glucose into the blood and the adipose cells to release free fatty acids so that muscles have no shortage of fuel. It also stimulates glycogen breakdown in skeletal muscle so that the cells produce ATP, at the maximal rate. The pattern has an overall logic of ensuring that muscles can react violently and instantly to escape the threat.

Problems for Chapter 5

1 Why do cells of the body store glucose as glycogen? Why not save trouble and store it as glucose?

2 Triacylglycerol (TAG) is stored in much greater quantities in the body than is glycogen. Why is this so?

3 Can glucose be converted to fat? Can fatty acids be converted in a net sense to glucose? Do amino acids have a special dedicated long-term storage form in the body? Give brief explanations.

4 In terms of food logistics, describe the chief metabolic characteristics of the liver.

5 Can the brain use fatty acids for energy generation?

6 What is the chief metabolic characteristic of fat cells?

7 What energy source do red blood cells use?

8 What are the main hormonal controls on the logistics of food movement in the body in different nutritive states?

Chapter 6

Biochemical mechanisms involved in food transport, storage, and mobilization

In this chapter we deal with the biochemical mechanisms involved in fuel transport, storage, and release processes described in Chapter 5.

Glucose traffic in the body

Mechanism of glycogen synthesis

The synthesis of glycogen from glucose is an **endergonic process**—it requires energy. Hydrolysis of the glycosidic bond joining glucose units in glycogen has a $\Delta G^{0'}$ value of about 16 kJ mol^{-1} which would give an equilibrium totally to the side of glucose for the simple reaction: glycogen plus $H_2O \rightarrow$ glucose. Therefore energy must be supplied from high-energy phosphoryl groups to form the glycosidic bond.

Glycogen synthesis occurs by enlarging pre-existing glycogen molecules (called 'primers') by the sequential addition of glucose units. The synthesis of a glycogen granule is initially primed by the protein, **glycogenin**, which transfers eight glucosyl residues to a tyrosine–OH on itself. The protein remains inside the glycogen granule. The new units are always added to the nonreducing end of the polysaccharide. Glucose is an aldose sugar. This is seen clearly in the open-chain form which is in spontaneous equilibrium in solution with the ring form, the latter predominating.

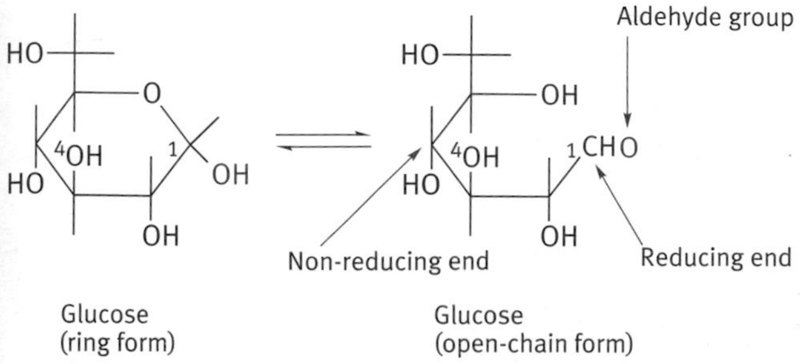

Glucose
(ring form)

Glucose
(open-chain form)

An aldehyde is a reducing agent, so that the carbon 1 end of the ring sugar is the reducing end and carbon 4 the non-reducing end. The glycogen chain thus always has a nonreducing end.

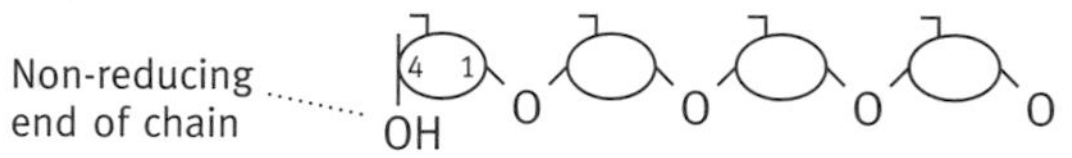

The process of glycogen synthesis in essence therefore is as follows.

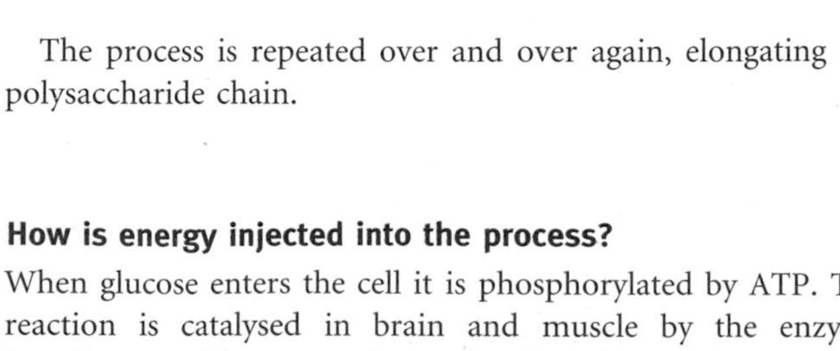

The process is repeated over and over again, elongating the polysaccharide chain.

How is energy injected into the process?

When glucose enters the cell it is phosphorylated by ATP. The reaction is catalysed in brain and muscle by the enzyme hexokinase. A **kinase** transfers a phosphoryl group from ATP to something else—in this case, glucose, which is a hexose sugar—hence the name **hexokinase**. In liver, a different enzyme catalysing the same reaction is called **glucokinase**.

Glucose

+ ATP

Hexokinase

(Glucokinase in liver)

$\Delta G^{0'} = -16.7$ kJ mol^{-1}

+ ADP

Glucose-6-phosphate

The ΔG of this reaction is strongly negative, making the reaction irreversible. The highly charged glucose-6-phosphate molecule is unable to traverse the cell membrane so that glucose phosophorylation has the effect of trapping the sugar inside the cell and facilitating entry of more glucose from the blood, since in facilitated diffusion (by which glucose enters) it can move only down a concentration gradient. The phosphoryl group is now switched around to the 1 position by a second enzyme, **phosphoglucomutase**, a freely reversible reaction.

Glucose-6-phosphate (G-6-P)

Phosphoglucomutase

$\Delta G^{0'} = -7.3$ kJ mol^{-1}

Glucose-1-phosphate (G-1-P)

The enzyme is called phosphoglucomutase because it mutates (changes) phosphoglucose.

The free energy of the phosphoryl ester group of glucose-1-phosphate (G-1-P; $\Delta G^{0'}$ for its hydrolysis = -21.0 kJ mol^{-1}) is about the same as that of the glycosidic bond in glycogen ($\Delta G^{0'}$ for hydrolysis of the same α $(1 \rightarrow 4)$-bond of maltose is -15.5 kJ mol^{-1}). You might expect that the glucose unit would be transferred from G-1-P to a glycogen primer and the synthesis would be done. It was thought years ago that this was what happened. But it's not so. We explained (page 3) that biochemical pathways are usually arranged to be thermodynamically irreversible—if necessary, energy is pumped in at the expense of —Ⓟ groups and the process given such a large negative ΔG value that the pathway is one-way only. If glycogen synthesis occurred directly from G-1-P, the process would be freely reversible and hence uncontrollable.

An extra step is inserted that renders glycogen synthesis thermodynamically irreversible. This extra step is somewhat curious. To understand it, first go back to the structure of ATP (Fig. 1.5). A similar compound, uridine triphosphate (UTP), exists in which adenine is replaced by uracil (U) whose structure doesn't matter in this context. (Uridine is uracil-ribose, a structure analogous to adenosine.) The cell synthesizes UTP (the energy coming from ATP). In glycogen synthesis, uridine diphosphoglucose (UDP-glucose) is made by a reaction between UTP and G-1-P as shown in Fig. 6.1.

This enzyme, for systematic nomenclature reasons, is named for the reverse reaction (which doesn't occur in the cell). This reaction (were it to occur) would cleave UDPG with pyrophosphate. The enzyme is therefore a pyrophosphorylase and is called **UDP-glucose pyrophosphorylase**; inorganic pyrophosphate is not available in the cell because it is rapidly destroyed and therefore the reverse reaction is inoperative.

UDPG is the 'activated' or reactive glucose compound that donates its glucosyl group to glycogen. Wherever in animals an 'activated' sugar has to be produced for a chemical synthesis, as a general rule it is a uridine diphosphate derivative. Whether there is a fundamental reason for this or whether it is an accidental quirk that evolution was locked into isn't known. (Plants use the corresponding adenine compound for starch synthesis.) Synthesis of glycogen using UDPG as glucosyl donor occurs as shown in Fig. 6.2. The enzyme that does this is called **glycogen synthase** ('synthase' is used wherever a synthesizing enzyme does *not* directly use ATP and 'synthetase' where ATP is involved). The UDPG pyrophosphorylase produces inorganic pyrophosphate. This is hydrolysed by inorganic pyrophosphatase, which, as explained on page 10, has a $\Delta G^{0'}$ value of -33.5 kJ mol^{-1}, and thus pulls the process of UDPG synthesis from G-1-P and UTP completely to the right. Glycogen synthesis in the cell is not therefore directly reversible by the route of its synthesis. The synthesis is summarized in the scheme shown in Fig. 6.3.

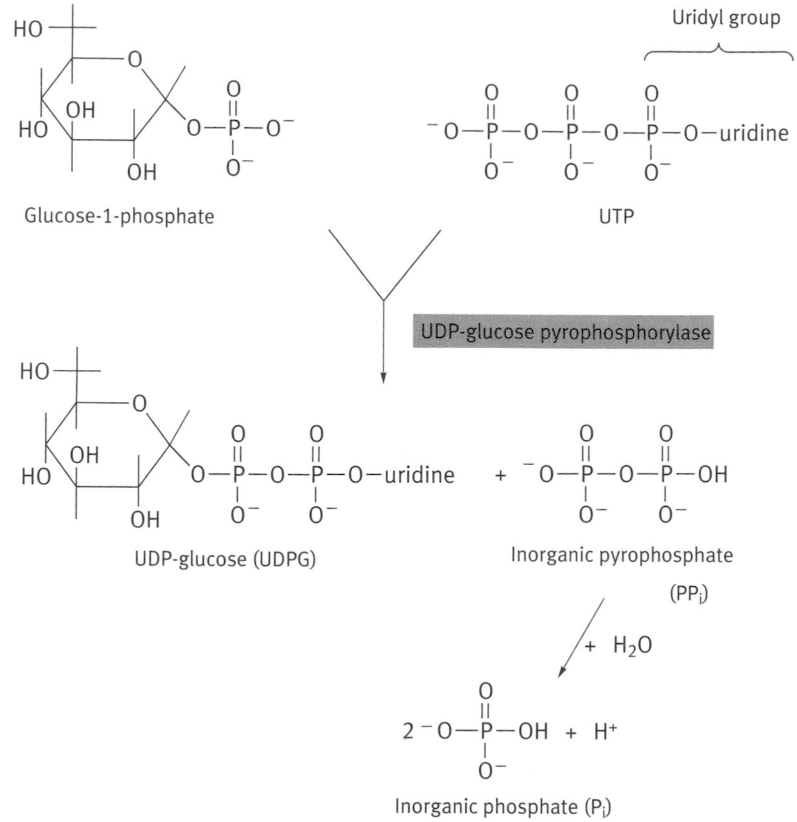

Fig. 6.1 Formation of UDP-glucose by a reaction between UTP and glucose-1-phosphate catalysed by UDP-glucose pyrophosphorylase.

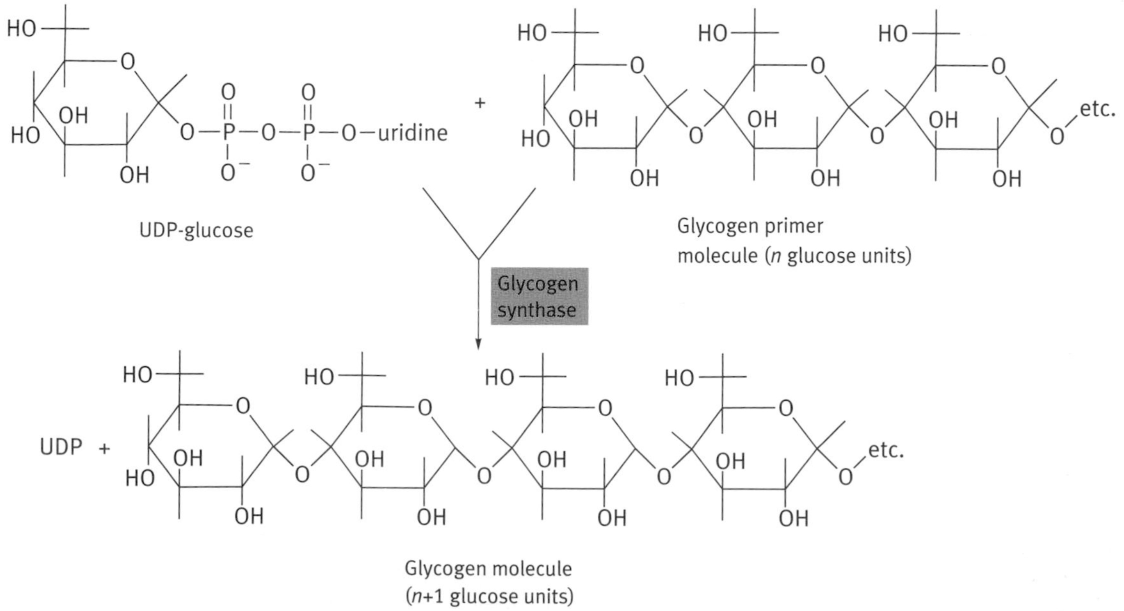

Fig. 6.2 Synthesis of glycogen by the elongation of a glycogen primer molecule using UDP-glucose (UDPG) as glucosyl donor.

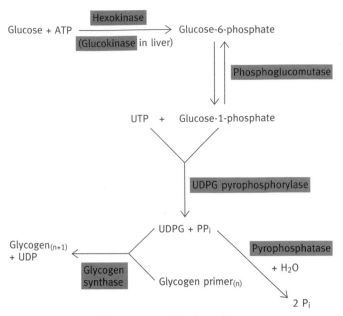

Glucose + ATP $\xrightarrow{\text{Hexokinase (Glucokinase in liver)}}$ Glucose-6-phosphate

Phosphoglucomutase

UTP + Glucose-1-phosphate

UDPG pyrophosphorylase

UDPG + PP$_i$

Glycogen$_{(n+1)}$ + UDP

Glycogen synthase

Glycogen primer$_{(n)}$

Pyrophosphatase

+ H$_2$O

2 P$_i$

Fig. 6.3 Summary of glycogen synthesis from glucose.

As with synthesis, **glycogen breakdown** occurs at the nonreducing end. Instead of splitting off (lysing) glucose by water (*hydro*lysis) it splits it off with phosphate (*phosphoro*lysis). The enzyme is called **glycogen phosphorylase.**

There's one more point about glycogen synthesis. If more and more glucosyl groups were added to primer molecules, the result would be very long polysaccharide chains—which is precisely what plants synthesize as the amylose starch component. Glycogen is different—instead of consisting of long chains of glucosyl units, it is a highly branched molecule. This structure is achieved by another enzyme, called the **branching enzyme.** When the glycogen synthase has made a 'straight chain' extension more than 11 units in length, the branching enzyme transfers a block of about seven terminal glucosyl units from the end of an $\alpha(1 \rightarrow 4)$-linked chain to the C(6)—OH of a glucose unit on the same or another chain (Fig. 6.4). The energy in the $\alpha(1 \rightarrow 6)$ link is about the same as in the $(1 \rightarrow 4)$ link so the reaction is a simple transfer one. It creates more and more ends both for glycogen synthesis and breakdown (see page 174).

Thus, to return to the starting point, blood glucose after a meal is taken up by tissues and converted into glycogen by the reactions outlined.

How does the liver release glucose?

The liver, you'll remember, stores glycogen, not so much for itself but for other tissues and especially the brain and red blood cells. In fasting, such as between meals, the liver breaks down glycogen (high glucagon/low insulin are the signals for this) and releases glucose to keep up the blood glucose level, for otherwise the brain would cease to function normally.

The G-1-P so formed, is converted by phosphoglucomutase working in the reverse direction, to glucose-6-phosphate and this is hydrolysed (in liver and kidney only) by the enzyme **glucose-6-phosphatase** to give free glucose which is released into the blood.

Glucose-1-phosphate $\leftrightarrow$ Glucose-6-phosphate,

Glucose-6-phosphate $+ H_2O \rightarrow$ Glucose $+ P_i$.

A summary of the production of glucose from glycogen is shown in Fig. 6.5.

Glucose-6-phosphatase is located in the membrane of the endoplasmic reticulum (page 306). It is a remarkable enzyme in that its active site is exposed to the lumen of the endoplasmic reticulum (ER), not the cytoplasm. Glucose-6-phosphate is transported through the ER membrane (by a transport protein) where it is hydrolysed. The products, glucose and P$_i$, are transported back into cytoplasm by separate transport systems.

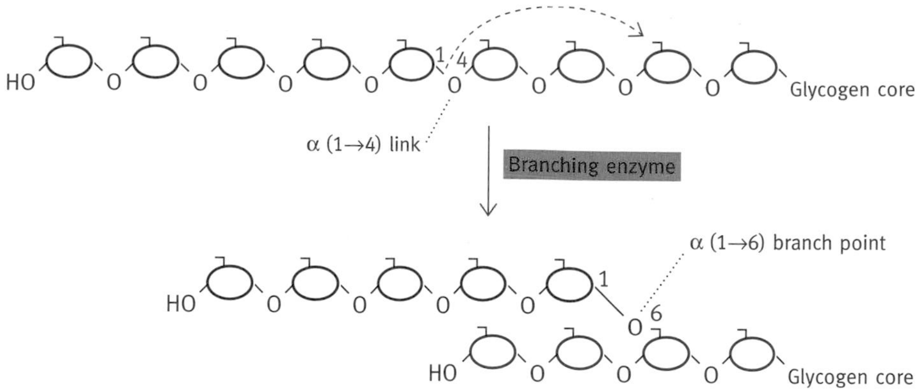

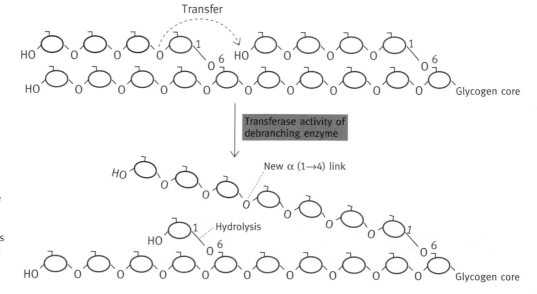

Fig. 6.4 Action of the branching enzyme in glycogen synthesis.

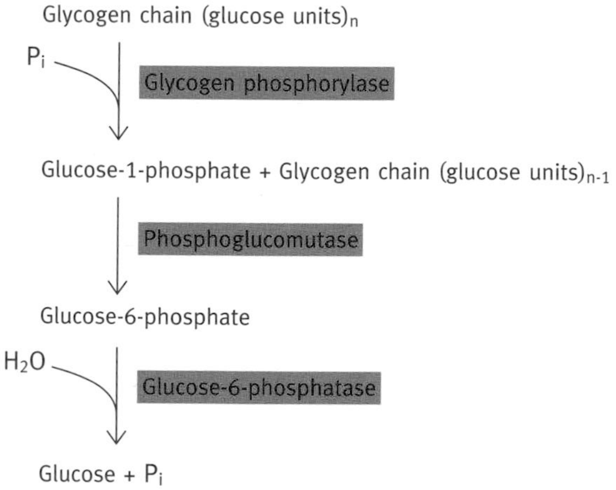

Glycogen chain (glucose units)$_n$

P_i → | Glycogen phosphorylase |

Glucose-1-phosphate + Glycogen chain (glucose units)$_{n-1}$

| Phosphoglucomutase |

Glucose-6-phosphate

H_2O → | Glucose-6-phosphatase |

Glucose + P_i

Fig. 6.5 Degradation of glycogen by phosphorolysis and the ultimate release of free glucose into the blood.

The branch points of glycogen create a problem, for phosphorylase cannot function within four glucose units of such points—the enzyme is big and presumably the branch gets in the way of the enzyme attaching to the site of the glucosyl bond to be attacked. A pair of enzymes takes care of this problem—so that the release of glucose can go on. The first of these, the **debranching enzyme**, transfers three of the glycosidic units of the branch to the 4-OH of another chain. This makes them part of a chain long enough for phosphorylase to work on. The last unit, in $1 \rightarrow 6$ linkage, is hydrolysed off by an $\alpha(1 \rightarrow 6)$-glucosidase activity of *the same enzyme* (which has two activities) and this opens up the chain for further attack by phosphorylase (Fig. 6.6). Glucose uptake, storage, and release in the liver is summarized in Fig. 6.7.

Fig. 6.6 Diagram of the debranching process. Before debranching, the structure above cannot be attacked by glycogen phosphorylase. After the transferase and hydrolysis actions of the debranching enzyme, both chains are now open to phosphorylase attack.

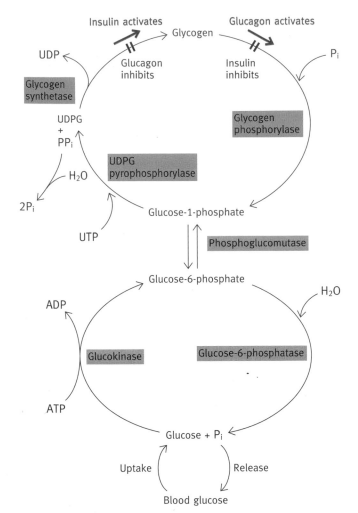

Fig. 6.7 Summary of glucose uptake, storage, and release in the liver. Note that glucagon and insulin do not act directly on glycogen metabolism enzymes (see Chapter 12).

There are some important points to note, especially the following.

1. The synthesis and degradation pathways are different and therefore independently controllable.

2. Glucose-6-phosphatase is found in liver but not muscle or adipose tissues, which therefore cannot release glucose into the blood. Only the kidney, among other tissues, has this enzyme.

3. In extrahepatic cells (as well as in liver) glucose-6-phosphate is on the path to glucose oxidation (Chapter 8).

4. Most importantly, the effect of insulin on cells is to activate glycogen synthesis and inactivate glycogen breakdown. The effect of glucagon on cells is the reverse.

Thus, insulin promotes glycogen synthesis; glucagon promotes release of glucose from liver. How these effects are achieved is the subject of later chapters when we come to deal with molecular signals to cells.

Why does liver have glucokinase and the other tissues hexokinase?

You will recall that once glucose enters a cell it is phosphorylated to glucose-6-phosphate. Glucokinase in liver and hexokinase in brain and other tissues catalyse the same reaction. Why is this so? In starvation, when glucose supply to the blood is the all-important problem, the entry of glucose into muscle and other cells is restricted because of the lack of insulin (the mechanism of this control is described on page 172), while the entry of glucose into brain, liver, and red blood cells is *not insulin-dependent*. A potentially illogical situation exists. The liver synthesizes glucose from amino acids supplied by the muscle so that it can keep the blood glucose level up to permit normal brain function and it would not make sense for the liver to take up glucose in competition with the brain. The liver glucokinase has a much lower affinity for glucose than has brain hexokinase (see Fig. 6.8). This means that the liver takes up glucose less readily when its concentration in blood is low since its passive entry depends on phosphorylation within the cell. After a meal, when blood glucose levels are high and when insulin signals the liver to make glycogen, glucokinase works maximally. Also hexokinase is inhibited by glucose-6-phosphate while glucokinase is not. Thus glycogen synthesis can proceed even at high glucose-6-phosphate levels in the cells. (In addition to this automatic control, the glucose transporter of liver has a higher Michaelis constant, K_m (see page 161) than that of many other cells which has the same effect. The same is true of the insulin-secreting cells of the pancreas, which coordinates insulin secretion with high blood sugar values.)

What happens to other sugars absorbed from the intestine?

The diet of animals usually results in large amounts of other sugars being absorbed into the portal bloodstream. For example, the sugar in milk is lactose (page 66), which on hydrolysis in the gut yields galactose and glucose. Sucrose yields fructose and glucose. So far as energy metabolism goes, the policy of the body

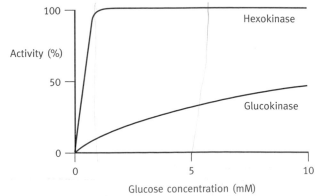

Fig. 6.8 The response of hexokinase and glucokinase to glucose concentration.

is simple—convert sugars such as galactose and fructose either to glucose or to compounds on the main glucose metabolic pathways in the liver. Thus, these sugars can be converted to anything to which glucose is converted—glycogen, fats, or to CO_2 and H_2O. What happens to galactose has special interest.

To convert galactose to glucose, the H and OH on carbon atom 4 must be inverted or **epimerized**.

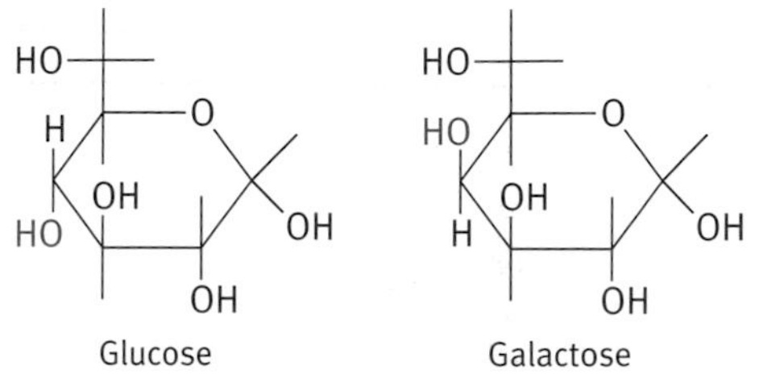

Glucose

Galactose

This is not done directly but by an epimerase that epimerizes not galactose itself, but UDP-galactose to UDP-glucose. UDP-glucose, you will recall, is formed in glycogen synthesis (page 81).

First, galactose is phosphorylated by galactokinase to galactose-1-phosphate.

+ ATP

Galactokinase

+ ADP

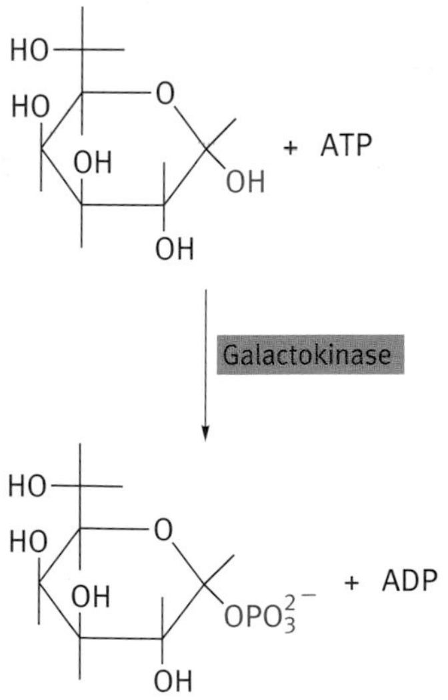

You might have expected this to react with UTP by analogy with UDP-glucose formation but that is not what happens. Instead, galactose-1-phosphate displaces glucose-1-phosphate from UDP-glucose forming UDP-galactose and glucose-1-phosphate as follows.

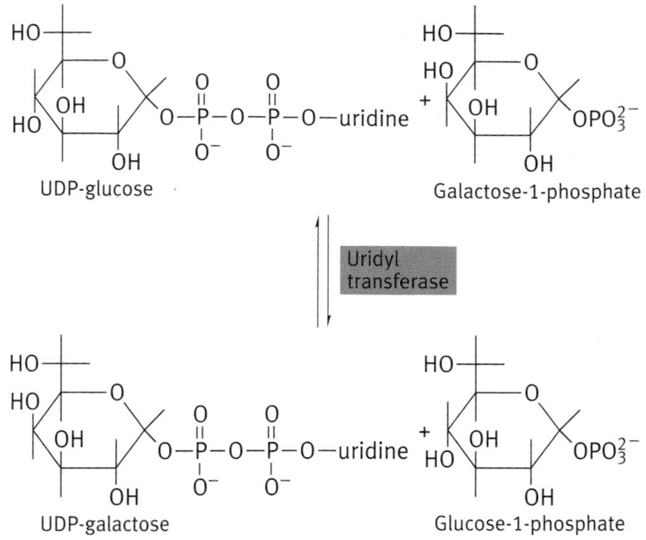

UDP-glucose

Galactose-1-phosphate

Uridyl transferase

UDP-galactose

Glucose-1-phosphate

You will see that what happens in this reaction is that the —P—uridine group (uridyl) is transferred from UDPG to galactose-1-phosphate and the enzyme is therefore a uridyl transferase. Now epimerization of UDP-galactose occurs.

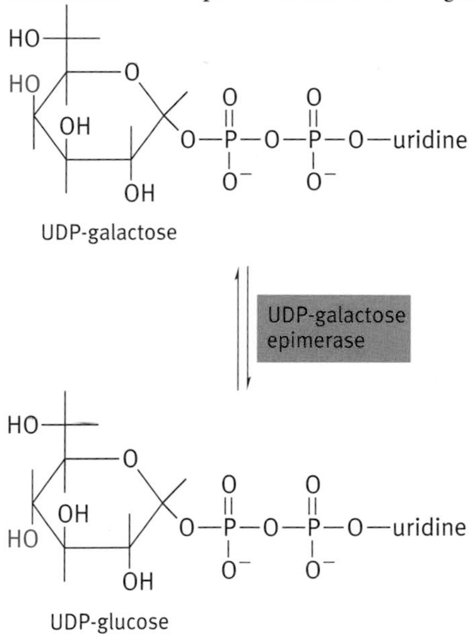

UDP-galactose

UDP-galactose epimerase

UDP-glucose

Putting the three reactions together, the following sequence occurs.

Galactose + ATP $\longrightarrow$ Galactose-1-P + ADP
Galactokinase

Galactose-1-P + UDP-glucose $\longleftrightarrow$ Glucose-1-P + UDP-galactose
Uridyl transferase

UDP-galactose $\longleftrightarrow$ UDP-glucose
UDP-galactose epimerase

The net effect is to convert galactose to glucose-1-phosphate (as summarized in Fig. 6.9).

In the infant genetic disease, galactosemia, uridyl transferase is missing and galactose cannot be converted as shown.

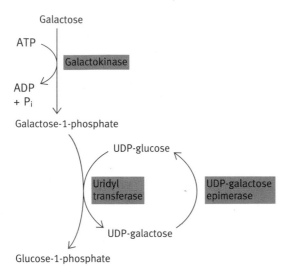

Fig. 6.9 Diagram showing how galactose enters the main metabolic pathways by conversion of galactose to glucose-1-phosphate. The green arrows represent the net effect of the reactions.

Accumulation of galactose and its metabolic products leads to impairment of brain development and blindness. Elimination of galactose from the diet avoids this. Since the epimerase is reversible, UDP-glucose can be converted to UDP-galactose so the patient can survive without any intake of galactose in the diet for essential syntheses of glycolipids (page 44) and glycoproteins (page 48). These require a donation of galactose residues (from UDP-galactose).

Amino acid traffic in the body (in terms of fuel logistics)

Since amino acids are not *deliberately* stored, there is no whole body story comparable to that of glycogen and fat traffic, but amino acid traffic does occur. The movement of greatest importance is that in starvation, when muscle proteins break down to produce amino acids that are transported to the liver to provide substrates for glucose synthesis. It will be more convenient to deal with this in detail when we come to the mechanisms of metabolic pathways. We will therefore move on to the traffic of fat in the body—a very important subject.

Fat and cholesterol traffic in the body

We need to explain why cholesterol comes into this story, since the chapter is concerned with the transport of energy-yielding fuels, and cholesterol is certainly not one of these. However, cholesterol is absorbed from the diet and transported by the same lipoproteins as is fat, no doubt because they both present the same problem of water insolubility. It is therefore necessary to discuss the two together.

Uptake of fat from chylomicrons into cells

After a fatty meal, the blood is loaded with **chylomicrons**. The chylomicron (page 68) has a shell of phospholipids, cholesterol, and proteins surrounding a core of neutral fat (TAG) and cholesterol esters.

Let's consider the fate of TAG first. It is taken up (but not as such) by the adipose cells for storage and the lactating mammary gland for secretion, by muscle and other tissues for energy production and also by liver.

Free fatty acids (FFA) readily pass through cell membranes; TAG cannot do so. The TAG in chylomicrons is hydrolysed in the capillaries by a lipase attached to the *outside* cells lining the blood capillaries to produce glycerol and FFA which immediately enter the adjacent cells.

$$
\begin{array}{l}
CH_2-O-\overset{\overset{\displaystyle O}{\|}}{C}-R_1 \\
\\
CH-O-\overset{\overset{\displaystyle O}{\|}}{C}-R_2 \quad + \ 3H_2O \longrightarrow \\
\\
CH_2-O-\overset{\overset{\displaystyle O}{\|}}{C}-R_3
\end{array}
\quad
\begin{array}{l}
CH_2OH \\
\\
CHOH \quad + \\
\\
CH_2OH
\end{array}
\quad
\begin{array}{l}
R_1-\overset{\overset{\displaystyle O}{\diagup}}{C}_{\diagdown O^-} \\
R_2-\overset{\overset{\displaystyle O}{\diagup}}{C}_{\diagdown O^-} \quad + \ 3H^+ \\
R_3-\overset{\overset{\displaystyle O}{\diagup}}{C}_{\diagdown O^-}
\end{array}
$$

TAG   Glycerol   Free fatty acids

The amount of lipase activity present in the capillaries of a particular tissue determines the amount of FFA release in that tissue and hence the amount of FFA uptake by the tissue. Adipose tissue is rich in lipase, as is lactating mammary gland, while tissues with less need for fat have less of the enzyme in the capillaries. The amount of lipase is varied according to physiological need; for example, a high-insulin, low-glucagon ratio causes increases in the amount of the enzyme. The chylomicron is a lipoprotein—a complex of protein and lipid. Hence the capillary lipase is called a **lipoprotein lipase** to distinguish it from other lipases. There is evidence that lipoprotein lipase causes the chylomicron to transitorily bind

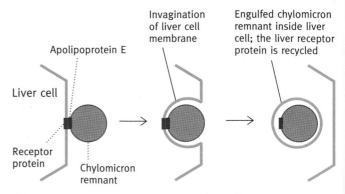

Fig. 6.10 Uptake of chylomicron remnants into a liver cell by receptor-mediated endocytosis. The receptor on the liver cell is specific for apolipoprotein E present on the chylomicron remnant. See Fig. 16.2 for further details of the fate of the engulfed particle and of the mechanism of endocytosis.

(a)

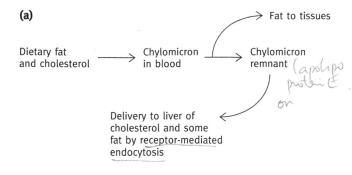

(b)

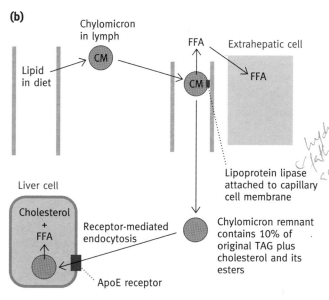

Fig. 6.11 (a) Transport of fat and cholesterol by chylomicron route. **(b)** Mechanism for the transport of fat and cholesterol from the intestine to the tissues by chylomicrons. CM, chylomicron; FFA; free fatty acids; HDL, high-density lipoprotein.

to the cell surface, presumably to facilitate TAG removal.

The progressive removal of triglyceride from chylomicrons reduces these in size to become **chylomicron remnants,** containing all the cholesterol and its esters and about 10% of the triglyceride of the original chylomicron. The remnants are taken up by the liver by receptor-mediated endocytosis (Fig. 6.10) to be destroyed inside the liver cell, thus delivering fat and cholesterol from the intestine to the liver.

A picture of chylomicron usage is shown in Fig. 6.11(a), (b).

Logistics of fat and cholesterol movement in the body

An overview

The movement of lipid and cholesterol around the body is a subject of great importance, particularly from a medical viewpoint.

The liver is the major source of lipid and cholesterol circulating in the blood in the form of lipoproteins, except for chylomicrons, which, as already described, carry absorbed fat and cholesterol from the intestine. The liver is active in synthesizing TAG from glucose and other metabolites and it receives about 10% of the absorbed fat due to uptake of chylomicron remnants. Nonetheless, it is not a storage organ for lipid—a 'fatty liver' in which there are extensive depositions of fat is pathological. The liver exports its TAG to peripheral tissues in the form of VLDL (very low-density lipoprotein), a lipoprotein similar in structure to a chylomicron. The rationale appears to be that liver, as a major source of fat, supplies this to major fat users such as muscle, which oxidize it for energy production, and to adipose cells for storage.

The liver is the major site of cholesterol synthesis in the body, and also receives dietary cholesterol via the chylomicron remnants. This also is exported to peripheral tissues in the VLDL resulting in an outward flow, from liver to peripheral tissues, of lipid and cholesterol.

There is also a reverse flow of cholesterol from peripheral tissues to the liver (Fig. 6.12). The physiological rationale for this apparent 'equilibration' of cholesterol between the liver and the rest of the body is not self-evident. It might be a way of ensuring that all cells have adequate cholesterol supplies with any excess returned to the liver, which excretes part of it into the intestine in the form of bile acids. Most cells are, however, capable of cholesterol synthesis.

A somewhat different speculative viewpoint has been put forward—that the outward flow of cholesterol may not be primarily a means of transporting this substance, but rather that cholesterol is needed for the lipoprotein structure used in lipid transport. On this view cholesterol transport is not the main aim of the outward flow, but is, as the authors of the view express it, the carrier bag that should be distinguished from the groceries (fat), the reverse flow of cholesterol being regarded as the return of carrier bag material to the liver. They suggest that perhaps this concept has been obscured by the intense interest in

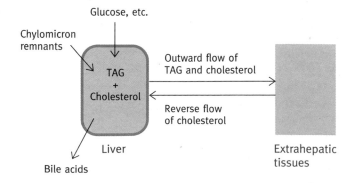

Fig. 6.12 Overview of the major movements of fat (TAG) and cholesterol to and from the liver. The liver synthesizes both components from metabolites and receives them from chylomicron remnants. It also converts cholesterol to bile acids.

cholesterol movement because of its very great medical interest in atherosclerosis (see below).

With that introduction we shall proceed to an account of the mechanisms involved, but it should be kept in mind that in detail it is an extraordinarily complex business, is incompletely understood, and is the subject of a massive continuing research effort.

Synthesis of cholesterol and its regulation

This is briefly dealt with on page 147.

Utilization of cholesterol in the body

Cholesterol is an important constituent of animal cell membranes (page 45). Because of the effects of its oversupply in causing cardiovascular disease, mechanisms for its removal from the body are of great interest.

The main route for disposing of cholesterol is bile acid formation by the liver. (The structures of cholesterol and of a bile acid are shown on page 67.) About 0.5 g per day is disposed of in humans by this route. Bile acids are reabsorbed from the gut and re-used. One therapeutic approach to lowering cholesterol levels in patients is to administer a compound that complexes bile acids in the intestine and prevents their reabsorption; a more recent approach is to inhibit its production, as described on page 147. Cholesterol is also used in the adrenal glands and gonads for the synthesis of steroid hormones. (Examples of the structures of the latter are given page 354.)

Cholesterol ester, whose structure and synthesis are given on page 68, is the storage form of cholesterol in cells and is the form in which much of the cholesterol in lipoproteins is carried.

Lipoproteins involved in fat and cholesterol movement in the body

There are two lipoproteins produced by the liver, **VLDL (very-low-density lipoprotein)** and **HDL (high-density lipoprotein)**. The production of lipoproteins involves the endoplasmic reticulum and the Golgi apparatus (see page 307), which packages them into membrane-bound vesicles for release by exocytosis. In the rat, about 80% of lipoprotein (other than chylomicrons) is made by the liver and the rest by intestinal cells. The VLDL is converted by the mechanism described below to **IDL (intermediate-density lipoprotein)** and then **LDL (low-density lipoprotein)** so that, including chylomicrons, five different lipoproteins are found in circulation. Fig. 6.13 shows electron micrographs of lipoproteins.

Apolipoproteins

Each type of lipoprotein has its own particular associated set of **apolipoproteins**. A dozen or more are already known. Their functions are not established in every case but they have several roles.

- Some are required as components for the production of lipoproteins. In the case of chylomicrons the main one is $apoB_{48}$ and, in that of VLDL, it is $apoB_{100}$.
- Some are required as destination-targeting signals—specific apoproteins designed to bind to specific receptors on the surface of cells. Binding leads to uptake of the bound lipoprotein by receptor-mediated endocytosis (see Fig. 6.10). In this way individual lipoproteins are taken up only by designated cells. Examples of this targeting function are $apoB_{100}$ on LDL which binds to LDL receptors, and apoE on chylomicron remnants which binds to liver receptors.
- Some apolipoproteins are required to activate enzymes. Thus apoCII on chylomicrons is necessary for lipoprotein lipase activity, which removes fatty acids from TAG. Not all of the apoproteins are present when secreted from the liver or intestine; some are delivered in the circulation. For example, apoCII produced by the liver is delivered to chylomicrons after release.

Mechanism of TAG and cholesterol transport from the liver and the reverse cholesterol transport in the body

Let us start with VLDL and the outward flow of TAG and cholesterol from the liver. In patients deficient in the synthesis of this lipoprotein, accumulation of TAG in the liver occurs due to the failure of its export.

Once VLDL is released, the TAG is progressively removed by lipoprotein lipase action in the way already described for chylomicrons (page 87). As the amount of fat diminishes, the percentage of cholesterol and its esters rises (see Table 6.1) causing an increase in density, and the lipoprotein structures shrink in size. The end-product of the process is LDL, formed via the intermediate state, IDL. LDL is taken up by peripheral tissue cells via receptor-mediated endocytosis so that the remaining TAG and cholesterol are delivered to extrahepatic tissues. The number of LDL receptors on cells is regulated to control LDL uptake. This outward flow of cholesterol from the liver is counterbalanced by a reverse flow; how does this occur?

The question brings us to the last of the lipoproteins, HDL. HDL is produced mainly by the liver in an 'immature' disc-like form, known as **nascent HDL**, made of phospholipid and apoproteins but containing little TAG or cholesterol. The nascent HDL pick up cholesterol from peripheral tissue cells; exactly how they do it is not established but they may

Table 6.1 Approximate content of fat and cholesterol as percentage dry weight of different lipoproteins

	Percentage dry weight of				
	Chylomicron	VLDL	IDL	LDL	HDL
Neutral fat (TAG)	80	50	30	10	8
Cholesterol + cholesterol ester	8	20	30	50	30

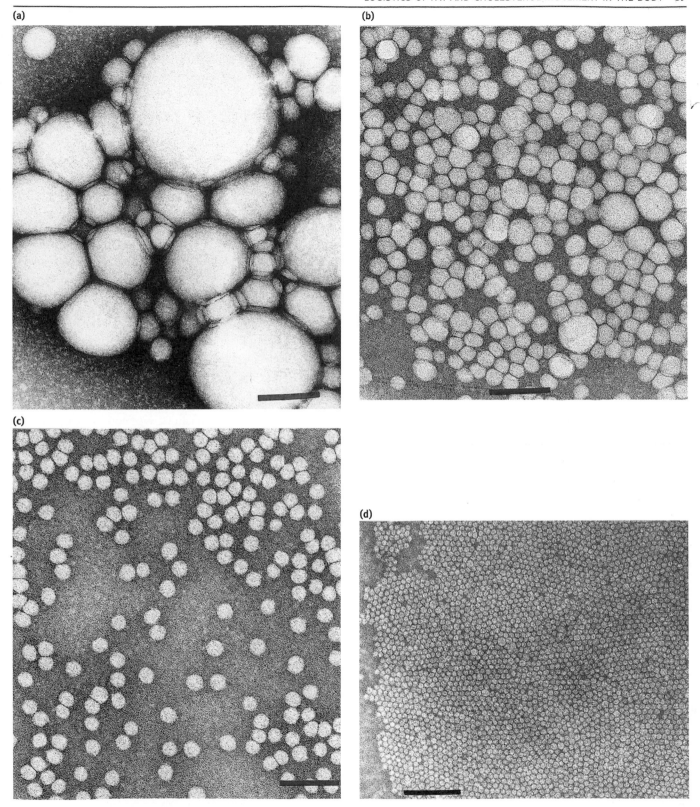

Fig. 6.13 Electron micrographs of: **(a)** chylomicrons (Chylo); **(b)** very low-density lipoproteins (VLDL); **(c)** low-density lipoproteins (LDL); and **(d)** high-density lipoproteins (HDL). Scale bar, 1000 Å. Kindly provided by Dr. Trudy Forte of the Lawrence Berkeley Laboratory, University of California.

transitorily attach to extrahepatic cell membranes and receive cholesterol by simple diffusion. Free cholesterol as such would remain in the peripheral layer of the lipoprotein with its —OH group in contact with water but, once in the HDL, it is converted to cholesterol ester. Hydrophobic force (page 14) causes the latter to migrate to the centre of the HDL, away from contact with water. The accumulation of the ester causes the HDL to fatten out into a spherical particle.

Cholesterol esterification *in cells* requires ATP (see page 68), but HDL has none of this available and an alternative strategy is used. In this, a fatty acyl group of lecithin, which is present in the hydrophilic shell, is transferred to cholesterol in an energy-neutral reaction by an enzyme associated with HDL, called **lecithin:cholesterol acyltransferase (LCAT)**.

$$CH_2-O-\overset{\overset{O}{\|}}{C}-R_1$$
$$CH-O-\overset{\overset{O}{\|}}{C}-R_2 \quad + \quad CHOL-OH$$
$$CH_2-O-\overset{\overset{O}{\|}}{\underset{\underset{O^-}{|}}{P}}-O-Choline$$

Lecithin   + cholesterol
(phosphatidylcholine)

$$CH_2-O-\overset{\overset{O}{\|}}{C}-R_1$$
$$CHOH \quad + \quad R_2-\overset{\overset{O}{\|}}{C}-O-CHOL$$
$$CH_2-O-\overset{\overset{O}{\|}}{\underset{\underset{O^-}{|}}{P}}-Choline$$

Lysolecithin +   cholesterol
                  ester

Reaction catalysed by lecitithin:cholesterol acyltransferase, where R_1 and R_2 are fatty acyl groups; CHOL—OH is cholesterol

The HDL transfers its cholesterol ester to chylomicrons (present after a fat-containing meal), VLDL, IDL, and LDL by means of a **cholesterol ester transfer protein (CETP)**. Although IDL and LDL deliver their contents to extra-hepatic tissues as described above, a proportion return to the liver. This is the explanation of the seemingly pointless recycling of IDL and LDL now enriched by cholesterol from HDL; it is the reverse flow of cholesterol. A summary of the steps in the reverse flow might be useful here.

1. Cholesterol is transferred from extrahepatic cell membranes to HDL.

2. The cholesterol is esterified by LCAT and the ester migrates to the centre of the HDL.

3. Cholesterol ester is transferred from HDL to chylomicrons (when present), VLDL, IDL, and LDL.

4. A proportion of the latter two lipoproteins recycle back to the liver, thus delivering cholesterol from extrahepatic cells to that organ (Fig. 6.14). When present after a meal, chylomicron remnants also participate in this reverse flow.

The above account emphasizes the reverse cholesterol flow but the transfer of cholesterol ester between lipoproteins occurs reversibly and a more complex traffic occurs between all of them; in addition, transfer of TAG takes place between the various components, catalysed by CETP which actually exchanges cholesterol ester for TAG. The precise reasons for this traffic complexity are not known.

LDL in popular medical terms is regarded as 'bad cholesterol' because excessive levels of cholesterol occur in a proportion of the population in this form, and these are associated with increased risk of atherosclerosis. The latter involves development of plaques in blood vessels, a complex process in which cholesterol deposition is involved. These block blood flow and, if present in coronary arteries, are the cause of heart attacks.

Patients with the genetic disease, familial hypercholesterolemia, are deficient in functional LDL receptors on cells, resulting in impaired removal of LDL from the blood and very high levels of circulating LDL-associated cholesterol. This leads to early cardiovascular disease. By contrast, HDL is popularly known as 'good cholesterol' because high levels are associated with decreased risk of atherosclerosis, possibly related to its role in

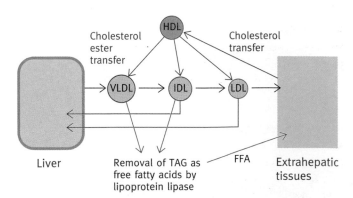

Fig. 6.14 Diagram showing the role of lipoproteins in the outward and reverse flow of cholesterol from, and to, the liver. The heavy red lines represent transfer of cholesterol esters. The transfer of cholesterol esters between lipoproteins is reversible and more complex than shown. In addition, TAG interchange occurs. The situation shown is that in the fasting condition in which chylomicrons are not present. After a meal containing fat, chylomicrons participate in the cholesterol ester transfer from HDL and, in delivering cholesterol to the liver via chylomicron remnants, also contribute to the reverse cholesterol flow. VLDL, very-low-density lipoproteins; IDL, intermediate-density lipoproteins; LDL, low-density lipoproteins; HDL, high-density lipoproteins.

reverse cholesterol flow and its resultant elimination from the liver, but the protective effect is incompletely understood. There is also the possibility, not proven, that HDL may reverse atherosclerosis by removing cholesterol from vascular cholesterol deposits.

How is fat released from adipose cells?

A lipase releases FFA from stored neutral fat. But there is a very important difference here. The lipase inside the adipose cell is *not* the same (lipoprotein) lipase as in the capillaries. It is a *hormone-sensitive lipase, activated by a glucagon signal to the cells and inhibited by an insulin signal to the cells.* The mechanism of how this occurs is described in Chapter 12 (page 178). You can see how it all fits in. After a meal, insulin is high and glucagon low. This instructs adipose cells to take up glucose from plasma and FFA from chylomicrons and convert them into neutral fat; release of FFA is inhibited. In fasting, the reverse occurs and tissues are supplied with FFA in the blood. In juvenile-onset diabetes, the low insulin/glucagon ratio means that the adipose cell hormone-sensitive lipase is activated and FFA pours out into the blood. The liver partially converts these to ketone bodies and this results in severe ketonemia (that is, ketone bodies in the blood).

A further point: the adipose cell hormone-sensitive lipase is also activated by epinephrine (adrenaline). Thus, in response to an epinephrine-liberating alarm, the adipose cells pour out FFA to supply the muscles with FFA to support any required violent muscular activity. How epinephrine does this is an important topic, but it is more suitable to deal with it in Chapter 12 (page 178). *Nor*epinephrine liberated from the sympathetic nervous system that innervates fat cells has the same effect.

How are free fatty acids carried in the blood?

It would be unsatifactory for relatively high levels of FFA to be carried in the blood since, at neutral pH, it would literally be a soap solution. Instead they are carried adsorbed to the surface of the blood protein, serum albumin. This protein has hydrophobic patches on its surface and is the carrier for many molecules with hydrophobic parts. The adsorbed FFA are in equilibrium with FFA in free solution so that, as cells take them up, more dissociate from the protein carrier into the serum where they are available to cells.

Further reading

Glycogen synthesis

Alonso, M. D., Lomako, J., Lomako, W. M., and Whelan, W. J. (1995). A new look at the biogenesis of glycogen. *FASEB J.*, **9**, 1126–37.
A beautiful article describing the role of glycogenin. Summarizes new concepts in a field many thought was a closed book.

Van Schaftingen, E., Detheux, M., and Da Cunha, M. V. (1994). Short-term control of glucokinase activity: role of a regulatory protein. *FASEB J.*, **8**, 414–19.
Describes the occurrence, properties, and control of this enzyme.

Lipid transport and lipoproteins

Nilsson-Ehle, P., Garfinkel, A. S., and Schotz, M. C. (1980). Lipolytic enzymes and plasma lipoprotein metabolism. *Ann. Rev. Biochem.*, **49**, 667–93.
Gives general summary of lipoproteins, including useful information at the physiological level.

Cannon, B. and Nedergaard, J. (1985). The biochemistry of an inefficient tissue. *Essays in Biochem.*, **20**, 110–64.
Deals with heat productton by brown fat cells.

Angelin, B. and Gibbons, G. (1993). Lipid metabolism. *Curr. Opin. Lipidology*, **4**, 171–6.
An excellent overview of lipoproteins.

Flier, J. S. (1995). The adipocyte: storage deposit or node on the energy information superhighway? *Cell*, **80**, 15–18.
Adipocytes have been viewed as a largely passive participant in the flux of lipid storage and utilization. This minireview puts the view that adipocytes can function as endocrine cells involved in a regulatory system of relevance to obesity.

Tall, A. (1995). Plasma lipid transfer proteins. *Ann. Rev. Biochem.*, **64**, 235–57.
Excellent review of the medically important area of the cholesterol ester transport protein (CETP), lipid transport, forward and revese cholesterol transport between liver and extrahepatic tissues.

Problems for Chapter 6

1   Starting with glucose-1-phosphate, explain how the synthesis of glycogen is made thermodynamically irreversible?

2   Have you any comments on the name of the enzyme UDP-glucose pyrophosphorylase?

3 Which tissues release glucose into the blood as a result of glycogen breakdown? Why is this so?

4 What reaction do the two enzymes glucokinase and hexokinase catalyse? Why should the liver have glucokinase while brain and other tissues have hexokinase?

5 In the genetic disease galactosemia, infants are unable to metabolize galactose correctly. Removal of galactose from the diet can prevent the deleterious effects of the disease. UDP-galactose epimerase is reversible; why is this important in connection with the above statement?

6 How is TAG removed from chylomicrons and utilized by tissues?

7 Explain what VLDL is and its likely role.

8 What is meant by reverse cholesterol flow?

9 What is the main route of cholesterol removal from the body?

10 Cholesterol is esterified in HDL. Conversion of cholesterol to its ester is an energy-requiring process but HDL has no access to ATP. How is the esterification achieved?

11 How is fat released from adipose cells and under what conditions?

Chapter 7

Energy production from foodstuffs—a preliminary overview

The generation of energy in the form of ATP from glucose, fats, and amino acids involves fairly long and somewhat involved metabolic pathways. There is the possibility that, if we go at once into these pathways, the overall strategy of energy generation by cells may be lost amongst the detail. We will therefore, at this stage, deal with major phases of the processes, picking out landmark metabolites only, and view the strategies on a broad basis. When this has been done, subsequent chapters will deal in more detail with the pathways; in short, we want you to see the wood before the trees. For convenience we will start with energy production from glucose.

Energy production from glucose

The main phases of glucose oxidation

Overall, glucose is oxidized as follows

$$C_6H_{12}O_6 + 6O_2 \rightarrow 6CO_2 + 6H_2O.$$

The $\Delta G^{0'}$ for this reaction is 2820 kJ mol^{-1}.

In the cell, this oxidation process is accompanied by the synthesis of more than 30 molecules of ATP from ADP and P_i. The entire process of glucose oxidation to CO_2 and H_2O can be divided up into three phases. If you cannot fully understand the summaries below don't be concerned—they will become clear shortly.

1. **Glycolysis**—which means lysis or splitting of glucose into two C_3 fragments in the form of pyruvic acid, accompanied by reduction of an electron carrier. This occurs in the **cytoplasm** of cells.

2. The **Krebs, citric acid, or tricarboxylic acid (TCA) cycle** (these are alternative names) in which carbon atoms two and three of pyruvate are converted to CO_2. Electrons are transferred to electron carriers. No oxygen is involved on this phase. Carbon atoms are released as CO_2. The cycle is located inside **mitochondria** in eukaryotes.

3. The **electron transport system** in which electrons are transported from the electron carriers to oxygen where, with protons from solution, water is formed. It is in stage 3 that most of the ATP is generated. This occurs in the inner mitochondrial membrane in eukaryotes.

Before we go into these stages we need to say a little about biological oxidation.

Biological oxidation and hydrogen transfer systems

Oxidation does not necessarily involve oxygen; it involves the removal of electrons. It may involve *only* electron removal such as in the ferrous/ferric system

$$Fe^{2+} \rightarrow Fe^{3+} + 2e^-$$

or it may be the removal of electrons accompanied by protons from a hydrogenated molecule

$$AH_2 \rightarrow A + 2e^- + 2H^+$$

In such chemical oxidation systems, the electrons must be transferred from the electron donor to an electron acceptor. Depending on the particular electron acceptor, the electron transferred may be accompanied by the proton in which case it is equivalent to a hydrogen atom being transferred or the proton may be liberated into solution.

The ultimate electron acceptor in the aerobic cell is oxygen. Oxygen is **electrophilic**—it has an avidity for electrons. When it

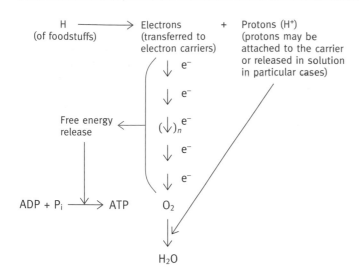

Fig. 7.1 Diagram illustrating the concept of electron transfer to oxygen and ATP generation.

accepts electrons, protons from solution join up to produce water

$$O_2 + 4e^- + 4H^+ \rightarrow 2H_2O.$$

However, oxygen is only the *ultimate* electron acceptor in the cell—there are other electron acceptors, which form a chain, carrying electrons from metabolites to oxygen. They accept electrons and hand them on to the next acceptor and thus are electron carriers. This is the electron transport chain, which plays a predominant part in ATP generation. The essential concept is given in Fig. 7.1. Two of the electron carriers involved in energy production are of such central importance in metabolism that we must now describe them. It is essential for you to be completely familiar with these.

NAD$^+$—an important electron carrier

The first electron carrier involved in the oxidation of most metabolites is NAD$^+$, which stands for **nicotinamide adenine dinucleotide**. You have already met a nucleotide in the form of AMP (Fig. 1.6); a **nucleotide** has the general structure, base–sugar–phosphate, the base in AMP being adenine. (The structures of adenine and other bases are dealt with later in Chapter 18 and need not concern us now.) In the case of NAD$^+$, a **dinucleotide** is formed by linking the two phosphate groups of two nucleotides (which is unusual and not how nucleotides are linked together in nucleic acids, as will be evident in later chapters). The general structure of NAD$^+$ is

Base–sugar–phosphate–phosphate–sugar–base.

The two bases are adenine (as in ATP) and nicotinamide, respectively—giving the structure of NAD$^+$ below.

```
Adenine        Nicotinamide
|              |
Ribose         Ribose
|              |
Phosphate——Phosphate
```

NAD$^+$ is a **coenzyme**—a small organic molecule that participates in enzymic reactions. It differs from an ordinary enzyme-substrate only in that the reduced compound leaves the enzyme and attaches to a second enzyme where it donates its reducing equivalents to a second substrate. NAD$^+$ thus acts catalytically by being continually reduced and re-oxidized, and in doing so transfers electrons from one molecule to another.

The 'business end' of the molecule is the nicotinamide group. It is derived from the vitamin, nicotinic acid or niacin. Niacin is an essential vitamin for animals, but humans can make it from the amino acid tryptophan. The rest of the molecule fits the coenzyme to appropriate enzymes but does not undergo any chemical change.

Nicotinamide has the structure

When linked in NAD$^+$ the structure is

where R is the rest of the NAD molecule.

NAD$^+$ can be reduced by accepting two electrons plus one proton as a hydride ion (H:$^-$), to give the following structure (the second proton being liberated into solution).

Reduced NADH + H$^+$

NAD$^+$ is the coenzyme for several **dehydrogenases** that catalyse this type of reaction

$$AH_2 + NAD^+ \leftrightarrow A + NADH + H^+.$$

The reduced NAD$^+$ can then diffuse to a second enzyme and participate in a reaction such as

$$B + NADH + H^+ \leftrightarrow BH_2 + NAD^+.$$

In this way NAD$^+$ acts as the carrier for the transfer of a pair of hydrogen atoms from A to B

$$AH_2 + B \rightarrow A + BH_2.$$

In biochemistry such reactions are often presented in the form of whirligigs.

In case it requires extra emphasis, NADH+H$^+$ can add *two* H atoms to a substrate since the second electron it transfers to an acceptor molecule may be joined by a proton from solution as shown above. In equations, reduced NAD$^+$ is written as NADH+H$^+$. In the text, the term NADH is used, but it always implies NADH+H$^+$.

Further study

FAD, FMN, and their reduction

FAD has the structure:

Isoalloxazine ring system—ribitol—phosphate—phosphate— ribose— adenine.

It is very unusual in having the linear ribitol molecule instead of ribose. The left-hand portion of the molecule is not, therefore, in the strict sense, a nucleotide, but nonetheless is always designated as such from long usage.

Oxidation–reduction reactions occur in the isoalloxazine structure.

FAD—another important electron carrier

Another electron (hydrogen) carrier is **FAD** or **flavin adenine dinucleotide**. It is derived from vitamin B$_2$ or riboflavin. The important feature of this molecule is that it can (in combination with appropriate proteins) accept two H atoms to become FADH$_2$. FAD is a prosthetic group—it is a permanent attachment to its apoenzyme unlike NAD$^+$ which moves from one dehydrogenase to another. Its role will become clear; for the present it is sufficient that FAD can be reduced. The chemistry of this reduction is given in the 'Further study' box that follows where a related carrier, FMN or flavin mononucleotide, is also described.

FAD (Oxidized form) FADH$_2$ (Reduced form)

Another carrier, FMN or flavin mononucleotide, has the structure isoalloxazine—ribitol—phosphate. It is reduced in the same way. The vitamin, riboflavin, lacks the phosphate. The flavins can exist in the semiquinone form and thus can participate in single-electron transfer reactions as well as in two-electron transfer ones.

Stages in the production of energy from glucose—first glycolysis

As stated earlier, glucose oxidation occurs in three main stages, the first of which is called **glycolysis**. It does not involve oxygen and very little energy is produced—in fact only two ATP molecules per molecule of glucose lysed. The end-products are pyruvate and NADH as shown in Fig. 7.2. In aerobic glycolysis, when the oxygen supply is plentiful, the NADH is re-oxidized to NAD$^+$ by mitochondria in eukaryote cells; the pyruvate is taken up by the mitochondria where it is oxidized to CO$_2$ and H$_2$O.

However, in the body, oxygen isn't necessarily always plentiful. In muscle, especially in the initial phase of emergency flight reactions before the heart has increased its pumping rate, the rate of glycolysis is dramatically increased with the aim of rapidly producing the ATP needed for muscle contraction. Since NAD$^+$ acts catalytically and is present in cells in small amounts, for glycolysis to proceed, the NADH must be recycled back to NAD$^+$; no NAD$^+$, no glycolysis. If the capacity of mitochondria to do this is inadequate at high glycolytic rates, glycolysis and the muscle's ability to maximally generate ATP, needed for contraction, would be impaired unless another way were found to regenerate NAD$^+$ from NADH. An 'emergency' system comes

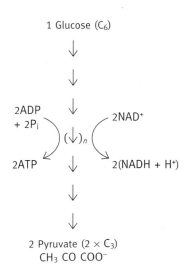

Fig. 7.2 Diagram illustrating the net result of aerobic glycolysis of glucose.

into play in this situation. The NADH is re-oxidized by reducing pyruvate to produce lactate. The reaction is

$$CH_3COCOO^- + NADH + H^+ \longleftrightarrow CH_3CHOHCOO^- + NAD^+ .$$

Pyruvate   Lactate dehydrogenase   Lactate

The enzyme catalysing this reaction is **lactate dehydrogenase**. The production of lactate from glucose is known as **anaerobic glycolysis**, as opposed to **aerobic glycolysis** which forms pyruvate, the NADH being oxidized by mitochondria. The object of anaerobic glycolysis is not to produce lactate but to re-oxidize NADH and thus permit continued ATP production from glycolysis (Fig. 7.3). Since there is a lot of lactate dehydrogenase in muscle, the NADH can be rapidly re-oxidized in this way and this in turn permits glycolysis to proceed at a very fast rate. The advantage of this is that, although only two ATP molecules are generated per molecule of glucose, relatively vast amounts of glucose can be broken down. The ATP so produced might make the difference between being eaten by a tiger and not being eaten if you happen to be chased by one. After a short time, the increased heart rate and dilatation of blood vessels supplying muscle increase the oxygen supply and the oxidative regeneration of NAD^+ increases. The lactate formed leaks out into the blood but is not wasted; it is used mainly by the liver as described later (page 153).

As a parenthetic note, yeast can live entirely on anaerobic glycolysis using an analogous trick to re-oxidize NADH. The pyruvate is converted to acetaldehyde+CO_2; a second enzyme, alcohol dehydrogenase, reduces the acetaldehyde to ethanol.

$$CH_3COCOO^- + H^+ \rightarrow CH_3CHO + CO_2$$
(Pyruvate decarboxylase)

$$CH_3CHO + NADH + H^+ \rightarrow CH_3CH_2OH + NAD^+$$
(Alcohol dehydrogenase).

Because of the low yield of ATP, vast quantities of glucose are broken down, producing alcohol and CO_2. The latter causes the explosion of bottles if, during ginger beer making, it is bottled too early.

An explanatory note
In the above account, for simplicity, we have talked about glucose being the carbohydrate that is glycolysed. It can indeed be free glucose, but in muscle it is mainly glycogen, the storage form of glucose, that is broken down. This does not affect the overall account given above except that, as described in later chapters, the initial steps of metabolism of glycogen and glucose are slightly different and the ATP yield from glycogen is three per glucosyl unit not two as from glucose. We also again remind you that in the next chapter the detailed mechanisms of glycolysis, the citric acid cycle, and the electron transport system will be given. At this stage the main thing is to get an overview.

Stage two of glucose oxidation—the citric acid cycle

The **mitochondria** are small organelles located in the cytoplasm of the cell (see the electron micrograph in Fig. 3.15). They are the main energy-generating units of aerobic cells where most of the ATP is produced. They are bounded by two membranes. The outer membrane is permeable and plays little role in energy generation. The inner membrane is highly impermeable to most substances, except where specific transport systems exist. Its surface area is increased by folds or cristae: the more energetic the tissue, the more cristae there are (Fig. 7.4). The mitochondrion is filled with a concentrated solution of enzymes—it is called the **matrix**. It is here that stage two of glucose metabolism mainly occurs, only one reaction being located in the inner membrane.

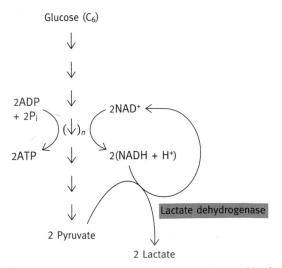

Fig. 7.3 Diagram illustrating the net result of anaerobic glycolysis of glucose.

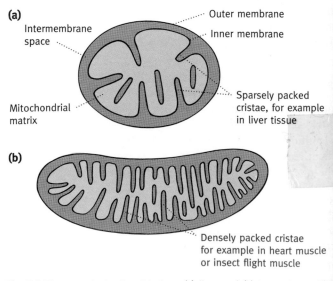

Fig. 7.4 Diagram of mitochondria from (**a**) liver and (**b**) heart tissue. The number of cristae reflects the ATP requirements of the cell.

Aerobic glycolysis, as stated, produces pyruvate and NADH in the cytoplasm. To be further oxidized the pyruvate must enter the mitochondria. A transport system in the inner mitochondrial membrane takes the pyruvate from the cytoplasm into the mitochondrion, but the cytoplasmic NADH cannot *itself* enter the mitochondrion for reoxidation. Instead, it hands over its reducing equivalent to either NAD^+ or an FAD enzyme *inside* the mitochondrion by what are known as shuttles, whose mechanisms are given in the next chapter. The end result is that NADH produced by glycolysis is re-oxidized by mitochondria leaving NAD^+ in the cytoplasm to participate further in glycolysis. The NADH and $FADH_2$ inside the mitochondrion are oxidized by the electron transport system (stage 3), but we must now deal with stage 2, the citric acid cycle, which metabolizes pyruvate, before we come to this. Some cells such as erythrocytes lack mitochondria and so must generate ATP by glycolysis only; they are glucose-dependent.

How is pyruvate fed into the citric acid cycle?

We now come to an enzyme reaction of major importance in which pyruvate, transported into mitochondria, is converted to a compound which is at the crossroads of metabolism. This is acetyl-CoA, a compound not previously mentioned in this book.

What is coenzyme A?

Coenzyme A is usually referred to as **CoA** for short, but is written in equations as CoA—SH because its thiol group is the reactive part of the molecule.

Unlike NAD^+ and FAD, coenzyme A is *not* an electron carrier, but an acyl group carrier (A for acyl). Like NAD^+ and FAD it is a dinucleotide and, as so often happens with cofactors, incorporates a water-soluble vitamin in its structure, in this case pantothenic acid. Pantothenic acid has an odd sort of structure (which, we suggest, it is not necessary to learn).

$$HO-CH_2-C(CH_3)(CH_3)-C(H)(OH)-C(=O)-NH-CH_2-CH_2-C(=O)OH$$

Pantothenic acid

You would normally expect the vitamin moiety of a coenzyme to play a part in the reaction for which the coenzyme is required (as with, for example, riboflavin in FAD and nicotinamide in NAD^+), but the pantothenic acid moiety is just 'there' and apparently quite inert. It presumably provides a recognition group to help bind the CoA to appropriate enzymes but why this particular structure is used is not obvious. It may be a quirk of evolution that was used by chance early in evolution and the cell became locked into its use.

The structure of CoA is as follows.

```
            Adenine      β-mercaptoethylamine
              |                |
Phosphate—Ribose          Pantothenic acid
              |                |
           Phosphate—Phosphate
```

The β-mercaptoethylamine part is the business end of the molecule.

$$RCO-NHCH_2CH_2-SH$$

β-Mercaptoethylamine moiety

The CoA molecule carries acyl groups as **thiol esters**. For example, acetyl-CoA can be written as $CH_3CO-S-CoA$. *The thiol ester is a high-energy compound* (unlike a carboxylic ester). It has a $\Delta G^{0'}$ of hydrolysis of approximately -31 kJ mol^{-1} as compared with about -20 kJ mol^{-1} for a carboxylic ester. The difference is due to the fact that the latter compound is resonance stabilized and so has a lower free-energy content than a thiol ester, which is not resonance stabilized. This is important in several areas of biochemistry. With that description of CoA we can now return to the question of what happens to pyruvate transported into mitochondria.

Oxidative decarboxylation of pyruvate

The pyruvate is subjected to an irreversible **oxidative decarboxylation** in which CO_2 is released (decarboxylation), a pair of electrons is transferred to NAD^+ (oxidation), and an acetyl group is transferred to CoA. Note that this is different from the nonoxidative decarboxylation carried out by yeast pyruvate decarboxylase, described earlier. In the latter there is, as implied, no oxidation, NAD^+ is not involved, and the product is acetaldehyde, not an acetyl group. The large negative free-energy change means that the reaction is irreversible. The reaction is as follows.

$$Pyruvate + CoA-SH + NAD^+ \rightarrow Acetyl-S-CoA$$
$$+ NADH + H^+ + CO_2 \quad \Delta G^{0'} = -33.5 \text{ kJ mol}^{-1}$$

The acetyl group of acetyl-CoA is now fed into the citric acid cycle. As indicated earlier, there are alternative names for this; it is known as the tricarboxylic acid (TCA) cycle since citric acid has three carboxyl groups, and as the Krebs cycle after its discoverer. The name citric acid cycle will be used here, but at this stage we will not be concerned with the reactions in it. The essential point is that the carbon atoms of the acetyl group of acetyl-CoA produced from pyruvate are converted to CO_2 while three molecules of NAD^+ are reduced to NADH. In addition, a molecule of FAD is reduced to $FADH_2$, the reducing equivalents coming partly from water. Almost as a sideline, the cycle generates one 'high-energy' phosphoryl group from P_i for each acetyl group fed in. This stage is summarized in Fig. 7.5.

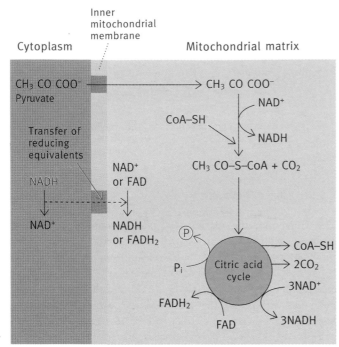

Fig. 7.5 Simplified diagram showing what happens to pyruvate and NADH generated in glycolysis and the generation of further NADH and FADH$_2$ inside mitochondria. The FAD is always attached to an enzyme. The source of the reducing equivalents is dealt with in the next chapter. The NADH and FADH$_2$ are oxidized by the electron transport pathway described shortly.

In summary, a molecule of pyruvate from the cytoplasm is converted in the mitochondria to three molecules of CO_2 and, in the process, three molecules of NAD^+ and one molecule of FAD are reduced but we have hardly started to make ATP—only two molecules in the glycolysis and two in the cycle (per starting glucose molecule) but there are almost 30 still to be made.

So far, it's been mainly preparation of fuel—now for the big return in the form of ATP generation.

Stage three of glucose oxidation—electron transport to oxygen

The oxidation of NADH and FADH$_2$ takes place in the inner mitochondrial membrane which contains a chain of electron carriers.

The electron transport chain—a hierarchy of electron carriers

The problem we are now concerned with, we remind you, is the transference of electrons from NADH and FADH$_2$ to oxygen with the formation of water

$$NADH + H^+ + \tfrac{1}{2}O_2 \rightarrow NAD^+ + H_2O.$$

The $\Delta G^{0'}$ of this reaction is $-220\ \mathrm{kJ\,mol}^{-1}$. To understand how the process occurs, we need to discuss the **redox potential** of compounds.

Compounds capable of being oxidized are, by definition, electron donors; in any oxido-reduction reaction there must be an electron acceptor (oxidant) and a donor (reductant). The reaction

$$X^- + Y \leftrightarrow X + Y^-$$

can be considered to occur by two theoretical half-reactions, as follows,

(a) $X^- \leftrightarrow X + e^-$;

(b) $Y + e^- \leftrightarrow Y^-$.

Each of these involves the reduced and oxidized forms of each reactant which are called **conjugate pairs** or **reduction–oxidation couples** or, the most convenient term, **redox couples**. X and X^- are such a couple and Y and Y^-, another couple. Clearly, any real-life oxidation–reduction reaction must involve a pair of redox couples for one must donate electrons and one must accept them.

Different redox couples have different affinities for electrons. One with a lesser affinity will tend to donate electrons to another of higher affinity. The **redox potential value** (E_0' is a measurement of the electron affinity or electron-donating potential of a redox couple). This is of importance in biochemistry for it is an indicator of the direction in which electrons will tend to flow between reactants. Equally important E_0' values are directly related to free-energy changes (see below).

E_0' values are expressed in volts—the more negative, the lower the electron affinity, the greater the tendency to hand on electrons, the greater the reducing potential, and the higher the energy of the electrons.

The reason why this chemical property is expressed in volts is due to the method of redox potential determination. As stated above, two redox couples must be involved in an oxido-reduction reaction but, since electron transfer is involved, they can be in two separate vessels (half-cells) if they are connected by a copper wire to conduct electron flow. The method is to use the $2H^+ + 2e^- \leftrightarrow H_2$ equilibrium (catalysed by platinum black) in one half-cell as the reference redox couple, and compare the unknown sample in the second half-cell with it. Electrons will flow through the wire according to the relative electron affinities of the two systems. The positive ions in each half-cell (H^+ in the hydrogen electrode half-cell and, for example, ferrous and ferric ions in the sample half-cell) are accompanied by anions. Change in the positive ions due to loss or gain of electrons in the half-cells necessitates a compensating anion migration from one half-cell to the other. The agar–salt bridge shown in Fig. 7.6 provides the route for this. The electrical potential between the half-cells is measured by a voltmeter inserted into the copper wire connecting electrodes in each half-cell. The reference (hydrogen electrode) half-cell is arbitrarily assigned the value of zero and the relative value of the sample cell is the redox potential of the

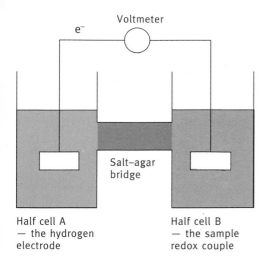

Half cell A
— the hydrogen
electrode

Half cell B
— the sample
redox couple

Fig. 7.6 Diagram of the apparatus for measurement of redox potentials. The reference hydrogen electrode in A contains the redox couple $H_2/2H^+$ catalysed by platinum black on the electrode. The sample half-cell B contains the redox couple whose redox potential is being determined. If B is more reducing than A, electrons will flow from B to A and reduce $2H^+$ to H_2 if the half-cells are connected by a connecting wire. The E_0' value is therefore more negative than that of the hydrogen electrode whose E_0' is assigned the value of −0.42 V. As protons in A are reduced to hydrogen atoms, anions will flow from A to B via the agar–salt bridge to preserve charge neutrality. If the sample in B is less reducing than the reference hydrogen electrode, a reverse series of events will occur and the sample redox couple will have an E_0' value more positive than −0.42 V.

redox pair in it. In physics, the convention is that electrical current flows in the opposite direction to electron flow and, therefore, the half-cell that is donating electrons has the more negative voltage. Thus, if the sample half-cell is more reducing than the reference half-cell, its redox potential value is more negative.

The E_0 values (written without a prime) are standard values measured at 1.0 M concentrations of components or H_2 gas at 1 atmosphere pressure. In biochemistry, values are adjusted to pH 7.0 instead of pH 0 and this brings the redox potential of the reference half-cell to a value of − 0.42 V. Redox potentials so

corrected are written as E_0' values. The half-reaction $NAD^+ + 2H^+ + 2e^- \rightarrow NADH + H^+$ has an E_0' value of − 0.32 V, and that of the half-reaction $\frac{1}{2}O_2 + 2H^+ + 2e^- \rightarrow H_2O$, an E_0' value of +0.82 V. The very large difference indicates that NADH has the potential to reduce oxygen to water but the reverse will not occur.

There is a direct relationship between the $\Delta G^{0'}$ value and the E_0' value of an oxido-reduction reaction, quantified by the Nernst equation,

$$\Delta G^{0'} = -nF\Delta E_0',$$

where n equals the number of electrons transferred in the reaction, F is the Faraday constant (96.5 kJ V^{-1} mol^{-1}), and $\Delta E_0'$ is the difference in redox potential between the electron donor and electron acceptor.

If we consider the oxidation of NADH,

$$NADH + H^+ + \tfrac{1}{2}O_2 \rightarrow NAD^+ + H_2O,$$

the half-reactions are

$$NADH + H^+ \rightarrow NAD^+ + 2H^+ + 2e^- \quad E_0' = -0.320V;$$

$$\tfrac{1}{2}O_2 + 2H^+ + 2e^- \rightarrow H_2O \quad E_0' = 0.816V.$$

For the overall reaction,

$$\Delta E_0' = -0.320V - 0.816V = -1.136V.$$

Therefore,

$$\Delta G^{0'} = -2(96.5\,\text{kJ V}^{-1}\,\text{mol}^{-1})(-1.136\,\text{V}) = -219.25\,\text{kJ mol}^{-1}.$$

The transport of electrons to oxygen does not happen in a single step. In the electron transport system of the mitochondria there is a chain of electron carriers of ever-increasing redox values (decreasing reducing potentials) terminating in the ultimate electron acceptor, oxygen. In effect, electrons from NADH and $FADH_2$ bump down a staircase, each step being a carrier of the appropriate redox potential and each fall releasing free energy (Fig. 7.7). The free energy is thus liberated in manageable parcels—manageable in the sense that it can be

Fig. 7.7 Diagram illustrating the principle of the electron transport chain. (The number of carriers shown is arbitrary; each is a different carrier.) With some carriers only electrons are accepted, protons being liberated into solution, while, in the case of other carriers, protons accompany the electrons. The final reaction with oxygen involves protons from solution. $FADH_2$ donates electrons to the chain at a point lower down than does NADH.

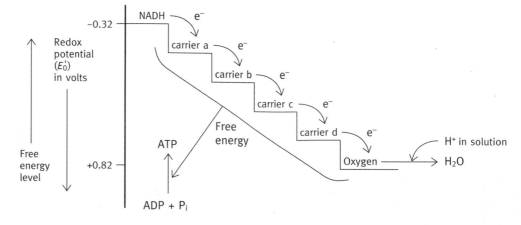

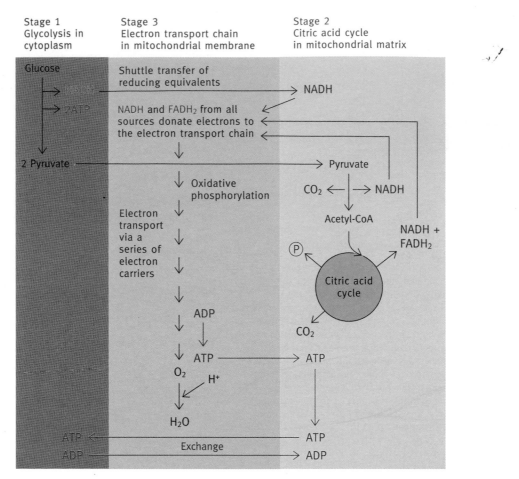

Fig. 7.8 Simplified diagram summarizing the oxidation of glucose. For ease of presentation only products are shown; obviously, reduced NAD is formed from NAD⁺, etc. The same scheme applies to glycogen except that three ATP molecules are produced per glucose unit. We indicate the production of one —Ⓟ from the citric acid cycle rather than as one ATP because, in this case, GDP rather than ADP is the acceptor. Energetically, it amounts to the same thing and is explained in the next chapter. The ATP generated is transported to the cytoplasm by a mechanism involving ADP intake. An important point to note is that NAD⁺ and FAD collect electrons from different metabolic systems including fatty acid oxidation and not just from glucose (see Fig. 7.10). NADH and FADH₂ donate electrons to different points in the electron transport chain. The glycerophosphate shuttle transfer to FAD in the mitochondrial membrane is not shown. See Figs 8.8 and 8.9 for details of shuttles.

harnessed into mechanisms that (indirectly) result in ATP generation from ADP and P_i rather than being wasted as heat, as occurs in simple burning of glucose. The oxidation of NADH and FADH₂ thus drives the conversion of ADP and P_i to ATP. Hence the complete process is called **oxidative phosphorylation**. Thirty or more ATP molecules are synthesized from the oxidation of one molecule or glucose (the exact number is discussed in the next chapter).

In Fig. 7.8, all three phases are put together in the one scheme.

Energy generation from oxidation of fat and amino acids

In addition to glucose and glycogen oxidation to supply the energy for ATP generation, the body oxidizes fat and excess amino acids for the same purpose. To start with fat; in terms of

energy production, it is the fatty acid components of neutral fat (TAG) that are quantitatively important, the glycerol portion being less significant. Chemically, fatty acids are quite different from glucose and you might well expect their oxidation to be correspondingly different. It is at this point that you might get your first glimpse of how, despite complexity in detail, the metabolism of different foodstuffs dovetails together with majestic simplicity. Glucose, as described, is manipulated so that acetyl-CoA is formed and fed into the citric acid cycle. Fatty acids are also manipulated so that carbon atoms are detached two at a time as acetyl-CoA which is fed into the citric acid cycle. Fatty acid oxidation, therefore, differs from glucose oxidation only in this preliminary formation of acetyl-CoA. Moreover, in this preliminary manipulation, NAD⁺ and FAD are reduced and the electrons they carry are fed into the electron transport system as in glucose oxidation. Figure 7.9 gives the relationship between glucose and fat oxidation.

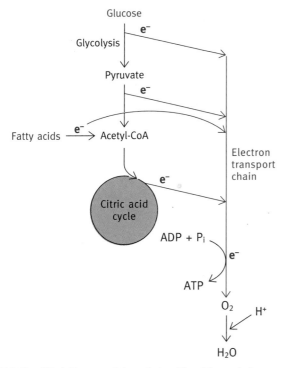

Fig. 7.9 Simplified diagram of the relationship of fat and glucose oxidation. Details given in Fig. 7.7 are omitted for clarity. This figure emphasizes the collection of electrons for the electron transport pathway.

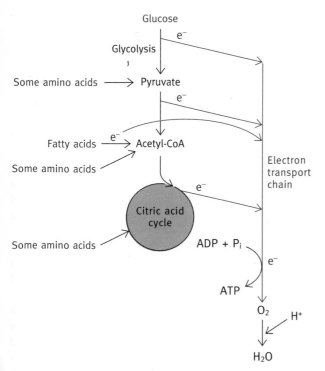

Fig. 7.10 Simplified diagram of the relationship of glucose, fat, and amino acid oxidation for energy generation.

The situation with regard to amino acid oxidation is more complex in detail but similar in concept. There are 20 different amino acids and, as explained earlier, if these are present in excess of immediate requirements they are deaminated and the carbon–hydrogen skeletons used as fuel. Once again, the metabolism of the latter shows a simplicity of concept. The carbon–hydrogen skeletons are converted either to pyruvate or to acetyl-CoA or to intermediates in the citric acid cycle, so that they also join the same metabolic path (as shown in Fig. 7.10). The citric acid cycle thus plays a central role in metabolism.

The interconvertibility of fuels

Although this chapter is concerned primarily with energy generation by food oxidation, a brief note here can throw light on the fuel logistics already described in earlier chapters. We described how glucose, in excess, can be stored as fat. This is because fatty acids can be synthesized from acetyl-CoA,

Glucose → pyruvate → acetyl-CoA → fatty acids.

However, in animals, fatty acids *cannot* be converted, in a net sense, to glucose, because to synthesize the latter (the topic of Chapter 11) pyruvate is needed. In animals, acetyl-CoA cannot be converted to pyruvate because the pyruvate dehydrogenase reaction is irreversible and, therefore, fatty acids cannot be converted to glucose but glucose can be converted to fatty acids and thence to neutral fat (Fig. 7.11). Those amino acids that give rise to pyruvate or a citric acid cycle acid can be converted to glucose (glucogenic amino acids). This is why, in starvation,

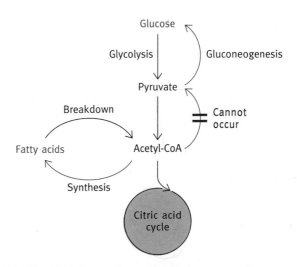

Fig. 7.11 Simplified diagram illustrating why glucose can be converted into fats but fats cannot be converted into glucose in animals. The reverse pathways in red are not completely the same as the forward pathways and are described in later chapters. In plants and bacteria, fat can be converted into glucose but not by the reversal of the pyruvate dehydrogenase reaction.

muscle proteins are destroyed and the amino acids so produced transported to the liver, which converts them to glucose. There are organisms such as bacteria and plants that do have the capacity to convert fat and C_2 compounds such as acetate to glucose, but this involves a special cycle known as the glyoxylate cycle (a modified citric acid cycle) that is not present in animals. This is also described later.

The directions of metabolism, whether fat and glucose are oxidized or are synthesized, are the result of control mechanisms to be described in Chapter 12.

Problems for Chapter 7

1 What are the three major phases involved in the oxidation of glucose and where do they occur?

2 Write down the overall structure of NAD^+ in words; show the structure of the electron-accepting group in the oxidized and reduced form. Explain how NAD^+ acts as a hydrogen carrier between substrates.

3 What is FAD and its role?

4 Explain the difference between aerobic and anaerobic glycolysis in muscle and the circumstances in which they occur. What is the point of anaerobic glycolysis?

5 Write down the structure of coenzyme A in words; also give the structure of its acyl-accepting group. What is the $\Delta G^{0'}$ of hydrolysis of a thiol ester? How does this compare with that of a carboxylic ester?

6 The pyruvate dehydrogenase reaction is of central importance. Write down the reaction and give its $\Delta G^{0'}$ value.

7 What normally happens to the acetyl-CoA generated in the pyruvate dehydrogenase reaction?

8 Glycolysis and the citric acid cycle produce NADH and $FADH_2$. What happens to these?

9 The redox couple: $FAD+2H^++2e^- \rightarrow FADH_2$ has an E'_0 value of -0.219 V. That of $\frac{1}{2}O_2+2H^++2e^- = 0.816$ V. Calculate the $\Delta G^{0'}$ value for the oxidation of $FADH_2$ by oxygen to water.

10 What is the major source of acetyl-CoA other than the pyruvate dehydrogenase reaction?

11 (a) Can glucose be converted to fat? Explain your answer.

(b) Can fatty acids be converted to glucose in animals. Explain your answer.

(c) Can fatty acids be converted to glucose in *E. coli*? Explain your answer.

Chapter 8

....

Glycolysis, the citric acid cycle, and the electron transport system: reactions involved in these pathways

In the previous chapter you saw the overall pattern of the way in which various foodstuffs are metabolized. We now want to fill in the mechanisms of the metabolic pathways, starting with carbohydrate oxidation. In subsequent chapters we will deal with how fatty acids are metabolized to join up to the glucose oxidation pathway at acetyl-CoA and then do the same with amino acid metabolism. A potential problem in studying these pathways is to forget their purposes—to get lost in the detail. If necessary, keep on going back to the previous chapter to refresh your memory on where pathways are heading.

....

Stage 1—glycolysis

This, we remind you, is the splitting of glucose or a glucosyl unit of glycogen into two molecules of pyruvate.

Glucose or glycogen?

So far, for simplicity, we have talked mainly of glucose metabolism. However, in normal circumstances most tissues will have glycogen stores (page 73), and glycolysis may be proceeding from this rather than from free glucose. There is a difference between the two. When glycogen is broken down (page 78), glucose-6-phosphate is produced via glucose-1-phosphate. In the liver this can be hydrolysed to release free glucose into the blood. However, glucose-6-phosphate is on the glycolytic pathway also and in all tissues is broken down to pyruvate. Free glucose is also converted to glucose-6-phosphate by phosphorylation using ATP. You have met this reaction already (page 78) because it is the same as that involved in

glycogen synthesis. Whether glucose-6-phosphate goes to glycogen or to pyruvate or to blood glucose in the case of liver depends on how the metabolic control switches are set, according to physiological needs, and this is a major later topic. The relationship between glycolysis, glucose, and glycogen is shown in Fig. 8.1. It costs a molecule of ATP to produce glucose-6-phosphate from glucose, but no ATP is used if we start with glycogen, because the energy of the glucosyl group in the latter (that is, the $\Delta G^{0'}$ of its hydrolysis) is similar to that of the phosphate ester and is preserved by phosphorolysis.

Why use ATP here at the beginning of glycolysis?

It may seem odd that a pathway designed to produce ATP should start by using up ATP. Why use it here? The reason is that glycolysis involves phosphorylated compounds and ATP must be used to phosphorylate glucose—it has the necessary energy potential. Glucose-6-phosphate is a low-energy phosphoryl compound so certainly we have lost a high-energy phosphoryl group in using ATP. Think of it as an investment for, as you'll see, a 100% profit is made on the ATP used in glycolysis of glucose.

Why is glucose-6-phosphate converted to fructose-6-phosphate?

The next step is to convert glucose-6-phosphate, an aldose sugar, to fructose-6-phosphate, its ketose isomer. The aim, you'll not forget, is to split glucose into two C_3 compounds. To (apparently) digress for a moment, organic chemistry textbooks describe a test-tube reaction called the **aldol condensation**. In this, an aldehyde and a ketone (or aldehyde) condense together

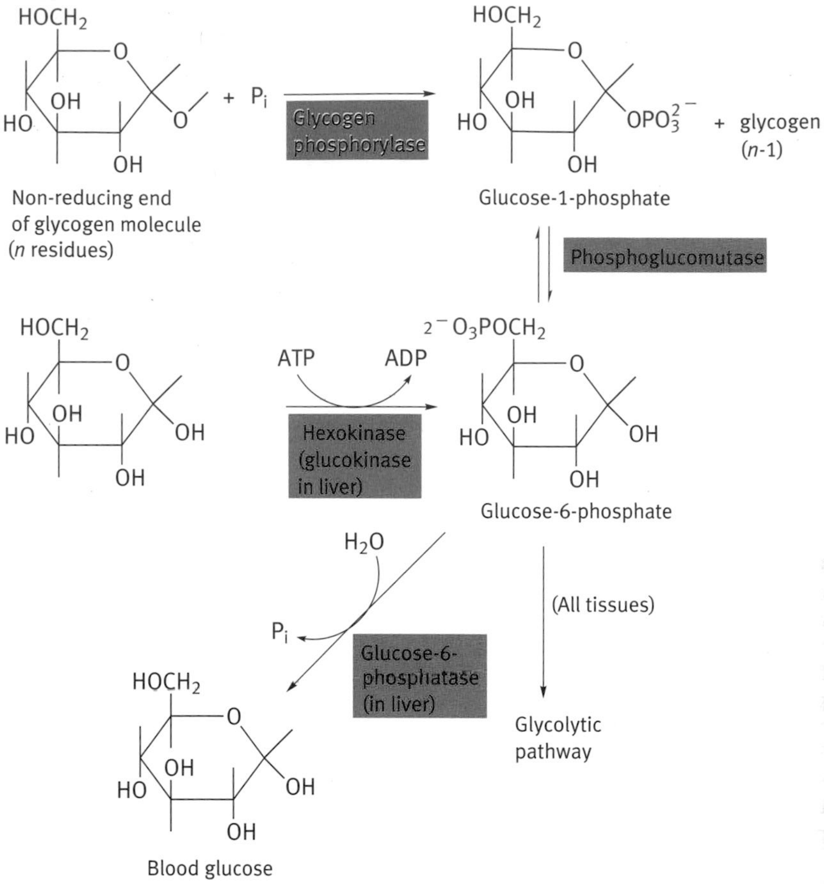

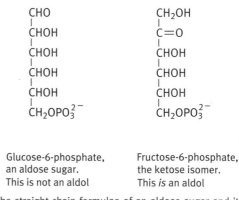

HOCH$_2$

Non-reducing end
of glycogen molecule
(n residues)

+ P$_i$

Glycogen
phosphorylase

HOCH$_2$

OPO$_3^{2-}$ + glycogen
(n-1)

Glucose-1-phosphate

Phosphoglucomutase

HOCH$_2$

ATP   ADP

Hexokinase
(glucokinase
in liver)

$^{2-}$O$_3$POCH$_2$

Glucose-6-phosphate

H$_2$O

P$_i$

Glucose-6-
phosphatase
(in liver)

(All tissues)

HOCH$_2$

Blood glucose

Glycolytic
pathway

Fig. 8.1 Production of glucose-6-phosphate from glycogen or free glucose and its fate. Which routes are operative depends on control mechanisms described in Chapter 12.

as shown in Fig. 8.2. The reaction is freely reversible and therefore can be used to split an aldol into two parts—an aldol being the β-hydroxycarbonyl compound shown. Now turn back to glucose-6-phosphate; this does not have the correct aldol structure for an aldol split, but fructose-6-phosphate does, as is easily seen in the straight-chain formulae in Fig. 8.3. By forming the fructose isomer, the sugar phosphate can be split into two. The glucose-6-phosphate is isomerized into fructose-6-phos-

phate by the enzyme **phosphohexose isomerase**. Before splitting, another phosphoryl group from ATP is transferred to the fructose-6-phosphate by the enzyme **phosphofructokinase** (PFK), yielding fructose-1:6-bisphosphate. The fructose-1:6-bisphosphate is now split by the enzyme **aldolase** catalysing the

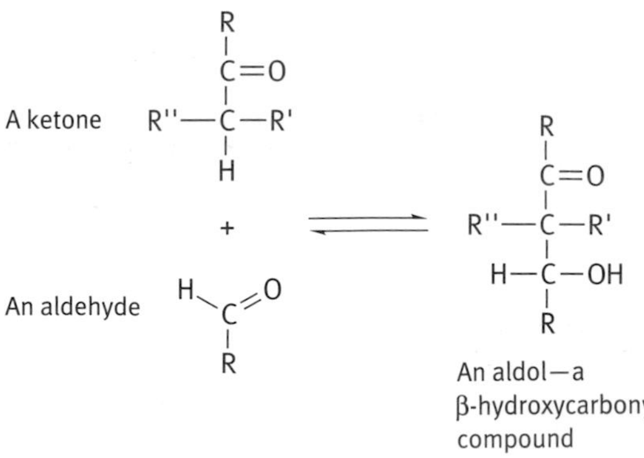

A ketone   R''—C—R'

R
|
C=O
|
C
|
H

+

An aldehyde

H
   \
    C=O
   /
  R

R
|
C=O
|
R''—C—R'
|
H—C—OH
|
R

An aldol—a
β-hydroxycarbonyl
compound

Fig. 8.2 The aldol condensation. A chemical reaction between an aldehyde and a ketone (or aldehyde).

CHO
|
CHOH
|
CHOH
|
CHOH
|
CHOH
|
CH$_2$OPO$_3^{2-}$

Glucose-6-phosphate,
an aldose sugar.
This is not an aldol

CH$_2$OH
|
C=O
|
CHOH
|
CHOH
|
CHOH
|
CH$_2$OPO$_3^{2-}$

Fructose-6-phosphate,
the ketose isomer.
This *is* an aldol

Fig. 8.3 The straight-chain formulae of an aldose sugar and its ketose isomer. The glucose-6-phosphate is in equilibrium with the six-membered ring (pyranose) form and the fructose-6-phosphate with the five-membered ring (furanose) form. The ring structures are shown in Fig. 8.5.

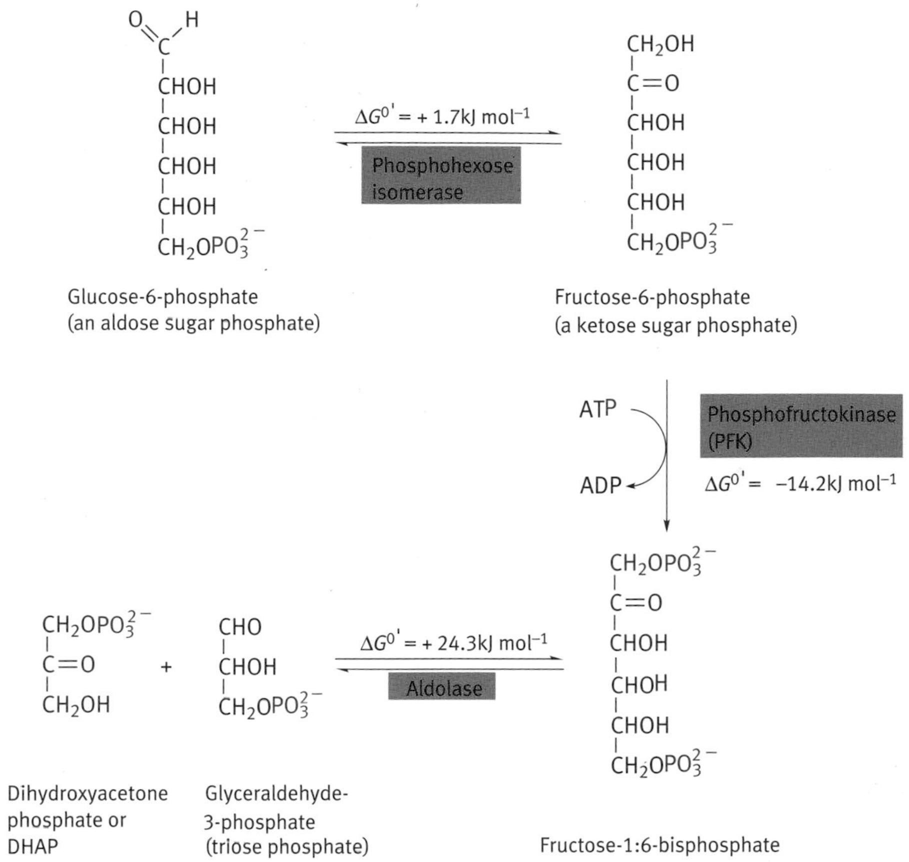

$$\text{Glucose-6-phosphate} \xrightleftharpoons[\text{Phosphohexose isomerase}]{\Delta G^{0'} = +1.7\,\text{kJ mol}^{-1}} \text{Fructose-6-phosphate}$$

Glucose-6-phosphate
(an aldose sugar phosphate)

Fructose-6-phosphate
(a ketose sugar phosphate)

Phosphofructokinase (PFK)

$$\text{ATP} \to \text{ADP}$$

$\Delta G^{0'} = -14.2\,\text{kJ mol}^{-1}$

Fig. 8.4 Conversation of glucose-6-phosphate into two C_3 compounds. The $\Delta G^{0'}$ value for the aldolase reaction in the forward direction would appear to preclude its occurrence, but see text for explanation of this. Straight-chain formulae for the sugars are used here for clarity; the reactions are commonly presented as in Fig. 8.5.

$$\text{Dihydroxyacetone phosphate or DHAP} + \text{Glyceraldehyde-3-phosphate (triose phosphate)} \xrightleftharpoons[\text{Aldolase}]{\Delta G^{0'} = +24.3\,\text{kJ mol}^{-1}} \text{Fructose-1:6-bisphosphate}$$

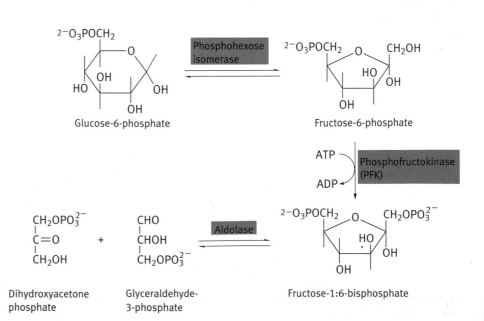

Fig. 8.5 This is the same scheme as in Fig. 8.4 but presented in the more usual form with sugars as ring structures.

Glucose-6-phosphate

Phosphohexose isomerase

Fructose-6-phosphate

ATP → ADP

Phosphofructokinase (PFK)

Dihydroxyacetone phosphate

Glyceraldehyde-3-phosphate

Aldolase

Fructose-1:6-bisphosphate

aldol reaction; the second phosphate means that each of the two C_3 products is phosphorylated to give glyceraldehyde-3-phosphate and dihydroxyacetone phosphate (Fig. 8.4). In Fig. 8.5, the same reactions are presented with the more commonly used ring structures for the sugars; the fructose-6-phosphate is in the five-membered ring (furanose) configuration. The $\Delta G^{0'}$ for the aldolase reaction is $+24.3$ kJ mol^{-1}, which would seem to preclude its ready occurrence. There are, however, special considerations applying to this reaction because one molecule of reactant gives rise to two molecules of product which means that the ΔG of the reaction is influenced by concentration to an unusual degree (see the 'Further study' section that follows). In cellular conditions, the ΔG is small and the reaction freely reversible.

Further study

A note on the $\Delta G^{0'}$ and ΔG values for the aldolase reaction

The reaction catalysed by aldolase has a $\Delta G^{0'}$ value of $+24.3$ kJ mol^{-1} and is freely reversible while reactions with smaller $\Delta G^{0'}$ values will not proceed in the cell. $\Delta G^{0'}$ values are determined at 1 M concentrations of reactants and products; since concentrations in the cell are more likely to be at 10^{-3}–10^{-4} M, ΔG values are always different from $\Delta G^{0'}$ values. Nonetheless, the latter are usually useful guides to metabolic events. This is not true in the case of aldolase where the correlation between $\Delta G^{0'}$ values and ΔG values in the cell is very poor.

The reason is that in the aldolase reaction, one molecule of reactant (fructose-1:6-bisphosphate) gives rise to two molecules of product (glyceraldehyde-3-phosphate and dihydroxyacetone phosphate). The relationship between $\Delta G^{0'}$ and ΔG values is given in the equation (page 12)

$$\Delta G = \Delta G^{0'} + RT \ln \frac{[\text{products}]}{[\text{reactants}]},$$

that is,

$$\Delta G = \Delta G^{0'} + RT \times$$
$$\ln \frac{[\text{glyceraldehyde-3-phosphate}][\text{dihydroxyacetone phosphate}]}{[\text{fructose-6-phosphate}]}.$$

Because there are two products of low concentration, the $RT \ln$ {[products]/[reactants]} moiety has a a large negative value, giving a ΔG value compatible with ready reversibility in the cell. To illustrate this, assume that all reactants are present in the cell at 10^{-4} M. Then

$$\Delta G = +24.3 \text{ kJ mol}^{-1}$$
$$+ (8.315 \times 10^{-3})(298)2.303 \log_{10} \frac{(10^{-4})(10^{-4})}{(10^{-4})}$$
$$= +24.3 - 22.8 = +1.5 \text{ kJ mol}^{-1}.$$

It can be seen that the actual ΔG value at these assumed concentrations is compatible with free reversibility.

Interconversion of dihydroxyacetone phosphate and glyceraldehyde-3-phosphate

Glyceraldehyde-3-phosphate and dihydroxyacetone phosphate are isomeric molecules. An enzyme, **triose phosphate isomerase**, interconverts these.

$$\Delta G^{0'} = +7.6 \text{ kJ mol}^{-1}$$

Glyceraldehyde-3-phosphate   Dihydroxyacetone phosphate

The two compounds are in equilibrium but, since glyceraldehyde-3-phosphate is continually removed by the next step in glycolysis, all of the dihydroxyacetone phosphate is progressively converted to glyceraldehyde-3-phosphate.

Glyceraldehyde-3-phosphate dehydrogenase—a high-energy-phosphoryl-compound-generating step

The aldehyde group of glyceraldehyde-3-phosphate is oxidized by NAD$^+$. You would expect this to produce a carboxyl group (and so it does, ultimately) but oxidation of a $-$CHO group to $-$COO$^-$ has a large negative ΔG value, sufficient in fact to generate a high-energy phosphate compound from P$_i$ on the way. The mechanism by which this is achieved is as follows: at the active site of the enzyme there is the amino acid cysteine which has a thiol or sulfhydryl group ($-$SH) on its side chain.

The aldehyde glyceraldehyde-3-phosphate condenses with the thiol to form a thiohemiacetal.

Enzyme with   Glyceraldehyde   Enzyme–thiohemiacetal
thiol group   -3-phosphate   complex

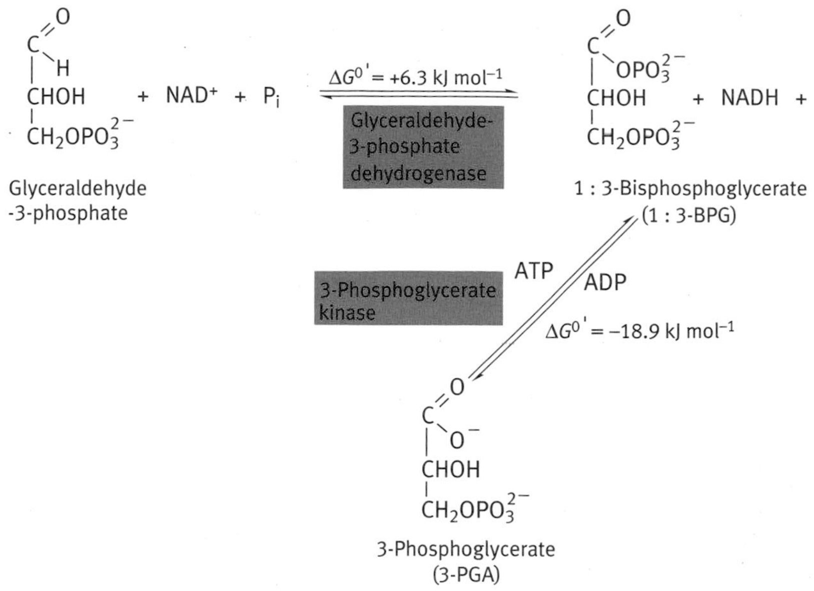

$$\Delta G^{0'} = +6.3 \text{ kJ mol}^{-1}$$

Glyceraldehyde-3-phosphate dehydrogenase

$$\text{Glyceraldehyde-3-phosphate} + NAD^+ + P_i \rightleftharpoons \text{1:3-Bisphosphoglycerate (1:3-BPG)} + NADH + H^+$$

3-Phosphoglycerate kinase

ATP / ADP

$$\Delta G^{0'} = -18.9 \text{ kJ mol}^{-1}$$

3-Phosphoglycerate (3-PGA)

Fig. 8.6 Conversion of glyceraldehyde-3-phosphate to 3-phosphoglycerate.

The enzyme substrate complex is now oxidized, the electrons being accepted by NAD^+, and a thiol ester is formed with the enzyme sulfhydryl group.

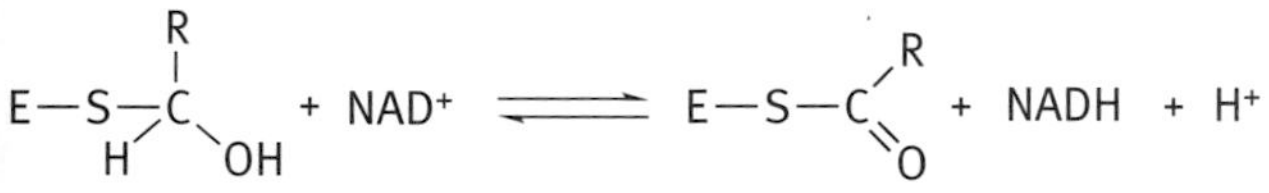

$$E-S-\underset{H}{\overset{R}{C}}-OH + NAD^+ \rightleftharpoons E-S-C\underset{O}{\overset{R}{\diagup}} + NADH + H^+$$

A thiol ester (R—CO—S—) as explained earlier (page 99) is a high-energy compound—of the same order as that of a high-energy phosphate compound. It is therefore thermodynamically feasible for P_i to react as follows.

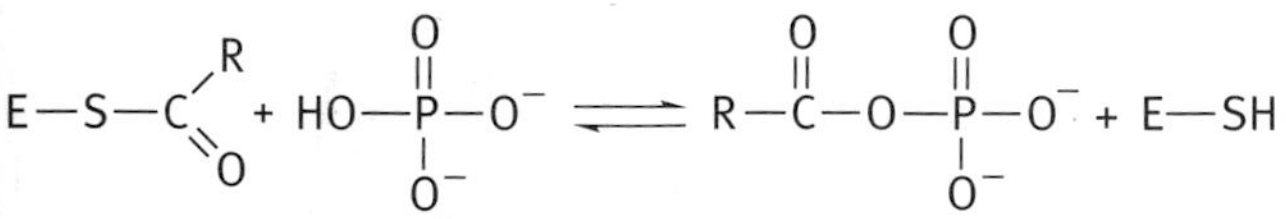

$$E-S-C\overset{R}{\underset{O}{\diagup}} + HO-\overset{O}{\underset{O^-}{\overset{\|}{P}}}-O^- \rightleftharpoons R-\overset{O}{\overset{\|}{C}}-O-\overset{O}{\underset{O^-}{\overset{\|}{P}}}-O^- + E-SH$$

The $\overset{O}{\overset{\|}{RC}}-PO_3^{2-}$ group is also high energy and so the phosphoryl group can be transferred to ADP forming ATP.

The responsible enzyme is **phosphoglycerate kinase** because, in the reverse direction, it transfers a phosphoryl group from ATP to 3-phosphoglycerate (3-PGA). Kinases are always named from the ATP side of the reaction (Fig. 8.6).

The generated phosphoryl group in this process is formed attached to the actual substrate of an enzyme. For this reason it is called **substrate-level phosphorylation**, a point to which we'll refer later. 3-Phosphoglycerate is a low-energy phosphate compound and cannot phosphorylate ADP. However, the next steps in glycolysis manipulate the molecule so that this low-energy phosphate ester becomes a high-energy phosphoryl

group, transferable to ATP. This is not energy for nothing—you'll see how it is done within the law.

The final steps in glycolysis

First, the phosphoryl group of 3-phosphoglycerate is transferred from the 3 to the 2 position as shown.

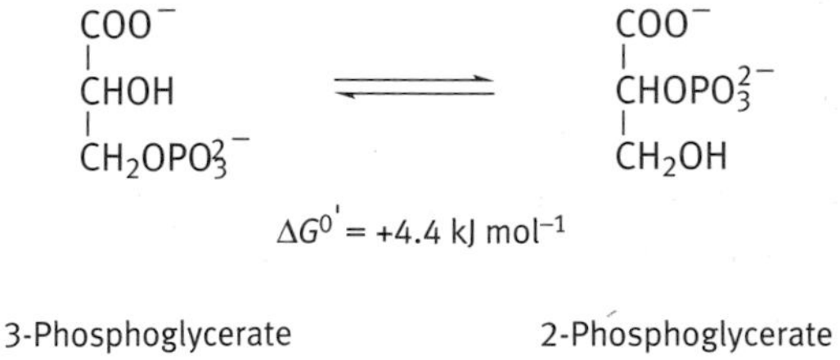

$$\begin{array}{ccc} COO^- & & COO^- \\ | & & | \\ CHOH & \rightleftharpoons & CHOPO_3^{2-} \\ | & & | \\ CH_2OPO_3^{2-} & & CH_2OH \end{array}$$

$$\Delta G^{0'} = +4.4 \text{ kJ mol}^{-1}$$

3-Phosphoglycerate   2-Phosphoglycerate

This is called the **phosphoglycerate mutase reaction**. The reaction is not really an *intra*molecular transfer of the phosphoryl group (though it is in the enzyme from plants). The enzyme from rabbit muscle contains a phosphoryl group that it donates to the 2-OH group of 3-phosphoglycerate forming 2:3-bisphosphoglycerate. The 3-phosphoryl group is now transferred to the enzyme to replace the donated phosphate, so that the net effect is the reaction shown above. The next step in glycolysis is that a water molecule is removed from the 2-phosphoglycerate. Enzymes catalysing such reactions are usually called dehydratases but in this particular case in glycolysis, the old established name is **enolase** (because it forms a substituted enol).

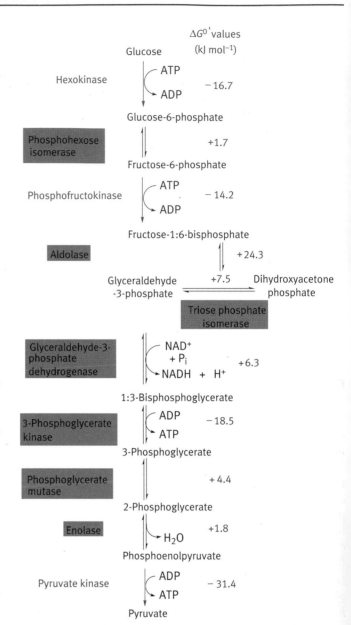

COO⁻
|
CHOPO₃²⁻
|
CH₂OH

2-Phosphoglycerate

Enolase
−H₂O

Phosphoenolpyruvate
(PEP)

Enolpyruvate

Pyruvate

+ ATP

Pyruvate kinase
ADP

Fig. 8.7 The glycolytic pathway. Irreversible reactions are indicated in red. The free reversibility of the aldolase reaction would appear to be inconsistent with such a large $\Delta G^{0'}$ value (see text for explanation).

The enolase reaction has a $\Delta G^{0'}$ of only +1.8 kJ mol⁻¹, but the enolphosphate compound is of the 'high-energy' type, with a $\Delta G^{0'}$ of hydrolysis of −62.2 kJ mol⁻¹; a reason for this is that the immediate product of the reaction, the enol form of pyruvate, spontaneously converts to the keto form, a reaction with a large negative $\Delta G^{0'}$ value. The conversion of a low-energy phosphate to a high-energy one by enolase is not a case of energy for nothing but rather that the energy of the molecule is rearranged. The phosphoryl group is transferred to ADP by the enzyme **pyruvate kinase**; this name might misleadingly imply that pyruvate can be phosphorylated by ATP by reversal of the reaction; the name of the enzyme derives from the convention mentioned earlier that a kinase is named from the reaction involving ATP even though, in this case, that reaction never occurs. The irreversibility of the conversion of PEP to pyruvate has important metabolic repercussions as you'll see when we come later to gluconeogenesis. (There is a potential source of confusion arising from the fact that in certain plants and microorganisms, pyruvate *is* directly converted to PEP by a quite different enzyme (described on page 195) that utilizes two phosphoryl groups from ATP. However, this reaction does *not* occur in animals.)

The complete glycolytic pathway is shown in Fig. 8.7.

The ATP balance sheet from glycolysis

Starting with glucose, two molecules of ATP were used to form fructose-1:6-bisphosphate. The phosphoglycerate kinase produced two ATP molecules per original glucose and the pyruvate kinase two—a total of four and a net gain of two. Starting from glycogen, only one molecule of ATP was consumed so the net gain is three molecules of ATP formed from ADP+P$_i$ per glucosyl unit.

Reoxidation of cytoplasmic NADH by electron shuttle systems

In the aerobic situation, the NADH generated in glycolysis is reoxidized by transferring its electrons into mitochondria. As already explained (page 98) NADH itself cannot enter the mitochondrion; there are two systems for transferring its electrons into mitochondria. The first involves dihydroxyacetone phosphate (DHAP, generated by the aldolase reaction). An enzyme in the cytoplasm transfers electrons from NADH to DHAP giving glycerol-3-phosphate (Fig. 8.8). The enzyme is

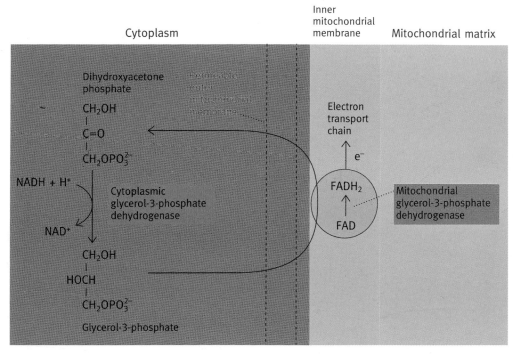

Fig. 8.8 Diagram of the glycerophosphate shuttle that transfers electrons from cytoplasmic NADH to the electron transport chain of mitochondria.

called **glycerol-3-phosphate dehydrogenase**, working in reverse in the above reaction.

Glycerol-3-phosphate can reach the inner mitochondrial membrane (the outer one being highly permeable) where a different type of glycerol-3-phosphate dehydrogenase, built into the membrane—this time with an FAD prosthetic group—transfers electrons from glycerol-3-phosphate to the mitochondrial electron transport chain. The DHAP so produced cycles (shuttles) back into the cytoplasm to pick up more electrons (Fig. 8.8). Note that in this, the electron carrier (glycerol-3-phosphate) does not have to enter the mitochondrial matrix but its pair of electrons gain access to the electron chain carrying electrons to oxygen, located in the inner mitochondrial membrane. The net effect is to transfer electrons from cytoplasmic NADH to the mitochondrial electron transport chain.

Another shuttle, the **malate–aspartate shuttle**, transfers electrons from cytoplasmic NADH to mitochondrial NAD^+. The mitochondrial NADH thus generated is then oxidized by the electron transport chain. This shuttle system involves transfer of electrons from NADH to oxaloacetate to form malate in the cytoplasm which is transported into the mitochondrion, by a specific carrier, where it is re-oxidized to oxaloacetate, mitochondrial NAD^+ being reduced (Fig. 8.9). The net effect is that NADH outside reduces NAD^+ inside. This shuttle is a little more complicated in that the oxaloacetate, so formed, can't traverse the mitochondrial membrane to get back to the cytoplasm. It is converted to aspartate, which is

transported, again by a specific carrier, to the cytoplasm and reconverted to oxaloacetate there; hence the name, the malate–aspartate shuttle. At this stage we'll not give the mechanism of aspartate $\leftrightarrow$ oxaloacetate interconversions since it will be more convenient to do this later when we deal with amino acid metabolism (page 200). Different tissues probably use the two shuttles to different extents. The two shuttles differ in a significant way—the glycerophosphate shuttle results in cytoplasmic NADH reducing mitochondrial FAD in the form of the prosthetic group of the mitochondrial flavoprotein, glycerol-3-phosphate dehydrogenase. The $FADH_2$ has a higher redox potential then NADH (lower energy); it hands on its electrons to the electron transport chain at a point that is further along the chain from that at which NADH hands on its electrons. The net ATP generation from the oxidation of a *cytoplasmic* molecule NADH by this $FADH_2$ route is 1.5 molecules (see below). The aspartate–malate shuttle starts with one molecule of cytoplasmic NADH and ends up with one molecule of mitochondrial NADH whose oxidation generates 2.5 molecules of ATP (explained later.

Transport of pyruvate into the mitochondria

The other product of glycolysis besides NADH is pyruvate. Unless it is reduced to lactate (see page 97) the pyruvate, as explained earlier, is transported into the mitochondrial matrix by an antiport type of membrane transport protein (page 52), which exchanges it for OH^- inside the matrix.

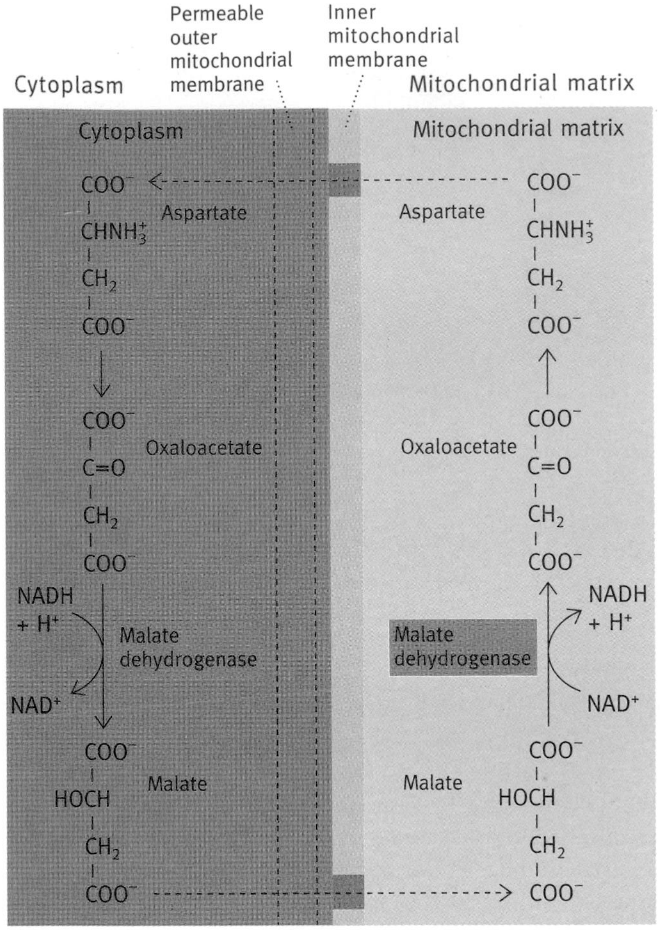

Fig. 8.9 Diagram of the malate–aspartate shuttle for transferring electrons from cytoplasmic NADH to mitochondrial NAD^+. The mechanism of the interconversion of oxaloacetate and aspartate is dealt with on page 200. This shuttle, unlike the glycerophosphate shuttle, is reversible, and can operate as shown, bringing NADH into the mitochondrion, only if the $NADH/NAD^+$ ratio is higher in he cytoplasm than in the mitochondrial matrix.

Stage 2—the citric acid cycle

Before we get to the cycle itself we will deal with the preparation of pyruvate to enter the cycle, by which we mean its conversion to acetyl-CoA.

Conversion of pyruvate to acetyl-CoA—a preliminary step before the cycle

As outlined earlier (page 99), pyruvate in the mitochondrial matrix is converted to acetyl-CoA, which feeds the acetyl group into the citric acid cycle (the structure of CoA is described on page 99).

The overall reaction catalysed by pyruvate dehydrogenase is repeated here

$$Pyruvate + NAD^+ + CoA-SH \rightarrow Acetyl-S-CoA$$
$$+ NADH + H^+ + CO_2.$$

Pyruvate dehydrogenase, the responsible enzyme, is a very large complex composed of many polypeptides. It essentially consists of three different enzyme activities aggregated together for efficiency, each catalysing one of the intermediate steps in the process (Fig. 8.10). The first step is decarboxylation of pyruvate to produce CO_2 and a hydroxyethyl group $CH_3\overset{H}{\underset{OH}{C}}$ attached to the cofactor **thiamin pyrophosphate (TPP)**. TPP is derived from thiamin, or vitamin B_1, deficiency of which impairs the ability to metabolize carbohydrate. The hydroxyethyl group, in a series of steps, is converted to the acetyl group of acetyl-CoA with the reduction of NAD^+. The process is known as an **oxidative decarboxylation** for obvious reasons. The structure of components of the reaction given in the 'Further study' section below (the cofactor TPP, and the lipoic acid that takes part in the reaction as part of the enzyme) are worth a look, but it is suggested that you need not learn these structures.

This conversion of pyruvate to acetyl-CoA is irreversible; the $\Delta G^{0'}$ of the reaction is -33.5 kJ mol^{-1}. As you will see later, this is of profound significance for irreversibility means that fatty acids can never be converted in a net sense to glucose in the animal body though, as mentioned, bacteria and plants have a special mechanism for achieving this. The acetyl-CoA now enters the citric acid cycle.

Further study

Components involved in the pyruvate dehydrogenase reaction

Thiamin pyrophosphate (TPP) has the structure

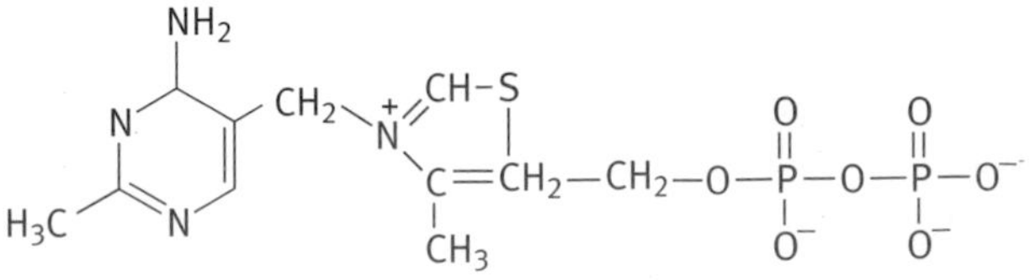

TPP-hydroxyethyl has the structure

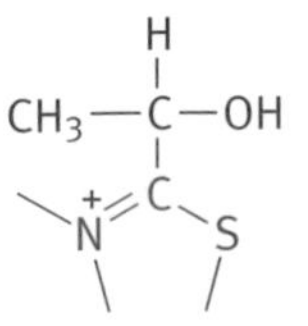

Lipoic acid in its reduced form has the structure

$$
\begin{array}{cc}
\text{SH} & \text{SH} \\
| & | \\
\text{CH}_2 & \text{CH}-\text{CH}_2-\text{CH}_2-\text{CH}_2-\text{CH}_2-\text{COO}^- \\
& \diagdown \\
& \text{CH}_2
\end{array}
$$

and in its oxidized form the structure is

$$
\begin{array}{cc}
\text{S}\!-\!\!-\!\!-\text{S} \\
| \quad\;\; | \\
\text{CH}_2 \quad \text{CH}-\text{CH}_2-\text{CH}_2-\text{CH}_2-\text{CH}_2-\text{COO}^- \\
\diagdown \\
\text{CH}_2
\end{array} \quad .
$$

Lipoyllysine consists of lipoic acid attached to a lysine side chain of the enzyme by —CO—NH— linkage. In Fig. 8.10 lipoyllysine is represented by the disulfide structure. The three enzymes represented by E_1, E_2, and E_3 are part of a very large protein complex.

Introduction—what is the real magic of the cycle?

The citric acid cycle, which looks like, and is, a series of ordinary enzymic reactions, is a device of astonishing ingenuity in that it produces a combustible fuel in the form of reducing equivalents of NADH and $FADH_2$, part of which, in effect, comes from splitting water. This fuel is burned in the next stage, the electron transport system, to produce ATP from ADP and P_i. Unlike schemes to run cars on water, the cycle is thermodynamically sound because it uses the free energy made available from the destruction of the acetyl group of acetyl-CoA to drive the process. It must be noted that the splitting of water is not a direct 'head on' process as occurs in photosynthesis (page 191) and oxygen is not liberated as such, but as CO_2. Rather it is an indirect splitting that can easily be overlooked and is seldom referred to in texts. We need to explain this aspect more clearly for it is central to appreciating what the cycle is all about.

Acetyl-CoA enters the cycle by a reaction with oxaloacetate to produce citrate. Do not bother at this stage with how this occurs—it is described later. The reaction involves the input of a

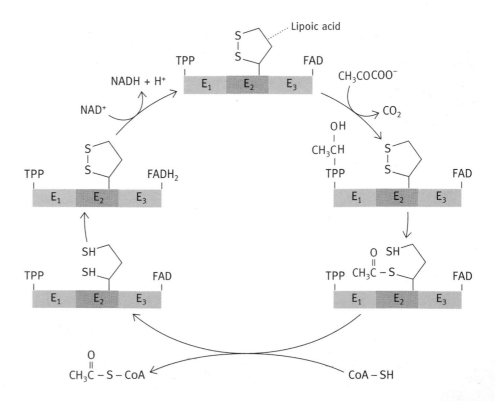

Fig. 8.10 Mechanism of the pyruvate dehydrogenase reaction. TPP, thiamin pyrophosphate; E_1–E_3, enzyme regions of complex.

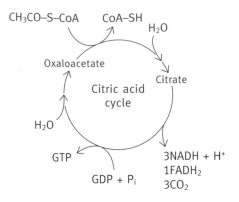

Fig. 8.11 Diagram of the inputs and outputs of the citric acid cycle. Individual cycle reactions are not indicated.

molecule of water. With one complete 'turn' of the cycle (again explained later) oxaloacetate is re-formed and the acetyl group of acetyl-CoA has disappeared. As a result of the cycle reactions, the products are as follows: two molecules of CO_2; the reduction of three molecules of NAD^+ to NADH; and that of one molecule of FAD to $FADH_2$. In addition, the CoA—S— of acetyl-CoA becomes CoASH (Fig. 8.11). (A single high-energy phosphoryl group, as GTP, is produced from P_i.) If you add up all the reducing equivalents of the three NADH and one $FADH_2$, there are eight. (Remember that NAD^+ accepts two electrons.) The formation of the thiol group of CoASH (from CoA—S—) requires one more—a total of nine reducing equivalents (effectively equivalent to nine H atoms).

The CH_3CO—S—CoA supplies three of these so that there is a shortfall of six. There is no involvement of oxygen in the cycle so there is also a shortfall of three oxygen atoms to produce the two molecules of CO_2. The source of these 'missing' atoms includes two molecules of H_2O. You will see below that, as well as the input of H_2O into citrate synthesis, a second molecule of water enters the cycle. But, this still leaves a shortfall of two hydrogen and one oxygen atoms. It will be more convenient to explain the source of these later (page 119).

Thus, we see the remarkable feat—electrons of H_2O are raised up the energy scale to become reducing equivalents of NADH and $FADH_2$ and, of course, electrons from the acetyl group also are utilized for this purpose. Again, it is emphasized that this does not mean that the components of H_2O go *directly* to these products but they are required for the chemical balance sheet and the net effect is that H_2O reduces NAD^+ and FAD and the oxygen is taken up by CO_2 production. As stated, the chemical reactions involved in converting the acetyl group to its products provide the free energy to convert H_2O into reducing equivalents. In fact, the whole cycle has a negative ΔG value and so proceeds thermodynamically 'downhill'.

With that preamble we can now turn to the reactions of the cycle.

A simplified version of the citric acid cycle

Possibly one obstacle to learning the cycle is that the progression of reactions doesn't make much sense until you have completed it, so we will first look at a simplified version devoid of detail.

Be sure you know the structure of oxaloacetic acid for that's where it all starts and finishes. Acetate is CH_3COO^-, the oxalo group is $^-OOC-C\overset{O}{\diagup}$ so oxaloacetate is

$$\overset{O}{\underset{H_2C-COO^-}{\overset{\|}{C}-COO^-}}.$$

The acetyl group of acetyl-CoA is joined to oxaloacetate to form citric acid; it is easy to see how citrate can be derived from oxaloacetate.

$$\begin{array}{c} CH_2COO^- \\ | \\ HO-C-COO^- \\ | \\ H_2C-COO^- \end{array}$$

It is helpful to remember that citrate is a C_6 *symmetric* tricarboxylic compound. This is converted to its asymmetric isomer, isocitrate, which, in the cycle, is progressively converted to α-ketoglutarate (C_5), succinate (C_4), fumarate (C_4), malate (C_4), and oxaloacetate (Fig. 8.12). This means that one turn of the cycle eliminates the acetyl group fed into it.

We suggest that you make yourself completely familiar with these acids of the cycle. With that preparation, a more detailed consideration of this superhighway of carbon compound metabolism can be given.

Mechanisms of the citric acid cycle reactions

We might, for convenience, divide the reactions into three groups. (1) the synthesis of citrate—this is the reaction feeding acetyl groups into the cycle; (2) the 'top part' of the cycle—involving conversion of C_6 citrate to C_5 α-ketoglutarate; and (3) the 'carbon 4' part-conversion of succinate to oxaloacetate.

The synthesis of citrate

The name of the enzyme involved here is **citrate** or **citryl synthase** (note, not synthetase, since ATP is not involved). The enzyme catalyses the condensation of acetyl-CoA with oxaloacetate to give citryl-CoA. This is unstable and hydrolyses to citrate. Citrate formation has a large negative ΔG value (-32.2 kJ mol^{-1}) and hence the reaction is irreversible.

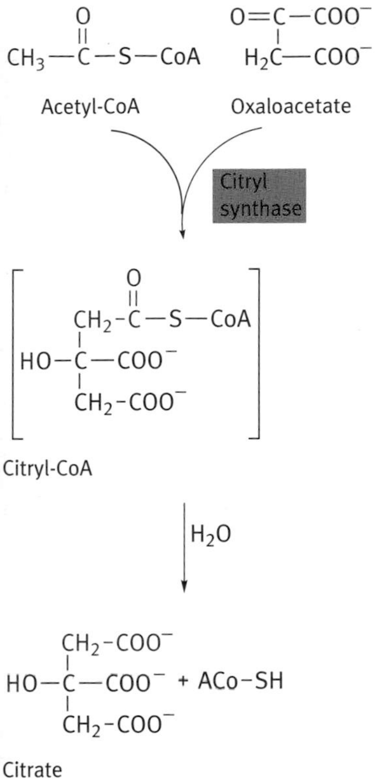

$$\underset{\text{Acetyl-CoA}}{CH_3-\overset{\displaystyle O}{\overset{\|}{C}}-S-CoA} \qquad \underset{\text{Oxaloacetate}}{\overset{\displaystyle O=C-COO^-}{H_2C-COO^-}}$$

Citryl synthase

$$\left[\begin{array}{l} CH_2-\overset{\displaystyle O}{\overset{\|}{C}}-S-CoA \\ HO-C-COO^- \\ CH_2-COO^- \end{array} \right]$$

Citryl-CoA

H_2O

$$\underset{\text{Citrate}}{HO-\overset{\displaystyle CH_2-COO^-}{\underset{\displaystyle CH_2-COO^-}{C-COO^-}}} + \text{ACo-SH}$$

Conversion of citrate to α-ketoglutarate

Citrate → Isocitrate.

The strategy of this reaction is to switch the hydroxyl group of citrate (a symmetric molecule) from the 3 position to the 2 position, giving isocitrate (an asymmetric molecule). The logic of this will become apparent. This isomerization of citrate is catalysed by a single enzyme which reversibly removes water and adds it back across the double bond in either direction.

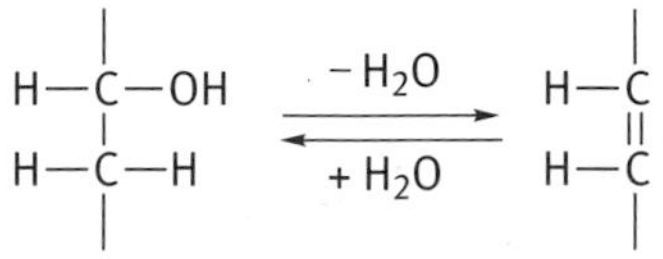

$$\underset{}{\overset{\displaystyle H-C-OH}{H-C-H}} \quad \underset{+H_2O}{\overset{-H_2O}{\rightleftharpoons}} \quad \overset{\displaystyle H-C}{\underset{\displaystyle H-C}{\|}}$$

Enzymes catalysing such reactions are usually called dehydratases but in this case, due to long tradition, its name is **aconitase** because the unsaturated product is *cis*-aconitate (found first in the plant genus *Aconitum*).

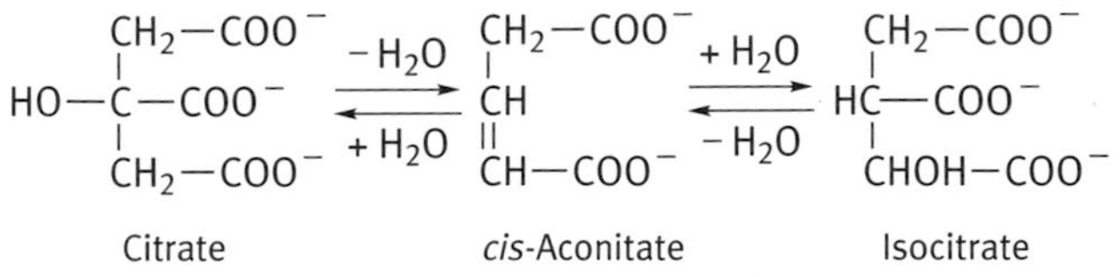

$$\underset{\text{Citrate}}{HO-\overset{\displaystyle CH_2-COO^-}{\underset{\displaystyle CH_2-COO^-}{C-COO^-}}} \underset{+H_2O}{\overset{-H_2O}{\rightleftharpoons}} \underset{\text{cis-Aconitate}}{\overset{\displaystyle CH_2-COO^-}{\underset{\displaystyle CH-COO^-}{\overset{\displaystyle CH}{\|}}}} \underset{-H_2O}{\overset{+H_2O}{\rightleftharpoons}} \underset{\text{Isocitrate}}{\overset{\displaystyle CH_2-COO^-}{\underset{\displaystyle CHOH-COO^-}{HC-COO^-}}}$$

Since isocitrate is now metabolized further in the cycle the net effect is that aconitase catalyses the reaction sequence

Citrate → *cis*-Aconitate → Isocitrate.

Isocitrate dehydrogenase.

You have already met the class of NAD^+-requiring dehydrogenases in lactate dehydrogenase (page 97). This again is a

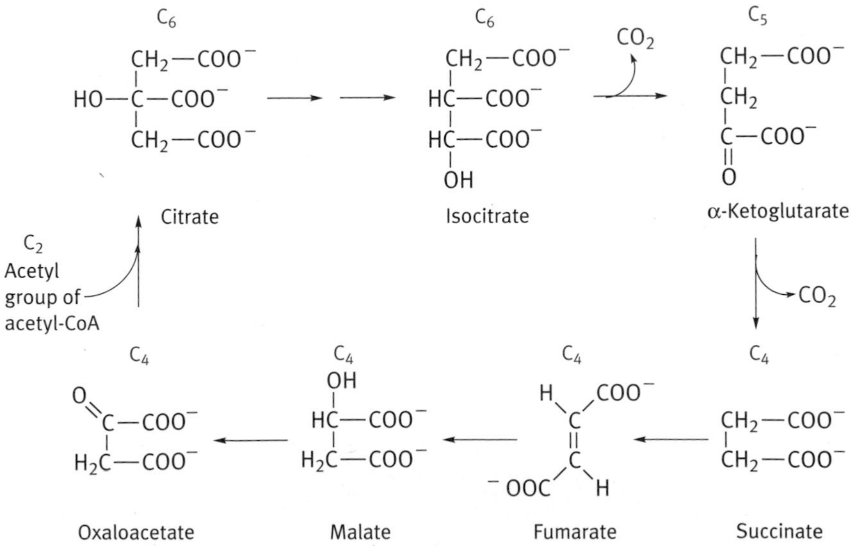

Fig. 8.12 Simplified citric acid cycle showing the component acids and their sequence. The purpose of this diagram is for you to learn the structures of the main cycle acids and their interrelationships without complicating details. When you are familiar with these it will be easier to appreciate the full cycle.

common reaction of the type

$$
\begin{array}{c}
\text{H}-\overset{|}{\underset{|}{\text{C}}}-\text{H} \\
\text{H}-\overset{|}{\underset{|}{\text{C}}}-\text{OH}
\end{array}
+ \text{NAD}^+ \longrightarrow
\begin{array}{c}
\text{H}-\overset{|}{\underset{|}{\text{C}}}-\text{H} \\
\overset{|}{\text{C}}=\text{O} \\
\end{array}
+ \text{NADH} + \text{H}^+ .
$$

In the cycle **isocitrate dehydrogenase** catalyses the reaction

$$
\begin{array}{l}
\text{CH}_2-\text{COO}^- \\
\overset{|}{\text{CH}}-\text{COO}^- \\
\overset{|}{\text{CHOH}}-\text{COO}^-
\end{array}
$$

Isocitrate

$\downarrow$ + NAD

$$
\left[
\begin{array}{l}
\text{CH}_2-\text{COO}^- \\
\alpha\overset{|}{\text{CH}}-\text{COO}^- \\
\beta\overset{|}{\underset{\|}{\text{C}}}-\text{COO}^- \\
\quad\ \text{O}
\end{array}
\right]
\longrightarrow
\begin{array}{l}
\text{CH}_2-\text{COO}^- \\
\overset{|}{\text{CH}_2} \\
\overset{|}{\underset{\|}{\text{C}}}-\text{COO}^- \\
\ \text{O}
\end{array}
+ \text{CO}_2 .
$$

+ NADH + H$^+$

Oxalosuccinate α-Ketoglutarate

The immediate product is oxalosuccinate which is a β-keto acid (the keto group is β to the centre-COOH group). Such acids are unstable and readily lose the carboxyl group as CO_2. This happens on the surface of the isocitrate dehydrogenase so the product is the C_5 acid α-ketoglutarate as shown.

The C_4 part of the cycle

α-Ketoglutarate is an analogue of pyruvate. We can write both as

$$
\begin{array}{l}
\text{R} \\
\overset{|}{\underset{\|}{\text{C}}}-\text{COO}^- \\
\ \text{O}
\end{array}
$$

For pyruvate $R=-CH_3$; for α-ketoglutarate $R=-CH_2CH_2COO^-$. We have already seen (page 99) that pyruvate dehydrogenase converts pyruvate to acetyl-CoA and CO_2. The reaction is repeated here.

$$
\begin{array}{l}
\text{R} \\
\overset{|}{\underset{\|}{\text{C}}}-\text{COO}^- + \text{NAD}^+ + \text{CoA}-\text{SH} \\
\ \text{O}
\end{array}
$$

$\downarrow$

$$
\begin{array}{l}
\text{R} \\
\overset{|}{\underset{\|}{\text{C}}}-\text{S}-\text{CoA} + \text{CO}_2 + \text{NADH} + \text{H}^+ \\
\ \text{O}
\end{array}
$$

An equivalent enzyme complex exists for α-ketoglutarate and precisely the same equation applies as above except that $R=-CH_2CH_2COO^-$; the enzyme complex attacks α-ketoglutarate rather than pyruvate. The product of α-**ketoglutarate dehydrogenase** is therefore succinyl-CoA, analogous to acetyl-CoA.

$$
\begin{array}{l}
\text{CH}_2-\text{COO}^- \\
\overset{|}{\text{CH}_2} \\
\overset{|}{\underset{\|}{\text{C}}}-\text{S}-\text{CoA} \\
\ \text{O}
\end{array}
$$

Succinyl-CoA

However, in the cycle, whereas acetyl-CoA is used to form citrate, succinyl-CoA is broken down to free succinate plus CoA—SH. In principle, the simplest way to do this would be to hydrolyse succinyl-CoA. However, the $\Delta G^{0\prime}$ of this hydrolysis of the thiol ester is -35.5 kJ mol^{-1}—enough energy to raise P_i to a high-energy phosphate compound. So, why waste this energy? Why not trap it instead? This is precisely what happens.

Generation of GTP coupled to splitting of succinyl-CoA

The reaction is

Succinyl-CoA + GDP + P_i $\leftrightarrow$ Succinate + GTP + CoASH;

$\quad \Delta G^{0\prime} = -2.9 \text{ kJ mol}^{-1}$.

The enzyme is named for the back reaction—hence **succinyl-CoA synthetase**—it synthesizes succinyl-CoA from succinate and GTP but, in the citric acid cycle, it works in the other direction, of course. In plants, ADP, not GDP, is used in this reaction; it is not known why GDP should be used in animals.

The mechanism of the reaction is as follows: P_i displaces CoA producing succinyl phosphate.

$$
\begin{array}{l}
\text{CH}_2-\text{COO}^- \\
\overset{|}{\text{CH}_2} \\
\overset{|}{\underset{\|}{\text{C}}}-\text{S}-\text{CoA} \\
\ \text{O}
\end{array}
+
\begin{array}{c}
\ \ \text{O} \\
\ \ \| \\
\text{HO}-\text{P}-\text{O}^- \\
\ \ \overset{|}{\text{O}^-}
\end{array}
\longrightarrow
\begin{array}{l}
\text{CH}_2-\text{COO}^- \\
\overset{|}{\text{CH}_2} \quad \text{O} \\
\overset{|}{\underset{\|}{\text{C}}}-\text{O}-\overset{\|}{\underset{\overset{|}{\text{O}^-}}{\text{P}}}-\text{O}^- \\
\ \text{O}
\end{array}
+ \text{CoA}-\text{SH}
$$

Succinyl-CoA P_i Succinyl phosphate

The phosphoryl group is now transferred to GDP, giving succinate and GTP.

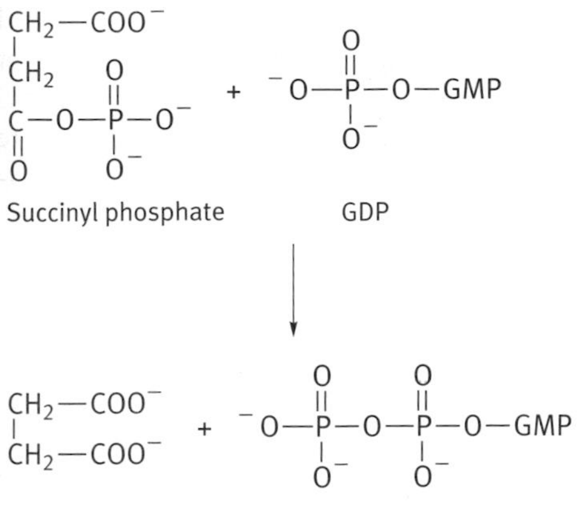

$$\text{CH}_2-\text{COO}^-$$
$$|$$
$$\text{CH}_2 \quad \text{O}$$
$$| \quad \quad ||$$
$$\text{C}-\text{O}-\text{P}-\text{O}^- \quad + \quad {}^-\text{O}-\overset{\overset{\text{O}}{||}}{\text{P}}-\text{O}-\text{GMP}$$
$$|| \quad \quad | \quad \quad \quad \quad \quad \quad \quad \text{O}^-$$
$$\text{O} \quad \quad \text{O}^-$$

Succinyl phosphate GDP

$$\downarrow$$

$$\text{CH}_2-\text{COO}^-$$
$$| \quad \quad \quad \quad \quad \quad \quad \quad \text{O} \quad \quad \text{O}$$
$$\quad \quad \quad \quad \quad \quad + \quad {}^-\text{O}-\overset{||}{\text{P}}-\text{O}-\overset{||}{\text{P}}-\text{O}-\text{GMP}$$
$$\text{CH}_2-\text{COO}^- \quad \quad \quad \quad \quad | \quad \quad \quad |$$
$$\quad \quad \quad \quad \quad \quad \quad \quad \quad \quad \text{O}^- \quad \quad \text{O}^-$$

Succinate GTP

Earlier (page 116), we pointed out that, to balance the output of CO_2 and reducing equivalents from the cycle with the input components, in addition to the two H_2O molecules entering the cycle we need two more hydrogen atoms and one oxygen. These arise from the involvement of inorganic phosphate in the breakdown of succinyl-CoA as described above. In case this is not clear, the *actual* reactions given above can be notionally regarded for balance sheet purposes as being *equivalent* to the two reactions

$$GDP + P_i \rightarrow GTP + H_2O;$$

$$Succinyl\text{-}CoA + H_2O \rightarrow Succinate + CoASH.$$

We emphasize that the reaction *does not* proceed in that way but it illustrates how the 'missing' elements of H_2O are supplied to the cycle to put the balance sheet in order.

Conversion of succinate to oxaloacetate

First, succinate is dehydrogenated but by **succinate dehydrogenase** whose electron acceptor is FAD (page 97), firmly bound to the enzyme and which can reversibly accept a pair of hydrogen atoms. Why do we use NAD^+ for the other dehydrogenation reactions of the cycle and FAD here? It's a question of redox potentials. On page 100, we described how electrons will, for thermodyamic reasons, flow from a lower redox potential (higher reducing potential or energy level) to electron acceptors of higher redox potential (lower reducing potential or energy level). In the case of succinate dehydrogenase the reaction is of the type

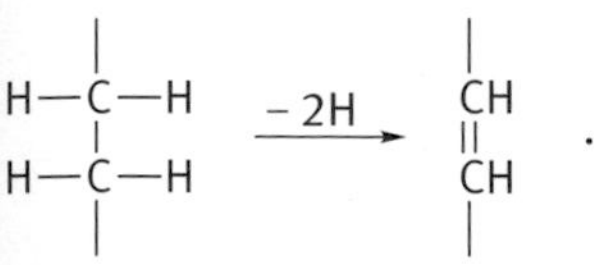

$$\text{H}-\overset{|}{\underset{|}{\text{C}}}-\text{H} \quad \xrightarrow{\;-2\text{H}\;} \quad \overset{|}{\underset{|}{\text{C}}}\text{H}$$
$$\text{H}-\overset{|}{\underset{|}{\text{C}}}-\text{H} \quad \quad \quad \quad \quad \overset{||}{\underset{|}{\text{C}}}\text{H}$$

The redox potential or reducing potential of this system is such that it cannot reduce NAD^+ but can reduce FAD (which is more

strongly oxidizing than NAD^+). The reaction is therefore

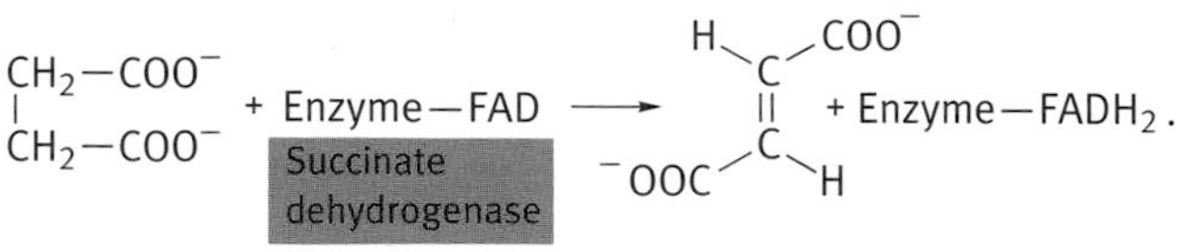

$$\text{CH}_2-\text{COO}^- \quad \quad \quad \quad \quad \quad \text{H}\diagdown_{\text{C}}\diagup\text{COO}^-$$
$$| \quad \quad \quad + \text{Enzyme}-\text{FAD} \longrightarrow \quad \quad || \quad \quad + \text{Enzyme}-\text{FADH}_2.$$
$$\text{CH}_2-\text{COO}^- \quad \quad \quad \quad \quad \quad {}^-\text{OOC}\diagup^{\text{C}}\diagdown\text{H}$$

Succinate Fumarate

The rest of the cycle, conversion of fumarate to oxaloacetate is plain sailing because you have already met the reaction types involved. A molecule of water is added to fumarate (cf. aconitase, above). The enzyme should logically be called fumarate hydratase but, by long usage, is called **fumarase**.

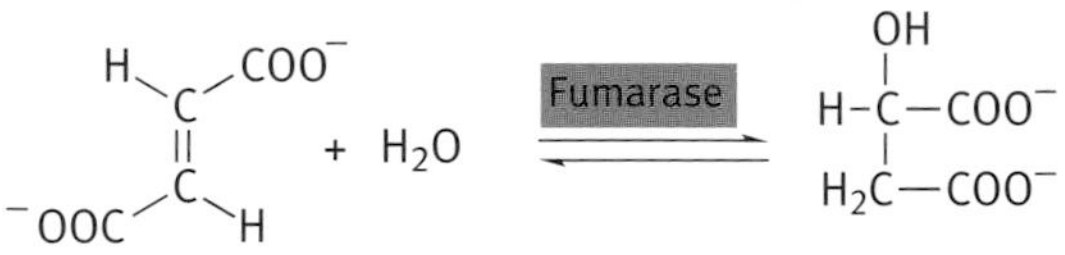

$$\text{H}\diagdown_{\text{C}}\diagup\text{COO}^- \quad \quad \quad \quad \quad \quad \quad \text{OH}$$
$$\quad \quad || \quad \quad \quad + \text{H}_2\text{O} \quad \xrightleftharpoons{\text{Fumarase}} \quad \text{H}-\overset{|}{\text{C}}-\text{COO}^-$$
$$\quad {}^-\text{OOC}\diagup^{\text{C}}\diagdown\text{H} \quad \quad \quad \quad \quad \quad \quad \text{H}_2\text{C}-\text{COO}^-$$

Fumarate Malate

The malate so produced is dehydrogenated by **malate dehydrogenase**, an NAD^+ enzyme, and so the cycle is back to the starting point, oxaloacetate.

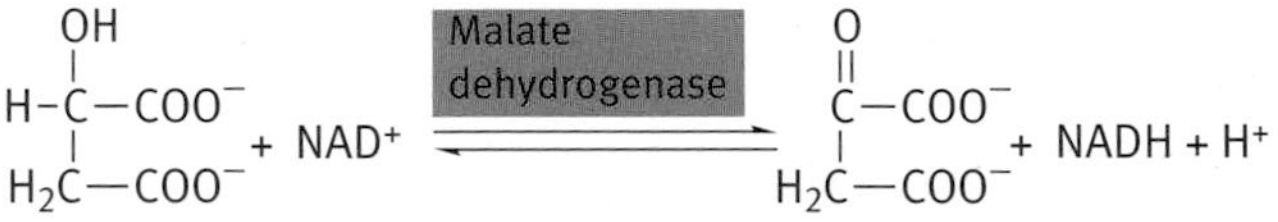

$$\text{OH} \quad \quad \quad \quad \quad \quad \quad \quad \quad \quad \quad \quad \quad \quad \text{O}$$
$$| \quad \quad \quad \quad \quad \quad \quad \text{Malate} \quad \quad \quad \quad || $$
$$\text{H}-\overset{|}{\text{C}}-\text{COO}^- \quad \quad \xrightleftharpoons{\text{dehydrogenase}} \quad \text{C}-\text{COO}^- \quad + \text{NADH} + \text{H}^+$$
$$\text{H}_2\overset{|}{\text{C}}-\text{COO}^- \quad + \text{NAD}^+ \quad \quad \quad \quad \quad \quad \text{H}_2\text{C}-\text{COO}^-$$

Malate Oxaloacetate

The $\Delta G^{0'}$ of this reaction is $+29.7$ kJ mol^{-1}, which is very unfavourable; the reaction proceeds because the next reaction, the conversion of oxaloacetate to citrate, is strongly exergonic and pulls the reaction over.

The complete cycle is shown in Fig. 8.13.

What determines the direction of the citric acid cycle?

The cycle operates in a unidirectional way as shown in Fig. 8.13. This is due to three of the reactions having a sufficiently large negative ΔG value to be irreversible. These are the synthesis of citrate from acetyl-CoA and oxaloacetate ($\Delta G^{0'} = -32.2$ kJ mol^{-1}), the decarboxylation of isocitrate to α-ketoglutarate ($\Delta G^{0'} = -20.9$ kJ mol^{-1}), and the α-ketoglutarate dehydrogenase reaction ($\Delta G^{0'} = -33.5$ kJ mol^{-1}). This results in the cycle operating in one direction even though the equilibrium of the malate dehydrogenase reaction is in favour of the reverse direction ($\Delta G^{0'} = +29.7$ kJ mol^{-1}). The overall operation of the cycle reactions has a negative ΔG value.

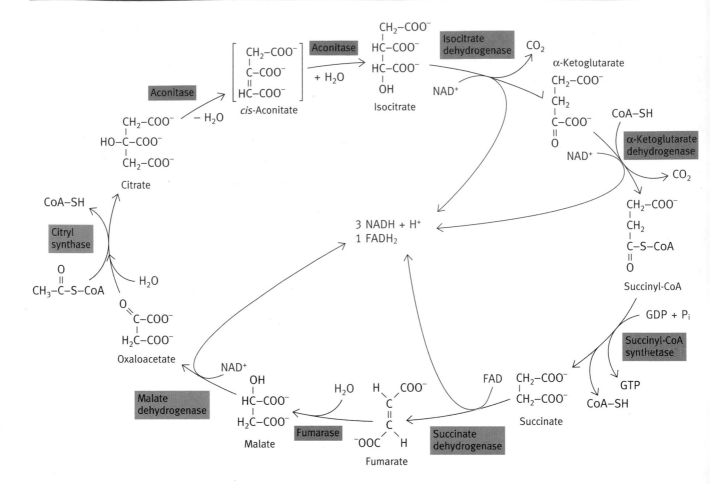

Fig. 8.13 Complete citric acid cycle. Red highlights the production of reducing equivalents from the cycle. Blue highlights the supply of the elements of H_2O to the cycle. (The conversion of citrate to isocitrate involves removal and addition of H_2O but there is no net gain.) Note that the FAD is not free but is attached to the succinate dehydrogenase protein. The involvement of water in the synthesis of citric acid is explained on page 117.

Stoichiometry of the cycle

The overall equation for the process is

$$CH_3CO-S-CoA + 2H_2O + 3NAD^+ + FAD + GDP + P_i \rightarrow$$
$$2CO_2 + 3NADH + 3H^+ + FADH_2 + CoA-SH + GTP;$$
$$\Delta G^{0'} = -40\,kJ\,mol^{-1}.$$

If you add up the hydrogen/oxygen atoms *as printed*, they don't balance on the two parts of the equation. The reason for the apparent shortfall of H_2O is explained on page 116.

Topping up the citric acid cycle

The cycle starts with oxaloacetate condensing with acetyl-CoA and ends with oxaloacetate so that the latter component is not used up. However, the cycle does not exist in isolation from the rest of metabolism and some of its acids are drawn off for other purposes. The cycle acids occupy a special place in metabolism in that they are not necessarily readily available from the diet in large amounts. Carbohydrates in the diet give rise to large

quantities of C_3 acid in the form of pyruvate production, and fats similarly provide large amounts of the C_2 (acetyl groups), but cycle acids (C_4, C_5, and C_6) are not similarly available in such quantity. It is true that certain amino acids can provide cycle acids but, by the same token, cycle acids are withdrawn to synthesize some amino acids (described in Chapter 15) and other metabolites. A complex equilibrium situation exists between metabolic systems. A means of topping up cycle acids to keep the energy-generating mitochondria reactions running properly is essential and there is such a provision in cells. An important reaction for this, called an **anaplerotic** or 'filling up' reaction is that of pyruvate plus CO_2 being converted to oxaloacetate, using energy from ATP hydrolysis.

$$ATP + \underset{CH_3}{\underset{|}{\overset{COO^-}{\overset{|}{C}}=O}} + HCO_3^- \longrightarrow \underset{H_2C-COO^-}{\underset{|}{\overset{O}{\overset{||}{C}-COO^-}}} + ADP + P_i + H^+$$

The enzyme is called **pyruvate carboxylase** (and quite different from pyruvate *de*carboxylase of yeast, please note). This is the crucial point at which C_3 acid can be converted to C_4 acid. Pyruvate carboxylase is an enzyme of central importance. It requires the vitamin, **biotin**, for its function. Wherever 'activated' CO_2 is needed for synthetic reactions catalysed by a group of carboxylase enzymes, biotin is the cofactor. Biotin is a water-soluble B group vitamin. It becomes covalently bound to its enzyme where it accepts a carboxy group from bicarbonate to form carboxybiotin, the reaction being thermodynamically driven by the conversion of ATP to ADP and P_i. Carboxybiotin is a reactive, but stable, form of CO_2 that can be transferred to another molecule that is to be carboxylated. The $\Delta G^{0'}$ for the cleavage of CO_2 from carboxybiotin is -19.7 kJ mol^{-1}. In this case pyruvate is the substrate, but other carboxylation enzyme systems are also biotin-dependent.

Pyruvate carboxylase has two catalytic sites—one to carboxylate the biotin and the other to transfer the carboxy group from biotin to pyruvate. (In some bacteria, the two activities reside on separate enzymes.) It is believed that the attachment of the biotin to the long lysyl side chain of the protein provides a flexible arm to permit the biotin to oscillate between the two sites (Fig. 8.14).

We will return to pyruvate carboxylase later, for it has metabolic importance other than its anaplerotic role for the citric acid cycle (page 151).

Stage 3—the electron transport chain that conveys electrons from NADH and FADH$_2$ to oxygen

Remember that what we are doing is to look at the three major phases involved in glucose (or glycogen) oxidation. Phase 1 was glycolysis, phase 2 was the citric acid cycle, and now we come to the final phase. Energy-wise we haven't achieved much yet—per starting molecule of glucose only a trivial yield of four ATP molecules—two from glycolysis and two from the cycle (via GTP), with one extra if a glucosyl unit of glycogen is the starting compound. Energy-wise, the other products per mole of starting glucose are 10 NADH, (two from glycolysis, two from pyruvate dehydrogenase, six from the cycle) and two FADH$_2$ from the cycle. (Don't forget that one molecule of glucose produces two pyruvate molecules and hence supports two turns of the cycle.)

The oxidation of the NADH and FADH$_2$ will produce most of the ATP (from ADP and P_i) generated by the oxidation of glucose.

The electron transport chain

For reasons that will become apparent, we will now discuss electron transport, pure and simple. Its purpose *is*, definitely, ATP production from ADP and P_i, but just for the moment forget all about phosphorylation of ADP to ATP and concentrate on electron movement to oxygen.

Where does it take place?

Electron transport carriers exist in or on the inner mitochondrial membrane. As we have already seen (Fig. 7.4), the inner membrane is folded into cristae, which increases the amount of inner membrane present, the density of cristae in a mitochondrion being related to the energy requirements of the cell.

Nature of the electron carriers in the chain

Heme is the prosthetic group of several electron carriers—called **cytochromes** because of their colour (red). The different cytochromes are called c_1, c, a, and a_3 (in order of their participation in the chain; the role of two b cytochromes is given later). The essentials of the heme structure are shown in Fig. 8.15, and its full structure in Fig. 27.3.

The important thing about the heme molecule is that, as the prosthetic group of the cytochrome electron carriers, the Fe atom oscillates between the Fe^{2+} and Fe^{3+} states as it accepts an electron from the preceding carrier or donates it to the next

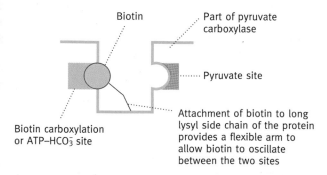

Fig. 8.14 Diagram of the role of biotin in the active site of enzymes catalysing carboxylation reactions. The example shown is pyruvate carboxylase.

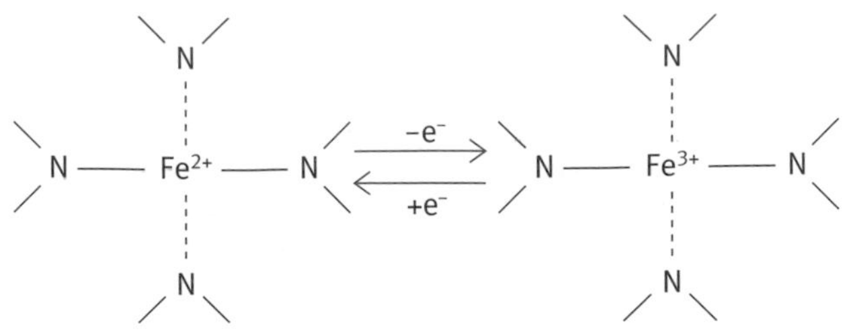

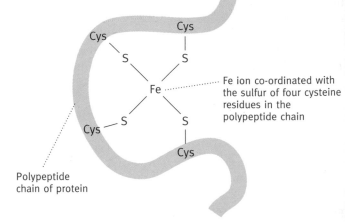

Fig. 8.15 Diagrammatic representation of heme (but take a look at the actual structure in Fig. 27.3).

Fig. 8.16 Diagram of iron–sulfur centre. There are several types of these, increasing in complexity and numbers of Fe and S atoms. The simplest form is shown here.

carrier in the chain. The characteristics of the heme molecule are modified by the specific protein to which it is attached, and variations in the heme side groups occur in different cytochromes and in their attachment to their apoproteins. Thus, it is not a contradiction that different cytochromes have different redox potentials and yet have heme as their prosthetic group.

Another type of electron carrier, based on iron, are the so-called non-heme iron proteins. In these, the iron is bound to the sulfhydryl side group of the amino acid cysteine of the protein and also to inorganic sulfide ions forming **iron–sulfur complexes**, or iron–sulfur centres. The simplest type is shown in Fig. 8.16. As with the cytochromes, the iron atom in these can accept and donate electrons in a cyclical fashion, oscillating between the ferrous and ferric state. Such iron–sulfur centres are associated with flavin enzymes. They accept electrons from FAD—enzymes such as succinate dehydrogenase and the dehydrogenase involved in fat oxidation (described in Chapter 9). Another type of carrier is an FMN-protein. FMN (flavin

adenine mononucleotide) consists of the flavin half of FAD (see page 97 (Ch. 7—Further study)). It carries electrons from NADH to an iron–sulfur centre. All of the iron–sulfur centres transfer electrons to ubiquinone (see below).

As well as these protein-bound electron carriers, there is one carrier not bound to a protein. This is a molecule, illustrated in Fig. 8.17, called **ubiquinone**, because it can exist as a quinone and is found ubiquitously. It is often referred to as CoQ, UQ, or Q. Q is an electron carrier because it can accept protons as shown in Fig. 8.17; it can exist as the free radical, semiquinone intermediate, thus permitting the molecule to hand over a single electron to the next carrier rather than a pair of electrons. The very long hydrophobic tail on the molecule (as many as 40 carbon atoms long in 10 isoprenoid groups; just have a look at

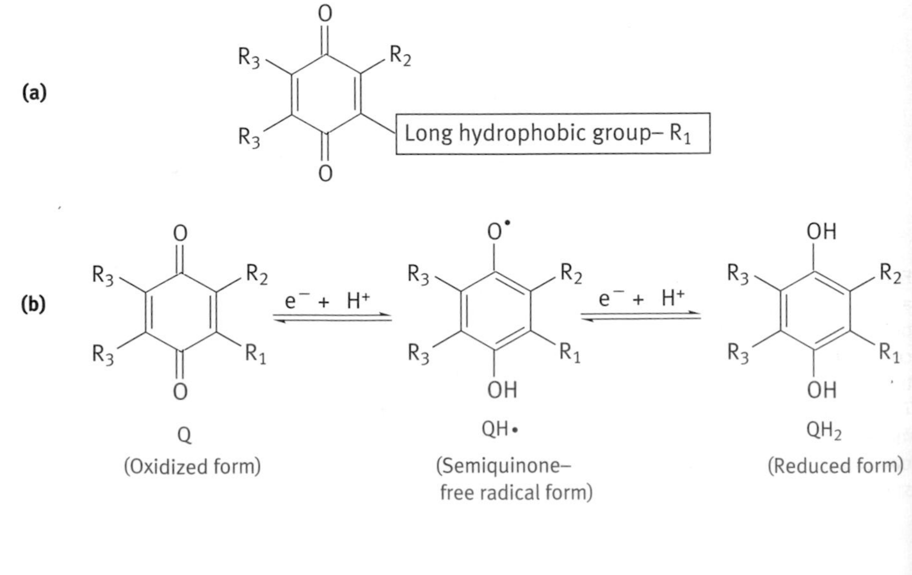

Fig. 8.17 (a) Diagram of ubiquinone or coenzyme Q structure (in the oxidized form). **(b)** Diagram of oxidized, semiquinone, and reduced forms of ubiquinone (Q). The semiquinone, QH·, can exist as the anion, Q·$^-$. R_1, Long hydrophobic group; R_2,–CH_3; R_3, –O–CH_3. The full structure of ubiquinone is given on page 123.

(a)

Long hydrophobic group– R_1

(b)

Q
(Oxidized form)

$e^- + H^+$

QH•
(Semiquinone–
free radical form)

$e^- + H^+$

QH_2
(Reduced form)

the structure below) make it freely soluble and mobile in the nonpolar interior of the inner mitochondrial membrane.

$(n = 5-9)$

Structure of ubiquinone (coenzyme Q)

To summarize, we have in the electron transport chain an FMN-protein, non-heme iron–sulfur proteins, Q not bound to a protein and freely mobile in the membrane, and heme proteins known as cytochromes. An important point is that one of the latter, cytochrome *c*, is a small water-soluble protein molecule (molecular weight ~ 12.5 kDa, just over 100 amino acids) that is loosely attached to the outside face of the inner mitochondrial membrane so that it also is free to move about. All the other proteins of the respiratory complexes are built into the membrane structure as integral proteins in fixed positions.

Arrangement of the electron carriers

In Chapter 7 (page 100) we discussed the redox potentials of electron acceptors and explained that electrons flow from a carrier of higher reducing potential (low redox potential) to one of lower reducing potential (more oxidizing, or higher redox potential). The electron carriers in the chain are of different redox potentials.

The redox potentials are directly related to $\Delta G^{0'}$ values, as already discussed on page 101. The electron carriers are arranged in the electron transport chain such that there is a continuous progression down the free-energy gradient (increasing redox potentials) with the corresponding release of free energy as the electrons move from one carrier to the next (Fig. 8.18). In considering glucose oxidation (the subject of this chapter) the task in this phase is to transfer electrons from NADH and FADH₂ to oxygen. The whole scheme involves a somewhat formidable list of steps but the carriers can be grouped into the four complexes shown in Fig. 8.19. These complexes are built into the structure of the inner mitochondrial membrane, interconnected by the mobile electron carriers, ubiquinone and cytochrome *c*. Ubiquinone takes electrons from complexes I and II and delivers them to complex III. Cytochrome *c* is the intermediary between complexes III and IV. Complex I carries electrons from NADH to Q. Complex II carries electrons from succinate via FADH₂ to Q; complex III uses QH₂ to reduce cytochrome *c*. Complex IV transfers electrons from cytochrome *c* to oxygen. Complexes I, III, and IV are, for convenience, referred to as NADH:Q reductase, QH₂:cytochrome *c* reductase, and cytochrome oxidase, respectively. Complex IV, cytochrome oxidase, is a multisubunit structure; electrons are donated to it by cytochrome *c* on the outer face of the inner mitochondrial membrane. The electrons are transported via two cytochromes, *a* and *a₃*, to oxygen. Cytochromes *a* and *a₃* have copper sites associated with them that oscillate between the Cu^+ and Cu^{2+} states. The final reduction catalysed by cytochrome oxidase is

$$O_2 + 4e^- + 4H^+ \rightarrow 2H_2O.$$

As explained later (Chapter 17), it is vital that the mechanism of cytochrome oxidase ensures that oxygen atoms are fully reduced by four electrons to form water without releasing any intermediate partially reduced states. These would be dangerous reactive free radicals or hydrogen peroxide.

How is the free energy released by electron transport used to form ATP?

If you remember how ATP is generated in glycolysis, you may,

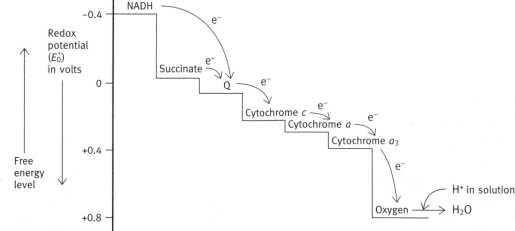

Fig. 8.18 Diagram showing the approximate relative redox potentials of some of the main components of the electron transport system in mitochondria. The arrows indicate electron movements. The role of cytochrome *b* components is shown in Fig. 8.22.

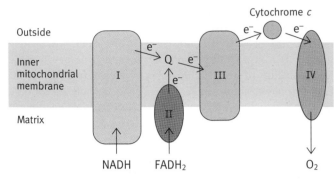

Fig. 8.19 Diagram of the electron transport chain with electron carriers grouped into four main complexes. Complex I, NADH–Q reductase; complex II, succinate–Q reductase; complex III, QH$_2$–cytochrome c reductase; complex IV, cytochrome oxidase. Q, ubiquinone or coenzyme Q. FADH$_2$ is generated in the cycle from succinate by succinate dehydrogenase. Note that the complexes are located in the inner mitochondrial membrane. Coenzyme Q and cytochrome c are mobile carriers capable of physically transporting electrons from one site in the membrane to another. Cytochrome c is surface-located. Note that FADH$_2$ exists attached to flavoprotein enzymes. The major ones are succinate dehydrogenase and fatty acyl-CoA dehydrogenases involved in fat oxidation; the latter is described in Chapter 9.

in fact should, be puzzled. In glycolysis there is substrate-level phosphorylation (page 111). In this, the phosphorylation (generation of ATP from ADP and P$_i$) is inseparably linked to the reactions in glycolysis. Either the reactions occur, involving ATP synthesis, or they don't occur. You can't have the relevant reactions without ATP generation since this is an intrinsic component of the reactions. The same is true of ATP generation in the citric acid cycle via GTP.

By contrast, we have described the electron transport reactions without even mentioning ATP production. The purpose of electron transport is to generate ATP but electron transport to oxygen proceeds very readily in broken up mitochondria without any ATP generation at all. This situation perplexed biochemists for decades. If electron transport generates ATP in the cell, why can it readily occur without this in broken up mitochondria? In short, in cells, or in intact 'healthy' mitochondria, electron transport results in ATP generation but, in broken mitochondria, electron transport occurs very happily without it. It conflicted with the known type of substrate-level phosphorylation that was known from glycolysis. The solution was discovered by the English biochemist, Peter Mitchell, who, in a private laboratory, in 1961 produced a concept of how electron transport causes ATP synthesis that is so novel that it was at first hardly taken seriously by most, and it took a long time for it to be accepted (and for a Nobel Prize to be awarded to Mitchell in 1978). The concept is based on the simple notion that gradients have the ability to do work. A gradient of water pressure can be used to generate electricity, a gradient of air pressure to drive a windmill, etc. A chemical gradient is no different. Molecules or ions will migrate from a high concentration to a low

concentration and, if a suitable energy-harnessing device is interposed, useful work can be done.

Applying this concept, two things are needed for ATP generation coupled to electron transport: firstly, electron transport must create a gradient of some sort; and, secondly, the gradient must be allowed to flow back through a device that uses the energy of the gradient to synthesize ATP from ADP and P$_i$. Mitchell's concept thirdly required the existence of intact vesicles across whose membrane a gradient could be established. It has to have an inside and an outside, as separate compartments, explaining why broken mitochondria don't make ATP. The membrane must therefore be impermeable to the solute of which the gradient consists.

Mitchell discovered that electron flow caused protons to be ejected from inside the mitochondrion to the outside, thus creating a proton gradient across the membrane; in other words, the pH of the external solution fell. A membrane (or charge) potential, negative inside and positive outside, is also generated by the proton expulsion and also contributes to the total energy gradient or **proton-motive force** available for ATP synthesis. The inner mitochondrial membrane itself is virtually impermeable to protons. This is a prerequisite for the system, but inserted itno the membrane are special proton-conducting channels. Protons flow from the outside through these channels back into the mitochondrial matrix and the energy of this flow is harnessed to the formation of ATP from ADP and P$_i$.

The proton-conducting channels are knob-like structures that completely cover the inner surface of the cristae and are made up of several proteins (Fig. 8.20). These are, in fact, ATP-synthase complexes that convert ADP and P$_i$ to ATP, the process being energetically driven by the proton flow. The complex is built into the membrane by the F$_0$ unit (Fig. 8.20), while the actual ATP synthesis occurs in the F$_1$ unit, both units containing multiple protein molecules. There is evidence that the actual proton-conducting channels consist of strategically placed charged amino acid residues in proteins and the protons hop from one to another.

The overall scheme in mitochondria is shown in Fig. 8.21. This scheme is known as the **chemiosmotic mechanism**. Two questions can be asked,

1 How does the flow of electrons from NADH and FADH$_2$ to O$_2$ cause protons to be pumped from the matrix side of the inner mitochondrial membrane to outside the membrane?

2 How does the flow of protons into the mitochondrion drive the synthesis of ATP from ADP and P$_i$?

How are protons ejected?

The mechanism by which protons are translocated as a result of electron flow is uncertain for complexes I and IV but is established in the case of complex III. The *principle* of Mitchell's original idea is as simple as it is ingenious. In essence, hydrogen *atoms* are assembled on the matrix side of the inner

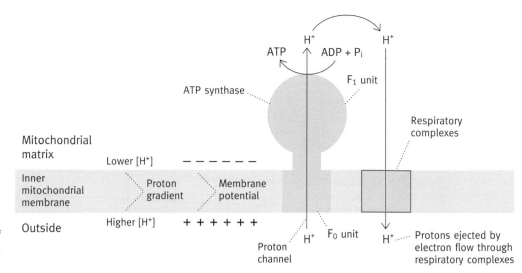

Fig. 8.20 The ATP-generating system of the inner mitochondrial membrane. The F$_1$ unit consists of nine subunits. The F$_0$ unit is also a multisubunit structure.

mitochondrial membrane, using protons from the matrix and electrons from the transport chain. The atoms are assembled on ubiquinone (Q), forming the reduced form, QH$_2$, which now diffuses to the opposite face of the membrane where the reverse happens—electrons are stripped off the hydrogen atoms and the resultant protons escape to the outside. The essentials of the process are the positioning of the responsible catalytic proteins on opposite sides of the membrane and a mobile carrier to transport the hydrogen atoms across the membrane from one face to another. It is called the **Q cycle**.

That is the principle for proton transport in complex III. The actual mechanism is shown in Fig. 8.22; it looks complicated but is in fact very simple with few reactions involved. Note first of all

that the red arrows simply represent physical movement of ubiquinone and its derivatives; the second point is that, in effect, two distinct processes are going on, both of which eject a pair of protons. In the membrane there are pools of Q and QH$_2$. Let us

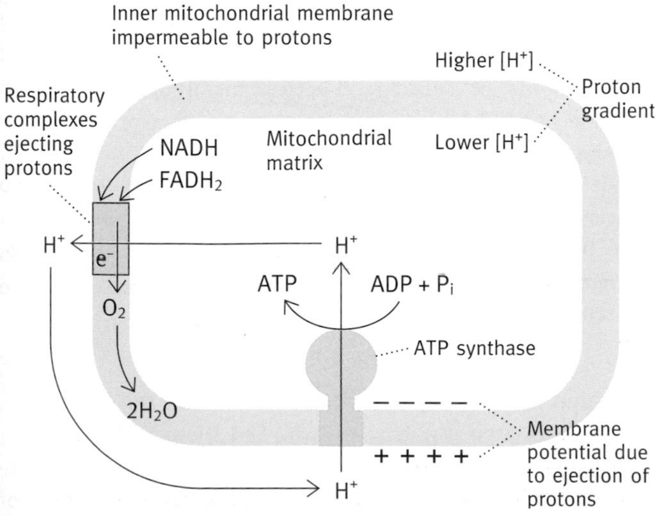

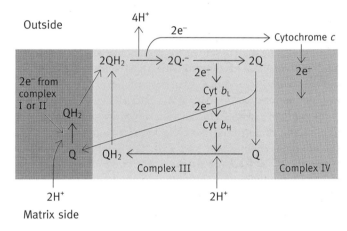

Fig. 8.22 Diagram of the mechanism of proton translocation by complex III as a result of electron transport in mitochondria. The red arrows represent physical diffusion of components rather than chemical transformations; the latter are indicated by black arrows and electron transport by blue arrows. (However, note that the blue arrow representing electron transport by cytochrome c is effected by the physical diffusion of the latter from complex III to complex IV and its return after oxidation.) Cytochrome b protein spans the membrane so that electrons are transferred from the outer face to the inner face. The molecules of Q and QH$_2$ are in equilibrium with a membrane pool of these, but this is not shown in order to simplify the diagram arrangement. The electrons from complex I (as QH$_2$) arise mainly from NADH generated in glycolysis and the citric acid cycle; those from complex II come from the succinate → fumarate step of the cycle via an FAD-protein. As will be described in the next chapter, electrons from fatty acid oxidation also enter complex III via complexes I and II. Q, Ubiquinone; Q·⁻, Semiquinone anion; Cyt, cytochrome. Cyt b_L and b_H refer to heme sites on cytochrome b of low and high redox potentials respectively.

Fig. 8.21 Generation of ATP in mitochondria by the chemiosmotic mechanism. Note that FADH$_2$, produced by the dehydrogenation of fatty acids (described in Chapter 9), enters the same pathway as that utilized by the oxidation of succinate.

start with QH_2, arriving from complexes I and II, and whose electrons originate from NADH and $FADH_2$, respectively. In the reduction of Q to QH_2 by complexes I and II, two protons are taken up from the matrix as shown to the left of the diagram. The QH_2 now migrates to a site, in complex III, *on the external face of the membrane*, where one electron is removed and passed on to reduce cytochrome *c*, which transports it to complex IV. Remember that cytochrome *c* is also mobile. The site on complex IV that accepts electrons from cytochrome *c* is exposed on the external face of the inner membrane and cytochrome *c* is also located on this face of the membrane. Two protons are ejected, leaving the half-oxidized quinone anion, $Q\cdot^-$. A further electron is now removed from the latter, but in this case the electron is handed on, not to cytochrome *c* but to cytochrome *b*, to which we'll return shortly. That is the end of that half of the story—a molecule of QH_2 from complexes I/II has been oxidized, two electrons passed (one to cytochrome *c* and one to cytochrome *b*), and two protons ejected from the matrix to the outside. The Q so formed now returns to the general pool. There is another part to the story. A *second* molecule of QH_2 is oxidized in the same way as the first, resulting in the ejection of two more protons. From the two molecules of QH_2 we thus have two electrons passing on to complex IV via cytochrome *c* and two to cytochrome *b*. The latter transfers the electrons to a different heme site on the same protein whose redox potential is greater (lower energy) and in this way transports the two electrons to the matrix side of the membrane. They are donated here to a molecule of Q and, with a pair of protons from the matrix, hydrogen atoms on Q are assembled as QH_2. The QH_2 now migrates back to the site of the external face and the merry-go-round starts again. (The diagram shows the same molecule of ubiquinone going round the cycle, for clarity, but molecules of Q and QH_2 will enter and exit the pool in a dynamic equilibrium—it amounts to the same thing.) This somewhat convoluted process actually oxidizes only one molecule of QH_2 but achieves the ejection of four protons. The *net* effect of the reactions *within complex III* can be summarized in the equation

$$QH_2 + 2H^+ \text{ (matrix)} + 2 \text{ cyt } c \text{ (Fe}^{3+}) \rightarrow$$
$$Q + 4H^+ \text{(outside)} + 2 \text{ cyt } c \text{ (Fe}^2).$$

Figure 8.23 illustrates the fact that the energy for proton pumping is derived from the free energy released as electrons are transported down the redox potential gradient.

How is ATP generated from proton flow?

The F_1 knob has an ATPase catalytic function in the test tube—it hydrolyses ATP to ADP and P_i but, in the mitochondrion, it works in the direction of ATP synthesis. The question is how the proton flow makes the reaction, $ADP + P_i \rightarrow ATP + H_2O$, possible. In the synthesis of ATP from ADP and P_i by the F_1 unit, there is no formation of an intermediate phosphoryl or ADP-group covalently bound to the protein. The condensation is a direct

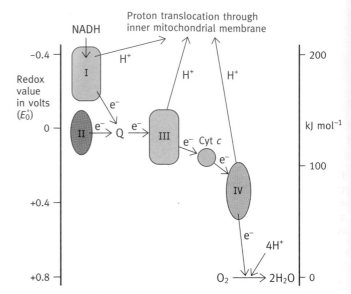

Fig. 8.23 Approximate positions on the redox potential scale of the electron transport complexes. The scale on the right gives the approximate free energy released when a pair of electrons is transferred from components to oxygen. Q, ubiquinone.

one between the two molecules. It appears that, when ADP and P_i are bound to the catalytic site, the formation of ATP *also bound to the catalytic site* involves little free-energy change but the release of that ATP is an energy-requiring step. It is believed that this is achieved by a conformational change in the protein, driven by the proton flow. The energy of the latter is thus harnessed into an energy-requiring conformational change that releases the ATP. It should be noted that the ease of formation of ATP on the enzyme surface does not negate the fact that conversion of ADP and P_i to ATP *in solution* requires the expected input of energy, suplied by proton flow.

By now you are used to the concept of ATP hydrolysis driving the transport of substances across membranes. The ATP synthase of mitochondria can be regarded as a proton pump working in reverse.

There is an awesome and majestic simplicity about the concept that, for billions of years, all aerobic life forms on earth have been driven by the creation of a small pH and charge gradient across a lipid bilayer membrane. It is one of the great concepts in biology.

Transport of ADP into mitochondria and ATP out

Most of the ATP synthesis in most eukaryote cells occurs in mitochondria while most of the ATP is used outside of the mitochondria. Hence ADP and P_i must enter the mitochondrion and ATP move out. The highly charged molecules cannot diffuse passively across the inner mitochondrial membrane and hence special transport mechanisms exist. **ATP–ADP translocase** exchanges ATP inside the mitochondrion for an ADP

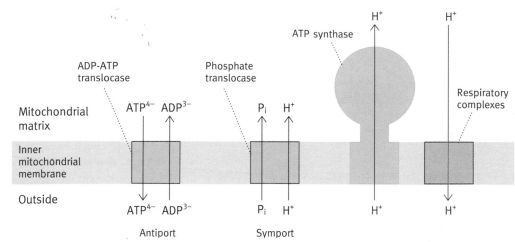

Fig. 8.24 Diagram of transmembrane traffic in mitochondria involved in ATP generation. All of the traffic is via specific transport proteins.

outside the mitochondrion (Fig. 8.24).

Where does the energy for this ATP:ADP exchange come from? As already explained, electron transport generates not only a pH gradient across the inner mitochondrial membrane but also a membrane potential, positive outside and negative inside, due to the ejection of H$^+$ ions. ATP carries four negative charges out, while ADP carries only three in. Thus the ATP:ADP exchange tends to neutralize this membrane potential or, in other words, the membrane potential drives the exchange. About 25% of the total energy yield of electron transfer is absorbed in the process. The transport of the P$_i$ needed along with ADP for ATP generation is catalysed by a phosphate translocase in the mitochondrial membrane which is also driven by the proton gradient (Fig. 8.24).

The balance sheet of ATP production by electron transport

It requires three protons to flow through the ATP synthase to generate one molecule of ATP from ADP and P$_i$, assuming that the latter two are already inside the mitochondrion. The transport of a molecule of ADP into the mitochondrion and that of one of ATP to the cytoplasm requires the energy equivalent of one proton entering the mitochondrion, since maintenance of the charge that drives the exchange requires the extra proton pumping. Hence four protons have to be pumped out of the matrix to drive the production of one molecule of ATP made available in the cytoplasm of the cell. For each pair of electrons transported from NADH to oxygen, the consensus is that 10 protons are pumped out of the mitochondrial matrix. Thus the oxidation of one molecule of NADH (located inside the mitochondrion) will produce 2.5 molecules of ATP. For the oxidation of one molecule of FADH$_2$ (that is, a pair of electrons from succinate), the value is 1.5. (Earlier estimates were 3 and 2, respectively.) The values are known as **P/O ratios** since a pair of electrons reduce one atom of oxygen.

The molecules of NADH produced in glycolysis, located in

the cytoplasm, require separate consideration; these, you will recall, donate their pairs of electrons either to NAD$^+$ located inside the mitochondrion or to mitochondrial FAD, depending on which shuttle mechanism (page 112) the cell happens to be using. Thus, a *cytoplasmic molecule* of NADH may give rise to either 2.5 or 1.5 molecules of ATP.

Yield of ATP from the oxidation of a molecule of glucose to CO$_2$ and H$_2$O

Starting from free glucose rather than glycogen, the net yield of ATP from the complete oxidation of the molecule is either 30 or 32 depending on which shuttle is used for the cytoplasmic NADH as already described above. To summarize: two from glycolysis at the substrate level (remember that, although four ATP molecules are generated in glycolysis, two were used at the start so the net gain is two). In the citric acid cycle, two molecules of ATP (via GTP in animals) are produced per molecule of glucose at the succinyl-CoA stage (one per turn of the cycle but two acetyl-CoA molecules are produced per glucose). Thus we have four molecules of ATP produced at the substrate level; all the rest come from electron transport.

Glycolysis produces, per molecule of glucose, two molecules of NADH, which are located in the cytoplasm. These will give rise either to a total of five or three molecules of ATP depending on the shuttle used. Per molecule of glucose, pyruvate dehydrogenase produces two molecules of NADH and the citric acid cycle, six. Oxidation of these will produce 20 molecules of ATP. Oxidation of the FADH$_2$ generated from the succinate → fumarate step produces a further three ATP molecules.

The total is therefore 2+5 (or 3) +2+20+3=32 (with the aspartic–malate shuttle used) or 30 (with the glycerol-3-phosphate shuttle used). Note that these values are estimates—with substrate-level phosphorylation, ATP generation is always a whole number, but there is no such whole-number relationship between proton ejection and ATP generation.

An *Escherichia coli* cell is equivalent in this context to a mitochondrion, the cell membrane equating to the inner mitochondrial membrane and the bacterial cytoplasm to the mitochondrial matrix. In such cells there is no need for shuttle mechanisms to transport NADH electrons to the respiratory pathway. In *E. coli* there is no transport of ATP and ADP needed into the cytoplasm. The ATP yield from the oxidation of a molecule of glucose in *E. coli* is therefore greater.

Is ATP production the only use that is made of the potential energy in the proton-motive force?

The answer is almost, but not quite. In newborn babies, heat production to maintain body temperature is helped by brown fat cells—brown because they are rich in mitochondria that contain the coloured cytochromes. The generation of ATP in mitochondria is dependent on the inner mitochondrial membranes being impermeable to protons, thus forcing the latter to enter the mitochondrial matrix only via the ATP-generating channels. If you made a hole in the membrane the protons would simply flood through it, effectively acting as a short circuit, no ATP would be generated, and the energy would be liberated as heat (which is why broken mitochondria cannot synthesize ATP). In brown fat cell mitochondria, this is essentially what happens, channels permitting non-productive (that is, no ATP synthesis) proton flow being made by a special protein, **thermogenin**. Chemicals that transport protons unproductively through membranes (dinitrophenol is the classical one) also 'uncouple' oxidation from ATP-generation.

Bacteria also generate energy by pumping protons across the membrane to the outside of the cell and generating ATP as described by reversed proton flow (the whole bacterial cell, as mentioned, in this regard, is equivalent to a mitochondrion). However, the proton gradient is also used in uptake of solutes into the cell, the H^+ gradient being used for cotransport (lactose uptake is an example) just as the Na^+ gradient is used in animal cells (page 51). Remarkably also, the cilia of bacteria are rotated by a flow of protons through the protein machinery that rotates the cilium; it runs on 'proticity' rather than the electricity used by an electric motor.

Further reading

Glycolysis

Boiteux, A., and Hess, B. (1981). Design of glycolysis. *Phil. Trans. Roy. Soc. Series B, London.*, **293**, 5–22.
An extensive general review of the pathway and its control.

Electron transport chain

Boyer, P. D., Chance, B., Ernster, L., Mitchell, P., Racker, E., and Slater, E. C. (1977). Oxidative phosphorylation and photophosphorylation. *Ann. Rev. Biochem.*, **46**, 955–1026.
An unusual article in which the leaders in the field wrote separate independent sections. Although dated in detail, it is worth reading for the general principles given, and for its historical interest in this central field of ATP generation.

Slater, E. C. (1983). The Q cycle, an ubiquitous mechanism of electron transport. *Trends Biochem. Sci.*, **8**, 239–42.
Describes the proton pumping system present in both mitochondria and chloroplast.

Mitchell, P. (1984). Chemiosmosis: A term of abuse. *Trends Biochem. Sci.*, **9**, 205.
An explanation by the discoverer of the chemiosmotic mechanism of why chemiosmosis is not an appropriate noun for the chemiosmotic process.

Babcock, G. T. and Wikstrom, M. (1992). Oxygen activation and the conservation of energy in cell respiration. *Nature*, **356**, 301–9.
In depth review of the cytochrome oxidases and associated proton pumping.

Capaldi, R. A., Aggeler, R., Turina, P., and Wilkens, S. (1994). Coupling between catalytic sites and the proton channel in F_1F_0-type ATPases. *Trends Biochem. Sci.*, **19**, 284–9.
Review of structural studies on the complex which synthesizes ATP from ADP and P_i, driven by proton flow.

Problems for Chapter 8

1 Explain the chemical rationale for the phosphohexose isomerase reaction.

2 What is meant by substrate-level phosphorylation? Give an example of such a system.

3 The reaction for which the enzyme pyruvate kinase is named never occurs. Discuss this.

4 How many ATP molecules are generated in glycolysis, from:
  (a) glucose;
  (b) a glycosidic unit of glycogen.

5 In calculating how many molecules of ATP are produced as a result of the oxidation of cytoplasmic NADH, we cannot be sure of the answer in eukaryotes. Why is this so?

6 In the citric acid cycle, one acetyl group is disposed of; the products are nine reducing equivalents (effectively nine hydrogen atoms) as NADH and $FADH_2$, and two molecules of CO_2. The acetyl group supplies three hydrogen atoms, and one oxygen atom, and two H_2O molecules enter the cycle. The total input is therefore seven hydrogen atoms, and three oxygen atoms. This leaves a deficit of the elements of H_2O. Where does this come from?

7 What is the chemical rationale for isocitrate being oxidized before loss of CO_2 occurs?

8 Explain how the citric acid cycle acids can be 'topped up' by an anaplerotic reaction.

9 What is the cofactor involved in carboxylation reactions? Explain how it works.

10 Outline the arrangement of respiratory complexes in the electron transport chain.

11 What characteristic do ubiquinone and cytochrome c have in common? What is their physical location in the cell?

12 What is the immediate role of electron transfer in the respiratory chain?

13 It is stated in the text that the yield of ATP from the complete oxidation of a molecule of glucose in eukaryote cells is either 30 or 32 molecules; in the case of *E. coli* it is stated that the yield is greater than this. Why does this difference in statements exist?

Chapter 9

Energy production from fat

In the previous sections you have seen how energy in the form of ATP is produced in the cell by the oxidation of glucose or glycogen.

Fat is another major source of energy for ATP production. It provides perhaps half of the total energy needs of heart and resting skeletal muscles. By far the largest amount of stored energy occurs as fat for, unlike the situation with glycogen, there appears to be no limit to the amount of neutral fat that can be stored in the body in the adipose cells. It is more highly reduced than carbohydrate and not hydrated as is glycogen, and therefore is a more concentrated energy store.

The subject of fat oxidation and concomitant ATP production is greatly simplified because fat oxidation involves the same citric acid cycle and electron transport system as in glucose oxidation. As already described, the two systems (glucose oxidation and fat oxidation) converge at acetyl-CoA so all we are mainly concerned with in fat oxidation is the relatively simple task of chopping up fatty acids by removing two carbon atoms at a time as acetyl-CoA. This involves NAD^+ and FAD reduction which are oxidized by the same pathways we have already discussed. The remarkable efficiency of using the same machinery for obtaining energy from all classes of foodstuffs should be noted. It is only the preliminary reactions for feeding the material into the cycle that vary between the foods being oxidized.

A few simple points first.

1. Before oxidation can occur, free fatty acids must be released by hydrolysis. This also produces a molecule of glycerol per TAG hydrolysed and this is metabolized separately. The glycerol is manipulated to enter metabolism in the glycolysis pathway—it is phosphorylated and oxidized to give dihydroxyacetone phosphate of glycolysis fame. Thus, in starvation, the glycerol moiety *can* give rise in the liver to glucose by the gluconeogenesis pathway (see Chapter 11).

2. Free fatty acids for oxidation are obtained by peripheral tissues from that released into the blood by adipose cells when the glucagon level is high. They also enter the cells as a result of lipoprotein lipase attack on chylomicrons or VLDL (page 88).

3. Free fatty acids from adipose cells are carried as ionized molecules attached in a freely reversible manner to serum albumin. They readily diffuse into cells so that the amount entering cells increases as their blood level rises as a result of release from adipose cells. As the fatty acids are taken up by cells, more will dissociate from the serum albumin carrier protein.

4. Fatty acids are broken down by removing two carbon atoms at a time as acetyl-CoA.

5. During conversion to acetyl-CoA the fatty acid is always in the form of an acyl-CoA. The first stage in oxidation of fatty acids is always to convert them to the fatty acyl-CoA compounds, a reaction known as fatty acid activation.

Mechanism of acetyl-CoA formation from fatty acids

'Activation' of fatty acids by formation of fatty acyl-CoA derivatives

The term 'activation' of a carboxylic acid refers to the fact that the thiol ester is a high-energy (or reactive) compound. The **activation reaction** is

$$R\,COO^- + ATP + CoA{-}SH \rightarrow R\,CO{-}S{-}CoA + AMP + PP_i;$$

$$\Delta G^{0'} = -0.9 \text{ kJ mol}^{-1}.$$

The free-energy change of this reaction is small (because of the high energy of the thiol ester) but hydrolysis of the PP_i by the ubiquitous enzyme, inorganic pyrophosphatase, makes the overall process strongly exergonic and irreversible ($\Delta G^{0'} = -32.5$ kJ mol^{-1}). (See page 12 if you've forgotten this point.)

There are three different fatty acid activating enzymes specific for short, medium, and long chain acids, respectively—called **fatty acyl-CoA synthetases**.

Transport of fatty acyl-CoA derivatives into mitochondria

Activation of fatty acids occurs on the outer mitochondrial membrane, which is effectively the same compartment as is the cytoplasm, but their conversion to acetyl-CoA occurs only in the mitochondrial matrix. The acyl group of fatty acyl-CoA is carried in, without the CoA, by a special transport mechanism and then handed over to CoASH *inside* the mitochondrion to become fatty acyl-CoA again. The high-energy nature of the acyl-bond is preserved during the transport—otherwise it could not re-form fatty acyl-CoA inside the mitochondrion without further energy expenditure. To achieve this, outside the mitochondrion, the acyl group is transferred to a rather odd hydroxylated molecule, **carnitine.**

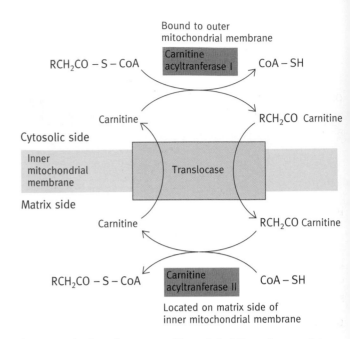

Fig. 9.1 Mechanism of transport of long chain fatty acyl groups into mitochondria where they are oxidized in the mitochondrial matrix. The acyl–carnitine bond is an unusual ester bond in that it has a high group transfer potential—the compound is of the high-energy type so that exchange of carnitine for CoASH inside the mitochondrion occurs without need for energy input. See text for structures.

$$CH_3 \overset{+}{\underset{\underset{CH_3}{|}}{N}}-CH_2-\overset{\underset{OH}{|}}{\underset{|}{C}}-CH_2-C\overset{O}{\underset{O^-}{\diagup}}$$

Carnitine

Although a carboxylic ester is usually of the low-energy type, the structure of carnitine is such that the fatty acyl–carnitine bond is of the high-energy type—the acyl group has a high group transfer potential. Presumably this is a reason for the evolution of carnitine as the carrier molecule in this transport system. The fatty acyl-carnitine so formed is transported into the mitochondrial matrix where the reverse reaction occurs—carnitine is exchanged for CoASH and the free carnitine is transported back to the cytoplasm to collect another fatty acyl group.

$$CH_3 \overset{+}{\underset{\underset{CH_3}{|}}{N}}-CH_2-\overset{\underset{O}{|}}{\underset{|}{C}}-CH_2-C\overset{O}{\underset{O^-}{\diagup}}$$
$$\underset{R}{\overset{|}{\underset{|}{C=O}}}$$

Fatty acyl-carnitine

The scheme is shown in Fig. 9.1.

The acyl transfer from fatty acyl-CoA to carnitine is catalysed on the cytoplasmic side of the mitochondrion by an enzyme called **carnitine acyltransferase I** and that on the matrix side by **carnitine acyltransferase II**. Genetic defects are known in which there is a carnitine deficiency or deficiency in the carnitine acyltransferase. In some cases this can manifest itself as muscle pain and abnormal fat accumulation in muscles.

Conversion of fatty acyl-CoA to acetyl-CoA molecules inside the mitochondrion

There are four separate reactions in this process, three of which are analogous in reaction types to those converting succinate to oxaloacetate in the citric acid cycle—namely, dehydrogenation by an FAD enzyme, hydration, and an NAD^+-dependent dehydrogenation. We suggest that you refresh your memory on these by having a quick look on page 120 at the reaction sequence, succinate → fumarate → malate → oxaloacetate. The corresponding reactions on fatty acyl-CoA derivatives are shown in Fig. 9.2. In fat oxidation, the keto acyl-CoA is cleaved by CoASH, splitting off two carbon atoms as acetyl-CoA but forming a shorter fatty acyl-CoA derivative. Because the molecule is split by the —SH group of CoASH, the enzyme is called a **thiolase**. The thiolase reaction preserves the free energy as thiol ester of fatty acyl-CoA. When the fatty acyl group has been shortened, by seven successive rounds of acetyl-CoA production, to the C_4 stage (butyryl-CoA) the next round of reactions produces acetoacetyl-CoA, which is finally split by CoASH into two molecules of acetyl-CoA. A specific thiolase in mitochondria performs this reaction.

$$CH_3\,CO\,CH_2CO-S-CoA + CoA-SH \rightarrow 2CH_3\,CO-S-CoA$$
$$\qquad\text{Acetoacetyl-CoA} \qquad\qquad\qquad \text{Acetyl-CoA}$$

The reaction sequence in each round involves conversion of saturated fatty acyl-CoAs to β-ketoacyl-CoAs. The process is therefore referred to as **β-oxidation of fatty acids**. The NADH and the $FADH_2$ (the latter on the acyl-CoA dehydrogenase) feed

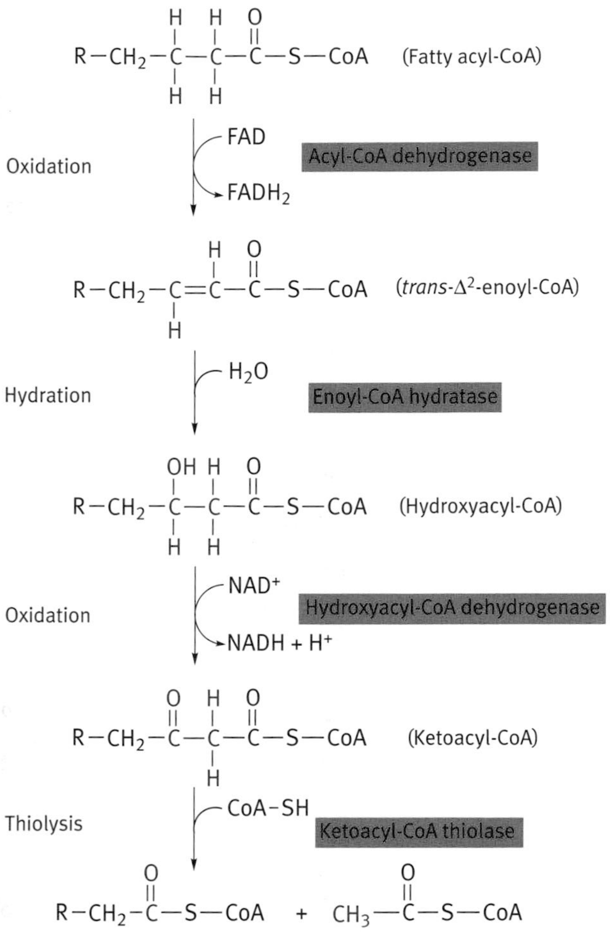

R—CH$_2$—C—C—C—S—CoA   (Fatty acyl-CoA)

Oxidation

FAD

| Acyl-CoA dehydrogenase |

FADH$_2$

R—CH$_2$—C=C—C—S—CoA   (*trans*-Δ^2-enoyl-CoA)

Hydration

H$_2$O

| Enoyl-CoA hydratase |

R—CH$_2$—C—C—C—S—CoA   (Hydroxyacyl-CoA)

Oxidation

NAD$^+$

| Hydroxyacyl-CoA dehydrogenase |

NADH + H$^+$

R—CH$_2$—C—C—C—S—CoA   (Ketoacyl-CoA)

Thiolysis

CoA–SH

| Ketoacyl-CoA thiolase |

R—CH$_2$—C—S—CoA   +   CH$_3$—C—S—CoA

Fig. 9.2 Diagram showing one round of four reactions by which a fatty acyl-CoA is shortened by two carbon atoms with the production of a molecule of acetyl-CoA. Note the similarity of reaction types in the desaturation, hydration, and ketoacyl formation with the succinate → fumarate → malate → oxaloacetate steps of the citric acid cycle.

electrons into the electron transport chain exactly as already described (see Fig. 8.19 for a summary of this).

Energy yield from fatty acid oxidation

A molecule of palmitic acid (C$_{16}$) is converted to eight molecules of acetyl-CoA and in this process we generate seven FADH$_2$ molecules and seven NADH molecules. The acetyl-CoA is oxidized by the citric acid cycle and the NADH and FADH$_2$ by the electron transport chain (Chapter 8). NADH oxidation generates 2.5 ATP molecules and FADH$_2$, 1.5 ATP molecules. (Since the NADH is generated in the mitochondrial matrix, it does not have to be carried in by a shuttle mechanism.)

If you count it all up (not forgetting the one GTP per acetyl-CoA from the citric acid cycle) the oxidation of one mole of palmitic acid generates 106 moles of ATP from ADP and P$_i$

(allowing for the two ATP consumed in formation of palmitoyl-CoA). This represents about a 33% efficiency in trapping the free energy in usable form. Put in another way, the oxidation of 256 g of palmitic acid would produce 45 kg of ATP from ADP and P$_i$. There isn't anywhere near that amount of ATP in the body, of course, for ATP recycles rapidly through ADP and P$_i$.

Oxidation of unsaturated fat

Olive oil is an oil rather than a solid because of its high content of **monounsaturated fat** (see page 45 for explanation of this). For example, palmitoleic acid has a double bond between carbon atoms 9 and 10. So far as oxidation goes, this is treated by the cell in exactly the same way as palmitic acid for three rounds of β-oxidation. At this point the product is *cis*-Δ^3-enoyl-CoA

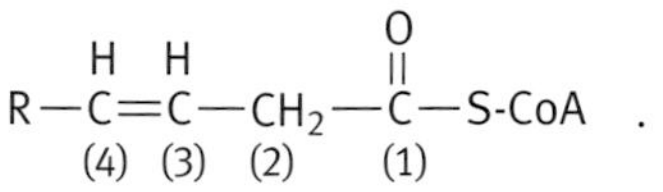

R—C=C—CH$_2$—C—S-CoA .
(4) (3) (2)   (1)

The double bond in this position prevents the acyl-CoA dehydrogenase from forming a double bond between carbon atoms 2 and 3, as is required in β-oxidation of a saturated acyl-CoA.

An extra isomerase enzyme takes care of this by shifting the existing double bond into the required 2–3 position; it generates the *trans*-isomer in doing so.

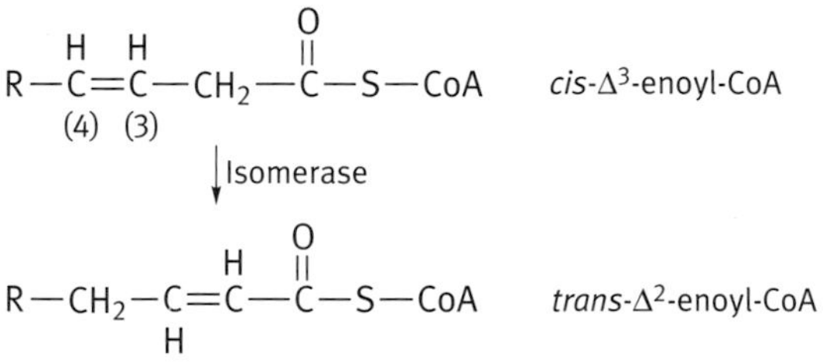

R—C=C—CH$_2$—C—S—CoA   *cis*-Δ^3-enoyl-CoA
(4) (3)

Isomerase

R—CH$_2$—C=C—C—S—CoA   *trans*-Δ^2-enoyl-CoA

This is now on the main pathway of fat breakdown and so the problem of monounsaturated fat oxidation is solved. (If you look at Fig. 9.2, you will see that, in oxidation of saturated acyl-CoAs, it is the *trans*-isomer that is generated.)

Polyunsaturated fatty acids pose additional problems—for example, linoleic acid has two double bonds (Δ^9 and Δ^{12}). In effect, one (Δ^9) is dealt with as described above while the second (Δ^{12}) is reduced to the saturated form by an additional enzyme at the appropriate stage, again putting the molecule on the normal metabolic path of β-oxidation (the sequence of the steps is not exactly as outlined and need not be given here).

Is the acetyl-CoA derived from fat breakdown always directly fed into the citric acid cycle?

So far we have explained that fatty acids are converted to acetyl-CoA, which then joins the citric acid cycle and is oxidized. This is true for all tissues in 'normal' circumstances but it has to be qualified for the liver in a particular circumstance.

As described in Chapter 5, the body can be in a physiological situation where fat metabolism is the main source of energy. This occurs in starvation after exhaustion of glycogen stores; the same can occur in diabetics where inability to metabolize carbohydrate effectively results in an almost analogous glucose 'starvation' irrespective of glucose availability. In this situation, the fat cells are pouring out free fatty acids (page 77) and the liver may produce excessive amounts of acetyl-CoA, 'excessive' in the sense that there is too much produced to be fed into the citric acid cycle. This is exacerbated because lack of carbohydrate metabolism leads to a relative shortage of oxaloacetate, which you will recall is needed to produce citrate from acetyl-CoA. (For an explanation of why oxaloacetate is in short supply in this situation, see page 120.)

As previously mentioned in Chapter 7, the liver cell, *in effect*, joins two acetyl groups together, by a mechanism to be described shortly, to form **acetoacetate** (CH_3 $COCH_2$ COO^-) which is partly reduced to **β-hydroxybutyrate** (CH_3 $CHOH$ CH_2 COO^-). The two (**ketone bodies**) are released into the blood.

How is acetoacetate made from acetyl-CoA?

Acetoacetyl-CoA is formed from acetyl-CoA by reversal of the ketoacyl-CoA thiolase reaction

$$2CH_3CO{-}S{-}CoA \leftrightarrow CH_3COCH_2CO{-}S{-}CoA + CoA{-}SH$$

One would have imagined that free acetoacetate would be formed by simple hydrolysis of the acetoacetyl-CoA (thus pulling the equilibrium over). However, it is not so. Instead, a third molecule of acetyl-CoA is used to form **3-hydroxy-3-methylglutaryl-CoA (HMG-CoA)**. This is now decomposed to form acetoacetate and acetyl-CoA. The scheme is shown in Fig. 9.3.

One cannot be sure why the synthesis of acetoacetate proceeds in this way. HMG-CoA is synthesized by many animal cells. It is a precursor of cholesterol, which is an essential constituent for their membranes (page 45). Ketone body formation occurs in mitochondria (where fat conversion to acetyl-CoA occurs). Formation of HMG-CoA for cholesterol synthesis takes place in the cytoplasm where a separate HMG-CoA synthase occurs attached to the endoplasmic reticulum membrane. Acetoacetate

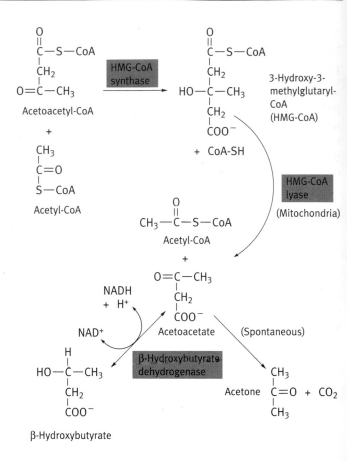

Fig. 9.3 Ketone body production in the liver during excessive oxidation of fat in starvation or diabetes. The process occurs in mitochondria. HMG-CoA is also the precursor of cholesterol but this occurs in the cytosol of rat liver where HMG-CoA synthase also occurs.

can be used by peripheral tissues to generate energy. In mitochondria it is converted to acetoacetyl-CoA by an acyl exchange reaction in which succinyl-CoA is converted to succinate.

Acetoacetate + succinyl-CoA $\xrightleftharpoons[\text{transferase}]{\text{CoA}}$ Acetoacetyl-CoA + succinate

Acetoacetyl-CoA $\xrightarrow{\text{CoA-SH} \atop \text{Thiolase}}$ 2-Acetyl-CoA

Acetoacetyl-CoA is cleaved by a thiolase using CoASH to form two molecules of acetyl-CoA. β-Hydroxybutyrate is also utilized by being dehydrogenated first to acetoacetate.

Oxidation of odd-numbered carbon chain fatty acids

A small proportion of fatty acids in the diet (for example those from plants) have odd-numbered carbon chains, β-oxidation of which produces, as the penultimate product, a five-carbon β-ketoacyl-CoA instead of acetoacetyl-CoA. Cleavage of this by thiolase produces acetyl-CoA and the three-carbon propionyl-CoA.

$$CH_3-CH_2-CO-CH_2-CO-S-CoA + CoA-SH \rightarrow$$

Propionyl-CoA

$$CH_3-CH_2-CO-S-CoA + CH_3-CO-S-CoA$$

Acetyl-CoA

Propionyl-CoA is converted to succinyl-CoA and hence on to the citric acid cycle mainstream by the following reactions. The epimerase catalyses the conversion of D- to L-methylmalonyl-CoA.

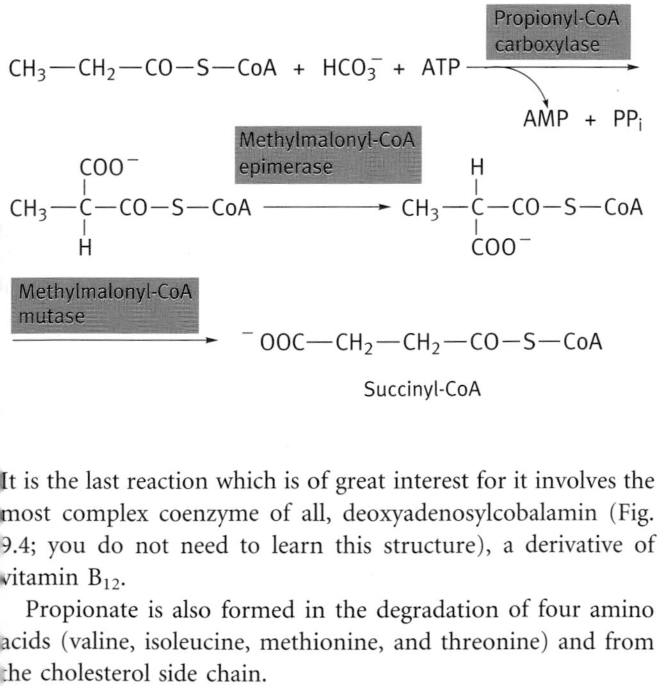

It is the last reaction which is of great interest for it involves the most complex coenzyme of all, deoxyadenosylcobalamin (Fig. 9.4; you do not need to learn this structure), a derivative of vitamin B_{12}.

Propionate is also formed in the degradation of four amino acids (valine, isoleucine, methionine, and threonine) and from the cholesterol side chain.

Deficiency of either the methylmalonyl-CoA mutase or ability to synthesize the required cofactor from vitamin B_{12} leads to methylmalonic acidosis, usually fatal early in life.

Peroxisomal oxidation of fatty acids

Peroxisomes are small membrane-bound vesicles found in many animal cells. They are discussed in Chapter 16. In the present context, they also oxidize fatty acids, though the bulk of this occurs in mitochondria. β-Oxidation of fatty acids in peroxi-

somes differs in that the $FADH_2$ produced by oxidation of saturated acyl-CoAs is directly oxidized by O_2, producing H_2O_2 (and heat). There is no electron transport system. The functions of peroxisomes are less well defined than those of other organelles but may be related to handling fatty acids longer than C_{18} which are not oxidized by the mitochondrial system.

We have so far dealt with energy production from glucose and from fat. The remaining third major food component is in the form of the amino acids. It would be logical to deal with energy production from amino acids next but, in fact, it is more convenient at this stage to remain with carbohydrate and fat metabolism and next to deal with glucose and fat synthesis and then with the control of metabolism. The reason for this is that energy production from individual amino acids essentially consists of converting them to compounds on the main glycolytic and citric acid cycle pathways so that there is very little information to be given in terms of energy production *per se*. The main biochemical interests of amino acid metabolism lie elsewhere, as will be seen when we deal with this in Chapter 15.

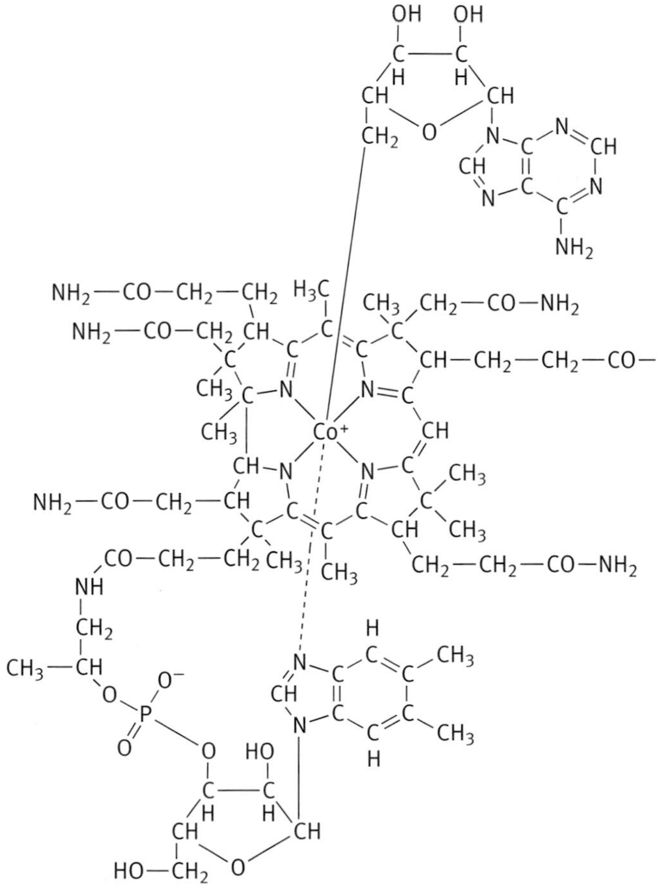

Fig. 9.4 Structure of deoxyadenosylcobalamin.

Further reading

McGarry, J. D. and Foster, D. W. (1980). Regulation of hepatic fatty acid oxidation and ketone body production. *Ann. Rev. Biochem.*, **49**, 395–420.
Despite the regulatory title, the article provides a general review of ketone body formation.

Lardy, H. and Shrago, E. (1990). Biochemical aspects of obesity. *Ann. Rev. Biochem.*, **59**, 689–710.
Very readable account relating obesity, hormones, and metabolism.

Problems for Chapter 9

1   Peripheral tissues obtain their free fatty acids for oxidation from the blood. Explain three ways in which free fatty acids become available to cells.

2   Which cells of the body do not use free fatty acids for energy supply?

3   In breaking down fatty acids to acetyl-CoA:
   (a) What is always the first step?
   (b) Where does this occur?
   (c) Where does fatty acid breakdown to acetyl-CoA occur in eukaryotes?
   (d) How do fatty acid groups reach this site of breakdown?

4   Illustrate similarities in the oxidation of fatty acids to acetyl-CoA with a section of the citric acid cycle.

5   What is the yield of ATP from the complete oxidation of a molecule of palmitic acid? Explain your answer.

6   Explain how a monounsaturated fat (Δ^9) is broken down to acetyl-CoA.

7   Is acetyl-CoA, derived from fatty acid breakdown, always fed into the citric acid cycle? Explain your answer.

8   HMG-CoA is an intermediate in both acetoacetate synthesis and cholesterol synthesis. Where do these two processes occur?

Chapter summary

Chapter 10

··

A switch from catabolic to anabolic metabolism—first, the synthesis of fat and related compounds in the body

In the previous chapters on metabolism we have been dealing with catabolic systems that break down fats and carbohydrates to yield energy. We are now starting on the other side of metabolism—the building up of molecules or anabolism. Since fat breakdown will be fresh in your minds from the previous chapter it seems a good idea to start with fat synthesis. We'll deal with carbohydrate synthesis in the next chapter.

Fat synthesis occurs mainly to convert carbohydrate in excess of that needed to replenish glycogen stores to a more suitable storage form, neutral fat. Alcohol and certain amino acids also give rise to fat. As already explained (page 74), if this weren't done, to store the amount of energy contained in our storage fat as glycogen we'd be much larger in size. Fat synthesis occurs in times of dietary plenty.

··

Mechanism of fat synthesis

General principles of the process

If you refer to the overall diagram of metabolism (Fig. 7.11), you will see that, as well as fatty acids being converted to acetyl-CoA, acetyl-CoA can be converted into fatty acids. The fact that fatty acids are synthesized from acetyl-CoA, two carbon atoms at a time, explains why natural fats almost always have even numbers of carbon atoms. Carbohydrate intake in excess can lead to fat deposition—glucose goes to pyruvate which goes to acetyl-CoA. However, since the latter step catalysed by pyruvate dehydro-

genase is irreversible, acetyl-CoA cannot be converted to pyruvate and hence fat cannot be converted into glucose in animals. (Bacteria and plants have, as you will see, a special trick for doing so—see page 154.)

You will by now be familiar with the concept that, in metabolic pathways, at least some reactions are different in the forward and reverse directions. This is true for fat breakdown and synthesis. The two pathways use, for some steps, essentially the same reactions, but there are others that are different in the two directions. These make both pathway directions thermodynamically favourable, irreversible, and separately controllable.

To render synthesis of fatty acids from acetyl-CoA thermodynamically favourable, we must have a reaction that injects energy into the process—a reaction that is irreversible. Refresh your memory on this, if necessary, by reading page 3 again, since it is essential to appreciate this for understanding the very first reaction in fat synthesis. In this, a molecule of CO_2 is added to the acetyl-CoA molecule, using ATP breakdown as energy source, to form malonyl-CoA, but, in the next reaction in fat synthesis, the CO_2 is released again. This seems pointless unless you remember that the process is thus made thermodynamically irreversible. It is to do with energy, rather than chemical change. Malonic acid has the structure $HOOC—CH_2—COOH$ so malonyl-CoA is $HOOC—CH_2—CO—S—CoA$. The enzymic reaction below is catalysed by **acetyl-CoA carboxylase**—it inserts a carboxyl group into acetyl-CoA forming malonyl-CoA.

$$CH_3-\overset{\overset{\displaystyle O}{\|}}{C}-S-CoA \ + \ ATP \ + \ HCO_3^-$$

$$\overset{\overset{\displaystyle O}{\diagdown}}{\underset{\displaystyle{}^-O}{}}C-CH_2-\overset{\overset{\displaystyle O}{\|}}{C}-S-CoA \ + \ ADP \ + \ P_i \ + \ H^+$$

The enzyme has a prosthetic group of biotin. All carboxylases using ATP to incorporate CO_2 into molecules, we remind you, have this feature. As an intermediate in the reaction, an activated CO_2–biotin complex is formed (page 121).

To synthesize fatty acids we have to add two carbon units at a time, starting with acetyl-CoA. *Malonyl-CoA is the active donor of two carbon atoms in fatty acid synthesis*, despite the fact that the malonyl group has three carbons. But before this, a few words of explanation about the **acyl carrier protein** or **ACP**.

The acyl carrier protein (ACP) and the β-ketoacyl synthase

As described in Chapter 9, when fatty acids are broken down to acetyl-CoA, all of the reactions occur, not as free fatty acids, but as thiol esters with CoA. When fatty acids are synthesized, all of the reactions also occur on fatty acyl groups bound as thiol esters, but in fatty acid synthesis, instead of CoA being used to bind the reactants, *half* of the CoA molecule is used. We remind you of the structure of CoA.

Phosphate—pantothenate—NHCH₂CH₂—SH
Phosphate—ribose—adenine
phosphate

Phosphopantotheine moiety in box

The half in the box is 4-phosphopantotheine and this is the 'carrier' thiol used in fatty acid synthesis. It is not free, as is CoA, but is bound to a protein called the acyl carrier protein or ACP. You can think of ACP as a protein with built in CoA or, alternatively, as a giant CoA molecule with the AMP part of CoA replaced by a protein.

Another enzyme or domain, **β-ketoacyl synthase**, also has a reactive thiol group in that of the amino acid cysteine that forms part of the structure of the active site. The ACP and β-ketoacyl synthase in mammals are part of a large multifunctional complex.

Mechanism of fatty acyl-CoA synthesis

We will start with the situation shown Fig. 10.1(a) in which the two thiol groups on the fatty acyl-CoA synthase complex are vacant.

$CH_3 CO—$ is transferred from acetyl-CoA to the ACP thiol by a specific transferase (Fig. 10.1(b)). This is then further transferred to the β-ketoacyl synthase thiol group (Fig. 10.1(c)) leaving the ACP site vacant. Another transferase transfers the malonyl group of malonyl-CoA to the latter, forming malonyl-ACP (Fig. 10.1(d)). The synthase now transfers the acetyl group to the malonyl group, displacing CO_2 and forming a β-ketoacyl-ACP; in this initial case it is β-ketobutyryl-ACP (Fig. 10.1(e)). The latter is reduced by activities on the same protein complex (described below) to form butyryl-ACP (Fig. 10.1(f)). The resultant situation is precisely analogous to that shown in Fig. 10.1(b) since both have a saturated acyl-ACP complex (acetyl and butyryl, respectively) and a vacant synthase site. A series of reactions now ensues, identical to those in Fig. 10.1 (starting with the reaction b → c), except that acetyl- is now butyryl- and the end-product is hexanoyl-. Five more rounds produces palmityl-. When C_{16} is reached, a hydrolase releases palmitic acid,

$$CH_3 (CH_2)_{14} CO—S—ACP + H_2O \rightarrow CH_3 (CH_2)_{14} COO^-$$
$$+ ACP—SH.$$

Note that the β-ketoacyl-ACP synthase reaction leading to stage (e) is irreversible because of the decarboxylation involved. This is the purpose of forming malonyl-CoA from acetyl-CoA: to make fatty acid synthesis pathway a one-way process.

If longer chain acids are required to be formed such as stearic acid (C_{18}), a separate system elongates the palmitate.

Organization of the fatty acid synthesis process

The sequence of reactions given above for the conversion of acetyl-CoA to palmitic acid is relatively simple or complex according to your viewpoint but, in animal tissues, the organization of the enzyme activities involved is remarkable from any viewpoint. In *E. coli*, the situation is straightforward, in that there is a series of separate enzymes, as is usual in metabolic pathways, and, after each reaction, the products dissociate and diffuse to the next enzyme. In animals, the situation is different; all of the catalytic activities of the steps following malonyl-CoA formation reside on one giant protein, the activities of which have not been separated. (A pair of such proteins collaborate as the functional dimer complex.) The growing fatty acid chain oscillates from the ACP-thiol group to the β-ketoacyl synthase thiol group but it is never released from the enzyme complex until palmitate synthesis is completed. The elongation step and reductive reactions all occur with the substrate attached to the ACP. The long flexible arm of the 4-phosphopantetheine is presumably needed to permit the different intermediates to interact with the appropriate catalytic centres of the complex. The advantage of a single large complex is that the process of synthesis can be more rapid since each intermediate is positioned to interact with the next catalytic centre rather than having to diffuse away and find the next enzyme.

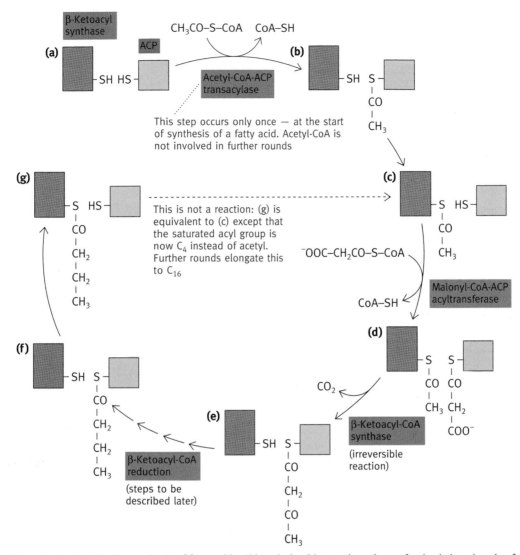

Fig. 10.1 Diagram of the steps involved in the synthesis of fatty acids. Although the β-ketoacyl synthase of animals is a domain of a single protein molecule, it is shown here as being separate. This is because the functional enzyme unit is believed to be a dimer arranged head to tail and it is also believed that the transfers between the two —SH groups occur between the domains on the two dimer-constituent protein molecules. Note that the thiol of the ACP is on the phosphopantetheine moiety and that of the synthase is on a cysteine residue of the protein. After seven successive rounds of reactions, the resultant palmityl-ACP is hydrolysed to release free palmitate.

The reductive steps in fatty acid synthesis

In the above sequence of events involved in saturated fatty acyl-ACP synthesis, the β-ketoacyl group attached to the ACP is reduced by a succession of three steps shown in Fig. 10.2. The reductant in these steps is **NADPH** (not NADH), an electron carrier, so far not mentioned in this book, and which will now be described.

What is NADP⁺?

NAD^+ is

P—ribose—nicotinamide
|
P—ribose—adenine

$NADP^+$ is

P—ribose—nicotinamide
|
P—ribose—adenine
|
P

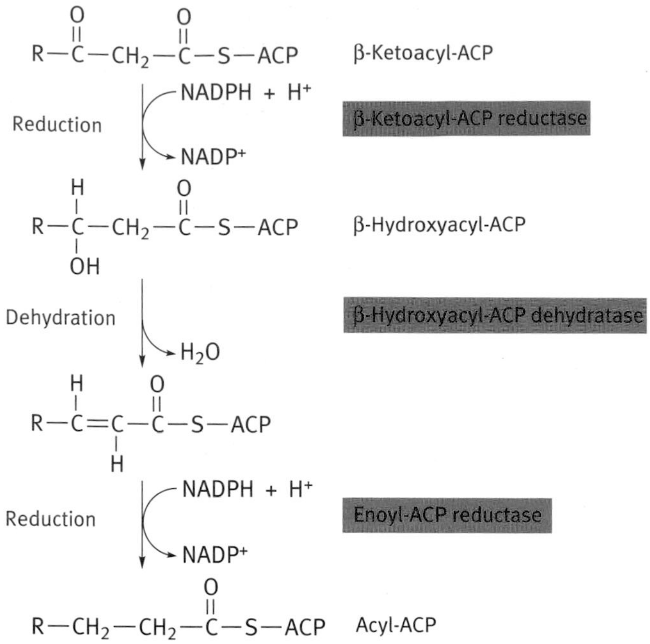

Fig. 10.2 Reductive steps in fatty acid synthesis.

The extra phosphoryl group is attached to the 2′ hydroxyl group of the ribose which is attached to adenine.

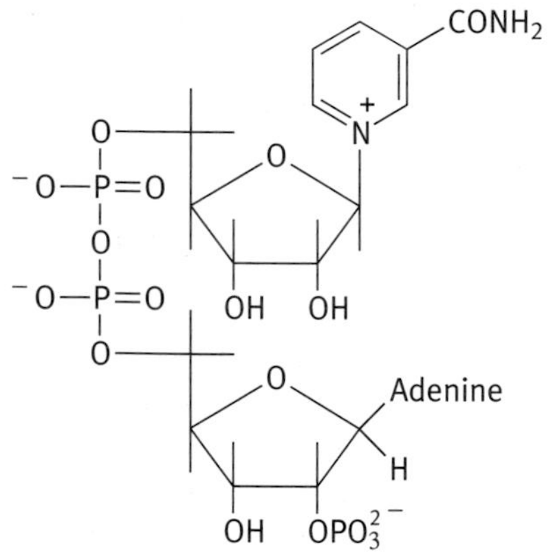

The extra phosphoryl group has no significant effect on the redox characteristics of the molecule—it is, in fact, purely an identification or recognition signal. An NAD^+ enzyme will not react with $NADP^+$ and vice versa (with the possible odd exception of little significance).

For a long time it was thought that this was just an odd quirk of nature, no more than a curiosity. However, eventually it was realized that the use of the two coenzymes constitutes an important form of metabolic compartmentation. To understand this, remember that in energy production the aim is to oxidize reducing equivalents, while in, for example, fat synthesis the aim is to use reducing equivalents for reductive synthesis. The aims are diametrically opposite. The cell keeps these metabolic activities separate by metabolic compartmentation—one, the oxidative part uses NAD^+ as electron carrier; the other, the reductive part, uses $NADP^+$.

Separate mechanisms exist to reduce NAD^+ and $NADP^+$. You already know how NAD^+ is reduced in glycolysis, the pyruvate dehydrogenase reaction, the citric acid cycle, and fat oxidation. How is $NADP^+$ reduced? This is explained in the context of the next question.

Where does fatty acid synthesis take place?

In terms of tissues, the main sites of fatty acid synthesis in the body are the liver and adipose cells (the former predominating) but many other tissues, especially the mammary gland during lactation, also produce fat.

Within a cell, **palmitate synthesis** from acetyl-CoA occurs in the **cytosol**. This is in contrast to **fatty acid oxidation** which occurs in the **mitochondria**. The major source of acetyl-CoA for fat synthesis is the pyruvate dehydrogenase reaction (page 114), which is located in the mitochondrial matrix. Acetyl-CoA cannot cross the mitochondrial membrane to the site of fatty acid synthesis in the cytosol so acetyl residues must be transported from mitochondria to the cytosol. What is done is part of an ingenious scheme. The acetyl-CoA in the mitochondrion is converted to citrate by the citric acid synthase reaction of the citric acid cycle (page 116). The citrate is transported by a mitochondrial membrane system into the cytosol where it is cleaved back to acetyl-CoA and oxaloacetate by a separate enzyme, called **citrate lyase** or the **citrate cleavage enzyme**. This reaction is coupled to hydrolysis of ATP to ADP and P_i, thus ensuring it goes to completion,

$$\text{Citrate} + \text{ATP} + \text{CoA—SH} + H_2O \rightarrow \text{acetyl-CoA} + \text{oxaloacetate} + \text{ADP} + P_i.$$

That is not the end of the story. The oxaloacetate cannot get back into the mitochondrion, whose membrane is impervious to it and for which no transporter exists. It is therefore reduced in the cytoplasm to malate by NADH (note, *not* NADPH); the malate so formed is oxidatively decarboxylated to pyruvate and CO_2 by an enzyme (the 'malic' enzyme) that uses $NADP^+$ (note, *not* NAD^+), thus producing NADPH, which is needed for fat synthesis (Fig. 10.3). The pyruvate is now transported back into the mitochondrion (Fig. 10.4). The pyruvate transported into the mitochondria can be converted back to oxaloacetate by the **pyruvate carboxylase reaction** (previously mentioned on page 120).

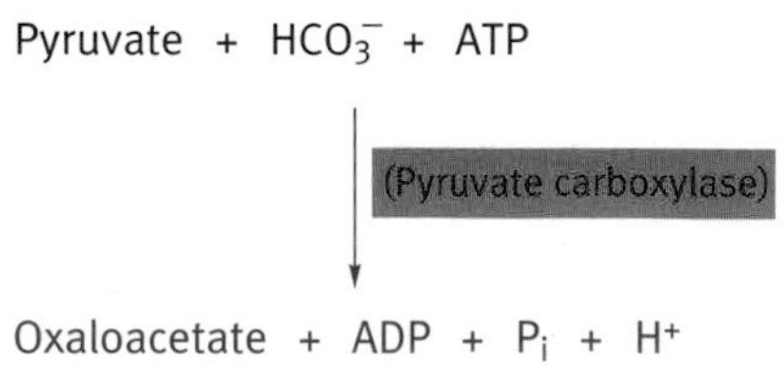

$$\text{Pyruvate} + HCO_3^- + \text{ATP}$$

(Pyruvate carboxylase)

$$\text{Oxaloacetate} + \text{ADP} + P_i + H^+$$

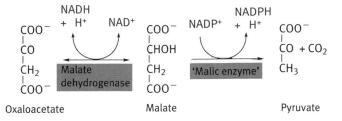

Fig. 10.3 Reduction of $NADP^+$ for fatty acid synthesis. The net effect of the two reactions is to transfer reducing equivalents from NADH to $NADP^+$. The pyruvate so produced in the cytoplasm enters the mitochondrion. The source of the oxaloacetate is shown in Fig. 10.4.

Citrate leaves the mitochondrion only when it is at a high concentration; this occurs when carbohydrate is plentiful. Citrate does not occur in the cytosol at other times.

Thus, the citrate mechanism not only transports acetyl-groups out of the mitochondrion, it also generates NADPH for fat synthesis. The reduction in the cytoplasm of oxaloacetate to malate by NADH and the oxidation of malate to pyruvate by $NADP^+$ constitutes a neat mechanism for transferring electrons from the NADH metabolic 'compartment' into the NADPH 'compartment' used for reductive synthesis reactions. In

addition, citrate *in vitro* activates the first reaction committed to fat synthesis—the acetyl-CoA carboxylase, producing malonyl-CoA, though the physiological significance of this activation has recently been questioned. Control of this enzyme is discussed in Chapter 12.

It will be apparent from the scheme given that, for every acetyl-CoA produced in the cytosol from citrate, we generate *one* NADPH molecule. However, each cycle of the fatty acid synthase process required *two* NADPH molecules (see Fig. 10.2). Yet another mechanism for producing the extra NADPH is required—this will be explained in Chapter 13 (page 181) for it belongs to a different metabolic topic, the pentose phosphate pathway.

Synthesis of unsaturated fatty acids

As explained (page 45), the body requires unsaturated fatty acids for the production of polar lipids for membrane synthesis and other roles. A separate enzyme system in liver can introduce a single double bond into the middle of stearic acid, generating oleic acid.

$$CH_3\ CH\ (CH_2)_7 CH{=}CH\ (CH_2)_7\ COOH$$

$$\text{Oleic acid, } 18{:}1(\Delta^9)$$

However, the system cannot make double bonds between this central double bond and the methyl end of the molecule. Linoleic acid which has two double bonds and linolenic acid with three thus cannot be synthesized by the body.

$$CH_3\ (CH_2)_4\ CH{=}CH\ CH_2\ CH{=}CH\ (CH_2)_7\ COOH$$

$$\text{Linoleic acid, } 18{:}2(\Delta^{9,12})$$

$$CH_3\ CH_2\ CH{=}CH\ CH_2\ CH{=}CH\ CH_2\ CH{=}CH\ (CH_2)_7\ COOH$$

$$\alpha\text{-Linolenic acid, } 18{:}3(\Delta^{9,12,15})$$

Since they are needed for membrane components and for synthesis of eicosanoid regulatory molecules (see below) these two acids must be obtained from the diet; plants have enzymes that can desaturate in the terminal half of the fatty acids. The liver can elongate linoleic acid and introduce an extra double bond to give the C_{20} arachidonic acid ($20{:}4$, $\Delta^{5,8,11,14}$, see below) with four double bonds. It can also elongate acids to the C_{22} and C_{24} acids involved in the lipids of nerve tissues. All of these transformations occur as the CoA derivatives.

Synthesis of triacylglycerol and membrane lipids from fatty acids

Neutral fat is synthesized for storage purposes. For attachment of a fatty acid in ester bond to glycerol, the acid must be activated to the form of acyl-CoA, a reaction catalysed by acetyl-CoA synthetase, as already described

$$RCOOH + CoA{-}SH + ATP \rightarrow R\ CO{-}S{-}CoA + AMP + PP_i.$$

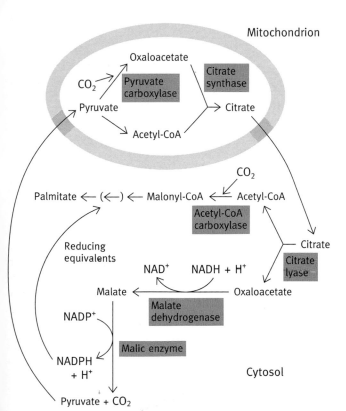

Fig. 10.4 Diagram illustrating source of acetyl groups (acetyl-CoA) and reducing equivalents ($NADPH + H^+$) for fatty acid synthesis. The citrate lyase reaction involves the breakdown of ATP to ADP and P_i.

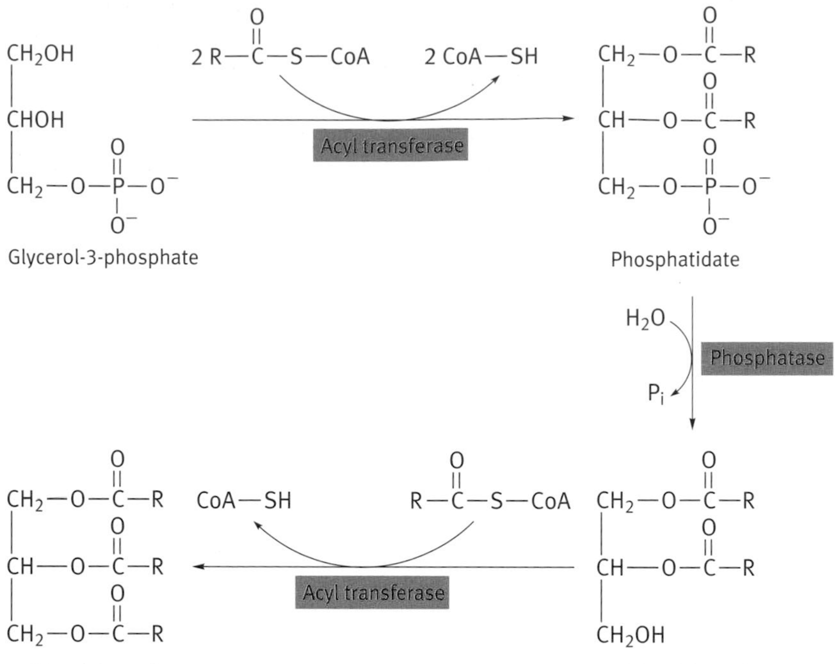

Fig. 10.5 Reactions involved in the synthesis of triacylglycerol (neutral fat) from glycerol-3-phosphate.

A carboxylic acid ester has a lower free energy of hydrolysis than has a thiol ester and hence activation of the acid makes triglyceride synthesis from the fatty acyl-CoA exergonic. Surprisingly perhaps, the acceptor of acyl groups is not glycerol but glycerol-3-phosphate, which arises mainly by reduction of the glycolytic intermediate, dihydroxyacetone phosphate (DHAP), though phosphorylation of glycerol can also occur in the liver but not in adipose cells, the latter point being of significance (page 154). The reaction is

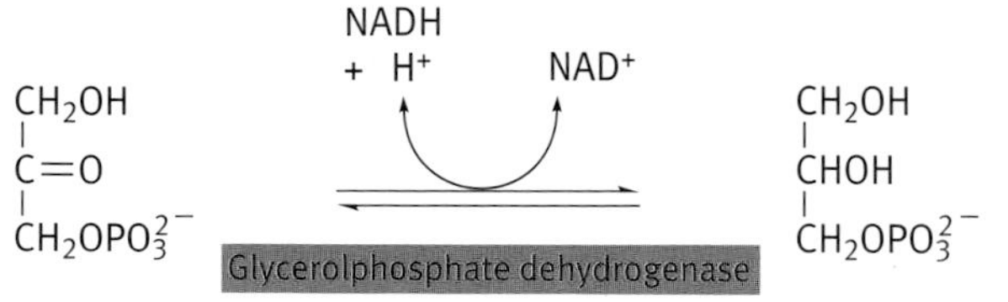

Dihydroxyacetone phosphate

Glycerol-3-phosphate

The steps in triacylglycerol synthesis are shown in Fig. 10.5.

Synthesis of glycerophospholipids

If you turn back to page 42 you will recall that membranes contain glycerol-based phospholipids in which a polar alcohol is attached to the phosphoryl group of phosphatidic acid.

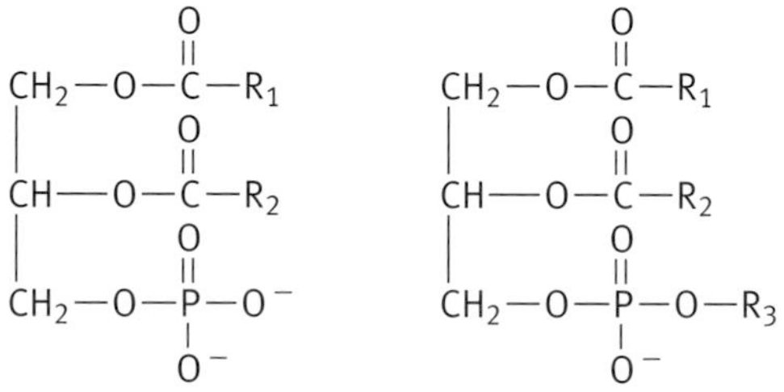

Phosphatidate

Phospholipid

R_3 may be serine, ethanolamine, choline, inositol, or diacylglycerol. The most prevalent membrane lipids in eukaryotes are phosphatidylethanolamine and phosphatidylcholine so we will consider their synthesis first; ethanolamine is $NH_2\,CH_2\,CH_2OH$ and choline is

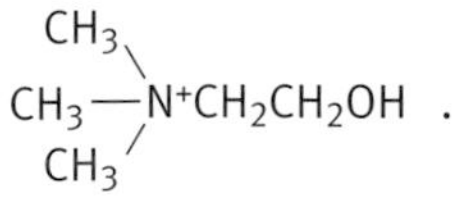

Both can be written as the alcohols 'R—OH'. The alcohols are 'activated' to participate in the synthesis; this occurs in two steps. First, a phosphorylation by ATP,

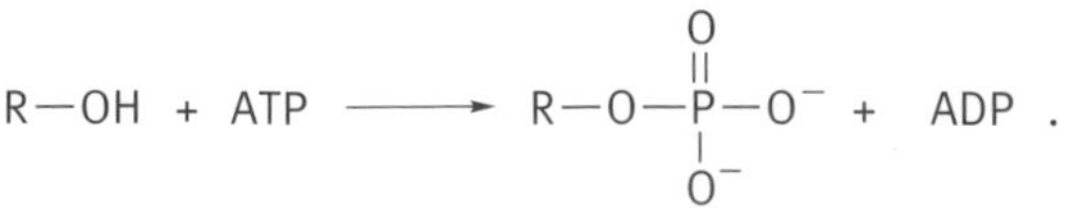

Secondly, a reaction with **cytidine triphosphate** or **CTP**. CTP is the same as ATP except that it has the base cytosine (C) in place of adenine (cytosine-ribose is called cytidine). We will come to its precise structure later in the book. All cells have CTP for it is needed for RNA and DNA synthesis (Chapter 21). Whether CTP involvement rather than ATP is due to an accidental quirk of evolution or whether there is a good reason for it is unknown, but the fact is, that CDP-compounds are always the 'high-energy' donors of groups in this area (just as for equally unknown reasons it is always UDP-glucose for processes such as glycogen synthesis in animals). The reaction is, in fact, very reminiscent of UDP-glucose formation from glucose-1-phosphate and UTP (page 81).

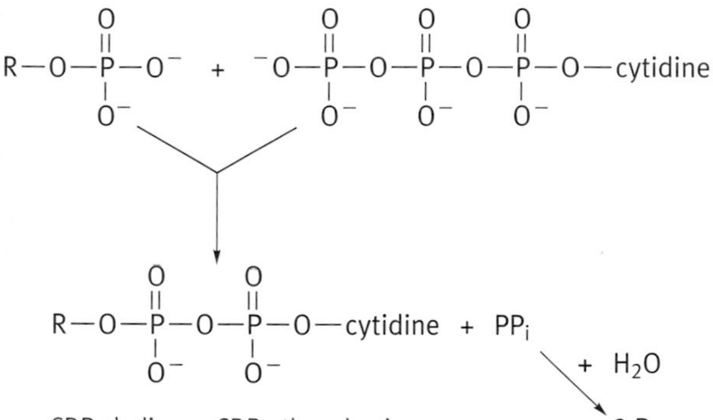

CDP-choline or CDP-ethanolamine

Hydrolysis of PP_i to P_i makes the reaction strongly exergonic.

The final reaction in phosphatidylethanolamine and phosphatidylcholine synthesis is as follows:

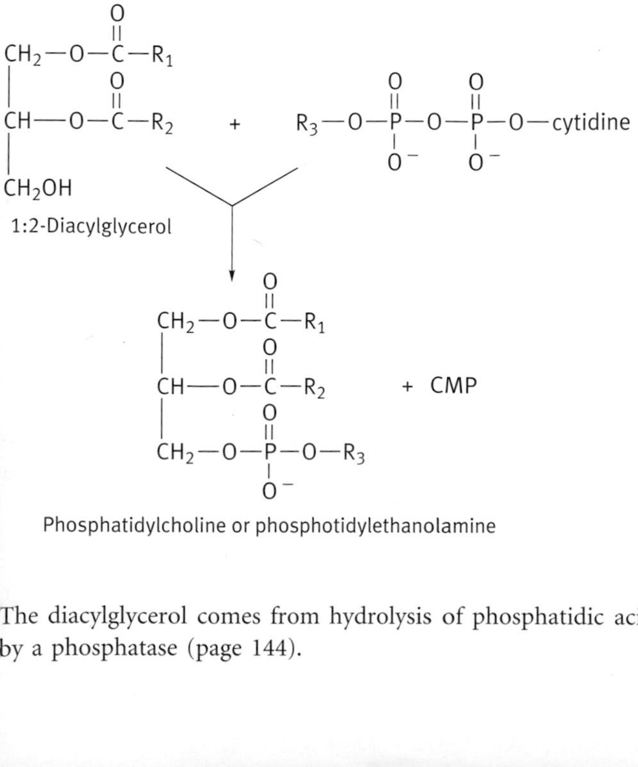

1:2-Diacylglycerol

Phosphatidylcholine or phosphotidylethanolamine

The diacylglycerol comes from hydrolysis of phosphatidic acid by a phosphatase (page 144).

In the reactions above we join an alcohol (ethanolamine or choline) to CDP and then transfer phosphoryl alcohol to another alcohol (diacylglycerol); the head group is 'activated'. Energetically, it would be the same to join diacylglycerol to CDP and transfer a diacylglycerol phosphoryl-group to ethanolamine or choline—that is, to activate the diacylglycerol instead of the polar alcohol or head group. This is what happens in phosphatidylinositol and cardiolipin (page 42) synthesis in eukaryotes. In the latter, instead of inositol, the reactant is another molecule of diacylglycerol.

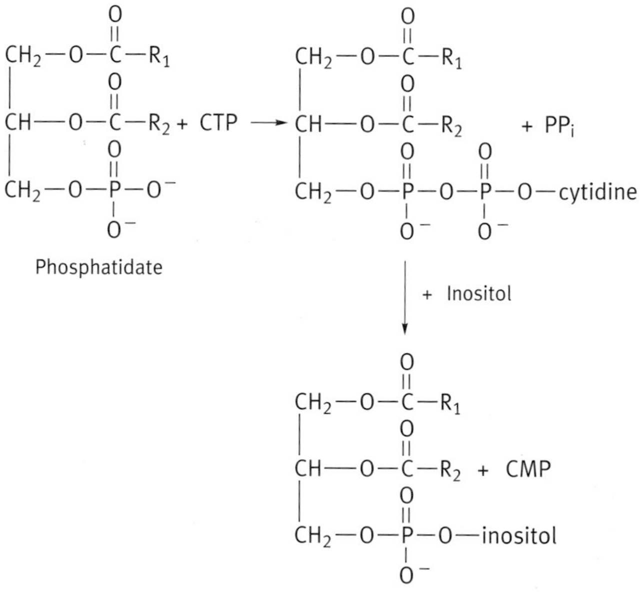

Phosphatidate

+ Inositol

Phosphatidylinositol

The situation is summarized in the scheme in Fig. 10.6. This scheme illustrates the principle of the two ways in which glycerophospholids can be synthesized. It should be emphasized that the field is a complex one in which extensive interconversion of the phospholipids occurs. For example, serine and ethanolamine can be exchanged, PE can be methylated to PC, PS can be decarboxylated to PE. The particular route of synthesis of a given phospholipid may vary in different organisms.

Site of membrane lipid synthesis

In animal cells, membrane lipid synthesis occurs on the smooth endoplasmic reticulum (ER) membrane. The newly synthesized lipids are transported to their various cellular destinations, probably as vesicles budded off from Golgi membranes or by a protein-mediated transport mechanism, not completely understood. We suggest that you simply note these statements and wait until we deal with the ER and Golgi apparatus in Chapter 22.

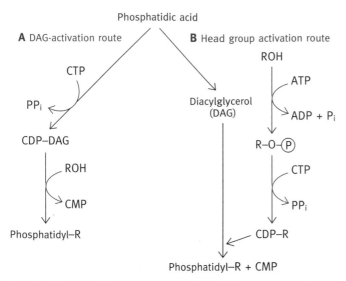

Fig. 10.6 Diagram of the two pathways (**A** and **B**) for synthesis of glycerophospholipid. (ROH, The polar head group to be attached to the phospholipid.) Different cells may use different routes for synthesizing a given phospholipid. In mammals, phosphatidylcholine (PC), phosphatidylserine (PS), and phosphatidylethanolamine (PE) are synthesized by route **B**, but the three are interconvertible by decarboxylation of PS to PE, methylation of PE to PC, and head-group exchange between PE and PS and free ethanolamine or serine. Synthesis of cardiolipin and phosphatidylinositol in mammals follows route **A**. In *E. coli* phospholipid synthesis occurs by route **A**.

Synthesis of prostaglandins and related compounds

The Greek word *eikosi* means 20; this is the basis for the name of a group of compounds called **eicosanoids**, all of which contain 20 carbon atoms and are related to polyunsaturated fatty acids. Although present in the body at very low levels, they have a wide range of physiological functions.

There are three main groups named after the cells in which they were first discovered. These are the **prostaglandins**, the **thromboxanes**, and the **leukotrienes**. Prostaglandins were first discovered in semen and so named because it was thought that they originated from the prostate gland (actually it is from the seminal vesicles). However, it is now known that very many tissues produce prostaglandins. Thromboxanes were discovered in blood platelets or thrombocytes and leukotrienes in leukocytes.

The prostaglandins and thromboxanes

A number of different prostaglandins exist, varying in their detailed structures, the main subclasses being PGA, PGE, and PGF. A subscript number indicates the number of double bonds in the side chains attached to the cyclopentane ring structure. If you look at Fig. 10.7(b), you will see the structure of PGE_2 as an

example. In humans the compounds with two double bonds are most important and are synthesized from arachidonic acid, whose structure is shown in Fig. 10.7(a). Other prostaglandins are derived from related unsaturated fatty acids. Arachidonic acid is present in cells, mainly as the fatty acid component of a phospholid. For prostaglandin synthesis it is released by a phospholipase.

The first step in synthesizing prostaglandins is carried out by a **cyclooxygenase**, which forms the ring structure from arachidonic acid (Fig. 10.7(a)). This enzyme is inhibited by **aspirin** (acetylsalicylic acid), which covalently modifies the enzyme by acetylating an essential serine group on the enzyme. Thromboxanes (Fig. 10.7(c)) are formed by further conversion of certain of the prostaglandins, so that aspirin inhibits this formation also.

The prostaglandins have a wide variety of physiological effects. They are released immediately after synthesis and act as local hormones on adjacent cells by combining with receptors. They cause pain, inflammation, and fever; they cause smooth muscle contraction and are involved in labour; they are involved

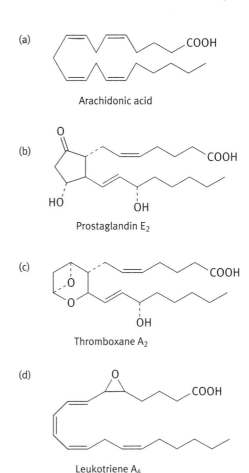

Fig. 10.7 Structure of prostaglandins and related compounds. (a) Arachidonic acid; (b) prostaglandin E_2 (PGE_2); (c) thromboxane A_2 (TXA_2); (d) leukotriene A_4 (LTA_4). Probably only those specially interested in this area would need to memorize these structures.

in blood pressure control; they suppress acid secretion in the stomach. Thromboxanes affect platelet aggregation and thus blood clotting.

Aspirin, by its inhibition of cyclooxygenase, suppresses many of these effects; in particular, a low dose (100 mg per day) inhibits platelet thromboxane formation resulting in decreased platelet aggregation. This therapy has been found to reduce the risk of heart attacks by reducing the danger of blood clot formation blocking the coronary arteries.

Leukotrienes

The structure of one of the leukotrienes is given in Fig. 10.7(d). Leukotrienes arise from arachidonic acid by a different route involving a lipoxygenase reaction. They cause protracted contraction of smooth muscle, are a factor in asthma symptoms by constricting airways, and play a role in white cell function.

Synthesis of cholesterol and its regulation

Most cells of the body are believed to be capable of cholesterol synthesis; cholesterol is an essential component of cell membranes and the precursor of bile salts and steroid hormones. The liver and, to a lesser extent, the intestine are the most active in its synthesis and from the liver a two-way flux 'equilibrates' the cholesterol of the liver and the rest of the body, as described in Chapter 6.

Remarkably perhaps, the sole starting material for synthesis of this large molecule is acetyl-CoA from which, by a long metabolic pathway, the details of which probably few biochemists carry in their heads, cholesterol is produced. The first few stages of the synthesis are of special interest for this is where control of cholesterol occurs and we will confine this account to the production of mevalonic acid, the first metabolite committed solely to cholesterol synthesis. This involves the reactions shown in Fig. 10.8 which occur in the cytosol.

HMG-CoA is also the precursor of acetoacetate, one of the ketone bodies (page 134), but for this purpose is synthesized inside the mitochondrion.

The synthesis of cholesterol is controlled at the HMG-CoA reductase step by three mechanisms. Cholesterol feedback inhibits synthesis of the reductase at the gene level, it results in destruction of the enzyme, and it also inactivates the enzyme by a phosphorylation mechanism. (The latter type of control is described on page 165.) Therapeutically, drugs have been developed to lower cholesterol levels in patients. These resemble mevalonic acid in structure but in the body act as transition-state (page 5) analogues of the HMG-CoA reductase reaction, which they inhibit very effectively by combining with the active site of the enzyme. The two drugs are known as Lovastatin and Simvastatin.

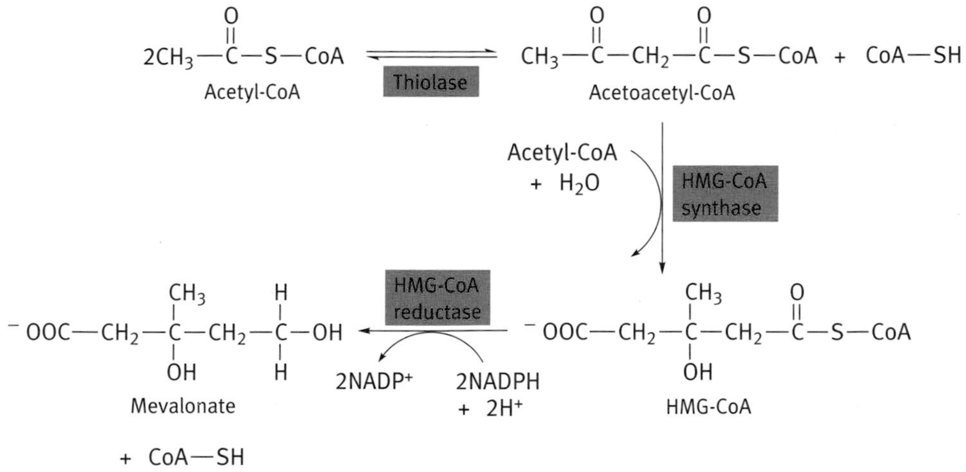

Fig. 10.8 The synthesis of mevalonate from acetyl-CoA. HMG-CoA, 3-Hydroxy-3-methylglutaryl-CoA.

....

Further reading

Fatty acid synthase

Smith, S. (1994). The animal fatty acid synthase: one gene, one polypeptide, seven enzymes. *FASEB J.*, 8, 1248–59.
Illustrated article on the structure and function of this remarkable protein, with each component activity discussed,

and how they are organized in the dimer complex. An interesting article on a classical topic.

Membrane lipid synthesis

Dawidowicz, E. A. (1987). Dynamics of membrane lipid

metabolism and turnover. *Ann. Rev. Biochem.*, **56**, 43–61.
Very readable review on membrane lipids, their site of synthesis, transport and assembly into membranes.

Rooney, S. A., Young, S. L., and Mendelson, C. R. (1994). Molecular and cellular processing of lung surfactant. *FASEB J.*, **8**, 957–67.

Reviews a more physiological role of phospholipid—that of lining the alveoli.

Kent, C. (1995). Eukaryotic phospholipid biosynthesis. *Ann. Rev. Biochem.*, **64**, 315–43.
A straightforward, clear account of the various pathways for the synthesis of the various membrane phospholipids.

Problems for Chapter 10

1 An early step in fatty acid synthesis is that acetyl-CoA is carboxylated, to be followed immediately by decarboxylation of the product. What is the point of this?

2 By means of a diagram, outline the steps in a cycle of elongation during fatty acid synthesis, omitting the reductive steps.

3 Discuss briefly the physical organization of the fatty acyl synthase complex in eukaryotes. How does this differ from that in *E. coli*? What is the advantage of the eukaryote situation?

4 Write down the structures of NAD^+ and $NADP^+$. Explain the reason for the existence of both NAD^+ and $NADP^+$.

5 What are the main tissues in which fatty acid synthesis occurs in eukaryotes?

6 Palmitate synthesis from acetyl-CoA occurs in the cytoplasm. However, pyruvate dehydrogenase, the main producer of acetyl-CoA used in fat synthesis, occurs inside the mitochondrion. Acetyl-CoA cannot traverse the mitochondrial membrane. How does the fatty acid synthesizing system in the cytoplasm obtain its supply of acetyl-CoA?

7 The synthesis of fatty acid involves reductive steps requiring NADPH. What are the sources of the latter?

8 Describe the synthesis of TAG from fatty acids.

9 What is the role of CTP in lipid metabolism?

10 (a) What are eicosanoids?

  (b) What are they made from?

  (c) Briefly describe their physiological significance.

  (d) What is the relevance of aspirin in this area of metabolism?

11 Describe a type of drug that is used to lower the rate of cholesterol synthesis.

Further reading

Problems

Chapter 11

. . . .

Synthesis of glucose (gluconeogenesis)

The body is able to synthesize glucose from any compound capable of being converted to pyruvate by the process of gluconeogenesis. As already stressed, this excludes fatty acids and acetyl-CoA but includes lactate (which is released into the blood by red blood cells and by muscle during exercise) as well as most of the amino acids. This has the role of permitting the storage of a variety of compounds as glycogen. However, in addition to this 'routine' metabolic role, gluconeogenesis can literally make the difference bertween life and death, despite the fact the glucose is such a common molecule prevalent in most diets. In case you have forgotten its significance (page 75), in starvation it assumes an overriding importance.

The brain, as stated several times earlier, requires a constant supply of glucose, which means that the blood glucose level must be kept within normal limits or coma and death will ensue. However, the liver's total store of glucose in the form of glycogen is small and, after about 24 hours without food, the entire store is exhausted. Nevertheless, people don't die as a result of a day's starvation. Supply of blood sugar in this situation is dependent on the liver—no other organ, other than the quantitatively unimportant kidney, can fill this need. If fasting is prolonged beyond about 24 hours, the liver must therefore synthesize glucose. Synthesis of about 100 g per day is needed in man.

Fat mobilization results in ketone body production by the liver which has a glucose-sparing effect (page 75) as the blood ketone level rises, but it cannot replace synthesis of glucose. Nor does the total energy requirement of the brain fall with starvation. Although the effects of glucose deprivation on brain are the most dramatic, cells such as those of the retina and kidney medulla and red blood cells are virtually (or completely in the latter, since they have no mitochondria) anaerobic and dependent on glycolysis of glucose for energy supply. The body normally has sufficient fat to supply energy for weeks of starvation but fatty acids do not pass through the blood–brain barrier and cannot be used by the brain.

The starting point for the gluconeogenesis pathway in liver is pyruvate. To re-emphasize a crucial point, whereas pyruvate can give rise to acetyl-CoA (page 103) and therefore lead to fat synthesis, the reverse cannot happen in animals. Acetyl-CoA cannot be converted to pyruvate and therefore fatty acids cannot be converted into glucose in a net sense.

We will first describe the pathway by which pyruvate is converted into glucose and after that discuss the broader biochemical implications.

Mechanism of glucose synthesis from pyruvate

If you examine the glycolytic pathway (Fig. 8.7), you will note that three reactions are irreversible because of thermodynamic considerations. These are: (1) phosphorylation of glucose by hexokinase (or glucokinase) using ATP; (2) phosphorylation of fructose-6-phosphate by phosphofructokinase; and (3) conversion of phosphoenolpyruvate (PEP) to pyruvate. Glucose is synthesized via the intermediate compounds found in glycolysis but a way has to be found to bypass the irreversible reactions.

The first thermodynamic barrier in the process is the conversion of pyruvate to PEP. Because the spontaneous conversion of the enol form of pyruvate to the keto form has a very large negative $\Delta G^{0'}$ value (page 112), the PEP → pyruvate reaction in glycolysis is irreversible in animal cells but, as mentioned earlier, because of the convention of naming kinases after the reaction involving ATP, it is called pyruvate kinase nonetheless. In animals, a roundabout route involving two reactions is used for the conversion of pyruvate to PEP. Two —Ⓟ groups are used in the process, making the process thermodynamically favourable. The route is as follows:

(1)   Pyruvate + ATP + HCO_3^- → Oxaloacetate + ADP + P_i + H^+; (catalysed by pyruvate carboxylase)

(2)   Oxaloacetate $+ \, GTP \rightarrow PEP + GDP + CO_2$;
(catalysed by PEP carboxykinase, or PEP-CK)

Sum. Pyruvate $+ \, ATP + GTP + H_2O \rightarrow$
PEP $+ \, ADP + GDP + P_i + 2H^+$.

The scheme is shown diagrammatically in Fig. 11.1.

It is not known why GTP is used in the second reaction rather than ATP. Energetically it is just the same. You may recall that reaction 1, catalysed by pyruvate carboxylase, which synthesizes oxaloacetate, has an important role in topping up the citric acid cycle (page 120) quite separate from its role in gluconeogenesis. The second enzyme is usually referred to as **PEP-CK**. Its full name is **phosphoenolpyruvate carboxykinase**—because in the reverse direction it carboxylates phosphoenolpyruvate and transfers a phosphoryl group.

Once PEP is formed, the reactions of glycolysis are all reversible until fructose-1:6-bisphosphate is reached. The formation of this in glycolysis from fructose-1-phosphate is irreversible, but the step is bypassed by the simple device of hydrolysing off the phosphoryl group from the 1:6 compound as shown.

Fructose-1:6-bisphosphate

+ H_2O

Fructose-1:6-bisphosphatase

Fructose-6-phosphate

Similarly, when glucose-6-phosphate is reached, the gluco-kinase (or **hexokinase**) reaction of glycolysis is irreversible but, in liver, **glucose-6-phosphatase** produces glucose which is secreted from the cell. The complete gluconeogenesis reactions are shown in Fig. 11.2. Note that whether at any given time glycolysis or gluconeogenesis is taking place depends on control mechanisms described in Chapter 12.

There are thus four enzymes involved in gluconeogenesis that do not participate in glycolysis: pyruvate carboxylase; PEP-CK; fructose-1:6-bisphosphatase; and glucose-6-phosphatase. The levels of these enzymes are about 20–50 times greater in rat

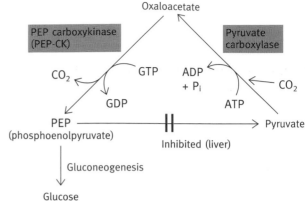

Fig. 11.1 Diagram showing the generation of phosphoenolpyruvate for gluconeogenesis from pyruvate in the liver. Note that this scheme makes sense only if the PEP → pyruvate reaction is prevented. How this is done is described in Chapter 12 on metabolic control (page 177).

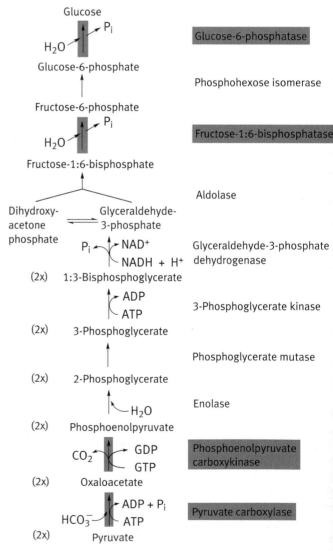

Fig. 11.2 The complete gluconeogenesis pathway from pyruvate to glucose. Reactions in yellow are different from those occurring in glycolysis.

liver than in rat skeletal muscle, in keeping with the importance of gluconeogenesis in liver.

What are the sources of pyruvate used by the liver for gluconeogenesis?

In starvation, after the glycogen reserves have been exhausted, the main source of pyruvate originates from the breakdown of muscle proteins. Hydrolysis of the latter produces 20 amino acids. Although the amino acid **alanine** represents only a small proportion of the latter, of the total amino acids leaving the muscle to be transported to the liver, more than 30% is in the form of this amino acid. Alanine is a **glucogenic amino acid**—its carbon skeleton is capable of being converted into glucose by the liver and the same is true of most of the other amino acids. (The metabolism of amino acids is separately discussed in Chapter 15, so do not be concerned with the mechanism of the reactions mentioned in this section.) How is alanine formed from the protein breakdown products in muscle? Several of the amino acids can give rise to extra oxaloacetate by reactions of the citric acid cycle. The oxaloacetate can be converted into pyruvate by the reactions shown in Fig. 11.1 and converted into alanine for release into the blood. The amino group of alanine comes from the other amino acids. Note, however, that this formation of alanine in muscle depends on the muscle not oxidizing the pyruvate to acetyl-CoA which, in starvation, would defeat the whole object of the exercise. In the situation of starvation the muscle will be preferentially utilizing fatty acids and ketone bodies for energy generation so that acetyl-CoA will be very plentiful. As will be described in Chapter 12, a high acetyl-CoA/CoA ratio leads to the inactivation of pyruvate dehydrogenase. The pyruvate produced from the amino acids is therefore reserved for alanine synthesis. In summary, the net effect is that many of the amino acids derived from muscle protein breakdown are converted to alanine, which migrates in the blood to the liver. (The amino acid, glutamine, is also synthesized in muscle for transport to the liver for glucose synthesis but the principle is exactly the same.) In the liver, the alanine is converted back to pyruvate and thence to glucose. The overall process is summarized in Fig. 11.3.

As the concentration of blood ketone bodies rises during starvation, the brain progressively uses more of these for energy generation and so reduces its utilization of glucose and lessens the demand for gluconeogenesis (which, however, always remains essential). This is important for it requires about 2 grams of muscle protein to be broken down for each gram of glucose made, a rate of loss that, if continued, would greatly reduce the period that could be survived in starvation.

It should be noted that, in Chapter 15 on amino acid metabolism, a glucose–alanine cycle is presented (Fig. 15.9). In this, alanine is synthesized in muscle and transported to the liver where it is converted to glucose as described here but with the difference that the pyruvate needed in muscle for alanine

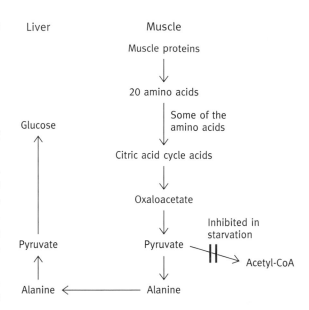

Fig. 11.3 Mechanism by which breakdown of muscle proteins supplies the liver with a source of pyruvate for gluconeogenesis during starvation. Note that the scheme is dependent on the pyruvate dehydrogenase not using the pyruvate (see text). Glutamine is also produced in muscle as a gluconeogenic substrate for the liver in addition to alanine.

synthesis comes from the breakdown of glucose. The sequence of events in this cycle is liver glucose → muscle glucose → muscle alanine → liver alanine → liver glucose. This does not result in a net gain of glucose and cannot therefore contribute to the blood glucose supply problem but may be relevant in the transport of amino nitrogen from muscle to liver.

A second source of pyruvate for gluconeogenesis, of less importance in starvation but important in more normal situations, is **lactate**, produced by the anaerobic glycolysis of glucose or glycogen (page 97). As stated, certain cells such as those of kidney medulla, and retina are virtually anaerobic and mature red blood cells, lacking mitochondria, have no oxidative generation of ATP. In normal nutritional situations, a major source of lactate is **muscle glycolysis**—in strenuous muscle activity, the glycolysis rate exceeds the capacity of mitochondria to reoxidize NADH and lactate is produced. This travels via the blood to the liver where it is converted to pyruvate and thence to glucose (or glycogen). This constitutes a physiological cycle, called the Cori cycle after its discoverers (Fig. 11.4). The cycle has two main effects; it 'rescues' lactate for further use and, secondly, it counteracts **lactic acidosis**. The introduction of large amounts of lactic acid into the blood may exceed its buffering power and cause a deleterious fall in pH of the blood. The synthesis of glucose from lactate involves the uptake of two protons (when $NADH+H^+$ is used for reduction of 1,3-bisphosphoglycerate).

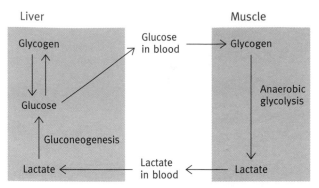

Fig. 11.4 The Cori cycle. Note that this is a physiological cycle involving muscle and liver. The muscle has very low levels of three enzymes essential for gluconeogenesis, which explains why the lactate has to travel from muscle to liver. Excess lactate is produced in muscle during violent action during which glycolysis outstrips the capacity of mitochondria to re-oxidize reduced NAD^+—that is, during anaerobic glycolysis.

Synthesis of glucose from glycerol

Another source of carbon compound for gluconeogenesis in the liver is **glycerol** released from **triglyceride hydrolysis** mainly in adipose tissue. This is taken up by the liver and converted to glucose by the route shown in Fig. 11.5. The enzyme glycerol kinase, which phosphorylates glycerol as the first step in the conversion, is very low in amount in adipose tissue and therefore glycerol is not converted to glucose in that tissue, but the enzyme is present in liver. This system is logical in that, in a situation where blood glucose supply is of vital importance, the

glycerol produced in fat cells is transported to the liver and converted to blood glucose. Adipose cells therefore do not use the glycerol themselves; supplying the brain with glucose takes priority over that.

During prolonged starvation, after a small initial drop in blood glucose level, the latter remains constant for several weeks and fatty acids supplied by adipose cells likewise remain at a constant level so that the mechanisms are extremely effective. It is, however, perhaps surprising that the whole business of muscle wastage during starvation could, on the face of it, have been avoided by animals using a simple metabolic trick present in bacteria and plants that permits fat to be converted to glucose. No doubt there are good evolutionary reasons for animals not using it; we will now describe this metabolic trick.

Synthesis of glucose via the glyoxylate cycle

E. coli can survive quite well with acetate as its sole carbon source. A net conversion of acetyl-CoA to C_4 acids of the citric acid cycle and thence to carbohydrate or any other component of the cell can occur, unlike the situation in animals. In plant seeds, energy stored as triglycerides is converted to glucose on germination. What allows bacteria and plants to do this?

These organisms possess the normal citric acid cycle but, in addition, can bypass some of its reactions by other reactions not present in animals. If you analyse the citric acid cycle, two carbon atoms are added to oxaloacetate from acetyl-CoA, giving citrate (C_6), but then two carbon atoms are lost as two CO_2 molecules in forming succinate (C_4), so that there is no *net* conversion of C_2 to citric acid intermediates.

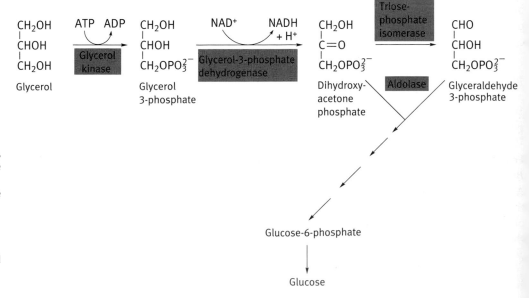

Fig. 11.5 Conversion of glycerol, released by neutral fat hydrolysis, to glucose in the liver. Much of the glycerol is produced in adipose cells but, since the glycerol kinase is not present there, gluconeogenesis from glycerol occurs in the liver. This is a sensible arrangement since it results, during starvation, in blood sugar production from the glycerol.

The **glyoxylate route** bypasses these losses. Isocitrate is directly split into succinate and glyoxylate—a sort of cycle shortcut.

$$\begin{array}{l} COO^- \\ | \\ CHOH \\ | \\ CHCOO^- \\ | \\ CH_2COO^- \end{array} \xrightarrow{\boxed{\text{Isocitrate lyase}}} \begin{array}{l} CH_2COO^- \\ | \\ CH_2COO^- \end{array} + \begin{array}{l} COO^- \\ | \\ CHO \end{array}$$

Isocitrate   Succinate   Glyoxylate

The C_2 glyoxylate now reacts with acetyl-CoA to produce malate—back on the citric acid cycle.

$$\begin{array}{l} CH{-}COO^- \\ \| \\ O \end{array} + \begin{array}{l} \quad\quad O \\ \quad\quad \| \\ CH_3{-}C{-}S{-}CoA \end{array}$$

Glyoxylate   Acetyl-CoA

$$+ H_2O$$

$$\boxed{\text{Malate synthase}}$$

$$\begin{array}{l} OH \\ | \\ CH{-}COO^- \\ | \\ CH_2{-}COO^- \end{array} + CoA{-}SH$$

Malate

The scheme is shown in Fig. 11.6. The net effect is that acetyl-CoA plus oxaloacetate convert to malate plus succinate. Succinate and malate can both be converted to oxaloacetate, one molecule then being available for conversion to glucose. In plants, this process occurs in membrane-bounded organelles, called **glyoxysomes**.

It should finally be mentioned that by far the greatest amount

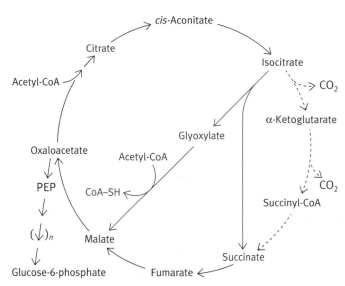

Fig. 11.6 Diagram of the glyoxylate cycle of plants and bacteria, which permits carbohydrate synthesis from acetyl-CoA. This does not occur in animals. The broken line represents the bypassed section of the citric acid cycle; the red lines are reactions peculiar to the glyoxylate cycle. The two CO_2 molecules are highlighted to emphasize that it is these two decaboxylation reactions that must be bypassed. It will be appreciated that, as a result of the glyoxylate reactions, two molecules of oxaloacetate are produced from citrate, one of which is used to form citrate again and one to produce PEP.

of carbohydrate synthesis on Earth occurs during photosynthesis in plants. This involves the fixation of CO_2 into another glycolytic intermediate, bisphosphoglycerate, from which glucose synthesis occurs. The mechanism of this is best left until we deal with photosynthesis in Chapter 14.

We have, in the last few chapters, dealt with energy production from glucose and fat, with fat and glucose synthesis. Now is the time to remind you that fat metabolism and carbohydrate metabolism do not go on independently. They form an integrated metabolic activity, each affecting the other, and the whole lot requiring regulation. This important topic will be dealt with in the next chapter.

Further reading

Felig, P. (1975). Amino acid metabolism in man. *Ann. Rev. Biochem.*, **44**, 933–55.
A most useful review which discusses amino acid metabolism at the organ level, together with the effects of starvation, diabetes, obesity and exercise on it. It also discusses gluconeogenesis at this level.

Snell, K. (1979). Alanine as a gluconeogenic carrier. *Trends Biochem. Sci.*, **4**, 124–8.

Reviews the role of muscle in supplying the liver with metabolites for glucose synthesis.

Pilkis, S. J., El-Maghrabi, M. R., and Claus, T. H. (1988). Hormonal regulation of hepatic gluconeogenesis and glycogen. *Ann. Rev. Biochem.*, **57**, 755–83.
Comprehensive review.

Problems for Chapter 11

1 After 24 h starvation, the glycogen reserves of the liver are exhausted but there are still relatively large stores of fat. What is the point of the body going to such trouble to make glucose by gluconeogenesis when virtually unlimited supplies of acetyl-CoA from fatty acids are available to supply energy?

2 Gluconeogenesis requires production of phosphoenolpyruvate (PEP), but pyruvate kinase cannot form this from pyruvate. Why is this and how is the problem overcome?

3 From PEP, gluconeogenesis in the liver produces blood glucose mainly by reversal of glycolytic enzymes. However, two enzymes are involved that do not participate in glycolysis. What are these?

4 Does muscle have glucose-6-phosphatase? Explain your answer.

5 What is the Cori cycle and its physiological role?

6 Adipose cells do not have glycerol kinase, the enzyme which converts glycerol to glycerol-3-phosphate, even though adipose cells produce glycerol from TAG hydrolysis. Liver does have the enzyme. Explain why this appears to be a logical situation.

7 Animals cannot convert acetyl-CoA to carbohydrate. Bacteria and plants can. Explain how they do this.

Chapter 12

Strategies for metabolic control and their application to carbohydrate and fat metabolism

This chapter should be read in conjunction with the preceding chapters on metabolism. We suggest you keep referring back to the relevant metabolic pathways to refresh your memory. What we will do in this chapter is to put together all the preceding metabolism to see how the various pathways work and interact. In many ways this is the pay-off for all the work you've done so far because it is where the many reactions make collective physiological sense.

We have collected together metabolic control mechanisms into a separate chapter rather than dealing with the topic when the pathways themselves were discussed, for three reasons: (1) the pathways themselves are complicated enough without the additional information on control; (2) metabolic control makes more sense when the different pathways are considered together; and (3) less importantly perhaps, it is easy enough to go back to the earlier chapters and indeed the revision may be beneficial.

Although control extends to all facets of metabolism, the control and integration of carbohydrate and fat metabolism are especially important because the flow of chemical change (flux) through these pathways is very large and, in the body, the direction of the fluxes is changed very frequently. In an animal with periodic meals, interspersed by periods of fasting, there is a need for change of direction of metabolic flows. Control of carbohydrate and fat metabolism illustrates all of the principles of metabolic control. The subject is a fairly complex one so let's be clear on how this chapter deals with it. Enzymes are responsible for metabolism and therefore metabolic control involves control of enzyme activities. We therefore first deal in a general sense with *how* enzyme activities are regulated. After this we describe how the pathways are kept in balance by the regulatory enzymes, which sense the levels of indicative metabolites in other pathways and automatically adjust their rates of activities to the information.

This is not the end because the metabolic activities of cells are controlled in a wider sense by circulating hormones and neural inputs that operate in the body as a whole according to current needs. We finally therefore describe how signals external to the cells—sometimes called extrinsic signals—regulate metabolism on top of the automatic controls first described above.

Before we deal with any of this we must consider the following question.

Why are controls necessary?

It is obvious when you think of the major pathways that we've dealt with—glycogen synthesis, glycogen breakdown, glycolysis, gluconeogenesis, fat breakdown, fat synthesis, the citric acid cycle, electron transport, etc.—that they can't all be running at full speed (or even necessarily at all) at the same time. If a metabolic pathway is proceeding in one direction the reactions in the reverse direction must be switched off. In a single pathway such as glycolysis, its required rate will vary enormously according to the energy needs at the time. The basal metabolic rate of someone playing squash is about six times greater than that of the person at rest and, in the most strenuous type of physical exercise, it may be about 15 times greater than the resting value.

In short, the metabolism of the main energy-yielding materials must be regulated for at least two broad purposes:

(1) to respond to energy production needs as energy expenditure varies;

(2) to respond to physiological needs—metabolic pathways need to work in different directions after a meal when metabolites are being stored (Chapter 5) as compared with intervals between meals when storage metabolites are being utilized. The needs are different again in prolonged starvation and in pathological situations such as diabetes where carbohydrate metabolism is abnormal.

There is another reason that is less self-evident—the potential problem of **futile cycles** or **substrate cycles** as they are now more properly called.

The potential danger of futile cycles in metabolism

The process of gluconeogenesis discussed in the preceding chapter provides a good example for illustrating the potential danger in metabolic systems of futile cycles. One wonders how the problem was coped with during the evolution of the pathways. Consider the phosphofructokinase (PFK) step in glycolysis and its reversal in gluconeogenesis by fructose-1:6-bisphosphatase (Fig. 12.1). In one direction, fructose-6-phosphate is phosphorylated by PFK to yield 1:6-bisphosphate, while in the reverse direction the bisphosphatase hydrolyses the product back to fructose-6-phosphate. This cycle of events could, unchecked, proceed and achieve nothing more than to rapidly destroy ATP with the generation of heat—more or less equivalent to an electrical short circuit. (Bumble-bees deliberately use this trick to warm up their flight muscles on cold mornings before take-off.)

Extending this to the whole of glycolysis and gluconeogenesis, these pathways, uncontrolled, could constitute a giant futile cycle again doing nothing but uselessly destroying ATP. The same applies to glycogen synthesis and breakdown, and fat synthesis and breakdown (Fig. 12.2), or, for that matter, to virtually any synthesis and breakdown pathways.

It is clear from this that the breakdown and synthesis directions of a metabolic pathway must be controlled in a

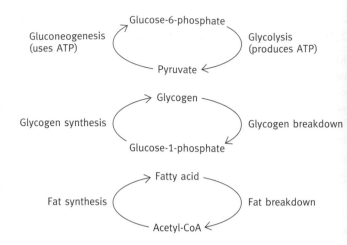

Fig. 12.2 Potential large-scale futile cycles if metabolism were not controlled.

reciprocal manner—activation of one and inhibition of the other. This independent control of the two directions can occur only, as mentioned earlier (page 3), at irreversible metabolic steps for it is here that there are separate reactions in the two directions catalysed by distinct enzymes that can be separately controlled. A freely reversible reaction is catalysed by the same enzyme in both directions and cannot be reciprocally controlled in the two directions. With the realization that controls are operative, the term 'futile cycle' has been replaced by 'substrate cycle' to describe situations such as the fructose phosphate one. Such substrate cycles are, in fact, crucially important control points since they provide the opportunity for independent controls on the forward and reverse metabolic pathways. Substrate cycling, if allowed to occur at all, may appear to be completely wasteful. However, a small amount of cycling may be advantageous in making controls on the forward and backward metabolic pathways more effective. Suppose you have a pathway in which A is converted to C via B and the reaction A ↔ B occurs via a substrate cycle.

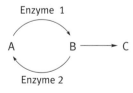

The rate of conversion of A to C can be reduced by inhibiting enzyme 1 or activating enzyme 2. If you do both, the control is much more effective; in fact, complete shutdown could be achieved without requiring excessively high levels of the inhibitor for enzyme 1. The same type of considerations apply to increasing the flux of metabolite from A to C. The fructose-6-phosphate ↔ fructose-1:6-bisphosphate cycle has dual controls of the type described.

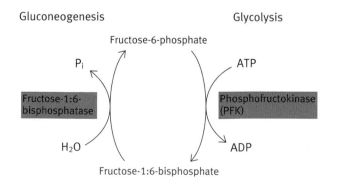

Fig. 12.1 Potential futile cycle at the phosphofructokinase (PFK) step in glycolysis if the reactions were uncontrolled.

How are enzyme activities controlled?

Metabolic regulation of a pathway means regulation of the rate of catalysis of one or more reactions in that pathway. The enzymic process could include translocation of a metabolite across a membrane as well as a chemical reaction. There are basically two ways of reversibly modulating the rate of an enzyme activity in a cell. These are:

(1) to change the amount of the enzyme;

(2) to change the rate of catalysis by a given amount of the enzyme.

There are, of course, ways of irreversibly activating enzymes such as the proteolytic conversion of trypsinogen to active trypsin (page 64). However, in this chapter we are dealing with control mechanisms that are reversible since this is essential to meaningful metabolic control.

Metabolic control by varying the amounts of enzymes

The level of a protein in a cell can be changed either by altering the rate of its production or the rate of its destruction. Proteins are relatively short-lived—the half-life of enzymes in, say, the liver, might range from about an hour to several days.

Metabolic control in animals by changes in enzyme levels is important. The control is a relatively long-term affair with effects being seen in hours or days rather than seconds. It is at the level of adaptation to physiological needs. There are many instances where enzyme levels are adjusted. As illustrations of this, the amount of lipoprotein lipase in capillaries (page 86) is adjusted to the fat demands of the tissues, its level increasing, for example, in lactating mammary glands. The liver changes its enzyme content within hours in response to dietary changes— whether it has to cope with high-fat or high-carbohydrate intake. In the well-fed state, hepatic enzymes involved in fat synthesis increase in amount, while a few hours of starvation reverses this situation. Intake of foreign chemicals such as drugs results in a rapid increase in hepatic drug-metabolizing enzymes (page 216). However, as already emphasized, the changes usually occur over hours or days in animal systems, although in specific cases they can be more rapid if the enzyme has a short half-life; nonetheless it is generally true that only slow responses of metabolic control can be obtained through changes in enzyme levels. Control at the level of enzyme synthesis is rapid in bacteria—for example, if *E. coli* is exposed to lactose as its sole carbon source, synthesis of the enzyme, β-galactosidase, needed to hydrolyse the sugar, starts up immediately and an increase in enzyme activity is detectable in minutes and a thousandfold increase can be measured within hours. In such rapidly dividing cells, changes in enzyme levels are much more dramatic than in the body but, nonetheless, we are still dealing with significant time periods. By the same token, after the stimulus for enzyme synthesis is no longer there, reversals of increases in enzyme levels occur relatively slowly since the levels are returned to lower values only by destruction of the proteins (or dilution by cell growth in bacteria) and this is related to their half-life. We will return to this subject later when we deal with control of gene expression (Chapter 21). Control of enzyme level does not come into the instant automatic metabolic control that exists over the activities of pre-existing enzymes in the cell.

Metabolic control by regulation of the activities of enzymes in the cell

For regulation of *activity*, the amounts of given enzymes are not altered, only the rate at which they work. The essential point about this method of control is that it can be instantaneous. It is necessary at this point to deal with some of the properties affecting the catalytic rate of an enzyme.

Basic kinetics of enzyme action

Hyperbolic kinetics of a 'classical' enzyme

You may recall (page 4) that in enzyme (E) catalysis, the substrate (S) combines reversibly at the active centre of the enzyme to form an enzyme–substrate complex (ES). The reaction then occurs, yielding product(s) (P) that dissociate(s) from the enzyme leaving it free to recycle. This is the model developed by Michaelis and Menten to fit the observed kinetics of enzyme catalysis,

$$E + S \leftrightarrow ES \rightarrow E + P.$$

At low substrate concentration [S], the rate of an enzyme-catalysed reaction is determined by the collision rate between S and the free enzyme E. At these low [S] values, ES formation is not favoured and hence the rate of reaction is low. As [S] increases so will the proportion of enzyme in the form of ES increase and, therefore, the rate of reaction will increase. However, as [S] further increases, the point is reached at which the rate of formation of ES ceases to be limiting and the catalytic activity of the enzyme becomes rate-limiting. The catalytic rate of the enzyme is an inherent property and varies from enzyme to enzyme. Some enzymes work extremely rapidly, and others more slowly. At high levels of [S], the enzyme is said to be saturated with substrate. The rate of the reaction is then called v_{max} for maximum velocity, and further increases in the substrate have no effect on the rate of catalysis. The result of this is a hyperbolic curve as shown in Fig. 12.3. These kinetics are referred to as Michaelis–Menten kinetics and an enzyme displaying them as a **Michaelis–Menten enzyme**. For a long time they were the only type of enzyme recognized and they are often referred to as 'classical' enzymes.

So far, we have used the term 'low substrate concentrations'. This is imprecise since what, in absolute molar terms, might be a low concentration of S for one enzyme may be a saturating concentration of substrate for another enzyme. It depends on

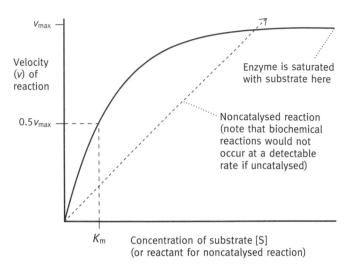

Fig. 12.3 Effect of substrate concentration on the reaction rate catalysed by a classical Michaelis–Menten type of enzyme. K_m, Michaelis constant. The dashed line shows, for comparison, the effect of reactant concentration on a noncatalysed chemical reaction. Note that the two lines are drawn to illustrate their shapes, not their relative rates. The reactions proceeding in the living cell would not proceed at a detectable rate unless catalysed.

how tightly the substrate binds to the enzyme or as it is usually expressed what the affinity of the enzyme is for its substrate. This will depend on the precise relationship of the substrate to its enzyme—the nature and number of the weak bonds established between them.

This relative affinity of an enzyme for its substrate in many cases can quantitively be measured as the K_m (m for Michaelis) of the enzyme. (Note however that it is a true affinity measurement only where the rate of dissociation of ES to E + S is much greater than that of the catalytic step of ES to E + P.) It is defined as that [S] at which the enzyme is working at half-maximal velocity as shown in Fig. 12.3 and is therefore independent of the total amount of E. (Other methods of plotting such graphs can be used to determine K_m values more precisely but no attempt is made here to deal with enzyme kinetics beyond the simple level needed to understand the basic principles of metabolic control.) The higher the K_m, the lower is the affinity of the enzyme for its substrate. K_m values can be used to compare the affinities of different enzymes for their substrates. It is impossible to make a precise statement about the values of K_ms since they can range from submicromolar to several millimolar but, as a generalization, K_ms for many enzymes are in the range of 10^{-4}–10^{-6} M. In the cell, concentrations of substrates are usually such that the general statement can be made that enzymes are not saturated with substrate but rather are working in subsaturation ranges of substrate concentrations. The important thing is that enzymes are unlikely to be saturated with their substrates and, indeed, their K_m values can usually be taken to reflect the cellular concentrations of their substrates.

Which enzymes in metabolic pathways are regulated?

The direct answer to this question is those enzymes designed by evolution to be regulated. In short, there are in metabolic pathways specific regulatory enzymes; they are strategically placed—for example, in reversible processes at irreversible steps in the reaction pathways (page 3).

The first enzyme of a pathway is usually a strategic place for control. Consider a metabolic pathway such as

A → B → C → D → E,

in which E is an end-product that is needed by the cell. It makes sense not to produce more E than is needed. An automatic control mechanism is for the reaction A → B to be controlled by the level of E as shown below.

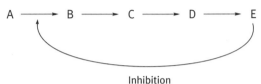

Inhibition

The control occurs through inhibition of enzyme activity and/or through the slower mechanism of lowering the level of the enzyme. This control is rational for it avoids build-up of the intermediates such as would occur if the block were at a later reaction. When little or no metabolite, E, is present, the inhibition is relieved and E is synthesized. This is analogous to a float-valve control on a water cistern: as the water rises, a float attached to an arm rises and closes the tap. It is known in biochemistry, though not in plumbing, as feedback or end-product inhibition. Numerous metabolic pathways exist in bacteria (such as for the synthesis of amino acids) in which such controls work with military precision. When it comes to fat and carbohydrate metabolism, the whole system is so complex that you can't always identify the end-products of pathways in quite the same precise way, so that 'end-product inhibition' is not such a relevant term. Nonetheless, the same principle applies, namely, that key intermediates (products) can control metabolically distant enzymic reactions. The control may be feedback as described, or 'feedforward' in which metabolites activate downstream enzymes that will be needed to cope with the flow of metabolites coming their way.

The nature of regulatory enzymes

There are essentially two main ways in which the catalytic activity of enzymes are modulated (as distinct from change in amount):

(1) by **allosteric control**;

(2) by **covalent modification** of the protein—mainly by phosphorylation and dephosphorylation.

These will now be described.

Allosteric control of enzymes

Allosteric control of enzymes is of central importance in metabolic regulation. The prefix *allo* means 'other'; it refers to the existence on an enzyme of one or more binding sites other than for substrate. Any small molecule that reversibly binds to a site on a protein is called a **ligand**; the definition is sometimes extended to include a small protein. The ligands that bind to the allosteric sites are called **allosteric effectors** or **moderators**. They need not, and usually don't, have any structural relationship to the substrates of the enzyme. *At a given substrate concentration*, a positive effector increases the activity of the enzyme when it combines at the allosteric site and a negative effector decreases it, so the terms allosteric activator and inhibitor are also used for allosteric ligands.

A few cases are known in which the allosteric effector reduces the catalytic rate of the enzyme so that the v_{max} is reduced. This is very uncommon as compared with the type in which the allosteric effector alters only the affinity of the enzyme for its substrate. Since, as mentioned before, enzymes typically work at subsaturating levels of [S], *increase in their affinity will increase their activity and decrease in their affinity has the reverse effect*. At saturating levels of [S], the v_{max} is unchanged by the allosteric effector even if the affinity is changed, but this situation does not (usually) occur in the cell.

The mechanism of allosteric control of enzymes

We will now discuss the nature of **allosteric enzymes** in which the substrate-binding affinity is moderated by binding of a ligand to an allosteric site, since this is the predominant type.

Such allosteric enzymes have a multi-subunit structure; they are made up of more than one catalytic protein molecule, which are assembled into a single enzyme complex by noncovalent bonds. The individual molecules are variably called **protein subunits**, **protomers**, or **monomers**. Although each is a separate catalytic entity, the subunits interact in the assembled enzyme molecule. A typical plot of reaction velocity versus substrate concentration for such an enzyme is shown in Fig. 12.4. It differs from a Michaelis–Menten enzyme in that the response of the enzyme velocity to changes in substrate concentration is S-shaped or sigmoidal, rather than hyperbolic. Why should this be, and how does it happen?

As to the first, when an allosteric activator (positive effector) binds to the enzyme, the S-shaped curve is moved to the left, and, with an allosteric inhibitor, to the right, as shown in Fig. 12.5. The former is due to increases in the binding affinity of the enzyme for its substrate; the latter, to a reduction. The sigmoidal shape of the response to substrate concentration means that over a range of substrate concentrations (centred at that giving half-maximum velocity), the rate of enzyme catalysis is more sensitive to substrate concentration change (and therefore to binding affinity change) than with a Michaelis–Menten type of

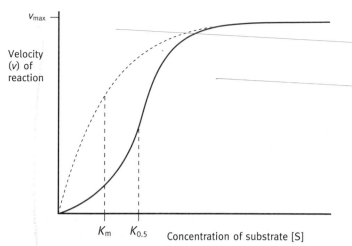

Fig. 12.4 Effect of substrate concentration on the rate of reaction catalysed by a typical allosteric enzyme. The dashed line shows, for comparison, the corresponding curve for a classical Michaelis–Menten enzyme. The term K_m is not strictly applicable to a non-Michaelis–Menten enzyme, and the term $K_{0.5}$ is used instead.

enzyme (the curve is steepest there).

What causes the sigmoidal response of velocity to substrate concentration? Binding of a substrate molecule to *one* of the monomers in an allosteric enzyme causes the binding of subsequent substrate molecules to the other monomers of the enzyme to occur more readily. This is known as **positive homotropic cooperative binding of substrate**—'homo', because only one type of ligand, the substrate, is involved. Since only the substrate binding site is involved in producing the sigmoid

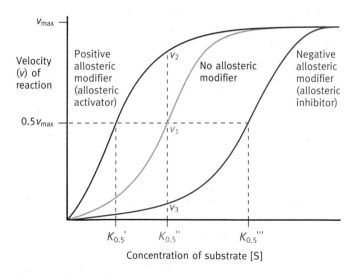

Fig. 12.5 Effect of substrate concentration on a typical allosterically regulated enzyme. v_1, v_2, and v_3 are, respectively, the reaction velocities observed with no allosteric modifier, with an allosteric activator, and with an allosteric inhibitor, at a fixed concentration of S ($K_{0.5}''$). The curves are drawn arbitrarily; actual shapes and positions will be a function of the particular system.

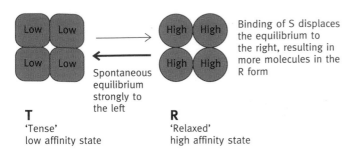

Binding of S displaces the equilibrium to the right, resulting in more molecules in the R form

T
'Tense'
low affinity state

R
'Relaxed'
high affinity state

Fig. 12.6 Diagram of concerted model of cooperative binding of substrate (S). In this model the enzyme exists in two forms, T and R, the two being in spontaneous equilibrium. Binding of a substrate molecule to the R state swings the equilibrium to the right, thus increasing the affinity of all of the subunits in that molecule. 'High' and 'low' refer to affinities of the enzyme for its substrate.

substrate concentration response, it does not strictly fit the name of allosteric conformational change involving another site, but it has become common to loosely call any conformational change produced by a ligand an allosteric one. Therefore, cooperative binding is often referred to as an allosteric effect, despite the fact that an allosteric site properly refers to a site other than that for substrate.

What is the mechanism of this? There are two theoretical models. In both of these, an increase in substrate concentration results in more and more catalytic subunits in a given collection of enzyme molecules being in a high-affinity state. The two models differ in the way in which the change occurs in a single enzyme molecule.

In the **concerted model** of Monod, Wyman, and Changeaux (MWC model), either *all* of the subunits in a molecule are low-affinity or *all* are high-affinity, the two being in spontaneous equilibrium (Fig. 12.6), but with the equilibrium strongly to the low-affinity side. Binding of substrate to a single subunit swings this equilibrium to the high-affinity state. Thus, as [S] increases, more and more enzyme molecules are in the high-affinity state. Note that what we have said so far is concerned with the cooperative binding of substrate; the effect of the allosteric modifier, or effector, which is what *controls* the enzyme, is a separate question. On this question, the model also assumes that allosteric modifiers alter the position of the equilibrium between the low-affinity (tense, or T) state and the high-affinity (relaxed, or R). If the allosteric modifier binds more tightly to the R state, it will displace the T ↔ R to the right and result in a higher proportion of molecules being in the R state—that is, it will, at a

given substrate concentration, activate the enzyme. As the positive allosteric modifier is increased and more of the enzyme converts to the high-affinity state, it will result in the substrate velocity curve tending towards the hyperbolic type. If, by contrast, the allosteric effector is of the negative type, it is assumed that it binds more tightly to the T state and thus results in more enzyme molecules being in this state and a decreased reaction rate in the collection of enzyme molecules as a whole.

The **sequential model** of Koshland, Nemethy, and Filmer (KNF, Fig. 12.7) differs in that it assumes that, in the absence of substrate, *all* enzyme molecules are in the low-affinity state; there is no equilibrium with high-affinity-state molecules. When one molecule of substrate binds to one of the subunits, it changes its conformation from tense (T) to relaxed (R). This has the effect of facilitating a similar change in an adjacent subunit so that a second molecule of substrate binds more easily. Binding of the second molecule of substrate increases still further the ease with which a third adjacent subunit can make the conformational change with a resultant increase of affinity for the substrate of that subunit and so on (Fig. 12.7). Both models explain observed effects; there is nothing to say that the actual mechanism of the conformational change cannot be something intermediate between the two models, and individual cases may differ in this respect.

A few enzymes such as glyceraldehyde phosphate dehydrogenase (page 110) exhibit cooperative substrate binding but do not have any allosteric modifiers. Such enzymes could be more sensitive to substrate concentration changes than are Michaelis–Menten-type enzymes, and this is possibly the reason for the cooperative binding effect. Also, to anticipate what follows, a different and important type of allosteric control involves an enzyme whose catalytic subunits are affected by the presence of an inhibitory or activating regulatory subunit. The latter responds to an allosteric modifier by detaching from the complex, thus freeing the enzyme from inhibition or activation by that subunit. Evolution is concerned with what works, not with the tidiness of the overall control system as presented on paper.

Reversibility of allosteric control

A most important point to note is that allosteric control is essentially instantaneous—both in its application and reversibility. The allosteric ligand attaches to its site by noncovalent

Fig. 12.7 Diagram illustrating the sequential model for the combination of an allosteric enzyme with its substrate. The binding of a single substrate molecule to a subunit causes a conformational change in the subunit. This facilitates the conformational change of a second subunit when it combines with a substate molecule and so on for the next subunit. The net effect is to make it 'easier' for successive subunits to undergo the conformational change, which is seen as increasing affinity for the substrate by successive subunits.

bonds and, when the concentration of ligand is reduced, it dissociates from combination with the enzyme and everything goes into reverse.

Allosteric control is a tremendously powerful metabolic concept

The vital point about allosteric control is that *the allosteric effector need not have any structural relationship whatsoever to the substrate of the enzyme regulated (and usually doesn't) or even belong to the same area of metabolism.* This means that any metabolic pathway or area of metabolism can be connected in a regulatory manner to any other metabolic area, the metabolite(s) of one pathway being regulator(s) of another. Moreover, an enzyme can have allosteric sites for more than one effector and thus can receive control signals from several metabolic areas, which increases the flexibility of control. The systems controlled are complex; glycogen breakdown feeds glycolysis, which feeds pyruvate into the citric acid cycle, and this, in turn, feeds electrons into the electron transport system. This feeds ATP into the energy-utilizing machinery of the cell. Fatty acid breakdown also feeds the citric acid cycle, supplying acetyl-CoA, while, in reverse, acetyl-CoA formed from pyruvate feeds fatty acid synthesis, and so on.

In such a complex system, each section has to 'know' what others are doing by sensing the levels of key metabolites. Is there enough or too little ATP? Is the citric acid cycle being supplied with too little or too much acetyl-CoA? Is glycolysis too fast or too slow? You can see the chemical chaos that would result unless there were constant second-to-second adjustment of pathways in the light of information reaching each pathway about all other pathways and, indeed, about parts of its own pathway. In case you have missed the point already stressed, this is why allosteric control is such an important concept— controlling enzymes can be designed, in the evolutionary sense, to receive signals from anywhere. Thus glycolysis can be 'informed' on how the citric acid cycle is doing and how the electron transport system is keeping up ATP supply, the information automatically adjusting activities to make the whole a harmonious chemical machine. As Monod, one of the discoverers of the allosteric concept, pointed out, you simply could not have anything as complex as a cell without it. He described it as the 'second secret of life' (DNA being the first).

The control network *is* complex. A given enzyme is not just switched on or off by one allosteric signal. As stated, it may receive multiple signals from all over the metabolic map, each partially inhibiting or stimulating its rate of reaction. It is a classic case of evolution being a tinkerer—one little adjustment here and another there, by yet another allosteric signal. This gives fine-tuning of enzyme activity, and natural selection of the control mechanisms ensures a smooth performance of the whole complicated mass of reactions.

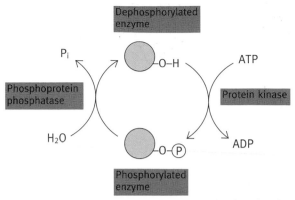

Fig. 12.8 Diagram of the control of an enzyme by phosphorylation. The phosphorylated enzyme may, in specific cases, be more active or less active than the unphosphorylated enzyme. Note that variations of this may occur in that the activity of an enzyme may be controlled by an inhibitor protein whose inhibitory activity is controlled by phosphorylation.

Control of enzyme activity by phosphorylation

So much for control of enzymes by allosteric modification. The second method is by phosphorylation. To be strictly accurate, this section ought to refer to covalent modification of enzymes rather than phosphorylation because other chemical modifications occur. However, phosphorylation is of such overwhelming importance and known alternatives so rare that we will confine ourselves to phosphorylation. The principle is very simple. Enzymes, called **protein kinases**, exist that transfer phosphoryl groups from ATP to specific proteins. When this happens, the target enzyme undergoes a conformational change such that the enzyme (or an enzyme inhibitor protein) changes its activity (or inhibitory effect). You can imagine that the addition of such a strongly changed group could have an effect on the structure of a protein molecule. To reverse the process there are protein phosphatases that hydrolyse the phosphate from the protein (Fig. 12.8).

The phosphorylation occurs on the hydroxyl group of a serine or threonine of the polypeptide chain of the enzyme, and is identified by the protein kinase by the neighbouring sequence of amino acids around the target —OH group. (In Chapter 26 we will describe tyrosine group phosphorylation but this is not relevant in this chapter.) We show serine as an example of the target group of the phosphorylation processes.

$$\begin{array}{c}|\\\mathrm{CH-CH_2OH}\\|\\\mathrm{CO}\\|\end{array} + \mathrm{ATP} \longrightarrow \begin{array}{c}|\\\mathrm{CH-CH_2-O-\overset{\displaystyle O}{\overset{\displaystyle \|}{P}}-O^-}\\|\qquad\qquad|\\\mathrm{CO}\qquad\quad O^-\\|\end{array} + \mathrm{ADP}$$

Serine residue
of polypeptide

This concept of control, as so far described, is very simple—it is a second general mechanism by which enzymes are controlled—allosteric control was the first, phosphorylation and dephosphorylation of proteins the second. We will return to the application of these to metabolism, soon.

Control of specific metabolic pathways

The two classes of controls—intrinsic and extrinsic (or extracellular)

Allosteric control gives an immediate regulatory mechanism, but control by phosphorylation requires that the phosphorylation and dephosphorylation processes themselves are controlled. The balance between phosphorylation and dephosphorylation determines the activity of the target enzyme. What, therefore, controls the protein kinases and phosphatases? The answer lies for the most part *not* in intracellular or intrinsic automatic controls but usually in controlling agents such as hormones, external to the cell. (Individual exceptions occur to this in which phosphorylation is part of the intrinsic metabolic controls, but the general point applies.)

Metabolic control is, as indicated, a complex business. To present the essentials of pathway regulation in a comprehensible manner we will deal with it in two broad sections. First, there are the internal or **intrinsic controls** that coordinate the different pathways and parts of pathways so that metabolic pile-ups and shortages don't occur. To repeat the point, these are the automatic controls in which metabolites 'signal' to other pathways, and other parts of their own pathway, so that a smoothly running chemical machine results. These controls are largely, but not completely, allosteric in nature. Imagine (again) the chaos there would be if glycolysis produced pyruvate irrespective of how much of it the citric acid cycle could handle, or if fatty acid synthesis took no notice of fatty acid oxidation, etc. In feedforward control, an early metabolite in a pathway 'activates' a distant enzyme in the same pathway to cope with the traffic coming its way.

A completely isolated cell could, in principle, carry out this intrinsic control. But, as already implied, in the body there is a second group of controls involving **extrinsic agents** (external to the cell, internal to the body) in the form of hormones and neurotransmitter substances that instruct cells on what their major metabolic direction at the time should be—such as whether to store fuel, or release it. This is essential if cells are to do whatever is beneficial for the body as a whole. Individual cells cannot make these decisions—they must receive signals from outside. Note that, after being so instructed, the intrinsic controls are still essential to keep the pathways coordinated to avoid bottlenecks and metabolic snarl-ups. It may be likened to the organization of a navy. Each individual ship has its own internal (allosteric) system of discipline that, in all circumstances, maintains it as an organized unit, but what it does as a

unit—where it sails and what it does—depends on external signals (hormonal and neurological) from higher naval authorities (endocrine glands and the brain).

With that introduction we will now deal with the intrinsic controls of individual metabolic pathways, picking out the major ones only. After that we will come to extrinsic controls.

The main instrinsic controls on carbohydrate and fat metabolism

As suggested earlier, it probably would be useful for you to refresh your memory on each pathway as we deal with it by referring to earlier chapters.

Instrinsic controls on glycogen metabolism

Glycogen metabolism involves synthesis of the polymer by glycogen synthase and its breakdown by glycogen phosphorylase. Glycogen synthesis occurs in times of dietary plenty and is controlled by extrinsic agents (see page 172). Glycogen breakdown has two physiological purposes and glycogen phosphorylase occupies a key strategic site in metabolism; it produces glucose-1-phosphate which is converted to the 6-phosphate by phosphoglucomutase. Glucose-6-phosphate is the entry point to the glycolytic pathway and hence to ATP production. Glycolysis is important for energy production in all tissues but it has especial importance in muscle because contraction makes large demands on the ATP supply (described in Chapter 28). In liver, glycogen breakdown has an additional special significance for the production of blood glucose.

The control of glycogen phosphorylase has been intensively studied, particularly in muscle, and it is now considered one of the 'classical' control enzymes. In **resting muscle**, the enzyme exists in a form known as phosphorylase *b* which, though inactive on its own, is allosterically partially activated by AMP. (Full activation requires the extrinsic controls described later.) Why is AMP used as the allosteric signal here? Glycogen phosphorylase needs to be activated when the energy reserves of the cell are low; that is, low ATP. Because of the reaction catalysed by adenylate kinase,

$$2ADP \leftrightarrow AMP + ATP,$$

AMP is indicative of the presence of ADP, and this, in turn, means that ATP is partially broken down (by energy-requiring reactions).

The equilibrium of the reaction catalysed by adenylate kinase is such that, if ATP is partially broken down to ADP, it results in a larger relative increase in AMP, which therefore is a sensitive indicator of the situation. Activation of phosphorylase by AMP, to provide more energy, is therefore an appropriate response. In keeping with this, ATP and glucose-6-phosphate both inhibit the enzyme (Fig. 12.9): if these are plentiful, why produce more? This simple intrinsic control situation is, however, overridden by extrinsic controls in different physiological situations (described on page 172). In liver, the situation is much the

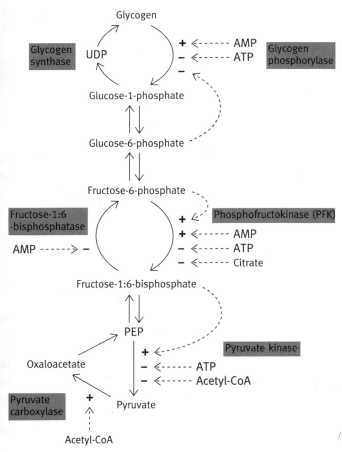

Fig. 12.9 Diagram of main intrinsic allosteric controls on glycogen metabolism, glycolysis, and gluconeogenesis. See text for explanation. Green broken lines, allosteric controls having positive effects on activities; red broken lines, allosteric controls having negative effects on activities. UDPG, Uridine diphosphoglucose.

same; AMP partially activates phosphorylase *b* (to about 20% of its potential maximum). As with muscle, this intrinsic control is overridden by extrinsic controls but for a different physiological reason (page 172). To summarize, in both muscle and liver, in the absence of the extrinsic controls described later, phosphorylase is in the inactive form *b*, unless allosterically activated by AMP, this activation being antagonized by ATP and glucose-6-phosphate.

Intrinsic allosteric controls on glycolysis and gluconeogenesis

Figure 12.9 shows the main systems. The rationale of controls is as follows: AMP indicates an increase in the ratio of ADP to ATP. As well as activating glycogen phosphorylase, AMP is an activator of phosphofructokinase (PFK). At the same time, it inhibits the fructose-1:6-bisphosphatase. Activation of glycogen breakdown increases the level of fructose-6-phosphate, which also activates PFK. Activation of PFK will, in turn, increase the level of fructose-1:6-bisphosphate, which activates pyruvate kinase, an example of feedforward control. Allosteric metabolic controls can be complex. Evolution is rarely satisfied by one-hit

controls. Thus, the activating effects of AMP on PFK are balanced by the inhibitory effects of ATP. The cell is thus constantly monitoring the ATP/ADP ratio (via AMP) and adjusting the glycolytic speed accordingly. In addition, when the ATP level is high, the citric acid cycle flux is diminished and citrate accumulates. The latter is transported out of the mitochondria to the cytoplasm where it allosterically inhibits PFK. This again makes sense since, if the citric acid cycle is partially shut down, glycolysis, which feeds the cycle, should likewise be inhibited. The same is true for the acetyl-CoA inhibition of pyruvate kinase.

The activation of pyruvate carboxylase by acetyl-CoA (Fig. 12.9) to produce oxaloacetate needs explanation. Accumulation of acetyl-CoA will occur if the citric acid cycle is deficient in oxaloacetate since the latter is needed to accept the acetyl moiety to form citrate. The acetyl-CoA thus automatically activates the cycle's anaplerotic or topping up reaction. It is also important in gluconeogenesis (see pages 120 and 151).

Intrinsic controls on pyruvate dehydrogenase, the citric acid cycle, and oxidative phosphorylation

Pyruvate dehydrogenase (page 114) occupies a highly strategic position in metabolism, for it is the irreversible reaction producing acetyl-CoA, by which the carbohydrate metabolic route feeds into the citric acid cycle and into the fat synthesis route. The pyruvate dehydrogenase enzyme complex is regulated in several ways (Fig. 12.10). Acetyl-CoA and NADH, both products of the reaction, allosterically switch off their own production while CoA—SH and NAD^+ substrates activate the enzyme. Thus the acetyl-CoA/CoA—SH ratio and the NADH/NAD^+ ratio control pyruvate dehydrogenase. The logic is obvious. If there is a great deal of acetyl-CoA and NADH, it means that the production of acetyl-CoA and NADH by pyruvate dehydrogenase should be reduced; the reverse is true if CoA—SH and NAD^+ are high.

Of major importance is the negative control by high ATP levels. This, in effect, is monitoring the 'energy charge'. If ATP is low, feed in more fuel; if high, cut off the fuel supply. The ATP control on pyruvate dehydrogenase is not a direct one. At high ATP/ADP ratios, a pyruvate dehydrogenase kinase is activated and this inactivates the dehydrogenase by phosphorylation of the enzyme; this antagonized by a phosphatase (Fig. 12.10). The kinase is actually part of the pyruvate dehydrogenase complex. The inactivating kinase is additionally allosterically activated by NADH and acetyl-CoA. The rationale of all this is as already stated: if the signals indicate adequate energy production, close off the fuel valve.

In the citric acid cycle and electron transport area, the intrinsic controls are somewhat different in that, while allosteric controls do exist, the major internal controls are the availability of NAD^+ and ADP as substrates. If much of the NAD^+ is in the form of NADH, then the dehydrogenases in the cycle are restricted in their activity. Since NADH will accumulate when

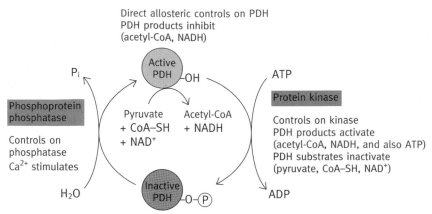

Fig. 12.10 Diagram of the intrinsic control of pyruvate dehydrogenase (PDH) by direct allosteric control and by phosphorylation and dephosphorylation in the mammalian enzyme complex. The multiplicity of these controls reflects the strategic position of PDH whose activity is the gateway for pyruvate to enter the citric acid cycle and the fat synthesis pathways. Despite their complexity in detail, they largely add up to activation by substrates and inhibition by products. (In this context ATP can be regarded as a product of the result of entry of acetyl-CoA into the citric acid cycle.) It is worth noting that control by phosphorylation of proteins is usually associated with extrinsic controls, rather than with the intrinsic ones here. As regards the stimulation by Ca^{2+}, the calcium control may be effected by catecholamines (see page 361).

electron transport to oxygen (or the supply of oxygen) cannot cope with available NADH produced by the cycle, this constitutes an important control. Similarly, if the ADP/ATP ratio is low, electron transport is inhibited because oxidation and phosphorylation are tightly coupled, again a desirable control. This tight coupling, called **respiratory control**, is of major importance though no precise mechanism for it is known. In short, if there is plenty of ATP the furnaces shut down automatically. In addition to this control by NAD^+ and ADP availability, the cycle is controlled at the citrate synthase step (ATP inhibits), at the isocitrate dehydrogenase step (ATP inhibiting and ADP activating), and at the α-ketoglutarate dehydrogenase step (succinyl-CoA and NADH inhibit). It is all very logical and the cycle metabolites are kept in balance one with another—no pile-ups, no shortages.

Intrinsic controls on fatty acid oxidation and synthesis

These are illustrated in Fig. 12.11. The rationale is that you want *either* fatty acid oxidation *or* fatty acid synthesis but not both at the same time in the same cell.

The two pathways mutually suppress each other. Fatty acyl-CoAs (the product of first step in fatty acid oxidation; page 131) allosterically inhibit acetyl-CoA carboxylase, the first committed step in fat synthesis. Malonyl-CoA, the first product irretrievably committed to fat synthesis (page 139), allosterically inhibits transfer of the fatty acyl groups to carnitine (page 132). This prevents the acyl groups being transported to the intramito-chondrial site of their oxidation. If you are synthesizing fats it is a good idea to stop fatty acids being fed into the mitochondrial furnaces. Acetyl-CoA carboxylase is activated *in vitro* by citrate (which also gives rise to acetyl-CoA, the substrate of the enzyme; see page 142). Citrate is transported out of the mitochondria only when its level is high and this happens only in times of food plenty when it is logical to store fat. There is no doubt that it

does activate acetyl-CoA carboxylase in the test tube, but some studies have raised questions of whether this activation is of significance in the cell. The major control on acetyl-CoA carboxylase is inactivation by phosphorylation. This is carried out by a protein kinase whose activity is dependent on the presence of AMP, but reversal of this by a phosphatase is hormone-controlled (see page 178).

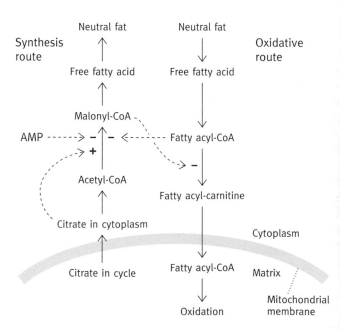

Fig. 12.11 Diagram showing major intrinsic control points in fat oxidation and synthesis by which the two routes mutually suppress each other. Note that, once acetyl-CoA is formed from fat breakdown, its further metabolism is subject to the controls of the citric acid cycle, etc. already described. Dashed lines indicate allosteric effects. While citrate activates acetyl-CoA carboxylase *in vitro*, whether it is significant in the cell has been questioned.

The above account, while not exhaustive, will have acquainted you with the main intrinsic controls operating inside cells on carbohydrate and fat metabolic pathways. From these you can see how all of the pathways interact so that metabolism is not a collection of isolated reaction sequences, but rather one complex network of chemical reactions.

What you cannot see from the above account, however, is how the cell's metabolism is directed according to physiological needs of the body as a whole, for this, we repeat, is the function of external control by hormones and other agents. They often regulate the same reactions as those controlled by intrinsic agents in metabolic pathways for, as explained (page 162), certain reactions are logical control points. The external controls operate on top of the intrinsic controls.

They are the subject of the next section.

Control of carbohydrate and fat metabolism by extracellular agents in the body

In this field, the hormones of especial importance are glucagon, insulin, epinephrine, and norepinephrine, and these are what we will deal with here. These are the ones that have immediate and rather dramatic effects and that are invoked on a daily basis as we oscillate between eating and fasting periods between meals (see page 73).

However, without going into any details, there is a great complexity of metabolic controls. Pituitary hormones, adrenal cortex hormones, and thyroid hormones (Table 26.1) all have their impact on metabolism in ways sometimes not fully, or hardly, understood and in general are of long-term indirect nature. You have only to remember the tendency for people with high thyroxine levels to be thin and overactive and those with low thyroxine levels to have reverse characteristics to appreciate the long-term impact of this hormone on fuel consumption in the body. But, as stated, certain hormones are of central importance in this field and have rapid, direct, effects. It is these we will now discuss.

Glucagon, you will recall, is the 'hunger' hormone produced by the pancreas when blood sugar is low; insulin has the opposite effect to glucagon (page 76). Epinephrine is the hormone liberated by the adrenal medulla and, along with norepinephrine, has a mobilizing effect on food storage reserves. Norepinephrine is also liberated from sympathetic nerve endings. The sympathetic system controls involuntary contraction of smooth muscle such as that of viscera but it also innervates adipose tissue. The sympathetic nerves are a means of rapidly liberating norepinephrine in the immediate locality of target organs—otherwise there is no fundamental difference from that of a hormone flooding the whole body and affecting target cells.

What regulates the levels of insulin, glucagon, and epinephrine in the blood?

Insulin is synthesized by the β cells of the pancreatic islets of Langerhans. They secrete insulin, a small protein, directly into the blood where it has a short half-life. Thus, the insulin signal to the body only persists as long as secretion continues. The β cells are very sensitive to blood glucose concentration—as soon as the level rises after food intake, insulin secretion starts. The response of β cells to raised blood glucose occurs within a minute. The glucose transporter by which glucose enters the cells has a low affinity for glucose so that it functions maximally only at high blood sugar levels, such as occur after a meal. The rate of insulin secretion plotted against glucose concentration is sigmoidal so that secretion rate is very sensitive to changes in glucose concentration above normal levels of about 90 mg dl^{-1} (or 5 mM). Glucagon, by contrast, is secreted by the pancreas in response to low blood glucose concentration. Epinephrine is released from the adrenal glands in response to nerve signals.

How do the hormones, glucagon, epinephrine, and insulin, work? **Hormones** are a major topic dealt with in Chapter 26 but a brief preliminary introduction to them is needed here. They are chemical signals released into the blood so that all tissues are exposed to them. However, only certain cells, called **target cells**, respond to a given hormone. Lipid-soluble hormones (steroids and thyroxine) go straight through cell membranes without resistance, but water-soluble ones such as glucagon, epinephrine, and insulin do not. Instead they combine with **receptor proteins**, specific for each hormone, displayed like aerials on the outside of cells. Hormones combine with their receptors with the precision of a substrate combining with an enzyme. Only target cells have the receptors to specific hormones—the rest have none and are 'blind' to the hormone.

The hormones are quickly eliminated from the blood so that, unless the source gland is releasing more, the concentration in the blood falls, hormone is no longer bound to the receptor, and the signal ceases. However, while the hormone is bound to the receptor, chemical changes occur inside the cell that lead to a response (Fig. 12.12). This brings us to the **second messenger concept**.

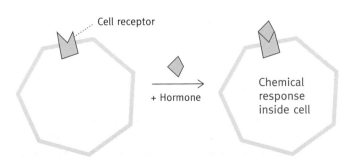

Cell receptor

+ Hormone

Chemical response inside cell

Fig. 12.12 Diagram illustrating the way in which hormone binding to surface receptor causes chemical events inside the cell.

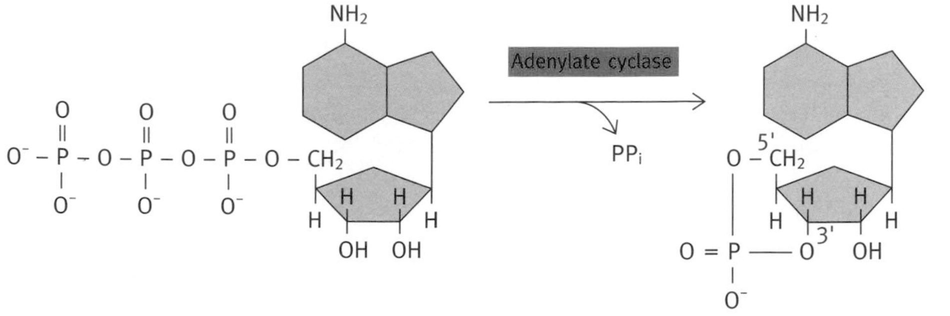

Adenosine triphosphate (ATP)

Adenosine-3',5'-cyclic monophosphate (cAMP)

Fig. 12.13 The synthesis of cyclic AMP from ATP by adenylate cyclase.

What is a second messenger?

If you regard the hormone as the first messenger, then binding to its external receptor on a cell causes a change in the level of a second messenger—an intracellular regulatory molecule that triggers cell responses.

What is the second messenger for glucagon, epinephrine, and norepinephrine?

In the case of glucagon and the epinephrines the second messenger is adenosine-3',5'-monophosphate, or **cyclic AMP (cAMP)** a molecule you haven't yet met in this book. It is synthesized from ATP by the enzyme **adenylate cyclase** (Fig. 12.13).

Now things start to come together because, inside the cell,

cAMP (note, not AMP) is the allosteric activator of a protein kinase. Thus, hormone attachment to an external cell receptor results in cAMP production inside the cell and this, in turn, allosterically activates a **cAMP-dependent protein kinase, PKA** (A for cAMP), which phosphorylates serine—or threonine—OH groups of specific enzymes thus modulating their activities. The enzymes affected will be given shortly when we deal with specific control systems. Thus the general mechanism is as in Fig. 12.14.

The way in which cAMP activates PKA is shown in Fig. 12.15. In the absence of cAMP, PKA is a tetramer with two catalytic subunits (C) and two regulatory subunits (R). When cAMP binds to the regulatory subunits, the enzymically active C monomers are released.

An enzyme, **cAMP phosphodiesterase**, hydrolyses cAMP to AMP (Fig. 12.16) inside the cell so that the activation of cellular PKA depends on continual production of cAMP, which occurs as long as the hormone is bound to the receptor. As mentioned, the hormones in the blood are also continuously destroyed or removed and their presence therefore depends on continuous production. Thus, everything has the required element of reversibility. If the hormone levels fall, cAMP production ceases and everything goes back to square one, phosphorylated proteins being dephosphorylated by protein phosphatases.

Hormone
(glucagon, epinephrine)

↓

Binds to receptor
of target cell

↓

Increases cAMP level
inside cell

↓

cAMP activates
a protein kinase (PKA)

↓

Protein kinase phosphorylates
specific proteins

↓ ↓ ↓ ↓

This phosphorylation leads to
changes in enzyme activites and
metabolic responses

Fig. 12.14 Steps in the hormonal control of metabolism. Not shown are: (1) that cAMP is continually destroyed; (2) that phosphorylation of proteins is reversed by phosphatases. The metabolic response shown occurs only as long as the hormones are bound to the receptor.

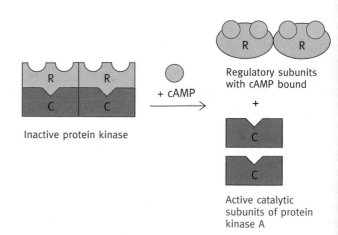

Fig. 12.15 Activation of cAMP-dependent protein kinase (PKA) by cAMP. R, regulatory subunit of PKA; C, catalytic subunit of PKA.

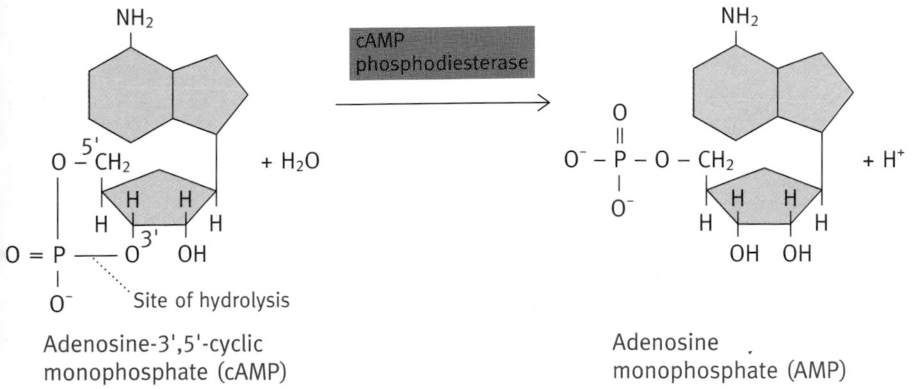

Fig. 12.16 Reaction catalysed by cAMP phosphodiesterase. The name arises from the fact that cAMP contains a doubly esterified phosphoryl group—that is, a phosphodiester.

What is missing so far in this account is an explanation of *how* the hormone binding switches on cAMP production and how the latter is switched off. We are deferring description of this until Chapter 26 (page 356) because it is part of the more general topic of signal transduction across cell membranes. For the moment then, we suggest that you just accept that glucagon and epinephrine binding to their cell receptors result in cAMP increase inside the cell.

Control of the number of receptors

The number of receptors on the surface of a cell specific for a given **agonist**, as the binding signal is called, can be regulated. Continuous exposure of a cell to an agonist can result in a temporary reduction in receptor number and desensitization of the cell to that agonist so that it responds less strongly. This is called **downregulation**. In addition, the receptor may be inactivated by phosphorylation on its cytoplasmic side so that, on combination of the agonist with the receptor, there is no stimulation of cAMP production. Since this phosphorylation is itself cAMP-stimulated, the system forms a logical cAMP feedback inhibition of its own production. Several hormone receptors, including those for epinephrine, insulin, and glucagon, display downregulation and desensitization.

How does insulin control metabolism?

In general, whatever glucagon does to energy metabolism, insulin does the opposite (Chapter 5). Insulin is a protein hormone that has been known for a long time and probably more research has been done on it than on any other hormone since it is of central importance in fat and carbohydrate metabolism and, because of diabetes, has a high profile. Despite this, we know less about its mechanism of action than those of the other hormones referred to. The existence of insulin receptors exposed to the outside of cell membranes is known. Quite recently, it has been shown that such receptors span the membrane and that the inside domain of the protein has a tyrosine protein kinase activity that is different from that already described and that is stimulated by the binding of insulin (see

Chapter 26). In the diagrams that follow, we show insulin as affecting specific enzymes, whereas for glucagon, epinephrine, and norepinephrine we take cAMP to be the second messenger. In short, it is not certain what the second messenger(s) for insulin is or are, or whether there are, indeed, any at all in this context. You will have a better idea of the possibilities of how insulin works after studying Chapter 26 but it would be inappropriate to go into these now.

Control of glucose and fat uptake into cells

Glucose cannot readily penetrate lipid bilayers and therefore **membrane transport proteins** are needed for its entry into cells. While absorption from the gut is active (page 66), transport into most other cells of the body is passive—the facilitated diffusion type in which a specific transporter protein is provided for glucose molecules to traverse the membrane, driven only by the glucose gradient across the membrane (page 49).

In brain and liver, this uptake of glucose is not insulin-responsive but in muscle and adipose cells it is, the rate being stimulated by the hormone. It appears that, inside insulin-responsive cells, there is a reserve of non-functional glucose transport proteins and insulin causes a recruitment of these into the cytoplasmic membrane, thus effectively increasing the rate of glucose transport (Fig. 12.17). It has been shown in adipose cells that the effect of insulin in causing transfer of the specific glucose transporter proteins into the cell membrane is complete in about 7 minutes. If the insulin is removed, the process reverses, the transporters being returned to the nonfunctional intracellular reserve in about 20–30 minutes.

The mechanism of this remarkable effect is unknown. It may be that the hormone initiates movement of the transporters, or it may be that they are continually recycling from the intracellular pool to the cytoplasmic membrane. In the latter situation, insulin might cause the transporters to accumulate in the membrane.

The energetically unassisted transport of glucose by itself would simply equilibrate cellular and blood levels. However,

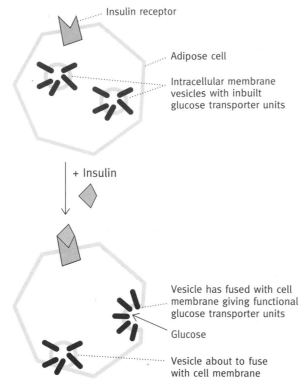

Fig. 12.17 Diagram of the effect of insulin in mobilizing glucose transporter units in adipose cells and presumably other cells with insulin-dependent glucose entry. The glucose transport is of the passive facilitated diffusion type. Note that liver and brain are not insulin-responsive in this respect.

once in the cell, the glucose is immediately removed by glycogen synthesis or other metabolism and this constitutes a driving force for uptake of the sugar. The first step is the phosphorylation of glucose

Glucose + ATP → glucose-6-phosphate + ADP.

The enzyme in liver doing this is glucokinase and elsewhere hexokinase, different enzymes carrying out the same reaction. As mentioned earlier, glucokinase has a much higher K_m for glucose (that is, a lower affinity) than has hexokinase (see page 84). What might seem to be a trivial 'stamp-collecting' fact is an important regulatory device. Given the glucostatic role of liver, there would be no point in it taking up glucose except when blood glucose is high. When blood glucose is low, the liver is turning out the sugar so, to take it up again, would not be logical. The high K_m of liver glucokinase means that this is prevented; only when the blood sugar is high does glucokinase work rapidly. Moreover, this enzyme is **induced** by insulin. (The term 'induced' has a technical meaning in biochemistry—it means that synthesis of the enzyme is increased by the inducer—in this case, insulin. The mechanism by which this is caused by the hormone is not known.)

In the case of fat uptake, as explained, only free fatty acids (not triglycerides) can enter cells, which they do freely because

the lipid bilayer poses no barrier to them. Therefore, control of fatty acid uptake is largely dependent on control of their blood level—which is determined by glucagon, epinephrine, norepinephrine, and insulin acting on adipose cells as described later in this section.

Having given general information on control by external agents, we now turn to the control of specific metabolic pathways by these agents.

Control of glycogen breakdown and synthesis by extracellular agents

Control of glycogen breakdown

We earlier described the regulation of glycogen phosphorylase b in resting muscle and liver involving the activation of the enzyme by AMP. This, as it were, represents the bread and butter control system to cater for basic needs. However, a quite different regulatory system exists in which phosphorylase b is converted to phosphorylase a, which is active without the presence of AMP. The conversion overrides the 'routine' metabolic control by AMP. Phosphorylase b is converted to the 'a' form by a protein kinase that phosphorylates a serine —OH group on the enzyme causing a conformational change to the activated form. (It is important to be clear on the difference between **phosphorolysis** which involves splitting off of glycogen residues by inorganic phosphate, the reaction catalysed by phosphorylase, and **phosphorylation** which involves transfer of a phosphoryl group of ATP to the serine —OH by the protein kinase.) In muscle, the signal for the latter reaction is the presence of epinephrine in the blood which attaches to receptors on the muscle cell membrane and causes the synthesis of cAMP inside the cell. The mechanism by which cAMP causes the phosphorylation of phosphorylase b is dealt with below. But before that we should look at the physiological rationale of the whole process.

Epinephrine is liberated by the adrenal glands in response to a neural signal from the brain that there is an emergency or threatening situation that may require a very rapid muscular response—known as the **fight or flight reaction**. Epinephrine presses the metabolic alarm button—the muscle cell is then no longer concerned with waiting for an intrinsic signal by AMP indicating that ATP supplies are not maximal, but rather with unrestricted pouring on the fuel for ATP-generating systems to cope with the threatening situation. Increasing the supply of glucose-6-phosphate (via glucose-1-phosphate) is especially appropriate since the instant flight response depends on glycolysis for ATP generation (explained on page 389). Because of its low yield of ATP, glycolysis must proceed on a relatively massive scale. Activation of phosphorylase permits this to occur.

Epinephrine also mobilizes the liver for the flight response so that it rapidly releases glucose into the blood, the rationale again

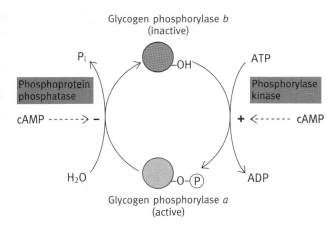

Fig. 12.18 Diagram of the conversion of glycogen phosphorylase *b* to glycogen phosphorylase *a* by phosphorylase kinase and the reverse by protein phosphatase. It is important to note that the effects of cAMP are not exerted *directly* on the reactions shown—see text. The —OH group is that of a serine residue in the protein.

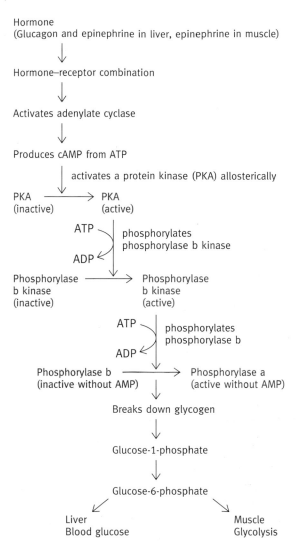

being that in a threatening situation you maximize the fuel supply to the muscles. Thus, the activation of phosphorylase by phosphorylation supplies the glucose-6-phosphate (via glucose-1-phosphate) for glucose production. It is interesting that potato phosphorylase (which breaks down starch by the same reaction) is a protein with a high degree of homology to muscle phosphorylase (the amino acid sequences of the two proteins are almost 50% identical) but it exhibits no such control properties, presumably related to the fact that potatoes have no flight or fight response and no blood sugar to control.

In addition to this, the liver produces blood glucose in response to glucagon. This hormone, which is released from the pancreas in response to low blood sugar, also raises the level of cAMP in the liver cell and has the same effect on phosphorylase as epinephrine. (Muscle does not have receptors for this hormone and therefore does not respond to it.)

With that background we will now turn to the question of how the conversion of phosphorylase *b* to *a* occurs.

Mechanism of cAMP control of glycogen breakdown and synthesis

As stated above, glycogen phosphorylase exists in two forms: *a* and *b*. '*a*' is phosphorylated on a serine hydroxyl residue; '*b*' is not, the two being interconverted as in Fig. 12.18.

The cAMP activation of the phosphorylase kinase (that is, a protein kinase), which phosphorylates glycogen phosphorylase, is not direct; as stated earlier, it activates a different protein kinase, called PKA (A for cAMP), which, in turn, activates phosphorylase kinase (also by phosphorylation) which then activates phosphorylase. The whole scheme from the hormone onwards is therefore as shown.

Why so complicated a mechanism? A cell will have only a relatively small number of hormone molecules attached to its receptors. Especially in the case of a life-threatening emergency, the body requires a huge response to epinephrine binding to cells. It is worthwhile recalling that a minute amount of glucose, such as 1 micromole (180 micrograms), contains 6.025×10^{17} molecules. The attachment of a relatively few molecules of hormones to cell receptors must result in astronomical numbers of glucose molecules being turned out by the liver and, if you are being chased by a tiger, preferably in an extremely short time. (The tiger's liver is doing this too.) The same applies to glucose-6-phosphate production in muscles for generating energy.

What is needed is a rapid, vastly amplified response, in which the minute signal (hormone binding) cascades into a flood of reaction (glycogen breakdown). The multiple steps in phosphorylase activation constitute an amplifying cascade. Suppose one molecule of hormone activates one molecule of adenylate cyclase and that the latter produces 100 molecules of cAMP per minute. In this time period the amplification is a hundredfold. If the cAMP activates a second enzyme that produces an activator at the same rate, the amplification becomes more than 100×100, and so on. In fact, glycogen phosphorylase activation involves four such amplification steps. It is known as a **regulatory cascade** and is used whenever a massive chemical response is needed from a minute signal. Phosphorylase attacks the end of glycogen chains (page 82) and it wouldn't be much use producing large numbers of active phosphorylase molecules if they couldn't find a glycogen end to work on. This is possibly one reason why glycogen is so highly branched compared with starch. Plants break down starch in a similar manner but they do not need the same emergency breakdown rate—hence, a more leisurely rate of starch degradation permits the molecule to have fewer ends to attack, or so this speculation might suggest.

Reversal of phosphorylase activation

All metabolic controls must be reversible—that is, there must be a mechanism for switching 'off' as well as 'on' and this provides a potential additional control point. In the case of phosphorylase, the 'a' form is converted back to the inactive 'b' form by a phosphatase, called protein phosphatase 1, that hydrolyses off the phosphoryl group from its serine residue

ENZ—O—P + H$_2$O $\rightarrow$ ENZ—OH + P$_i$ + H$^+$.
('a' form)   ('b' form)

In liver this is stimulated by free glucose due to an allosteric effect on the 'a' form.

It is advantageous during the conversion of phosphorylase b to the 'a' form, to switch off the opposing phosphatase action since this will result in more rapid and more complete activation of the enzyme, an important consideration in a life-threatening situation. There exists in many cells a protein called phosphatase inhibitor 1; when phosphorylated by the same protein kinase A that is activated by cAMP, it inhibits the phosphatase. Thus, cAMP simultaneously switches on the activation of phosphorylase and switches off its deactivation.

Finally, it is to be noted that in *normal* muscle contraction (that is, not in the emergency situation involving cAMP), phosphorylase b kinase is allosterically activated by the Ca^{2+} that triggers the contraction following a motor nerve signal. (The contraction mechanism is dealt with later in Chapter 28.) Since *this* activation of phosphorylase kinase does not involve phosphorylation, it is instantly reversed by the removal of Ca^{2+} that occurs after cessation of the neural signal for contraction. The Ca^{2+} activation is mediated by a protein called **calmodulin**, which is found in almost all eukaryote cells as part of enzyme complexes. It acts as a Ca^{2+} detector—when the ion binds to the subunit it undergoes a conformational change which activates the enzyme. Many enzymes are controlled in this way.

A summary of the main facts of this somewhat complicated control system may be useful at this stage (see also Fig. 12.19).

1. On terminology, phosphorylase is the enzyme catalysing glycogen breakdown; phosphorylation is a completely distinct matter involving transfer of a —Ⓟ group from ATP to a protein serine—OH group by a protein kinase; phosphatase splits off the phosphoryl group from proteins.

2. In resting muscle and in liver, phosphorylase is in the unphosphorylated 'b' form, which is inactive unless allosterically (partially) activated by the presence of AMP. This activation does not involve phosphorylation of the protein.

3. Epinephrine in muscle and liver cells and glucagon in liver cells increase the level of cAMP.

4. cAMP allosterically activates a protein kinase, which, in turn, activates phosphorylase b kinase. The latter phosphorylates the 'b' form of phosphorylase, producing the active 'a' form. The process is an amplifying cascade.

5. In normal muscle contraction, Ca^{2+} ions are released into the cell by the motor neural signal and these *allosterically* partially activate phosphorylase b kinase; this gives a partial activation of phosphorylase. Unlike the cAMP-induced activation of the phsophorylase b kinase, the Ca^{2+} activation of this enzyme does not involve phosphorylation.

6. Conversion of the 'a' form back to the inactive 'b' form of phosphorylase is effected by protein phosphatase 1. The latter is controlled by protein phosphatase inhibitor 1 which

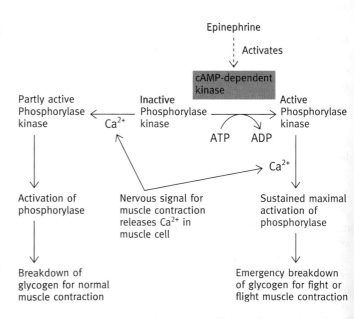

Fig. 12.19 Control of muscle phosphorylase kinase. This has, as well as the mechanism already described for both liver and muscle (to right of diagram), an additional Ca^{2+} activator (cAMP-independent) of phosphorylase kinase. This is to provide energy for normal contraction, which is triggered by Ca^{2+} release. The active (phosphorylated) kinase similarly requires Ca^{2+} for full activity. The kinase contains a protein subunit, calmodulin, to which Ca^{2+} binds, causing the calmodulin to activate the kinase.

inhibits the phosphatase only when it (the inhibitor) is phosphorylated. The phosphorylation is effected by the same cAMP-stimulated protein kinase that acts on phosphorylase *b* kinase. Thus, activation of phosphorylase to the '*a*' form is automatically accompanied by inactivation of the phosphatase. Importantly, the conversion of the '*a*' to the '*b*' form in liver is activated by glucose.

How is glycogen synthase controlled?

Glycogen synthesis and glycogen breakdown need to be controlled in a reciprocal manner—when one is switched on the other has to be switched off to avoid futile cycling. We have seen that phosphorylase *b* is activated (indirectly) by phosphorylation by PKA, the latter being activated by cAMP. Glycogen synthase is directly phosphorylated by the same PKA but in this case phosphorylation *inactivates* the enzyme. The cascade for this process is shorter, possibly reflecting the fact that there is not the same panic to synthesize glycogen as there can be to break it down and therefore a smaller amplification will suffice.

Hormone
(glucagon and epinephrine in liver, epinephrine in muscle)
↓
Hormone-receptor combination
↓
Activates adenylate cyclase
↓
Produces cAMP from ATP
↓ activates a protein kinase (PKA) allosterically
PKA   →   PKA
(inactive)   (active)
      ATP ⤵
      ADP ⤴   phosphorylates glycogen synthase
Glycogen synthase → Glycogen synthase
(active)   (inactive)

We have also seen that phosphorylase *a* inactivation by dephosphorylation is also controlled—the phosphatase doing this is inactivated by PKA phosphorylating the phosphatase inhibitor and activating the latter as described above. A somewhat similar control exists for the synthetase but with these important differences: (1) dephosphorylation of the synthase *activates* it; (2) a protein kinase *inactivates* the phosphatase inhibitor. The latter kinase is not PKA but one that is specifically activated by insulin. By this means, insulin *activates* phosphatase 1, which activates glycogen synthase. A remarkable fact is that the same phosphatase inhibitor 1 is involved in both systems but PKA activates the inhibitor while the insulin-activated kinase inactivates it. A lot of tinkering has gone on. These mechanisms are illustrated in Fig. 12.20 and the overall net effect on metabolism of glycogen in Fig. 12.21.

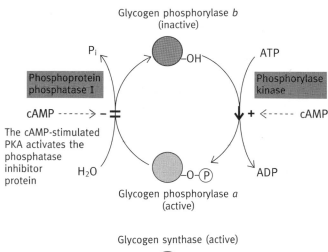

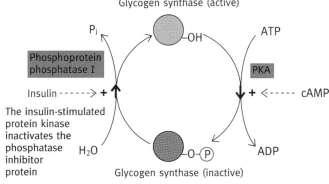

Fig. 12.20 Diagram illustrating the reciprocal controls operating on glycogen phosphorylase and glycogen synthase. The coloured circles represent enzyme molecules: red, inactive; green, active.

Control of glycolysis and gluconeogenesis by extracellular agents

It is logical to deal with these two together since they involve most of the glycolytic enzymes in common. The signal for the liver to produce blood glucose by either glycogenolysis (or glycogen breakdown) or gluconeogenesis is glucagon. Thus, it is logical for cAMP *in liver* to switch on glycogen breakdown *and* gluconeogenesis. It is *not* therefore logical for cAMP to switch on glycolysis, the reverse of gluconeogenesis, in liver (Fig. 12.22(a)).

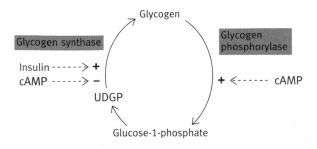

Fig. 12.21 Overall scheme of external control of glycogen metabolism. Note that the effects of insulin and cAMP on the enzymes are indirect.

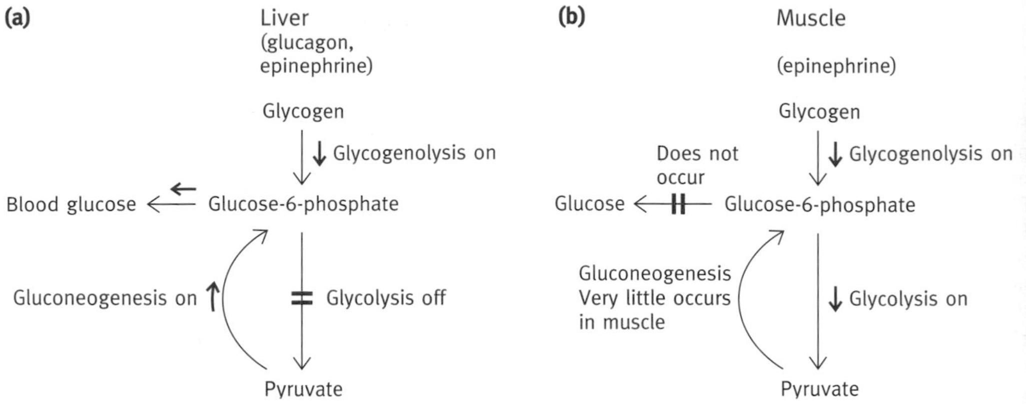

(a) Liver
(glucagon,
epinephrine)

Glycogen

↓ Glycogenolysis on

Blood glucose ⟵ Glucose-6-phosphate

Gluconeogenesis on ↑ ⟺ Glycolysis off

Pyruvate

(b) Muscle

(epinephrine)

Glycogen

Does not ↓ Glycogenolysis on
occur

Glucose ⟵‖⟶ Glucose-6-phosphate

Gluconeogenesis
Very little occurs ↓ Glycolysis on
in muscle

Pyruvate

Fig. 12.22 Diagram illustrating the different control requirements of **(a)** liver and **(b)** muscle in response to glucagon and/or epinephrine. Both hormones have a cAMP as second messenger. The term glycogenolysis used here is a synonym for glycogen breakdown.

In muscle, the situation is quite different. There, epinephrine stimulation of cAMP production is meant to increase generation of energy. It is, in contrast to liver, logical in muscle for cAMP to switch on glycogen breakdown but it is *not* logical to also inhibit glycolysis for this would result in the situation of producing glucose-6-phosphate from glycogen but preventing any usage of it (see Fig. 12.22(b)). In fact, the reverse (glycolysis speed-up) is required. We thus see that cAMP needs to block glycolysis in liver but activate it in muscle. The most important control point in glycolysis is the phosphofructokinase (PFK) step. (Refresh your memory with Fig. 8.7.) In *liver*, cAMP (glucagon increased) causes inhibition of PFK, thus inhibiting glycolysis; in muscle, cAMP (epinephrine increased) results in increased glycolysis. The effect of cAMP on PFK in liver is not direct, which leads to the next question.

How does cAMP control PFK?

The mechanism of this in liver is known and is not difficult to understand but it can be slightly tricky at first to sort out who is phosphorylating what, so you might have to read this section a few times. When blood sugar is low, glucagon is high and therefore liver cAMP is high. This causes a decrease in the cells of **fructose-2:6-bisphosphate**. This is *not* a misprint for **fructose-1:6-bisphosphate**, which is produced in glycolysis by PFK, but a new compound not mentioned previously. It is the most powerful allosteric activator of PFK and thus of glycolysis. It is

solely a regulator molecule. Raise its level and glycolysis increases; reduce its level and glycolysis decreases. Destruction of the PFK reaction product—that is, the hydrolysis of the fructose-1:6-bisphosphate by the bisphosphatase is reciprocally *inhibited* by the 2:6 compound as shown in Fig. 12.23. Recall that gluconeogenesis requires the fructose-1-6-bisphosphatase reaction (page 152).

You can easily deduce that, in liver therefore, cAMP must *reduce* the 2:6 level—thus inactivating glycolysis and activating gluconeogenesis. How is this achieved? The 2:6 compound is produced by a second PFK, called PFK$_2$. After its discovery, the first PFK had to be called PFK$_1$. cAMP causes *inhibition of PFK$_2$* (Fig. 12.24). It does so by causing activation of a protein kinase that phosphorylates the PFK$_2$ protein (Fig. 12.25). This has a double-hit effect, because the *phosphorylated* PFK$_2$ not only does not synthesize the 2:6 compound but actively destroys it by hydrolysis; the unphosphorylated PFK$_2$ does not destroy the 2:6 compound, but it synthesizes it. Thus, the PFK$_2$ protein is a double-headed enzyme in which the synthesis and destruction of the 2:6 compound are reciprocally controlled by a single phosphorylation; it is an astonishing enzyme.

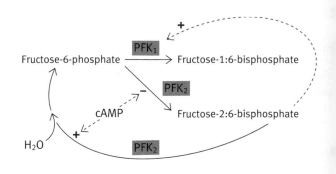

Fig. 12.24 Synthesis of fructose-2:6-bisphosphate by a second form of PFK and its inhibition *in liver* by cAMP. The inhibition is indirect: it activates a protein kinase that phosphorylates PFK$_2$, inactivating synthesis of the 2:6 compound. The phosphorylated PFK$_2$ now destroys the latter. Thus cAMP reduces the fructose-2:6-bisphosphate level in liver.

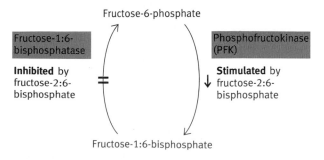

Fig. 12.23 Fructose-2:6-bisphosphate and the control of glycolysis.

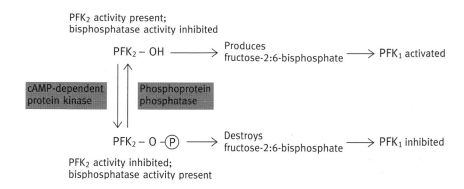

Fig. 12.25 Control of liver PFK$_2$ by phosphorylation by cAMP-dependent protein kinase. The PFK$_2$ is a double-headed enzyme: (1) it synthesizes fructose-2:6-bisphosphate when unphosphorylated; (2) it destroys the 2:6-compound when phosphorylated.

This is fine for liver but not what is required in muscle where PFK$_1$ must not be inhibited when epinephrine is released, since maximum glycolysis is needed in emergency. There is less information on this. It is reported that, in the presence of epinephrine, the level of the 2:6 compound increases. It may be that the increase is due to the increased level of substrate for PFK$_2$ (fructose-6-phosphate) resulting from cAMP-induced glycogen breakdown, though this is speculative. Fructose-6-phosphate is known to activate PFK$_2$.

Pyruvate kinase at the far end of glycolysis (Fig. 8.7) is another enzyme responding to glucagon. In the liver, but not in muscle, cAMP (and therefore glucagon) causes phosphorylation of the enzyme, resulting in its inactivation. The rationale of this is that, if blood glucose is low, the liver needs to produce glucose, by the gluconeogenesis pathway, not to use it. This is a glycolytic switch-off, additional to that at the PFK step. Once again, muscle has different needs—it does not synthesize glucose, glycolysis must not be switched off, and its pyruvate kinase is not phosphorylated in response to cAMP.

Gluconeogenesis is, as stated, a function of the liver mainly in response to glucagon. In Chapter 11 we described how gluconeogenesis from pyruvate requires the steps,

(1)   Pyruvate + ATP + HCO$_3^-$ → Oxaloacetate + ADP + P$_i$ + H$^+$

(2)   Oxaloacetate + GTP ↔ PEP + GDP + CO$_2$.

You can see that there is again a potential futile cycle, if pyruvate kinase is catalysing the reaction

PEP + ADP → pyruvate + ATP.

The futile cycle would be as shown below.

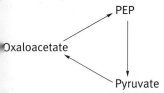

For gluconeogenesis, the pyruvate kinase in liver needs to be inactivated so that the PEP is sent back up the gluconeo-

genic path. Inactivation of the liver enzyme by cAMP described above achieves this (see Fig. 12.26). In liver, the PFK in the presence of glucagon is inhibited and the fructose-1:6-bisphosphatase is active, so gluconeogenesis is activated to supply blood sugar.

Control of fat metabolism by extracellular agents

The most important decisions in this area are whether fat cells should store fat as triglyceride or release free fatty acids from stored triglycerides into the blood. Insulin is the signal for the

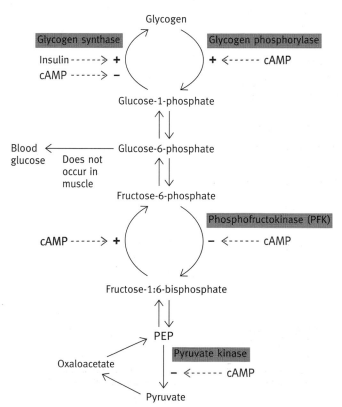

Fig. 12.26 Diagram of the main external controls in this area of metabolism in liver. Note that 'cAMP' represents the action of glucagon and epinephrine. Its action is not directly on the enzyme controlled (see page 170).

former. Glucagon is a signal for the latter, but in some animals, including humans, epinephrine and norepinephrine are important in this regard. Adipose tissue is innervated by norepinephrine-releasing sympathetic nerves and is subject to epinephrine hormone control also. All three agents, glucagon, epinephrine, and norepinephrine, use cAMP as second messenger. Fat cells contain a hormone-sensitive lipase that carries out the reaction

Triacylglycerol $+$ H_2O $\rightarrow$ Diacylglycerol $+$ free fatty acid.

The diacylglycerol is further degraded to glycerol and free fatty acids. The hormone-sensitive lipase is controlled by phosphorylation, the phosphorylated enzyme being active. A cAMP-activated protein kinase carries this out and a protein phosphatase reverses it (Fig. 12.27). Insulin antagonizes these effects when it combines with receptors on the cells.

The effect of insulin in promoting the uptake of glucose is an important aspect of the way in which fat cells are directed to convert excess glucose to fat. The key control enzyme in fat synthesis is acetyl-CoA carboxylase, which, you will recall, catalyses an irreversible reaction. It is inactivated by a protein kinase that is activated by AMP (not cAMP this time—see page 168). This inactivation by phosphorylation is reinforced by the fact that, in the presence of glucagon or epinephrine, the phosphatase that activates acetyl-CoA carboxylase by removing

the phosphate group is itself inactivated by cAMP. Thus, if ATP levels are low, AMP causes inactivation of the carboxylase; if blood glucose is low, glucagon keeps it inactive. Insulin has the opposite effect, as usual. It somehow activates the enzyme and promotes fat synthesis.

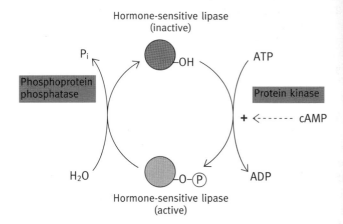

Fig. 12.27 Activation of adipose cell hormone-sensitive lipase by cAMP-dependent phosphorylation. Insulin antagonizes the effects of the catecholamine hormones and glucagon, which increase the cAMP level. Note also that glucagon and epinephrine cause inhibition of fat synthesis in liver and adipose cells, respectively, by preventing dephosphorylation of acetyl-CoA carboxylase. Insulin activates the carboxylase.

Further reading

Substrate cycles

Newsholme, E. A., Challiss, R. A. J., and Crabtree, B. (1984). Substrate cycles: their role in improving sensitivity in metabolic control. *Trends Biochem. Sci.*, **9**, 277–80.
Describes the regulatory roles of these cycles.

Glycogen metabolism—regulation

Johnson, L. N. (1992). Glycogen phosphorylase: control by phosphorylation and allosteric effectors. *FASEB J.*, **6**, 2274–82.
A comprehensive review of intrinsic and extrinsic controls on glycogen breakdown.

Walsh, D. A. and Van Patten, S. M. (1994). Multiple pathway signal transduction by the cAMP-dependent protein kinase. *FASEB J.*, **8**, 1227–36.

An advanced but readable account of the protein kinase structure and mechanism.

Ketogenesis—regulation

McGarry, J. D. and Foster, D. W. (1980). Regulation of hepatic fatty acid oxidation and ketone body production. *Ann. Rev. Biochem.*, **49**, 395–420.
Despite the regulatory title, the article also provides a general review of ketone body formation.

Quant, P. A. (1994). The role of mitochondrial HMG-CoA synthase in regulation of ketogenesis. *Essays in Biochem.*, **28**, 13–25.
Summarizes the control of ketogenesis and the physiological repercussions of the process.

Problems for Chapter 12

1 What are the two main ways by which the activities of enzymes may be reversibly modulated?

2 Compare the relationship between substrate concentration and rate of enzyme catalysis in a nonallosteric enzyme and a typical allosteric enzyme.

3 What is the advantage of the allosteric enzyme substrate–velocity relationship?

4 If a typical allosterically controlled enzyme is exposed to saturating levels of substrate, what would be the effects of allosteric effectors on the reaction velocity?

5 Describe the two models that explain homotropic cooperative binding of substrate to an enzyme.

6 What is the main feature of allosteric control that makes it such a tremendously important concept?

7 What are the salient features of intrinsic and extrinsic regulation?

8 By means of a diagram, illustrate the main *intrinsic* controls on glycogen metabolism, glycolysis, and gluconeogenesis and explain the rationale.

9 Pyruvate dehydrogenase is a key regulatory enzyme. In general, products of the reaction inhibit the reaction. There are three mechanisms of control involved; what are these?

10 Explain the intrinsic controls on fat synthesis and oxidation.

11 What controls the release of insulin and glucagon from the pancreas?

12 What is a second messenger? Name the second messenger for epinephrine and glucagon and explain how it exerts metabolic effects.

13 How does insulin control the rate of glucose entry into fat cells?

14 Explain how cAMP activates glycogen breakdown.

15 Glucagon activates liver phosphorylase via cAMP as its second messenger. Muscle does the same with epinephrine stimulation. However, cAMP has quite different effects on liver and muscle glycolysis. Explain these.

16 Several hormones that elicit completely different cellular responses nonetheless use cAMP as their second messenger. How can one compound be used for these?

17 Phosphofructokinase is a key control enzyme. What is the major allosteric effector for this enzyme and how is its level controlled in liver?

18 To synthesize glucose in the liver, PEP is needed. However, production of PEP would achieve little in this regard if it were dephosphorylated to pyruvate by pyruvate kinase. How is this futile cycle avoided? Why would this mechanism not be appropriate in muscle?

19 How does glucagon cause fatty acid release by fatty cells?

Chapter 13

. . . .

Why should there be an alternative pathway of glucose oxidation—the pentose phosphate pathway?

A completely different pathway of glucose oxidation exists, called the pentose phosphate pathway but sometimes called the 'direct oxidation pathway' or the 'hexose monophosphate shunt'. The need for another pathway will probably puzzle you, since in the preceding chapters we have dealt with a most extensive pathway (glycolysis, the citric acid cycle, and the electron transport pathway) for oxidizing glucose. Hence the question in the chapter title.

Paradoxically, it isn't really a pathway to provide for a different method of glucose oxidation nor is it primarily for energy production, but rather it is to cater for some specialized metabolic needs.

1. It supplies ribose-5-phosphate for nucleotide and nucleic acid synthesis (the latter dealt with in later chapters).

2. It suplies NADPH for fat synthesis (see page 143).

3. It provides a route for excess pentose sugars in the diet to be brought into the mainstream of glucose metabolism.

The oxidative steps

The pentose phosphate pathway has two main parts. First, glucose-6-phosphate is converted to ribose-5-phosphate and CO_2, during which $NADP^+$ is reduced to NADPH. The **6-phosphogluconate dehydrogenase** generates a β-keto acid which is decarboxylated to a keto-pentose (ribulose); an isomerase converts the latter to the aldose isomer, ribose-5-phosphate. This is the oxidative section (Fig. 13.1). It produces the two components, ribose-5-phosphate and NADPH. Why then does there follow another, nonoxidative series of reactions, which when written down in detail may appear somewhat formidable?

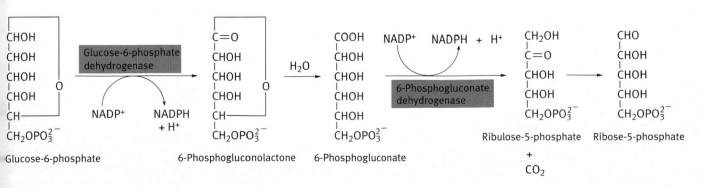

Fig. 13.1 Oxidative reactions of the pentose phosphate pathway.

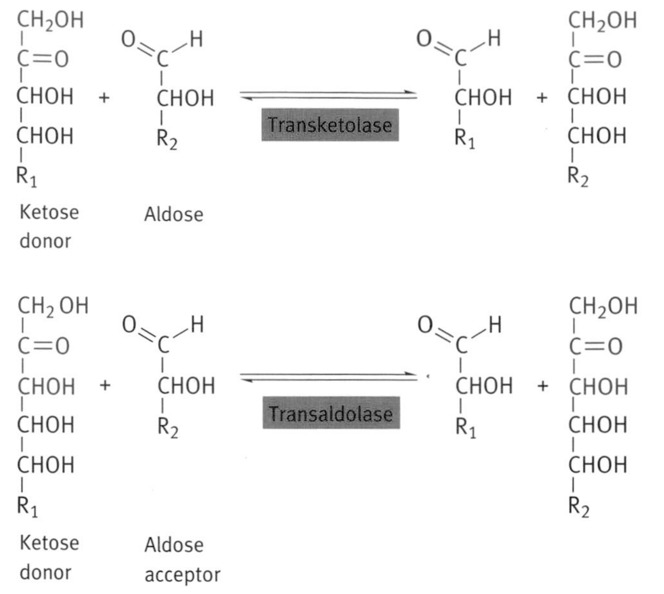

Fig. 13.2 Reactions catalysed by transketolase and transaldolase.

The nonoxidative section and its purpose

Tissue demands for the two products, ribose-5-phosphate and NADPH, vary greatly. For example, fat synthesis requires large supplies of NADPH (page 143). Fat cells and liver cells have large amounts of the pentose phosphate pathway enzymes to supply this, while muscle has very little of these. Production of NADPH for fat synthesis also produces large amounts of ribose-5-phosphate by the reactions given in Fig. 13.1, which may be far more than the cell needs for nucleotide synthesis. Conversely, in a rapidly dividing cell that is not synthesizing fat, large amounts of ribose-5-phosphate are needed to synthesize nucleotides, but there is little requirement for NADPH. The requirements for ribose-5-phosphate and NADPH may, in other cells, be anywhere in between this—equal amounts of the two products or more of one than the other.

The nonoxidative branch is a team act of a pair of enzymes, **transaldolase** and **transketolase**, that, together, interconvert sugars in accordance with the metabolic needs of the cell. These two enzymes detach, from a ketose sugar phosphate, C_3 and C_2 units, respectively, and transfer them to other aldose sugars (Fig. 13.2). It is interesting that transketolase is a thiamin pyrophosphate enzyme as is pyruvate dehydrogenase (page 114). In the latter case the enzyme also acquires a C_2 unit from pyruvate which it transfers to another acceptor, CoA—SH.

Between them, transketolase and transaldolase can work an almost bewildering range of sugar interconversions (which probably few biochemists carry in their heads in detail). If you put together the reactions of the *oxidative* section given above, the following balance sheet emerges.

Glucose-6-phosphate $+ 2NADP^+ + H_2O \rightarrow$
   Ribose-5-phosphate $+ 2NADPH + 2H^+ + CO_2$.

In situations where the needs for ribose-5-phosphate and NADPH are *balanced* the nonoxidative section is not required since the oxidative part produces both products in appropriate amounts.

However if, say, a nondividing fat cell requires *more NADPH than ribose-5-phosphate* the nonoxidative section reconverts (recycles) the latter into glucose-6-phosphate according to the stoichiometry

6 Ribose-5-phosphate $\rightarrow$ 5 Glucose-6-phosphate $+ P_i$.

The mechanism of this process in broad terms is given below, but we suggest you also follow the steps in the detailed reactions in the 'Further study' section that follows.

First, part of the ribose-5-phosphate (R-5-P), an aldose sugar, is converted into xylulose-5-phosphate (X-5-P), a ketose sugar, since both transaldolase and transketolase use only ketose sugars as group donors. The remaining fraction of the ribose-5-phosphate is the aldose sugar acceptor. The following transformations then occur, reaction 1 being between X-5-P and R-5-P.

(1) $2 C_5 \longleftrightarrow C_3 + C_7$ (Transketolase)

(2) $C_7 + C_3 \longleftrightarrow C_4 + C_6$ (Transaldolase)

(3) $C_5 + C_4 \longleftrightarrow C_3 + C_6$ (Transketolase)

The final C_3 compound is glyceraldehyde-3-phosphate, two molecules of which are converted to glucose-6-phosphate by the reversal of glycolysis and loss of P_i. The net effect of the above three reactions is that three molecules of C_5 (red type) are converted into 2.5 molecules of C_6 (blue type). Note that this set of reactions can also convert dietary ribose to glucose-6-phosphate following its conversion to ribose-5-phosphate by an ATP-requiring kinase.

The pentose phosphate pathway is extremely flexible. Consider another situation where a cell needs to make ribose-5-phosphate for nucleotide synthesis, but has little demand for NADPH. The following overall process caters for this

5 Glucose-6-phosphate $+$ ATP $\rightarrow$
   6 Ribose-5-phosphate $+$ ADP $+ H^+$.

The mechanism is as shown in Fig. 13.3. In this, glucose-6-phosphate is converted by the glycolytic pathway partly to fructose-6-phosphate and partly to glyceraldehyde-3-phosphate. These are the C_6 and C_3 products of reaction 3 above. Reversal of the three steps produces xylulose-5-phosphate which is isomerized into ribose-5-phosphate. This scheme does not involve the oxidative part of the pentose phosphate pathway at all.

Further study

Reactions involved in the conversion of ribose-5-phosphate to glucose-6-phosphate

```
    O
    ‖
    C—H          Phosphopentose        CH₂OH      Phosphopentose       CH₂OH
    |                                   |                               |
    CHOH            isomerase           C=O           epimerase         C=O
    |                                   |                               |
  H—C—OH          ⇌                   H—C—OH        ⇌                HO—C—H
    |                                   |                               |
  H—C—OH                              H—C—OH                          H—C—OH
    |                                   |                               |
    CH₂—O—PO₃²⁻                        CH₂—O—PO₃²⁻                     CH₂—O—PO₃²⁻

Ribose-5-phosphate              Ribulose-5-phosphate            Xylulose-5-phosphate
(R-5-P)                                                          (X-5-P)
```

Only part of the R-5-P undergoes the above conversion. The remainder is needed for reactions 1–3 below.

```
CH₂OH          CHO                          CH₂OH         CHO
|              |                            |             |
C=O            CHOH                         C=O           CHOH
|              |                            |             |
CHOH     +     CHOH                         CHOH    +     CHOH
|              |                            |             |
CHOH           CHOH                         CHOH          CHOH
|              |                            CH₂—O—PO₃²⁻   CH₂—O—PO₃²⁻
CH₂—O—PO₃²⁻    CH₂—O—PO₃²⁻

   X-5-P          R-5-P                       X-5-P         E-4-P
```

Reaction 1 — Transketolase

```
            CH₂OH
            |
            C=O
            |
            CHOH
            |
CHO   +     CHOH
|           |
CHOH        CHOH
|           |
CH₂—O—PO₃²⁻ CH₂—O—PO₃²⁻

Glyceraldehyde-  Sedoheptulose-
3-phosphate      7-phosphate
(G-3-P)          (S-7-P)
```

Reaction 2 — Transaldolase

```
            CH₂OH
            |
            C=O
            |
CHO         CHOH
|           |
CHOH   +    CHOH
|           |
CHOH        CHOH
|           |
CH₂—O—PO₃²⁻ CH₂—O—PO₃²⁻

Erythrose-4-    Fructose-6-
phosphate       phosphate
(E-4-P)         (F-6-P)
```

Reaction 3 — Transketolase

```
CH₂OH
|
C=O
|
CHOH
|
CHOH    +    CHO
|            |
CHOH         CHOH
|            |
CH₂—O—PO₃²⁻  CH₂—O—PO₃²⁻

   F-6-P         G-3-P
```

Two rounds of reactions 1–3 will produce two molecules of G-3-P.

```
      CHO
      |
  2   CHOH         ⟶ ⟶ ⟶      1   F-6-P  +  Pᵢ .
      |          (Gluconeogenesis
      CH₂—O—PO₃²⁻   reactions)

      G-3-P
```

Fructose-6-phosphate is converted to glucose-6-phosphate by phosphohexose isomerase.

The scheme can be summarized as

$$2C_5 + 2C_5 \rightleftharpoons 2C_3 + 2C_7$$
$$2C_7 + 2C_3 \rightleftharpoons 2C_4 + 2C_6$$
$$2C_5 + 2C_4 \rightleftharpoons 2C_3 + 2C_6$$
$$2C_3 \longrightarrow 1C_6 .$$

The net effect of these reactions is

$$6 \text{ Ribose-5-phosphate} \longrightarrow 5 \text{ Glucose-6-phosphate} + P_i .$$

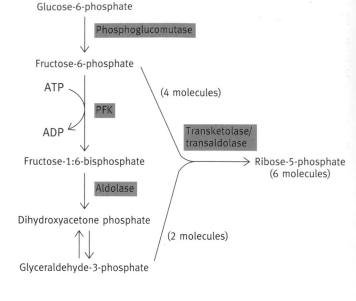

Fig. 13.3 Scheme by which glucose-6-phosphate is converted to ribose-5-phosphate without the production of NADPH. The oxidative part of the pentose phosphate pathway is not involved. See text for the transketolase/transaldolase steps.

Where does the complete oxidation of glucose come into all of this?

The oxidative part of the pentose phosphate pathway (see Fig. 13.1) is sometimes referred to as being capable of the direct oxidation of glucose to CO_2, and NADPH + H^+. The overall stoichiometry of the overall process is

$$6 \text{ Glucose-6-phosphate} + 12 \text{ NADP}^+ + 7H_2O \rightarrow$$
$$5 \text{Glucose-6-phosphate} + 6CO_2 + 12NADPH + 12H^+ + P_i.$$

In fact, it never occurs, as such, as a metabolic route. A *single* molecule of glucose-6-phosphate is *not* converted to six CO_2 molecules by the pathway. What happens is that six molecules of glucose-6-phosphate can *each* give rise to one molecule of CO_2 and one molecule of ribose-5-phosphate by the reactions given above. If, now, nonoxidative reactions convert the six ribose-5-phosphate molecules back to five glucose-6-phosphate molecules, as described above, then on a balance sheet it looks as if six glucose-6-phosphate molecules have been converted to five molecules of glucose-6-phosphate plus six molecules of CO_2 but, as stated, it hasn't really oxidized one molecule of glucose completely. This set of events generates NADPH without production of ribose-5-phosphate—a situation required in a cell with rapid fat synthesis but no cell division.

Why do red blood cells have the pentose phosphate pathway?

Erythrocytes do not divide so they have no need for ribose-5-phosphate for nucleic acid synthesis, nor do they synthesize fat. Their energy is derived from anaerobic glycolysis—anaerobic, for they have no mitochondria.

Nonetheless, in patients whose red blood cells lack the first enzyme in the pentose phosphate pathway (glucose-6-phosphate dehydrogenase), a massive hemolytic anaemia may be induced by the antimalarial drug, **pamaquine**. The reason for this relates to the fact that NADPH, generated by the pentose phosphate pathway, is needed for a protective mechanism described on page 219, namely the reduction of glutathione, a thiol compound, which maintains hemoglobin in its reduced (ferrous) state.

An interesting sidelight on glucose-6-phosphate dehydrogenase deficiency is that mutations leading to a defective enzyme confer a selective advantage in areas where a lethal type of malaria is endemic. Possible explanations for this are that the parasite has a requirement for the products of the pentose phosphate pathway and/or that the extra stress caused by the parasite causes the deficient red blood cell host to lyse before the parasite completes its development. It is interesting to compare the selective advantage conferred in this case with that conferred by the sickle-cell trait (page 383) where a potentially lethal disease can give a survival advantage because it gives protection against a more lethal disease, malaria.

Further reading

Beutler, E. (1983). Glucose-6-phosphate dehydrogenase deficiency. In *The metabolic basis of inherited disease* (ed. Stanbury, J. B., Wyngaarten, J. B., Fredrickson, D. S., and Brown, M. S.), pp. 1629–53. 5th edn. McGraw-Hill.
An account of the medical aspects. Comprehensive but very readable.

Problems for Chapter 13

1 What are the functions of the pentose phosphate pathway?

2 What is the oxidative part of the pentose phosphate pathway?

3 What enzymes are involved in the nonoxidative part?

4 A nondividing fat cell requires large amounts of NADPH for fat synthesis but very little ribose-5-phosphate. However, the oxidative section produces equal amounts of the two products. Explain how the nonoxidative reactions cope with this problem. (The answer need not give actual reactions.)

5 The pentose phosphate pathway was sometimes referred to as the direct oxidation pathway for glucose. Why was this and why was it a somewhat misleading term?

6 Mature red blood cells have no need for nucleotide or fat synthesis. Why then do they have glucose-6-phosphate dehydrogenase?

Chapter summary

Chapter 14

Raising electrons of water back up the energy scale—photosynthesis

From Chapter 8 you will have learned that ATP generation in aerobic cells depends on transporting electrons of **high energy potential** present in food down the energy scale to end up as the electrons present in the hydrogen atoms of water.

Since the amount of food on Earth is limited, if life in general is to continue indefinitely, a way must exist to kick those electrons back up the energy scale. A minor qualification to this statement is that life forms have been discovered in deep oceans around cracks in the Earth's crust from which compounds such as H_2S emerge. H_2S is a reducing agent of low redox potential and its electrons could be transported down the energy gradient releasing energy, provided appropriate biochemical systems are there. Such life could presumably exist as long as H_2S and other such agents are generated in the Earth's crust but, for continuation of the vast bulk of life, electron recycling is necessary. Early life forms must have survived on ready-made 'food', such as H_2S, in the primeval oceans but, given the exponential nature of biological reproduction, such resources would have been exhausted relatively quickly. For that matter, suitable electron 'sinks' might also have been exhausted in the anaerobic atmosphere believed to exist before the onset of photosynthesis.

Overview

Photosynthesis is the biological process that recycles electrons. It has several crucial advantages—water is an inexhaustible source of electrons, sunlight an inexhaustible source of energy, and, in releasing oxygen, an inexhaustible supply of electron sinks is provided to allow the energy to be extracted back from the high-energy electrons of food. The onset of photosynthesis was arguably the most important biological event following the establishment of life. The global energy cycle is shown in Fig. 14.1.

You are probably familiar with the concept of photosynthesis producing carbohydrate (usually starch or sugar) from CO_2 and H_2O, The overall equation (written for glucose production) is

$$6CO_2 + 6H_2O \rightarrow C_6H_{12}O_6 + 6O_2 . \Delta G^{o'} = 2820 \text{ kJ mol}^{-1}$$

To synthesize glucose from CO_2 and H_2O there are two basic essentials looked at from the point of view of energy. First, there must be a reducing agent of sufficiently **low redox potential (high energy)**. If you need to refresh your memory on redox potentials, turn to page 100. In photosynthesis, the reducing agent is NADPH (page 141). (Gluconeogenesis in animals, as described in Chapter 11, uses NADH as the reductant, but, in

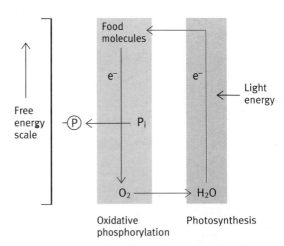

Fig. 14.1 Diagram of global 'electron cycling' by oxidative phosphorylation and photosynthesis. Note that the fixation of CO_2 is a process secondary to the raising of electrons from water to a higher energy potential.

photosynthesis, note that it is NADPH.) Secondly, there must be ATP to drive the synthesis.

Light energy is directly involved only in transferring electrons from water to $NADP^+$ and in the generation of ATP.

Site of photosynthesis—the chloroplast

Photosynthesis occurs in the **chloroplasts** of green plant cells. They are reminiscent of mitochondria in being membrane-bound organelles in the cytoplasm with an outer permeable membrane and an inner one impermeable to protons. Like mitochondria they have their own DNA coding for part of their proteins. Their protein-synthesizing machinery is prokaryotic in type (page 303) and it is believed that they arose by a symbiotic colonization of eukaryote cells by primitive prokaryotic photosynthetic unicellular organisms.

Unlike mitochondria, however, chloroplasts contain yet another type of membrane-bound structure—the **thylakoids**—membrane sacs within the chloroplast. The thylakoids are flattened sacs piled up like stacks of coins into grana, these being connected at intervals by single layer extensions. Inside the thylakoids is the thylakoid lumen; outside is the chloroplast stroma (Fig. 14.2). All of the light-harvesting chlorophyll and the electron transport pathways are in the thylakoid membranes. The conversion of CO_2 and H_2O into carbohydrate molecules is not light-dependent and occurs in the chloroplast stroma. The latter processes are referred to as 'dark reactions', not to imply that they only occur in the dark, but rather that light is not involved in them. In fact, the dark reactions occur mainly in the light, when NADPH and ATP generation is occurring in the thylakoid membranes. This is summarized in Fig. 14.3.

An overview of the photosynthetic apparatus and its organization in the thylakoid membrane

It would be useful for you to refresh your memory of the electron transport chain in mitochondria for there are considerable similarities between this and the photosynthetic machinery. In the inner mitochondrial membrane there are four complexes (see Fig. 8.18) that transport electrons. Connecting

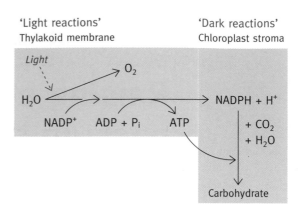

Fig. 14.3 Summary diagram of the processes in photosynthesis.

these complexes by ferrying electrons between them are ubiquinone, the small lipid-soluble molecule, and cytochrome *c*, the small water-soluble protein.

In the thylakoid membrane there are three complexes (Fig. 14.4); connecting the first two is the electron carrier plastoquinone whose structure is very similar to that of ubiquinone. Connecting the second two complexes is plastocyanin, a small water-soluble protein that has a bound copper ion as its electron-accepting moiety; this oscillates between the Cu^+ and Cu^{2+} states as it accepts and donates electrons.

The three complexes are named photosystem II (PSII), the cytochrome *bf* complex, and photosystem I (PSI). PSII comes

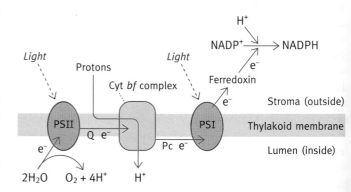

Fig. 14.4 The diagram gives an overall view of the light-dependent part of the photosynthetic apparatus, without reaction details. The feature to note is that the purpose is to take electrons from water and transfer them to $NADP^+$ and in the process to create a proton gradient across the thylakoid membrane that can drive ATP synthesis by the chemiosmotic mechanism. The gradient is created from two sources—the splitting of water (photosystem II, PSII) and the proton pumping of the Q:cytochrome *bf* complex. Electrons are transported from PSII to the cytochrome *bf* complex by plastoquinone (Q). The cycle is analogous to that in mitochondria (where Q is used for ubiquinone) that results in four protons translocated per QH_2 molecule oxidized, but there is some uncertainty still in the number transported here. Note also that plastocyanin (Pc) is a mobile carrier analogous in this respect to cytochrome *c* in mitochondria and that ferredoxin is located on the opposite side of the membrane.

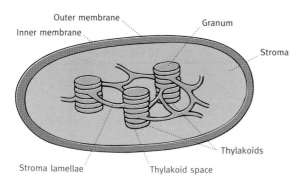

Fig. 14.2 Diagram of chloroplasts. Grana are stacks of thylakoids.

before PSI in the scheme of things; the numbers refer to the order in which they were discovered. The function of the whole array shown in Fig. 14.4 is to carry out the overall reaction

$$2H_2O + 2NADP^+ \rightarrow O_2 + 2NADPH + 2H^+.$$

This involves a very large increase in free energy. What is unique in photosynthesis, so far as biochemical reactions are concerned, is that this energy is supplied by light. For each molecule of NADPH produced, two photons are absorbed. These provide sufficient energy, not only for a molecule of $NADP^+$ to be reduced by water but also there is some to spare and this is used for ATP production. The arrangement is rather beautiful for it supplies the two requirements for carbohydrate synthesis from CO_2 and water—a reducing agent and ATP. We now turn to the light-harvesting machinery present in the photosystems II and I.

What is chlorophyll?

In green plants, chlorophyll is the light receptor. Other receptor pigments exist in bacteria and algae but we shall confine this account to higher plants.

Chlorophyll is a tetrapyrrole much as is heme (page 374) except that it has a magnesium atom at its centre instead of iron and the substituent side groups are different. One of the latter is a very long hydrophobic group, that anchors it into the lipid layer. As in heme, there is a conjugated double bond system (alternate double and single bonds right round the molecule) resulting in a strong colour—that is, strongly light-absorbing. In green plants there are two chlorophylls (*a* and *b*) differing in one of the side groups. They both absorb light in the red and blue ranges, leaving the intermediate green light to be reflected. The two chlorophylls have slightly different absorption maxima, which complement each other so that in the red and blue ranges between them they absorb a higher proportion of the incident light. We suggest that you memorize the structure of chlorophyll *a* shown below only in broad outline without bond or side-chain details. (Chlorophyll *b* is the same except for the $-CH_3$ side chain (in red), which is $-CHO$ in chlorophyll *b*.)

Structure of chlorophyll *a*

When a chlorophyll molecule absorbs light, it is excited so that one of its electrons is raised to a higher energy state; it jumps into a new atomic orbit. In an isolated chlorophyll molecule, after such excitation, the electron drops back to its ground state liberating energy as heat or fluorescence in doing so (and nothing is thereby achieved). But, when chlorophyll molecules are closely arranged together, a process known as **resonance energy transfer** transfers that energy from one molecule to another. In green plants, chlorophyll molecules are packed in functional units called **photosystems**, such that this resonance transfer occurs readily. When a chlorophyll molecule is excited by absorption of a photon, it passes on its excitation to a neighbour and drops back to the ground state itself and so the process continues. The excitation wanders at random from one chlorophyll molecule to another (Fig. 14.5).

This has a function, for amongst large numbers of 'ordinary' chlorophyll molecules there is a special **reaction centre** (probably an arrangement of a pair of chlorophyll molecules in association with proteins). The properties of this reaction centre are such that the excitation of a constituent chlorophyll molecule by resonance transfer results in the excited electron being at a somewhat reduced energy level, as compared with that of other excited chlorophyll molecules, so that resonance transfer *from* this molecule to other chlorophyll molecules does not occur. The excitation energy is, in this sense, trapped in an energetic hole—a shallow hole because the 'trapped' electron is still at a higher energy level than an unexcited electron, sufficient for the excited reaction centre to hand on an electron to an electron acceptor of appropriate redox potential. The latter is the first carrier of an electron transport chain in photosynthesis (to be described shortly).

A photosystem is therefore a complex of light-absorbing chlorophylls, reaction centre chlorophyll, and an electron

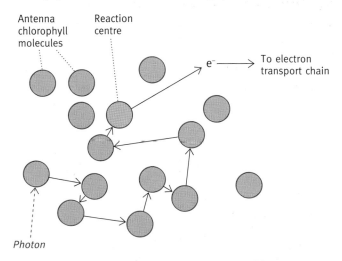

Fig. 14.5 Diagram of activation of reaction centre chlorophyll molecule (red circle) by resonance transfer of energy from activated antenna chlorophyll molecules (green circles). The reaction centre of photosystem II is called P680 and of photosystem I, P700.

transport chain. In the case of photosystem II (the first one) the reaction centre chlorophyll is called P680 because it absorbs light up to that wavelength and photosystem I is called P700 for an equivalent reason. The chlorophyll molecules feeding excitation energy to the centres are called **antenna chlorophylls** (Fig. 14.5).

Mechanism of light-dependent reduction of NADP$^{+\cdot}$

Figure 14.6 shows the 'Z' arrangement of the two photosystems II and I, with the redox potentials of the components indicated. Why two photosystems? A familiar analogy might help at the outset. An electric torch with a bulb requiring 3 volts to light it up, usually employs two 1.5-volt batteries in series. In photosynthesis, lighting of the bulb is represented by NADP^{+} reduction, and the batteries by the two photosystems. In fact, as stated, the latter supply more energy than is needed and some of it is sidetracked into ATP generation.

Let us start with a P680 chlorophyll of a reaction centre of photosystem II. In the dark, it is in its ground, unexcited state in which it has no tendency to hand on an electron. When energy

of a photon reaches it via antenna chlorophylls, it is excited such that it has a strong tendency to hand on its excited electron. It is, in fact, a reducing agent and it reduces the first component (a chlorophyll-like pigment lacking the Mg^{2+} atom, called pheophytin) of the PSII electron transport chain.

Two molecules of reduced pheophytin then hand on an electron each (one at a time) to reduce **plastoquinone**, the lipid-soluble electron carrier between PSII and the cytochrome *bf* complex.

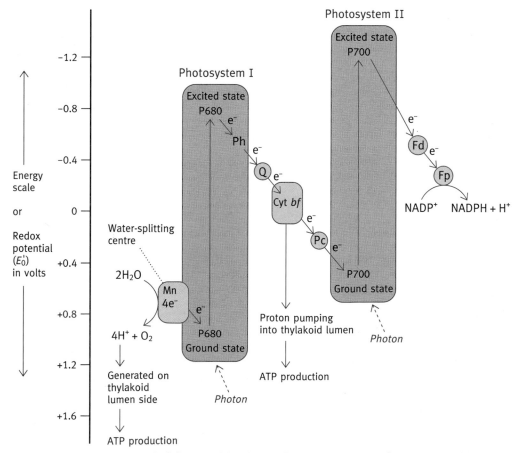

The latter complex contains two cytochromes and an iron sulfur centre (see page 122 if you need to be reminded what this is); the

Fig. 14.6 The Z-scheme of photosynthesis. Simplified diagram of the electron flow from H$_2$O to NADP^{+}. In viewing the diagram, start with a photon of light activating P680 and the transfer of an electron to P700. The electron-depleted P680 then accepts an electron from water and awaits a new round of activation. P700 behaves similarly except that it accepts an electron from the photosystem II transport chain and its donated electron is transferred to NADP^{+}. Ph, pheophytin; Q, plastoquinone; Pc, plastocyanin; Fd, ferredoxin; Fp, ferredoxin–NADP reductase.

complex transports electrons from QH_2 to plastocyanin to give the reduced form of the latter. At this point we'll leave PSII and move on to PSI but we will return to the reduced plastocyanin very soon.

The chlorophyll at the reaction centre of PSI is P700; when activated by a photon arriving from the antenna chlorophylls it becomes a reducing agent. It passes on its electron to a short chain of electron carriers (we will not give the details of this), which reduces **ferredoxin**, a protein with an iron cluster electron acceptor; ferredoxin is a water-soluble, mobile protein residing in the chloroplast stroma (that is, outside of the thylakoid membrane). It reduces $NADP^+$ by the following reaction, catalysed by the FAD-enzyme called **ferredoxin–NADP reductase**

$$2 \text{ Ferredoxin}_{RED} + NADP^+ + 2H^+ \rightarrow$$
$$2 \text{ Ferredoxin}_{OX} + NADPH + H^+.$$

If we summarize what has taken place in PSI, an electron has been excited out of P700 and transported to ferredoxin which, in turn, has reduced $NADP^+$. However, this has left P700 an electron short; it is now $P700^+$, an oxidizing agent. It accepts an electron from plastocyanin (Pc), which, you remember, we left after it had been reduced by PSII. The reaction is

$$P700^+ + Pc_{Cu^+} \rightarrow P700 + Pc_{Cu^{2+}}.$$

To go back further, remember that we started with light exciting an electron out of P680 (Fig. 14.6), the reaction centre pigment of PSII; this leaves $P680^+$ which must have its electron restored so that it can revert to the ground state, ready for another photon to start a new round of reactions. The electron comes from water.

The water-splitting centre of PSI

$P680^+$ is a very strong oxidizing agent—it has a very strong affinity for an electron (greater than that of oxygen) so that it can even extract electrons from water. Four electrons are extracted from two molecules of H_2O with the release of O_2 and four H^+ into the thylakoid lumen. It is necessary to extract all four electrons so as not to release any intermediate oxygen free radicals, which are dangerous to biological systems, just as, in mitochondria, the addition of electrons to oxygen to form H_2O must be complete. In PSII there is a complex of proteins with Mn^{2+}, known as the **water-splitting centre**, which extracts the electrons from water, with the release of oxygen and protons, and passes them on to $P680^+$ molecules, thus restoring them to the ground state (Fig. 14.8(a)). The P680 is now ready for further rounds of reaction.

How is ATP generated?

The **cytochrome *bf* complex** of PSII, which uses QH_2 to reduce plastocyanin, resembles complex III of mitochondria (see Fig.

8.19 for the latter) in that electron transport through the complex causes translocation of protons from the outside of the thylakoid membrane to the inside. By analogy with the Q cycle in mitochondria it would be expected that the cytochrome *bf* complex would translocate four protons per QH_2 oxidized (see page 125). However, experimental data has not yet confirmed this number so there is some degree of uncertainty on this point. In addition, the water-splitting centre generates protons inside the thylakoid lumen, the two effects producing a proton gradient that lowers the pH of the thylakoid lumen to about 4.5. The uptake of a proton in the reduction of $NADP^+$ by ferredoxin in the stroma further contributes to the proton gradient across the membrane.

There is a device that under certain circumstances, leads to an increased proton translocation and therefore a greater potential for ATP synthesis. When virtually all of the $NADP^+$ has been reduced by ferredoxin, the latter donates electrons to the cytochrome *bf* complex (Fig. 14.7) instead. The passage of these through the complex to plastocyanin leads to increased proton pumping by that complex. The extra ATP production is referred to as **cyclic photophosphorylation** driven by **cyclic electron flow**.

The proton gradient is used to generate ATP from ADP and P_i by the chemiosmotic mechanism described for mitochondria. (The whole process of the events described so far is summarized in Fig. 14.8.)

An explanatory note:
In mitochondria, the proton gradient is from the outside (high) to the inside (low). The thylakoid membrane is formed from invaginations of the inner chloroplast membrane (cf. the inner mitochondrial membrane) which explains why the proton gradients and the ATP synthases of mitochondria and thylakoid discs look as if they are in opposite orientation.

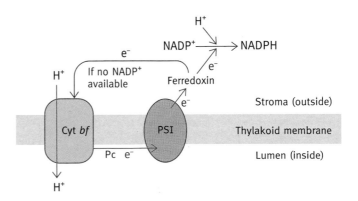

Fig. 14.7 Diagram of cyclic electron flow. When all of the $NADP^+$ is reduced, ferredoxin transfers electrons back to the cytochrome *bf* complex. Flow of electrons through this leads to increased proton pumping and hence increased ATP synthesis.

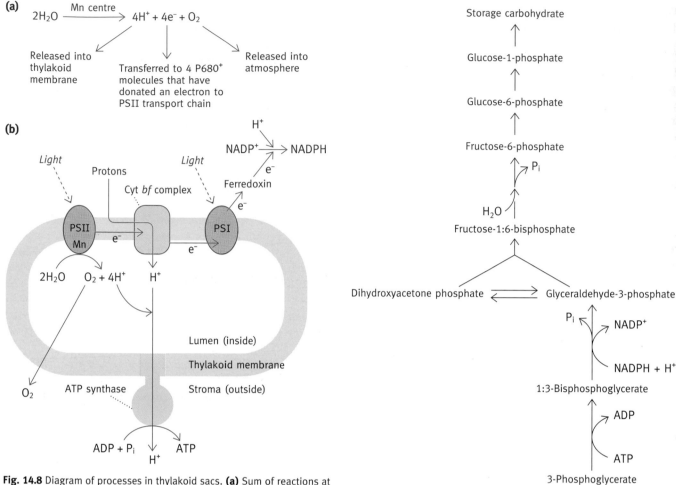

Fig. 14.8 Diagram of processes in thylakoid sacs. **(a)** Sum of reactions at the Mn water-splitting centre. **(b)** Routes followed by protons and electrons in thylakoids.

How is CO$_2$ converted to carbohydrate?

As already emphasized, the aspect of photosynthesis that is fundamentally different from other biochemical processes is the harnessing of light energy to split water and reduce NADP$^+$. From there, although the actual metabolic pathway by which glucose and its derivatives are synthesized is unique to plants, it is nonetheless 'ordinary' enzyme biochemistry and quite secondary to the light-dependent process.

We can dispose of most of the pathway of carbohydrate synthesis very quickly because from the metabolite 3-phosphoglycerate onward it is the same as the process in the liver (Fig. 11.5) except that NADPH is used instead of NADH as the reductant (Fig. 14.9). This leaves the question of how 3-phosphoglycerate is produced in photosynthesis.

Where does the 3-phosphoglycerate come from?

The most plentiful single protein on Earth is the enzyme called 'Rubisco' for short, which stands for ribulose-1:5-bisphosphate carboxylase. It is the enzyme that utilizes CO$_2$ to produce 3-phosphoglycerate.

Fig. 14.9 Pathway of starch synthesis in photosynthesis, starting with 3-phosphoglycerate. The process is the same as in gluconeogenesis in liver except that NADPH is the reductant rather than NADH. In starch synthesis, the activated glucose is ADP-glucose rather than the UDP-glucose involved in glycogen synthesis. The special question in photosynthesis is the mechanism by which 3-phosphoglycerate is produced (see text for this).

To remind you of terminology, ribose is an aldose sugar and ribulose is its ketose isomer. Ribulose-1:5-bisphosphate is cleaved by Rubisco into two molecules of 3-phosphoglycerate; this fixes one molecule of CO$_2$.

$$
\begin{array}{ccc}
\begin{array}{l}
CH_2OPO_3^{2-} \\
|\\
C{=}O \\
|\\
CHOH \\
|\\
CHOH \\
|\\
CH_2OPO_3^{2-}
\end{array}
&
\xrightarrow[\substack{\text{Ribulose-1:5-}\\\text{bisphosphate}\\\text{carboxylase}}]{CO_2 \ + \ H_2O}
&
\begin{array}{l}
CH_2OPO_3^{2-} \\
|\\
CHOH \\
|\\
COO^- \\
\quad + \quad\quad + \ 2H^+ \\
COO^- \\
|\\
CHOH \\
|\\
CH_2OPO_3^{2-}
\end{array}
\\[2pt]
\text{Ribulose-1:5-bisphosphate} & & \text{Two molecules of 3-phosphoglycerate}
\end{array}
$$

The 3-phosphoglycerate is converted to carbohydrate as already described. This leads to the next question.

Where does the ribulose-1:5-bisphosphate come from?

The answer to this is very simple, in principle. From six molecules of ribulose-bisphosphate (30 carbon atoms in total) and six molecules of CO_2 (six carbon atoms) we get 12 molecules of 3-phosphoglycerate (36 carbon atoms) produced. Two of these latter molecules ($2C_3$) are used to make storage carbohydrate (C_6) by the pathway already outlined in Fig. 14.9.

The remaining 10 phosphoglycerate molecules (30 carbon atoms in total) are manipulated to produce six molecules of ribulose-bisphosphate (30 carbon atoms in total). The manipulations are complex and involve C_3, C_4, C_5, C_6, and C_7 sugars, aldolase reactions, and transfer of C_2 units by transketolase reactions (page 182) reminiscent of the pentose phosphate pathway. The reactions here are rather involved and are not presented here; they are summarized below.

$$C_3 + C_3 \longrightarrow C_6 \qquad \boxed{\text{Aldolase}}$$
$$C_6 + C_3 \longrightarrow C_4 + C_5 \qquad \boxed{\text{Transketolase}}$$
$$C_4 + C_3 \longrightarrow C_7 \qquad \boxed{\text{Aldolase}}$$
$$C_7 + C_3 \longrightarrow C_5 + C_5 \qquad \boxed{\text{Transketolase}}$$

In sum,

$$5C_3 \longrightarrow 3C_5.$$

The outcome of it all is that six molecules of ribulose bisphosphate, plus six molecules of CO_2 and six of H_2O, are converted to 12 molecules of 3-phosphoglycerate. From there, the six molecules of ribulose bisphosphate are regenerated, plus the dividend of one molecule of fructose-6-phosphate which is converted to the storage carbohydrate. This whole process known as the Calvin cycle is shown Fig. 14.10. The stoichiometry of the whole business is given in the following rather lengthy equation which it is suggested need not be memorized.

$$6CO_2 + 18ATP + 12NADPH + 12H^+ + 12H_2O \longrightarrow$$
$$C_6H_{12}O_6 + 18ADP + 18P_i + 12NADP^+ + 6H^+.$$

Has evolution slipped up a bit?

The chemical devices in life so constantly astonish one with their elegance and efficiency that it comes almost as a relief to find what (at present, anyway) looks almost like a slip-up in the 'design' of the CO_2-fixing enzyme, 'Rubisco'.

At the dawn of photosynthesis there was no oxygen and much higher CO_2 levels than now, but, as photosynthesis occurred, oxygen accumulated. It so happens that Rubisco, in addition to

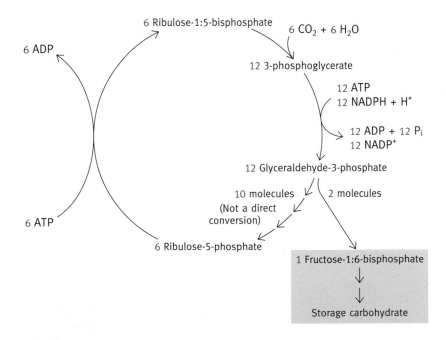

Fig. 14.10 Diagram of the net effect of the reactions in the Calvin cycle. *Note.* The diagram is simplified in that the conversion of 3-phosphoglyceraldehyde to ribulose-5-phosphate involves part of it being converted to fructose-bisphosphate as an intermediate. The presentation is intended to show the net effect of all the reactions involved. Dihydroxyacetone phosphate, which is in equilibrium with glyceraldehyde-3-phosphate, is also omitted for simplicity.

using CO_2, can also react with oxygen, the two competing with one another, so that strictly the enzyme should be called ribulose-bisphosphate carboxylase/oxygenase. The oxygenation reaction, so far as is known at present, serves no useful purpose and is apparently wasteful in the sense that it destroys ribulose-1:5-bisphosphate and wastes ATP in a reaction pathway known as **photorespiration**, the details of which we are not concerned with here. In high light intensity and temperatures, CO_2, in the immediate neighbourhood of the leaf, is rapidly removed by photosynthesis and the wasteful oxygen reaction is therefore maximized to the detriment of photosynthetic efficiency. This can reduce CO_2 assimilation by about 30%. Presumably, in earlier evolutionary times, when there was little oxygen and higher CO_2 levels, this would not have been significant, but, today, in the presence of high oxygen levels, the situation is different.

It might have been expected that a new Rubisco that excluded the oxygen reaction would have been evolved but this has not occurred, for reasons that are not apparent. It might be a rare evolutionary slip-up (which seems improbable in a system of such importance), or there may be good reasons for the apparent inefficiency that are not yet appreciated. However, in some plants, a biochemical device has evolved to raise the CO_2 level in cells where Rubisco operates, and thus minimize the oxygenase reaction. This occurs in species of plants that live in high light and high temperature environments where the problem of photorespiration would be maximized—plants such as maize (corn) and sugar cane.

The C_4 pathway

C_3 plants are so called because the first stable labelled product that is experimentally detectable, if they are allowed to photosynthesize in the presence of $^{14}CO_2$, is the C_3 compound, 3-phosphoglycerate, produced by the Rubisco reaction. Some plants, however, *initially* fix CO_2 from the atmosphere into oxaloacetate (Fig. 14.11). The latter is a C_4 compound, the process is referred to as C_4 **photosynthesis**, and the plants as C_4 **plants**.

The anatomical or cellular structure of C_4 plant leaves differs from that of C_3 leaves. In the former, the mesophyll cells at the surface, which are exposed to the atmospheric CO_2, do not contain the Rubisco enzyme but do fix CO_2 very efficiently into oxaloacetate by the carboxylation of phosphoenolpyruvate (PEP) by the enzyme PEP carboxylase (which is not found in animals)

$$CO_2 + H_2O + PEP \rightarrow Oxaloacetate + P_i.$$

PEP carboxylase has an effectively higher affinity for CO_2, and there is no competition from oxygen. The oxaloacetate is reduced to malate which is transported into neighbouring bundle sheath cells where photosynthesis occurs. Here the CO_2 is released from malate by the 'malic enzyme' (which you have met before in fatty acid synthesis, page 143)

$$Malate + NADP^+ \rightarrow Pyruvate + CO_2 + NADPH + H^+.$$

This raises the concentration of CO_2 in the bundle sheath cells

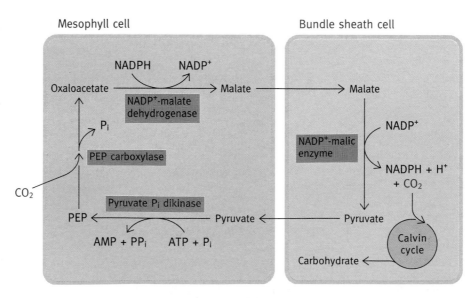

Mesophyll cell   Bundle sheath cell

Fig. 14.11 Diagram of the C_4 pathway for raising the CO_2 concentration for photosynthesis in bundle sheath cells. *Important note*. The pathway of oxaloacetate formation from pyruvate is quite different from that in animals. Direct conversion of pyruvate to phosphoenolpyruvate (PEP) does not occur in animals nor does the carboxylation of PEP. It is important to be clear on this for it has a major effect on metabolic regulation in animals. Note also that the malate dehydrogenase that reduces oxaloacetate uses NADPH, unlike that in the citric acid cycle that is NAD^+-specific.

10–60-fold, resulting in a more efficient operation of the Rubisco reaction. The pyruvate returns to the mesophyll cell where it is reconverted to PEP by **pyruvate phosphate dikinase**. This is an unusual reaction in which two $-\text{P}$ groups of ATP are liberated.

$$CH_3-CO-COO^- + ATP + P_i$$

$$\downarrow$$

$$CH_2{=}C-COO^- + AMP + PP_i + H^+$$
$$\quad\ |$$
$$\quad OPO_3^{2-}$$

In animals, PEP can be made from pyruvate only via oxaloacetate by a quite different route (see page 151). Once phosphoglycerate is made in the bundle sheath cells, the Calvin cycle operates exactly as in C_3 plants.

The C_4 route incurs a price for raising the CO_2 concentration in bundle sheath cells, since ATP is consumed in making PEP and transporting acids. However, at higher temperatures, the C_4 route becomes a considerable advantage. C_4 plants such as corn or sugar cane grown in areas of high temperatures and light intensities are prolific producers of carbohydrates.

There is great diversity in the biochemical strategies used by different C_4 plants. For example, the C_4 acid labelled in the presence of $^{14}CO_2$ may be aspartate in some species, and not malate. These plants have high levels of aspartate aminotransferase (Chapter 15) instead of $NADP^+$ malic dehydrogenase. C_4 plants have also evolved three distinct options for decarboxylating C_4 acids in bundle sheath cells, two being located in the mitochondria unlike the $NADP^+$-malic enzyme shown in Fig. 14.11, which is located in the cytoplasm. They are all mechanisms to achieve the same end—increase of CO_2 levels where Rubisco operates.

....

Further reading

The light-harvesting pathway

Slater, E. C. (1983). The Q cycle, an ubiquitous mechanism of electron transport. *Trends Biochem. Sci.*, **8**, 239–42.
Describes the proton pumping system present in both mitochondria and chloroplast.

The C_1 pathway

Rawsthorne, S. (1992). Towards an understanding of C_3–C_4 photosynthesis. *Essays in Biochem.*, **27**, 135–46.
Discusses the distribution of the biochemical systems among plant species as well as the pathways themselves.

....

Problems for Chapter 14

1 Explain, in general terms, what is meant by the terms 'light' and 'dark' reactions in photosynthesis.

2 What is meant by the term 'antenna chlorophyll'?

3 What is meant by:
   (a) photophosphorylation?
   (b) cyclic photophosphorylation?

4 The oxidation of water requires a very powerful oxidizing agent. In photosynthesis what is this agent?

5 Proton pumping due to electron transport in photosystem II (PSII) causes movement of protons from the outside of thylakoids to the inside. In mitochondria protons are pumped from the inside to the outside. Comment on this.

6 If a photosynthesizing system is exposed for a very brief period to radioactive CO_2, the first compound to be labelled is 3-phosphoglycerate. Explain why this is so.

7 Explain the Calvin cycle in simplified terms.

8 Starting with 3-phosphoglycerate, outline the pathway of starch synthesis.

9 The enzyme Rubisco can react with oxygen as well as CO_2, the oxygen reaction being, so far as we know, an entirely wasteful one. At low CO_2 concentrations such as can occur particularly in intense sunlight, the wasteful oxygenation reaction is maximized. What mechanisms have evolved to ameliorate this problem?

10 In C_4 plants, pyruvate is converted to phosphoenolpyruvate by an ATP-requiring reaction. Have you any comments on this?

Chapter summary

Chapter 15

···

Amino acid metabolism

Digestion of proteins in a normal diet results in relatively large amounts of the 20 different types of amino acids being absorbed from the intestine into the portal blood, which immediately traverses the liver. All cells, except terminally differentiated ones such as erythrocytes, use amino acids for protein synthesis and for the synthesis of a variety of essential molecules such as membrane components, neurotransmitters, heme, and the like. All cells, therefore, take up amino acids by selective transport mechanisms since, in their free, ionized form, they do not readily penetrate the membrane lipid bilayer.

As already stated, in the body there is no dedicated storage of amino acids in the sense that there is no polymeric form of amino acids whose function is simply to be a reserve of these compounds to be called upon when needed. The only 'reserves' are in the form of functional proteins and, since the biggest mass is in muscle proteins, the latter are the main reserves. But, these muscle proteins are part of the contractile machinery and, if broken down, for instance, to provide amino acids for gluconeogenesis in the liver, muscle wasting ensues (see page 153).

This biochemical situation, again as pointed out earlier, can in certain human populations today be an unfortunate one because, during evolution, humans have lost the ability to synthesize 10 of the amino acids. Interestingly enough, those that we cannot synthesize are those with many steps in the synthesis and therefore require many enzymes and many genes to code for them—they are, in essence 'expensive' to manufacture

Why this evolutionary loss occurred is an interesting question that can only be answered in speculative terms. One possibility is that it was simply a measure of economy—that it was more advantageous to 'obtain' 10 of the more complicated amino acids ready-made in the diet and to make only the simpler ones. When the human diet was from the 'wild', if sufficient food could be obtained to sustain life, energy-wise, it would probably contain all the necessary amino acids. In this situation, loss of the ability to synthesize certain amino acids would be pure gain, since you would have been more likely to die from insufficient energy sources than from lack of essential amino acids in the food that was available.

With the development of agriculture, however, vast production of chemical energy in the form of the carbohydrate of cereals permitted large populations to survive but did not necessarily supply adequate amounts of the essential amino acids, for some plant proteins, such as those in corn, lack lysine and tryptophan. What exacerbates the situation is that, when humans have a rich source of protein, containing enough essential amino acids to provide for an extended period, there is no method for storing them. After immediate needs are satisfied, the surplus amino acids are simply oxidized or converted to glycogen or fat. In some circumstances, it is somewhat like burning diamonds to keep warm.

The result of a diet inadequate in even a single amino acid is the condition known as **kwashiorkor**. Since virtually all human proteins contain all 20 amino acids, if a single one of the latter is deficient in amount, the necessary functional proteins cannot be made. This leads to wasting, apathy, inadequate growth, and lowered levels of serum proteins which, by reducing the osmotic pressure of the blood, are probably one cause of edema of the tissues. The condition is a vicious circle since the cells lining the intestine are constantly renewed and, in kwashiorkor, this may be inadequate, as might also apply to the production of digestive enzymes. The result is that what food is available may be inadequately digested and absorbed. The disease affects developing children more than adults because of their greater demand for protein to support growth. This evolutionary loss of the ability to synthesize certain amino acids and the lack of provision for their storage have turned out to be, apparently, considerable human disadvantages. The conclusion is speculative, however, for it may be that unknown compelling reasons dictated the evolutionary loss of synthetic capability. One suggestion has been made that the intermediates in the synthesis of essential amino acids may have been toxic to higher organisms (brain function impairment has been suggested), but there is little evidence on the matter.

Nitrogen balance of the body

The nutritive aspects of amino acids can be treated on a generalized level through the concept of **nitrogen balance**. If the total intake of nitrogen (mainly as amino acids) equals the total excretion, the individual is in a state of nitrogen balance. During growth or repair, more is taken in than is excreted—this is positive nitrogen balance. Negative nitrogen balance occurs in wasting, where excretion exceeds intake, when, for example, amino acids of muscle proteins are converted to glucose and the nitrogen excreted. The proteins of animals are continually 'turning over'; they are continuously broken down and resynthesized. Although much of the derived amino acids is recycled back into proteins, something of the order of about 0.3% of total body protein nitrogen per day is converted to urea and excreted.

An essential amino acid is one that, when omitted from an otherwise complete diet, results in negative nitrogen balance or fails to support the growth of experimental animals. Some amino acids are essential without qualification such as lysine, phenylalanine, and tryptophan (see Table 15.1 for a list of essential and nonessential amino acids; we suggest that you do not need to memorize this table). Tyrosine is not essential provided sufficient phenylalanine is available since the latter is convertible to tyrosine; similarly, cysteine synthesis in mammals requires the availability of methionine, another essential amino acid. The nutritive picture, in this respect, is not completely neat and tidy. 'First-class' proteins are rich in essential amino acids. Plant proteins may be poor in certain essential amino acids. Since the amino acid compositions of proteins vary, a mixture of plant proteins is needed to ensure adequate amino acid nutrition in vegetarian diets.

Table 15.1 Classification of dietary amino acids in humans

Nonessential	Essential
Alanine	Arginine‡
Asparagine	Histidine
Aspartic acid	Isoleucine
Cysteine*	Leucine
Glutamic acid	Lysine
Glutamine	Methionine
Glycine	Phenylalanine
Proline	Threonine
Serine	Tryptophan
Tyrosine†	Valine

* Cysteine is produced only from the essential amino acid, methionine.
† Tyrosine is produced only from the essential amino acid, phenylalanine.
‡ Arginine is required only in the growing stages.

General metabolism of amino acids

The general situation of amino acid metabolism is shown in Fig. 15.1. It is essentially a repeat of that given in Chapter 5 on the broad aspects of metabolism (page 75).

Aspects of amino acid metabolism

In considering how amino acids in excess of the immediate requirements for protein synthesis and other specific needs are broken down for energy production or fuel storage, there are a number of separate questions. We need also to consider how the nonessential amino acids are synthesized and how certain amino acids are metabolized to some other small molecules of physiological importance. The questions we deal with in this chapter are the following:

1. How are the amino groups of amino acids removed?—in other words, how does deamination occur? Although there are 20 different amino acids most of them are deaminated by a common mechanism and this is relatively easy to deal with.

2. What happens to the keto acids—the carbon–hydrogen 'skeletons'—of the amino acids after deamination? In this case, each amino acid has its own special metabolic route and we will give only a limited treatment of this aspect, mentioning features of special interest or of broader relevance.

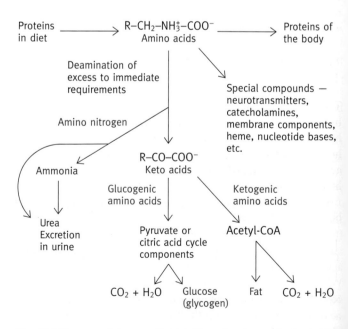

Fig. 15.1 Diagram of the overall catabolism of amino acids. Note that some amino acids are partly ketogenic and partly glucogenic (see page 200 for explanation of these terms).

3. How is the amino group nitrogen, which is removed from amino acids, converted to urea? This is an important area and will be dealt with.

4. How are amino acids synthesized? In the animal body this concerns only the nonessential amino acids. If the carbon skeleton (the keto acid) is available, synthesis is often the reverse of deamination. We will deal with only a few examples of these. In bacteria and plants, all amino acids are synthesized by individual pathways deriving their starting products from general metabolic intermediates. Animals depend on this synthetic activity to supply their essential amino acids. The pathways for the synthesis of all amino acids constitute a formidable amount of detailed information which probably few would carry in their heads without a specific need to do so. We will deal only with particular cases of amino acid biosynthesis of special interest.

5. Not a question, but there are special aspects of amino acid metabolism. Certain amino acids have metabolic reactions of special interest and more general significance which we collect into this section.

There are also major topics of biochemistry, such as protein synthesis, heme synthesis, and nucleotide synthesis, that have amino acids as their reactants. These are major topics and are dealt with in the relevant later chapters. After this rather long preliminary, let us now start.

Deamination of amino acids

First, a piece of simple chemistry. It concerns **Schiff bases**. A Schiff base results from a spontaneous equilibrium between a molecule with a carbonyl group (aldehyde or ketone) and one with a free amino group.

$$\underset{|}{\overset{|}{C}}=O \ + \ H_2N-R \ \rightleftharpoons \ \underset{|}{\overset{|}{C}}=N-R \ + \ H_2O$$

The reaction is freely reversible so a Schiff base can readily hydrolyse. If R = H in the above equation, the Schiff base will hydrolyse to give NH_3. The importance of this is that it shows how the removal of a pair of hydrogen atoms from an amino acid can result in **deamination**, and the supply of a pair of hydrogen atoms can result in the synthesis of an amino acid from a keto acid and ammonia (nonionized structures are used below for clarity).

$$\underset{|}{\overset{|}{C}}HNH_2 \ \xrightarrow{-2H} \ \underset{|}{\overset{|}{C}}=N-H \ \xrightarrow{H_2O} \ \underset{|}{\overset{|}{C}}=O \ + \ NH_3$$

Now back to biological deamination. Glutamic acid is an amino acid of central importance in metabolism. It is deaminated by **glutamate dehydrogenase**, unusually working with either NAD^+ or $NADP^+$. (The NH_3 is protonated to NH_4^+ at physiological pH.)

$$\begin{array}{l} COO^- \\ | \\ CH_2 \\ | \\ CH_2 \\ | \\ CHNH_3^+ \\ | \\ COO^- \end{array} + NAD(P)^+ + H_2O \longrightarrow NAD(P)H + \begin{array}{l} COO^- \\ | \\ CH_2 \\ | \\ CH_2 \\ | \\ C=O \\ | \\ COO^- \end{array} + NH_4^+$$

Glutamate   α-Ketoglutarate

As is described below, glutamate dehydrogenase plays a major role in the deamination, and therefore in the oxidation of many amino acids. In keeping with this, the enzyme is allosterically inhibited by ATP and GTP (indicators of a high energy charge) and activated by ADP and GDP which signal that an increased rate of oxidative phosphorylation is needed.

The α-ketoglutarate can feed into the citric acid cycle, and that disposes of glutamic acid oxidation to CO_2 and H_2O. Since α-ketoglutarate is converted to oxaloacetate in the cycle (Fig. 8.12), glutamic acid can give rise to glucose synthesis in appropriate physiological situations (page 153). It is a glucogenic amino acid. However, there are no corresponding dehydrogenases for the other amino acids. How are they deaminated? A few have special individual mechanisms (see below) but most of them transfer their amino groups to α-ketoglutarate forming glutamate and the corresponding keto acid. The former is then deaminated by the reaction given above. The amino acids are thus deaminated by a two-step process.

The first reaction is called **transamination**.

$$\begin{array}{l} R \\ | \\ CHNH_3^+ \\ | \\ COO^- \end{array} + \begin{array}{l} COO^- \\ | \\ CH_2 \\ | \\ CH_2 \\ | \\ C=O \\ | \\ COO^- \end{array} \longrightarrow \begin{array}{l} R \\ | \\ C=O \\ | \\ COO^- \end{array} + \begin{array}{l} COO^- \\ | \\ CH_2 \\ | \\ CH_2 \\ | \\ CHNH_3^+ \\ | \\ COO^- \end{array}$$

Let us suppose R = CH_3; the amino acid is then alanine. Deamination of alanine proceeds as follows.

1. Alanine + α-ketoglutarate → Pyruvate + glutamate
2. Glutamate + NAD^+ + H_2O → α-ketoglutarate + NADH + NH_4^+

Net reaction: Alanine + NAD^+ + H_2O → Pyruvate + NADH + NH_4^+

The two-step process involving transamination and then deamination of glutamate is called **transdeamination** for obvious reasons. The enzymes involved in reaction 1 above are called **transaminases** or **aminotransferases**. A number of these exist with specific substrate specificities and most amino acids can be deaminated by this route.

The reversibility of transamination means that, provided a keto acid is available, the corresponding amino acid can be synthesized. (The keto acids for essential amino acids are not, however, synthesized in the body.) An example of the importance of this is in the malate–aspartate shuttle (page 114) in which the following reversible reaction occurs.

$$
\begin{array}{c}
\text{COO}^- \\
| \\
\text{CH}_2 \\
| \\
\text{CH}_2 \\
| \\
\text{CHNH}_3^+ \\
| \\
\text{COO}^-
\end{array}
\quad + \quad
\begin{array}{c}
\text{COO}^- \\
| \\
\text{C}=\text{O} \\
| \\
\text{CH}_2 \\
| \\
\text{COO}^-
\end{array}
\quad
\boxed{\text{Aspartate aminotransferase}}
\quad \rightleftharpoons \quad
\begin{array}{c}
\text{COO}^- \\
| \\
\text{CH}_2 \\
| \\
\text{CH}_2 \\
| \\
\text{C}=\text{O} \\
| \\
\text{COO}^-
\end{array}
\quad + \quad
\begin{array}{c}
\text{COO}^- \\
| \\
\text{CHNH}_3^+ \\
| \\
\text{CH}_2 \\
| \\
\text{COO}^-
\end{array}
$$

Glutamate   Oxaloacetate   α-Ketoglutarate   Aspartate

Mechanism of transamination reactions

All transaminases have, tightly bound to the active centre of the enzyme, a cofactor, **pyridoxal-5′-phosphate (PLP)** that participates in the transaminase reaction.

Pyridoxal phosphate is a remarkably versatile cofactor, it participates as an electrophilic agent in a wide variety of reactions involving amino acids. In simple terms, PLP acts as an intermediary, accepting the amino group from the donor amino acid and then handing it on to the keto acid acceptor, both phases occurring on the same enzyme. As so often is the case, the cofactor is a B vitamin derivative. Vitamin B_6 in the diet consists of three interrelated compounds, pyridoxin, pyridoxal, and pyridoxamine (Fig. 15.2). They can all be converted to pyridoxal phosphate (Fig. 15.2) in the cell. The business end of the molecule is the −CHO group. The general reaction catalysed in transamination is

ENZ−PLP + amino acid 1 ↔ ENZ−PLP−NH$_2$ + keto acid 1
ENZ−PLP−NH$_2$ + keto acid 2 ↔ ENZ−PLP + amino acid 2

where PLP represents pyridoxal phosphate and PLP−NH$_2$, pyridoxamine phosphate. The mechanism of the reactions is as shown in Fig. 15.3. Both parts of the reactions occur at the active site of the transaminase, the pyridoxamine phosphate remaining attached to the enzyme.

In addition to the general transdeamination reactions, certain amino acids have their own particular way of losing their amino groups.

Special deamination mechanisms

Serine is a hydroxy amino acid; cysteine is the corresponding thiol amino acid. Serine can be deaminated by a **dehydratase reaction** (dehydratases have been described in several places— glycolysis, the citric acid cycle, and fat metabolism). Cysteine can be deaminated by a somewhat analogous reaction in which H_2S is removed instead of H_2O. In both cases the resultant keto acid is pyruvate (Fig. 15.4).

Fate of the keto acid or carbon skeletons of deaminated amino acids

As far as general metabolism is concerned, some amino acids are glucogenic and some are ketogenic. The term, 'ketogenic', might seem to imply that an amino acid giving rise to acetyl-CoA results in formation of ketone bodies in the blood, which would conflict with the earlier explanation (page 77) that ketone bodies arise only in conditions of excess fat metabolism. Acetyl-CoA in normal metabolic situations does *not* give rise to ketone bodies.

The universally used term **ketogenic** for certain amino acids is an old one, resulting from the use of starving animals as a test system for the metabolic fate of amino acids since, in such animals, any increase in acetyl-CoA production causes increases in blood ketone bodies that are easily measured. It does not mean that ketogenic amino acids produce these exclusively in normal animals, but simply that they *can* give rise to acetyl-CoA but not to pyruvate. **Glucogenic** amino acids were detected by increased levels of blood glucose or glucose excretion when administered to diabetic animals, and similar qualifications apply to this term.

Aspartate, like glutamic acid, is converted to a metabolite of the citric acid cycle (oxaloacetate) by the transamination reaction already described and is therefore glucogenic. Alanine and glutamate, producing pyruvate and α-ketoglutarate, respectively, on deamination, are also glucogenic as are serine, cysteine, and others.

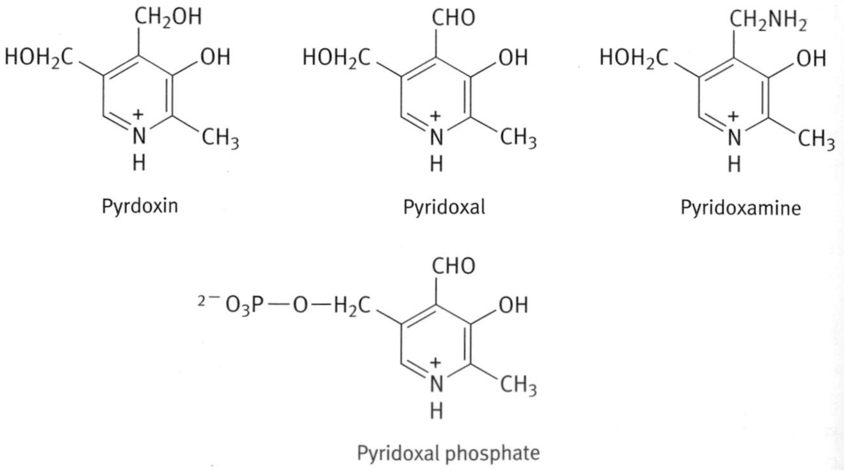

Fig. 15.2 Structures of vitamin B_6 components and of the transaminase cofactor, pyridoxal phosphate.

Pyrdoxin   Pyridoxal   Pyridoxamine

Pyridoxal phosphate

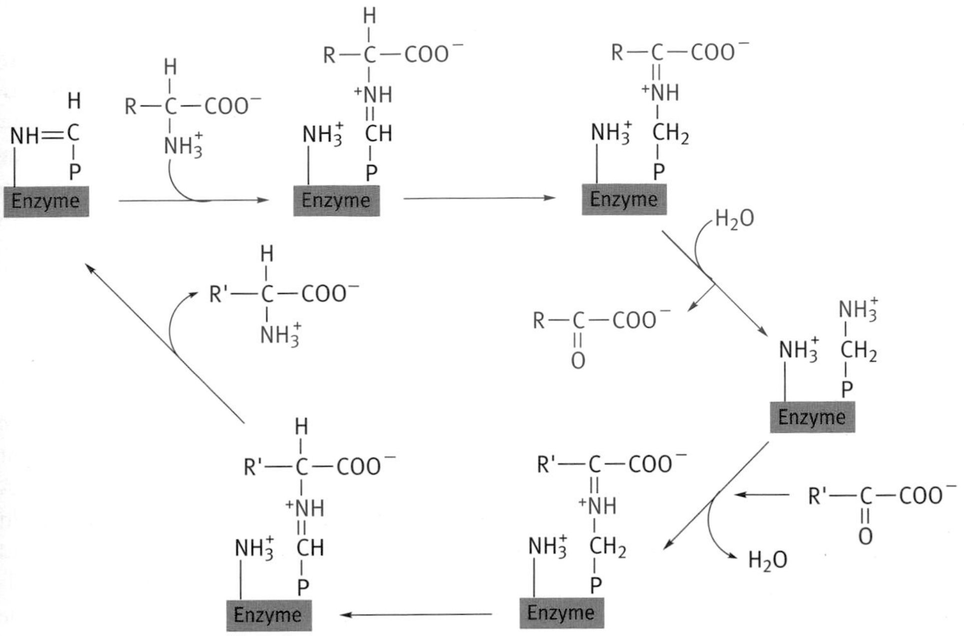

Fig. 15.3 Simplified diagram of the mechanism of transamination. P—CH=NH— in the first figure represents pyridoxal phosphate complexed with a lysine-amino group of the protein. P—CH$_2$—NH$_3^+$ represents pyridoxamine phosphate. The forward reaction (red arrows) results in the conversion of an amino acid to a keto acid and of pyridoxal phosphate to pyridoxamine phosphate. The reverse reaction (blue arrows) reacting with a different keto acid results in transamination between the (red) amino acid and the (blue) keto acid.

Some amino acids are both ketogenic and glucogenic—for example, phenylalanine produces fumarate (an intermediate in the citric acid cycle) and acetyl-CoA. Of the 20 amino acids only two (leucine and lysine) are solely ketogenic. The degradation of four amino acids (isoleucine, methionine, threonine, and valine) has been referred to on page 135.

Phenylalanine metabolism has special interest

Phenylalanine is an aromatic amino acid, an excess of which is normally converted to tyrosine by an enzyme, **phenylalanine hydroxylase** (Fig. 15.5). This enzyme is interesting in that a pair of hydrogen atoms are supplied by an electron donor—a coenzyme molecule called tetrahydrobiopterin (RH$_4$ in Fig 15.5).

It may seem odd that a reaction requires *both* oxygen and a reducing agent but it is a mechanism used in other reactions also as you'll see later. The trick is that one *atom* of oxygen is used to form an —OH group on the aromatic ring, but this leaves the other oxygen atom to be taken care of. It is reduced to H$_2$O by the two hydrogen atoms donated by the tetrahydrobiopterin. A separate enzyme system reduces the dihydrobiopterin formed back to the tetrahydro form, using NADH as reductant, and so the cofactor acts catalytically.

The phenylalanine hydroxylase belongs to a class of enzymes known as **monooxygenases** (because one atom of O appears in the product) or, alternatively, mixed function oxygenases because two things are oxygenated—the amino acid and a pair of hydrogen atoms. (See page 216 for a discussion of the difference between an **oxidation** that removes electrons and an **oxygenation** that adds oxygen.)

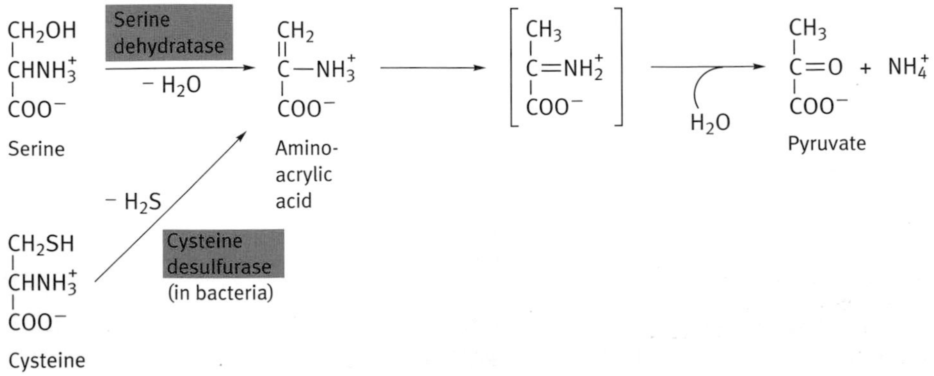

Fig. 15.4 Conversion of serine and cysteine to pyruvate.

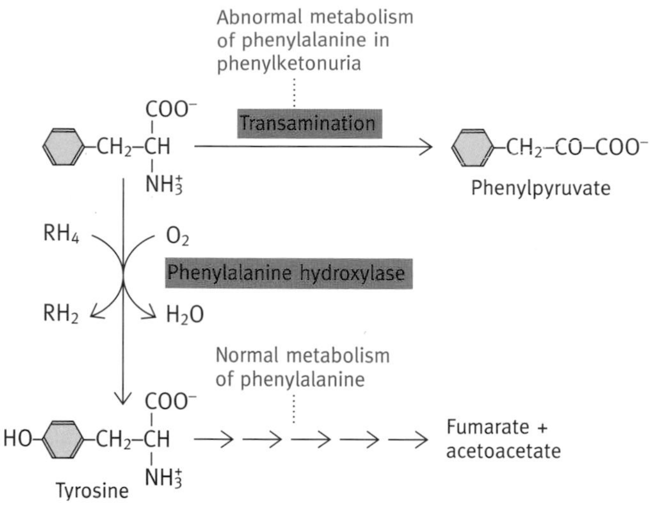

Fig. 15.5 Normal and abnormal metabolism of phenylalanine. RH$_4$, tetrahydrobiopterin; RH$_2$, dihydrobiopterin. Structures are given below for reference purposes—have a look at these.

(a)

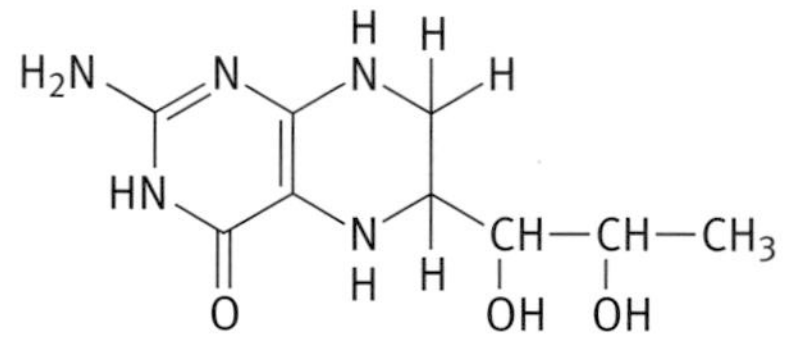

(b)

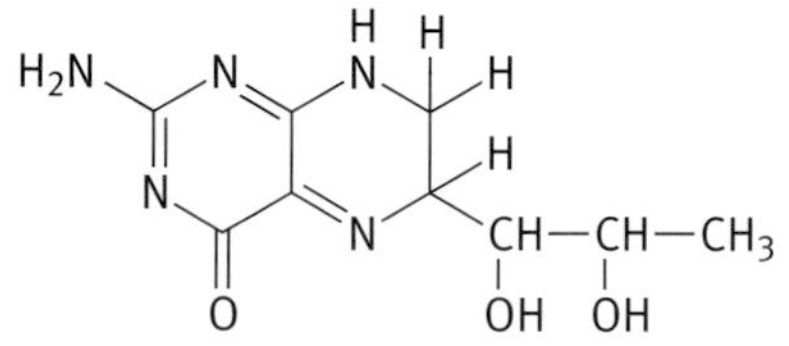

Phenylalanine as such is not normally deaminated, being converted to tyrosine, and only then does deamination occur (Fig. 15.5). However, there is a relatively common genetic abnormality in which the phenylalanine conversion to tyrosine is impaired or blocked due to enzyme deficiency or, rarely, to lack of tetrahydrobiopterin. This causes excess phenylalanine to accumulate and, in this situation, it abnormally participates in transamination producing phenylpyruvate (an abnormal metabolite), which spills out in the urine (Fig. 15.5). The disease is called **phenylketonuria** or **PKU**. The consequences of phenylpyruvate in babies are irreparable mental impairment and early death. If diagnosed at birth (by urine analysis, or preferably, blood analysis, for phenylpyruvate) a child with the disease can be given a diet limited in phenylalanine (but adequate in tyrosine) and development is normal.

It is not known how phenylpyruvate causes such deleterious brain damage. Curiously enough, another genetic condition, called maple syrup disease, involves accumulation of the keto acids of three aliphatic amino acids, valine, isoleucine, and leucine (whose structures are given on page 25), and this

also involves brain impairment. (The name of the disease comes from the keto acids, in the urine, having a characteristic smell.) This disease is much rarer than PKU. Another much-quoted genetic condition is alcaptonuria in which the urine turns black on exposure to air, but this is a benign condition. It is due to a block in the tyrosine degradation pathway in which a diphenol intermediate metabolite, homogentisate, is excreted. The diphenol in air oxidizes to form a dark pigment.

Methionine and transfer of methyl groups

Methionine is one of the essential amino acids. It has the structure

$$CH_3-S-CH_2-CH_2-CH-COO^- .$$
$$\underset{NH_3^+}{|}$$

The interesting part is the **methyl group**. Methyl groups are very important in the cell—a variety of compounds are methylated, and methionine is the source of methyl groups that are transferred to other compounds. Methionine is a stable molecule—the methyl group has no tendency to leave; however, if the molecule is converted into **S-adenosylmethionine** (**SAM**), a sulfonium ion is created and the methyl group is 'activated'—it has a strong group transfer potential making it thermodynamically favourable for it to be transferred (by transmethylase enzymes) to other compounds. ATP supplies the energy for SAM synthesis—in this case three $-ⓟ$ groups are converted to PP$_i$ + P$_i$ and then the PP$_i$ is cleaved to two P$_i$ (Fig. 15.6).

Transfer of the methyl group of SAM to other compounds generates *S*-adenosylhomocysteine. The latter is hydrolysed to produce homocysteine—this is methionine with −SH instead of S−CH$_3$. A reaction sequence exists for transferring the thiol group to serine, producing cysteine (Fig. 15.6). The intermediate compound, cystathionine—a complex between the two amino acids—is mentioned here because a defect in its hydrolysis, to form cysteine, results in the disease cystathionuria.

What are the methyl groups transferred to?

In the body the methyl group of creatine (page 390), phosphatidylcholine (page 42), and epinephrine (page 204) come from *S*-adenosylmethionine and, in addition so do the methyl groups attached to the bases of nucleic acids. Note, however, that the latter do not include that of thymine which is separately synthesized and is an important topic to be dealt with later (page 233) as also is the subject of nucleic acid base methylation.

Synthesis of amino acids

In the body, as explained, only the nonessential amino acids can be synthesized. We will not include the details of all of these, for

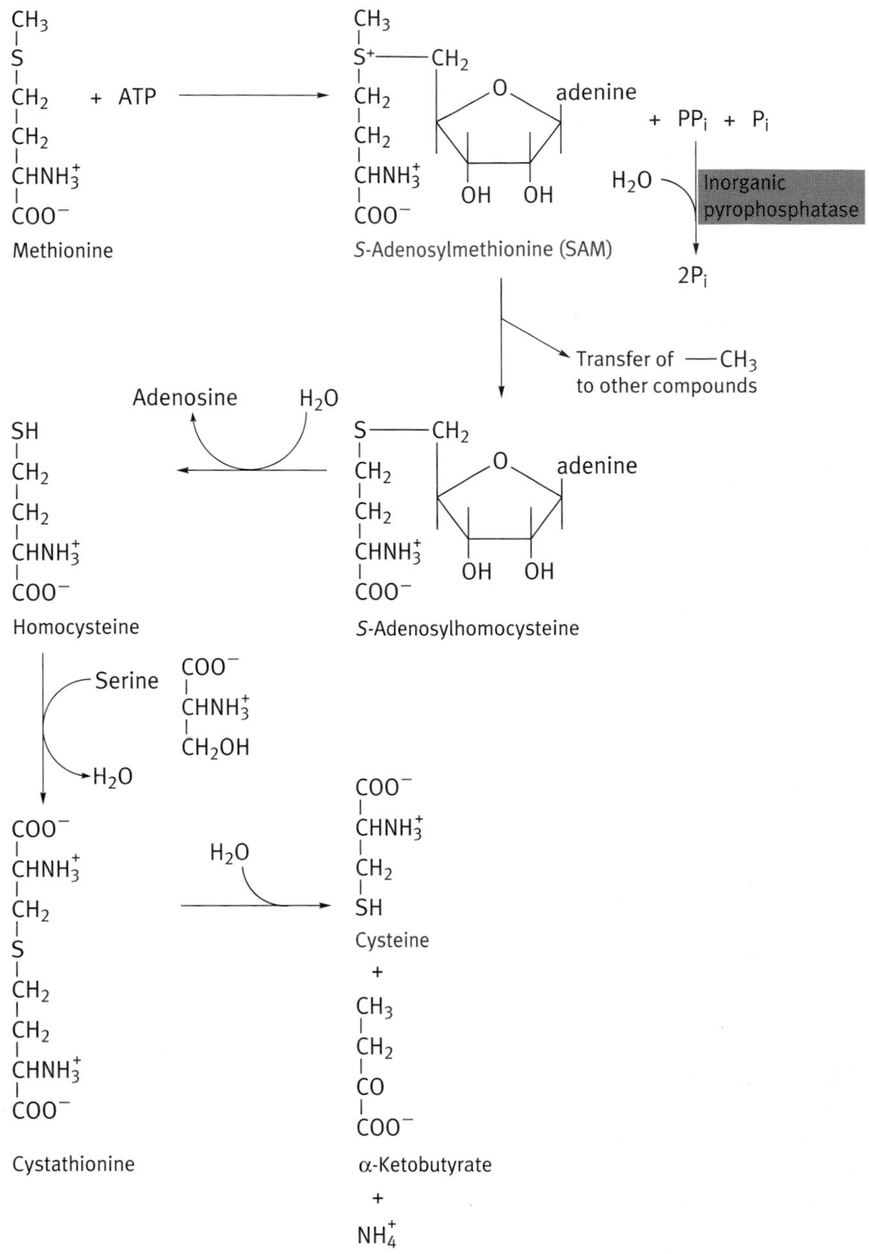

Fig. 15.6 The synthesis of S-adenosylmethionine (SAM) from methionine and the synthesis of cysteine from homocysteine.

the chemistry is quite extensive and much of it relevant only to itself. If needed, pathway details are readily available elsewhere. Our aim here is to deal only with aspects of special interest and to illustrate how amino acids are synthesized from glycolytic and citric acid cycle intermediates. In fact, five of these intermediates (3-phosphoglycerate, phosphoenolpyruvate, pyruvate, oxalo-acetate, and α-ketoglutarate) together with two sugars of the pentose phosphate pathway are the precursors of all 20 amino acids in those organisms, such as plants and bacteria, that synthesize all of them.

Synthesis of glutamic acid

You will recall that deamination of this amino acid occurs via glutamate dehydrogenase, an NAD^+ or $NADP^+$ enzyme of central importance. This is reversible, but probably is less important in glutamate formation in animals than is transami-nation of α-ketoglutarate using other amino acids such as alanine or aspartate as the donor of the amino group. See the reaction for aspartate aminotransferase on page 200. A different energy-requiring route of α-ketoglutarate amination is also used in prokaryotes in situations where the NH_4^+ concentration is very low.

Synthesis of aspartic acid and alanine

These amino acids come from transamination of oxaloacetate

$$\left(R = \begin{array}{c} COO^- \\ | \\ CH_2 \\ | \end{array} \right)$$

and pyruvate ($R = CH_3$) respectively.

$$\begin{array}{c} R \\ | \\ C=O \\ | \\ COO^- \end{array} + glutamate \longrightarrow \begin{array}{c} R \\ | \\ CHNH_3^+ \\ | \\ COO^- \end{array} + \alpha\text{-ketoglutarate}$$

Synthesis of serine

This is formed from a glycolytic intermediate, 3-phosphogly-cerate, which is first converted to a keto acid, 3-phospho-hydroxypyruvate.

$$\begin{array}{c} COO^- \\ | \\ CHOH \\ | \\ CH_2OPO_3^{2-} \end{array} \xrightarrow[]{NAD^+ \quad NADH} \begin{array}{c} COO^- \\ | \\ C=O \\ | \\ CH_2OPO_3^{2-} \end{array}$$

This keto acid is transaminated by glutamic acid to give 3-phosphoserine which is hydrolysed to serine and P_i.

$$\begin{array}{c} COO^- \\ | \\ C=O \\ | \\ CH_2OPO_3^{2-} \end{array} \xrightarrow[]{\boxed{Transamination}} \begin{array}{c} COO^- \\ | \\ CHNH_3^+ \\ | \\ CH_2OPO_3^{2-} \end{array} \xrightarrow[]{\boxed{Hydrolysis}} \begin{array}{c} COO^- \\ | \\ CHNH_3^+ \\ | \\ CH_2OH \end{array} + P_i$$

Synthesis of glycine

Glycine is the simplest amino acid of all ($CH_2\,NH_3^+\,COO^-$). It is formed by a reaction that is completely new, so far as this book is concerned, involving withdrawal of a methylene group ($-CH_2$) from serine and adding it to a coenzyme, tetrahydro-folate (again, not yet metioned in this book), whose function is to act as a one-carbon unit carrier. The one-carbon unit transfer area is of importance in nucleotide synthesis. It will be more appropriate to go into this more thoroughly later (page 228).

Synthesis of other molecules from amino acids

There is a whole range of physiologically active small molecules made from amino acids. **Amines** are produced by decarboxyla-tion of amino acids.

$$RCH_2NH_3^+COO^- \rightarrow RCH_2NH_3^+ + CO_2$$

Catecholamines includes the hormones dopamine, epinephr-ine and norepinephrine, whose synthesis is given in Fig. 26.2.

The collective term derives from their structural relationship to catechol.

Epinephrine   Catechol

Several neurotransmitters whose role in nerve conduction is described in Chapter 26, are derived from amino acids. They include γ-aminobutyrate (GABA) and 5-hydroxytryptamine as well as the catecholamines. The hormone, thyroxine, is derived from tyrosine (see Fig. 26.4(a)). This list is not exhaustive but is illustrative of the general principle that amino acids are the precursors of many compounds in the body.

What happens to the amino groups when they are removed from amino acids?—the urea cycle

The amino groups of catabolized amino acids are excreted in mammals as urea, which is a highly water-soluble, inert, and nontoxic molecule. It is produced in the liver from the guanidino group of arginine by the hydrolytic enzyme arginase. The other product is ornithine, an amino acid not found in proteins.

Arginine   Ornithine

The amino nitrogen of the catabolized 20 amino acids is used to convert ornithine back to arginine. The extra carbon comes from CO_2. This forms a metabolic cycle (Fig. 15.7), the first ever discovered. Krebs (who also discovered the citric acid cycle) together with Henseleit observed that, when arginine was added to liver cells, the increased urea formation caused by this addition far exceeded in amount that of the arginine added—that is, it was acting catalytically and ornithine did the same, suggesting a cyclical process. A cycle in which CO_2 and two nitrogen atoms were added to ornithine to produce arginine required more than one step and it was discovered that

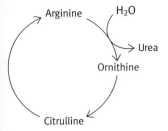

Fig. 15.7 Outline of the arginine–urea cycle. The input of CO_2 and nitrogen into the cycle will be dealt with in the next section.

citrulline, an amino acid intermediate between ornithine and arginine, is involved, since it also acts catalytically on urea synthesis when added to liver cells. This discovery led to the famous **urea cycle**. Citrulline is not one of the 20 amino acids used for synthesizing proteins.

$$
\begin{array}{c}
H_2N \\
\searrow C {=} O \\
| \\
NH \\
| \\
CH_2 \\
| \\
CH_2 \\
| \\
CH_2 \\
| \\
CHNH_3^+ \\
| \\
COO^-
\end{array}
$$

Citrulline

Mechanism of arginine synthesis

We need now to see how ornithine is converted to arginine. Ammonia, CO_2 and ornithine are the reactants for the first step, that of citrulline synthesis which occurs in the mitochondrial matrix. Energy is needed and ATP supplies it. Firstly, ammonia and CO_2 are converted to a reactive intermediate, **carbamoyl phosphate**, which then combines with ornithine to give citrulline. Carbamic acid has the structure NH_2COOH and carbamoyl phosphate is therefore $NH_2{-}\overset{\overset{\displaystyle O}{\|}}{C}{-}PO_3^{2-}$. It is a high-energy phosphoryl compound, being an acid anhydride. It is synthesized by an enzyme carbamoyl phosphate synthetase, catalysing the following reaction.

$$NH_4^+ \; + \; HCO_3^- \; + \; 2ATP$$

$$\downarrow$$

$$NH_2{-}\overset{\overset{\displaystyle O}{\|}}{C}{-}O{-}\overset{\overset{\displaystyle O}{\|}}{\underset{\underset{\displaystyle O^-}{|}}{P}}{-}O^- \; + \; 2ADP \; + \; P_i \; + \; 2H^+$$

Two ATP molecules are used, the first ATP being broken

down to ADP and P_i to drive the production of carbamate from ammonia and CO_2 and the second being used to phosphorylate the carbamate. It all happens on the surface of the one enzyme. The carbamoyl group of carbamoyl phosphate is now transferred to ornithine by ornithine transcarbamoylase giving citrulline. If we represent ornithine as $R{-}NH_3^+$ the reaction is

$$RNH_3^+ \; + \; NH_2{-}\overset{\overset{\displaystyle O}{\|}}{C}{-}O{-}\overset{\overset{\displaystyle O}{\|}}{\underset{\underset{\displaystyle O^-}{|}}{P}}{-}O^- \longrightarrow HN{-}\overset{\overset{\displaystyle O}{\|}}{\underset{\underset{\displaystyle R}{|}}{C}}{-}NH_2 \; + \; P_i$$

Ornithine   Carbamoyl phosphate   Citrulline

Conversion of citrulline to arginine

The final step in arginine synthesis is to convert the $C{=}O$ group of citrulline to the $C{=}NH$ of arginine. This occurs in the cytosol. Ammonia is *not* used here but, instead, the amino group of aspartate is added directly. (You will appreciate that ammonia can be converted to glutamate and that this can generate aspartate by transamination with oxaloacetate. In this way, ammonia and also the amino groups of most amino acids can also be converted to urea via this second stage of the cycle.)

First, aspartate condenses with citrulline as follows:

$$
\begin{array}{ccc}
& \overset{\displaystyle C{=}O}{R{-}NH\diagup\quad \diagdown NH_2} & +\quad
\begin{array}{c}
COO^- \\
| \\
H_3\overset{+}{N}{-}CH \\
| \\
CH_2 \\
| \\
COO^-
\end{array}
\end{array}
$$

Citrulline   Aspartate

$$\Big\downarrow \text{ATP}$$

$$\Big\downarrow \text{AMP} \; + \; PP_i$$

$$
\begin{array}{c}
COO^- \\
| \\
\overset{\displaystyle C{-}NH{-}CH}{R{-}NH\diagup\quad \diagdown NH_2^+}\quad
\begin{array}{c}
\\ | \\ CH_2 \\ | \\ COO^-
\end{array}
\end{array}
$$

Argininosuccinate

The molecule formed is **argininosuccinate**. The name derives from the fact that the molecule structurally is like an arginine derivative of succinate. It might be somewhat confusing, given this name, that the molecule is now 'pulled apart' by argininosuccinate lyase to yield arginine and fumarate (not arginine and succinate).

$$COO^-\!\!-CHNH_3^+\!\!-(CH_2)_3\!\!-NH \overset{C-NH-CH}{\underset{NH_2^+}{\Big\backslash}} \begin{array}{l} COO^- \\ | \\ CH \\ | \\ CH_2 \\ | \\ COO^- \end{array}$$

Argininosuccinate

$$COO^-\!\!-CHNH_3^+\!\!-(CH_2)_3\!\!-NH\!\!-\overset{NH_2}{\underset{NH_2^+}{\overset{|}{C}}} \quad + \quad \begin{array}{l} COO^- \\ | \\ CH \\ \| \\ CH \\ | \\ COO^- \end{array}$$

Arginine   Fumarate

The whole cycle is therefore as shown in Fig. 15.8. Major control of urea synthesis is exercised by the adjustment of the level of enzymes. These increase with a rich intake of amino acids and in starvation when muscle proteins are degraded. In addition, carbamoyl phosphate synthetase is allosterically activated by *N*-acetyl glutamate whose level reflects that of amino acids.

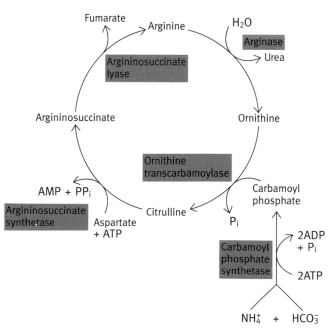

Fig. 15.8 The enzymes of the urea cycle. The levels of urea cycle enzymes are coordinated with the dietary intake of protein. The cycle is allosterically controlled at the carbamoyl phosphate synthetase step. The positive allosteric effector of the enzyme is *N*-acetylglutamate. The conversion of ornithine to citrulline takes place inside mitochondria while the rest of the cycle occurs in the cytoplasm.

How is the amino nitrogen transported from extrahepatic tissues to the liver to be converted into urea?

Transport of ammonia in the blood as glutamine

There are two main mechanisms. Ammonia produced from amino acids is toxic, and blood ammonia levels are kept very low since abnormally high levels can impair brain function and cause coma. Free ammonia is not therefore transported, as such, from peripheral tissues to the liver, but is first converted to the nontoxic amide glutamine. This is synthesized from glutamate by the enzyme, glutamine synthetase.

$$\begin{array}{l} COO^- \\ | \\ CH_2 \\ | \\ CH_2 \\ | \\ CHNH_3^+ \\ | \\ COO^- \end{array} + NH_4^+ + ATP \longrightarrow \begin{array}{l} CO-NH_2 \\ | \\ CH_2 \\ | \\ CH_2 \\ | \\ CHNH_3^+ \\ | \\ COO^- \end{array} + ADP + P_i + H^+$$

Glutamate   Glutamine

The reaction involves the intermediate formation of an enzyme-bound γ-glutamyl phosphate. This is a high-energy phosphoryl-anhydride compound with sufficient free energy to react with ammonia.

The glutamate for this synthesis can be formed from α-ketoglutarate generated in the citric acid cycle followed by transamination with other amino acids. The glutamine is carried in the blood to the liver where it is hydrolysed to release ammonia which is used for urea synthesis.

$$\begin{array}{l} CO-NH_2 \\ | \\ CH_2 \\ | \\ CH_2 \\ | \\ CHNH_3^+ \\ | \\ COO^- \end{array} + H_2O \xrightarrow{\text{Glutaminase}} \begin{array}{l} COO^- \\ | \\ CH_2 \\ | \\ CH_2 \\ | \\ CHNH_3^+ \\ | \\ COO^- \end{array} + NH_4^+$$

Glutaminase in the kidney also liberates ammonia to be excreted along with excess acids from the blood. Glutamine is one of the 20 amino acids found in proteins and is involved in the synthesis of several other metabolites. In the latter, the glutamine amide is used as a nitrogen souce. You will meet examples of this later in the book.

Transport of amino nitrogen in the blood as alanine

From muscle, as much as 30% of the amino nitrogen produced by protein breakdown is sent to the liver as alanine (as well as glutamine). The amino acids transaminate with pyruvate to yield alanine which is released into the blood. This is taken up by the liver; the amino group is used to form urea (via ammonia and/or aspartic acid). The released pyruvate is converted to blood glucose which can go back to the muscle. The sequence of events is referred to as the **glucose–alanine cycle** (Fig. 15.9).

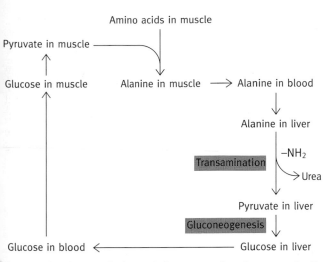

Fig. 15.9 The glucose–alanine cycle for transporting nitrogen to the liver as alanine, and glucose back to the muscles. See also page 153 for further comments on this cycle.

This alanine transport from muscle to liver has another important physiological role in starvation. As explained already (page 151), after glycogen reserves are exhausted the liver *must* make glucose to supply the brain and other cells with an obligatory requirement for the sugar. The main sources of metabolites for this hepatic gluconeogenesis are the amino acids derived from muscle protein breakdown. Protein breakdown in cells is a controlled process described in Chapter 22. Many of the amino acids can give rise to pyruvate in the muscle which is sent to the liver as alanine. As mentioned on page 153, however, the glucose–alanine cycle itself gives no net increase in glucose and in starvation the amino acids liberated from muscle protein breakdown must be converted to alanine without utilizing glucose as a source of pyruvate.

Further reading

Felig, P. (1975). Amino acid metabolism in man. *Ann. Rev. Biochem.*, **44**, 933–55.
A most useful review which discusses amino acid metabolism at the organ level together with the effects of starvation, diabetes, obesity, and exercise on it. It also discusses gluconeogenesis at this level.

Snell, K. (1979). Alanine as a gluconeogenic carrier. *Trends Biochem. Sci.*, **4**, 124–8.

Reviews the role of muscle in supplying the liver with metabolites for glucose synthesis.

Newsholme, E. A. and Leach, A. R. (1983). *Biochemistry for medical sciences*, pp. 417–29. Wiley.
A very useful description of amino acid metabolism in the different tissues of the body, and its physiological relevance.

Problems for Chapter 15

1 Explain how an oxidation can result in the deamination of an amino acid.

2 Which amino acid is deaminated by the mechanism referred to in question 1?

3 How are several of the amino acids deaminated, where the reaction in question 2 is involved? Use alanine as an example.

4 What is the cofactor involved in transamination? Give its structure and explain how transamination occurs.

5 Explain how serine and cysteine are deaminated.

6 What is meant by the terms glucogenic and ketogenic amino acids? Which amino acids are purely ketogenic?

7 Explain the genetic disease phenylketonuria.

8 What is the role of tetrahydrobiopterin in phenylalanine hydroxylation?

9 Methionine is the source of methyl groups in several biochemical processes. Explain how methionine is activated to donate such groups.

10 Outline the reactions of the urea cycle.

11 Why should the level of urea cycle enzymes be increased both in the situation of a high intake of amino acids and in starvation?

12 How are (a) ammonia and (b) amino nitrogen in peripheral tissues transported to the liver for conversion to urea?

Chapter summary

Chapter 16

Garbage disposal units inside cells

This chapter deals mainly with **lysosomes** whose job it is to destroy unwanted molecules or structures already in the cell or which are imported into the cell by endocytosis. It may seem an unglamorous metabolic role but its importance is highlighted by the existence of about 40 known genetic diseases associated with individual enzyme defects in lysosomes.

In addition to lysosomes we will here deal with the physically very similar **peroxisomes** though they cannot be regarded as garbage disposal units in the same way as lysosomes can be for they have special roles—in effect, they deal with the chemical conversions that the rest of the cellular metabolic machinery cannot cope with.

Lysosomes

These are membrane-bounded organelles in the cytoplasm of all eukaryote cells (see the electron micrograph in Fig. 3.15(a). There may be hundreds within a liver cell. They are literally bags of enzymes of great potential destructiveness. Inside lysosomes, which exist in the cytoplasm of cells, is a mixture of hydrolytic enzymes capable of taking almost any cellular component to pieces. The cell is protected from destruction by the lysosomal membrane. The lysosomal enzymes require an acid pH for activity; ATP-dependent proton pumps in the lysosomal membrane maintain the pH inside the organelle between pH 4.5 and 5.0 (Fig. 16.1). If a lysosome were to rupture, the buffering of the cytoplasm would maintain the pH at 7.3 or so, at which pH lysosomal enzymes are almost inactive.

Primary lysosomes are produced by budding off from the Golgi apparatus. Newly synthesized lysosomal enzymes are packaged into these vesicles during their formation, but we will leave the topic of how this is done until Chapter 22 (page 307), which deals with protein synthesis. The lysosomal enzymes include phosphatases, proteinases, esterases, DNase and RNase, and enzymes that destroy polysaccharides and mucopolysac-

charides. In short, and as stated, almost all biological molecules can be degraded by them. Hydrolytic enzymes catalyse irreversible reactions, so they proceed to completion.

How does material to be destroyed come Into contact with the lysosomal enzymes?

As implied, it would not do for lysosomal enzymes to be released for they could destroy the cell; they must always be segregated from the cytoplasm by a membrane. So, somehow, material to be destroyed must be incorporated into a membrane-bound vesicle that contains those enzymes.

Material for disposal may originate inside or outside the cell. In the former category are mitochondria. These are believed to last in a liver cell for about 10 days and then, in an unknown

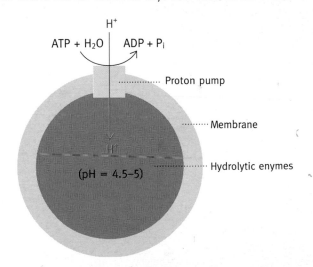

Fig. 16.1 Diagram of a primary lysosome budded off from the Golgi apparatus. This is essentially a transport vesicle that delivers its load of hydrolytic enzymes to endosomes. See text for the role of v-snares, which are believed to target vesicles to their destinations. The Golgi apparatus and its roles are described in Chapter 22.

way, they presumably must be selected for destruction, at which point they become enclosed in a membrane-bound vesicle, the membrane originating from the endoplasmic reticulum; the latter is described on page 308. The resultant vesicle is called an autophagic vesicle or **autophagosome** (Fig. 16.2). This contrasts with a **phagosome** that is formed by a phagocytic white cell engulfing a bacterium or other foreign particle.

Phagocytosis of bacteria is a specialized function of white cells but most cells of the body take in external particles by receptor-mediated endocytosis. The uptake of LDL (page 88) and chylomicron remnants (page 87) are good examples. The mechanism is shown in Fig. 16.2; the receptors for these, along with others, are present in clathrin-coated pits in the plasma membrane (see page 316 for more about endocytosis). On the binding of for example, an LDL particle, the pit invaginates and forms a clathrin-coated vesicle. The clathrin immediately returns to the membrane leaving a vesicle known as an **endosome** containing the ingested particle.

Another good example of endocytosis is the destruction of aged red blood cells by the liver and spleen. The major carbohydrates on red blood cells terminate in the sugar derivative sialic acid (page 44). If this end grouping is lost (which presumably happens with time), it exposes a galactose residue that marks the cell as ready for disposal. The average red blood cell circulates for 120 days and then is destroyed. Lysosomes also have an important role in releasing the contents of lipoproteins such as LDL taken up by endocytosis (see page 88).

The autophagosomes, phagosomes, and endosomes fuse with the primary lysosomes released by the Golgi (and containing the hydrolytic enzymes you will recall), forming secondary lysosomes in which digestion occurs. (The terminology in this area is somewhat varied. Sometimes the term lysosome is restricted to what is here called the secondary lysosome. We have omitted the secondary endosome believed to occur between the endosome and secondary lysosome for simplicity.) It is believed that transport vesicles produced by the Golgi may be targeted to the membrane with which they are to fuse by special recognition proteins ('snares') called **v-snares** (v for vesicle) that bind to complementary proteins in the target membrane (**t-snares**). Once the vesicle is attached to the target membrane, a separate energy-requiring process, probably involving GTP hydrolysis, is required for the actual fusion. The production of transport vesicles by the Golgi will be discussed in Chapter 22 when protein synthesis and delivery are dealt with.

In addition to the destruction of organelles and ingested structures described above, many cell components such as certain proteins, nucleic acids, lipid components, and even glycogen (see below) are believed to be selectively destroyed by lysosomal digestion. How they are taken up into the digestive vesicle isn't known but possibly specific membrane transport systems exist for this purpose. Specific membrane transporter proteins probably also exist to return digestion products back to the cytoplasm for reuse.

It seems that certain membrane components of cells are constantly degraded by lysosomes. A whole family of genetic diseases exists, called sphingolipidoses. In these, specific lysosomal enzymes are missing that impair the degradation of gangliosides (page 44) of the cell membranes. A classical example is Tay–Sachs disease (page 45).

The essential nature of lysosomal function

The very great importance of this is underlined by the existence of genetic diseases, already referred to, known as **lysosomal storage disorders**, that arise because of the lack of one or more specific lysosomal enzymes. Material taken into the lysosomes is not destroyed and the organelles become overloaded with it.

A rather puzzling role of lysosomes is their part in the breakdown of glycogen. You will recall (Chapter 6) that glycogen is degraded by glycogen phosphorylase but, in Pompe's disease, one of a series of glycogen storage diseases, the deficiency is of a lysosomal enzyme that hydrolyses the α-1:4 links of glycogen. In the absence of the enzymes there is a massive accumulation of glycogen in the cells usually causing death in infancy. Lysosomes do not come into schemes of

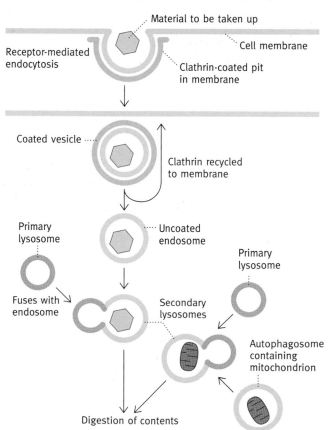

Fig. 16.2 Diagram of the digestion of the contents of coated vesicles produced by endocytosis and of auto-phagosomes. A similar mechanism occurs with the digestion of bacteria engulfed by phagocytes. See text for a note on terminology in this area.

Material to be taken up

Cell membrane

Receptor-mediated endocytosis

Clathrin-coated pit in membrane

Coated vesicle

Clathrin recycled to membrane

Primary lysosome

Uncoated endosome

Primary lysosome

Fuses with endosome

Secondary lysosomes

Autophagosome containing mitochondrion

Digestion of contents

glycogen metabolism. It may be that degraded glycogen molecules are taken up by lysosomes for final breakdown but very little is known as to exactly what their role is in this regard.

Peroxisomes

These are present in mammalian cells as membrane-bounded vesicles containing enzymes (oxidases) that oxidize a number of substrates using molecular oxygen generating not water, but hydrogen peroxide,

$$R'H_2 + O_2 \rightarrow R' + H_2O_2.$$

The H_2O_2 is further used to oxidize other substrates (including part of ingested alcohol) as

$$RH_2 + H_2O_2 \rightarrow R + 2H_2O.$$

The substrates include very long chain fatty acids that are not oxidized by mitochondria until shortened, some phenols, and D-amino acids. There is also evidence that the cholesterol side chain oxidation required for conversion to bile acids occurs in perixosomes and that the synthesis of some complex lipids requires peroxisomes. It is intriguing that the cholesterol level-reducing drug, **clofibrate**, causes proliferation of peroxisomes.

The believed roles of peroxisomes are rather a mixed bag, and sometimes not very well defined. However, their essential role is underlined by the existence of a rare fatal genetic disease in children lacking peroxisomes in some tissues. The disease is characterized by abnormally high levels of C_{24} and C_{26} fatty acids and of bile acid precursors.

Further reading

Neufeld, E. F. (1991). Lysosomal storage diseases. *Ann. Rev. Biochem.*, **60**, 257–80.
Reviews general principles and specific diseases.

Mallabiabarrena, A. and Malhotra, V. (1995). Vesicle biogenesis: the coat connection. *Cell*, **83**, 667–9.
A minireview that discusses why different coat proteins are used. For the transport from endocytotic vesicles to endosomes, and Golgi to endosomes, a clathrin coat is used. For others, two different coatamers are used. A complex but very interesting story, with more questions than answers.

Problems for Chapter 16

1   What are the purposes of lysosomes?

2   What are primary lysosomes and where are they produced?

3   How do the lysosomal enzymes come into contact with their target substrates?

4   What evidence is there to indicate that lysosomal digestion is an essential process?

5   Why is Pompe's disease somewhat of a biochemical puzzle?

6   What are peroxisomes and their function?

Chapter 17

Enzymic protective mechanisms in the body

The immune system (Chapter 24) is the major protective mechanism against attack by disease-causing pathogens. However, the body is threatened by other things. We will, in later chapters, describe DNA repair to cope with damage caused by ionizing radiation, UV light, and mutations (page 263), and the production of heat shock proteins (page 306) to cope with heat and other stresses. These fit into specific chapters but there exists, in addition, a variety of other protective mechanisms involving enzymic reactions and these are the subject of the present chapter. The fact that they are, in a biochemical sense, rather a varied collection should not obscure the fact that all have the same biological role—protection—and are essential to life.

Blood clotting

Coagulation of blood is needed to form blood clots, which plug holes in damaged blood vessels and so prevent bleeding. To be effective, the response has to be very rapid and relatively massive while the initial signal in chemical terms is exceedingly small. A massive amplification of the signal is therefore needed—the response must, in quantitative terms, be vastly greater than the signal.

We have seen earlier (see page 173) that biochemical amplification is achieved by means of a **reaction cascade**. In this, an enzyme is activated that then activates another enzyme and so on. The fact that the enzymes activated are themselves catalysts means there is an amplification at each step. In case this is not clear, if a single enzyme molecule activates 1000 molecules of the next enzyme in 1 minute and each molecule of the second enzyme does the same for the next enzyme and so on, in a cascade of four steps in a very short time, vast numbers of active molecules of enzyme number four are created. This enzyme at the end of the cascade can rapidly catalyse a massive response.

Blood clotting can conveniently be divided into two parts.

There is the cascade resulting in the activation of the enzyme that forms the clot, and there is the mechanism of clot formation itself by that enzyme. The cascade process is based on **proteolysis**—hydrolysis of peptide bonds of inactive precursor proteases activates them (cf. trypsinogen activation to trypsin in digestion, page 64). All of the necessary inactive precursor proteins are present in the blood waiting for the signal of a damaged blood vessel. They all have to be normally inactive.

What are the signals that clot formation is needed?

When a wound occurs, in terms of blood clotting, two things result.

1. The endothelial cell layer lining the blood vessels is damaged, exposing the structures underneath, such as collagen fibres that have a negatively charged or 'abnormal' surface. The blood-clotting response is a strictly localized reaction around the site of damage. Initially, a temporary plug is formed by the aggregation of blood platelets around the hole. For the formation of a clot, a small group of proteins in the blood absorb to the abnormal surface, the net result of which is that two proteases mutually activate each other in an autocatalytic manner. One of these is called factor XII. (The nomenclature is slightly confusing in that some 'factors' are enzymes, some are cofactors for enzymes, and they are not numbered according to the sequence of their appearance in the process.) Factor XII activates a cascade of three steps resulting in active factor X (another protease). Factor X activates **prothrombin** to the active protease, **thrombin**. Thrombin is what actually causes clotting (or thrombus formation). (We usually expect enzyme names to end in 'ase' but several of the classical proteases end in 'in'—for example, thrombin, pepsin, trypsin, chymotrypsin, and others.)

   The particular pathway of blood clotting is called the **intrinsic pathway** because, if blood is put into a glass vessel, the negatively charged glass surface triggers off clotting. Nothing has to be added; therefore the process is intrinsic.

2. Now for the **extrinsic pathway**. This is triggered by the release of a protein complex called tissue factor from damaged cells and tissues. Since something has to be added to blood, it is called the *extrinsic* pathway. This is a very short pathway. A protease is activated and this activates the same factor X as occurs in the intrinsic pathway resulting again in active thrombin formation. The two pathways are set out in Fig. 17.1.

The intrinsic pathway, being longer, is slower to cause clot formation when measured *in vitro*, than the extrinsic pathway, also measured *in vitro*. However, in the disease hemophilia A, in which blood clotting fails to occur, it is the intrinsic pathway that is deficient due to the absence of factor VIII required for factor X proteolytic activation. This seems paradoxical for the rapid extrinsic pathway, not requiring factor VIII, might seem able to do the job. However, it seems that, for normal physiological clotting, both pathways function as one, both being essential. Interactions between the two pathways are possibly involved.

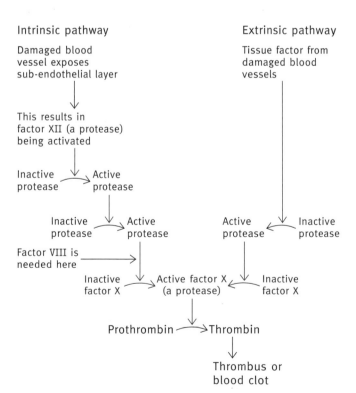

Fig. 17.1 Simplified diagram of intrinsic and extrinsic pathways of blood clotting. The names of the various proteases and other factors involved are omitted for simplicity. (Thirteen factors, numbered I–XIII, are, in fact, known.) The proteases listed are specific for their particular substrate. Factor VIII is the protein missing in patients with hemophilia A.

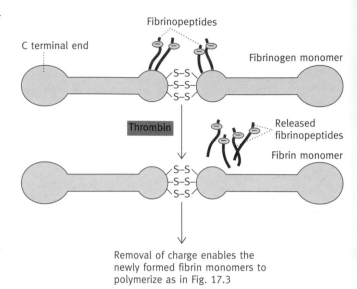

Fig. 17.2 Diagram of fibrinogen monomer and its conversion to fibrin monomer. Each half of the fibrinogen monomer is composed of three polypeptide chains, two of which terminate in the negatively charged fibrinopeptides.

How does thrombin cause thrombus (clot) formation?

In the circulating blood there is a protein called **fibrinogen**. The basic molecular unit consists of short rods made up of three polypeptide chains; two of these rods are joined together by S—S bonds near their *N*-terminal ends, forming the fibrinogen monomer. As shown in Fig. 17.2, at their joining points two of the three chains in each short rod project as negatively charged peptides, called fibrinopeptides. The negative changes mutually repel the monomers and prevent association.

Thrombin cleaves these off, giving a **fibrin monomer**. The fibrin monomer is now able to polymerize spontaneously by noncovalent bond formation. The sites at the end of the fibrin monomer are complementary to sites at the centre of adjacent molecules so that a staggered arrangement forms from the polymerisation (Fig. 17.3). This gives a so-called 'soft clot'. A more stable 'hard clot' is formed by covalent crosslinking between the side chains of adjacent fibrin molecules.

The covalent crosslinks are curious in that a glutamine side chain on one monomer is joined to a lysine side chain on the next in an enzymic transamidation reaction,

$$-CONH_2 + H_3N^+ - \rightarrow -CO-NH- + NH_4^+.$$

(Glutamine (Lysine   (Crosslink)
side chain)   side chain)

The fibrin strands entangle blood cells forming a blood clot.

Keeping clotting in check

Blood clotting is potentially dangerous unless strictly limited to local sites of bleeding. Once started, there is the danger of an

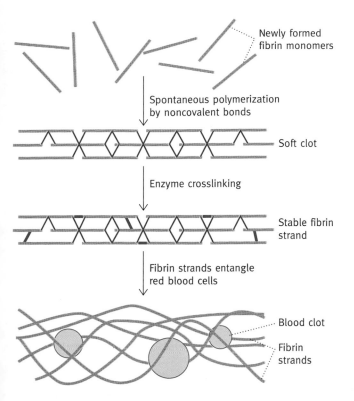

Newly formed fibrin monomers

Spontaneous polymerization by noncovalent bonds

Soft clot

Enzyme crosslinking

Stable fibrin strand

Fibrin strands entangle red blood cells

Blood clot

Fibrin strands

Fig. 17.3 Diagram of spontaneous polymerization of fibrin monomers and their enzymic crosslinking to form a stable fibrin strand. The fibrinogen cannot polymerize until thrombin hydrolyses off the fibrinopeptides because (1) the negative charges prevent association and (2) the central sites with which the ends of the fibrin monomers associate are masked by the fibrinopeptides. (Note that the covalent crosslinks in red are arbitrarily represented in position and number.)

autocatalytic process like this getting out of hand, causing inappropriate clotting. An elaborate series of safeguards exists. Proteinase inhibitors (for example, antithrombin) in blood 'dampen down' and prevent the clotting reactions from spreading; heparin, a sulfated polysaccharide present on blood vessel walls, increases this inhibitory effect; another protease, plasmin, dissolves blood clots. Plasmin itself is formed from inactive plasminogen; activation of this occurs by a protein, tissue plasminogen activator (TPA) released from damaged tissues. TPA is one of the new therapies for coping with blood clotting. Although present in tissues in minute amounts, its gene and the corresponding cDNA have been isolated and used for massive commercial production of the protein for injection (the technology is described in Chapter 24). This, and other inhibitory mechanisms, limit the reaction to the damaged surface area, the latter being required for initiation of the process. Blood clotting control is complex; inappropriate clotting is responsible for large numbers of deaths.

Rat poison, blood clotting, and vitamin K

The widely used rat poison, warfarin, kills by preventing blood clotting so that the rodents die from unchecked internal bleeding from minor lesions that continually occur. Warfarin is used clinically, for example, after strokes, to minimize the danger of clotting. It is structurally similar to vitamin K (K for *Koagulation*) and acts as a competitive inhibitor of vitamin K function (have a look at the two structures in Fig. 17.4 but you do not have to memorize these). Vitamin K is needed for prothrombin conversion to thrombin; it acts as a cofactor in an unusual enzyme reaction that adds an extra —COOH group, using CO_2, to a glutamic acid side chain of prothrombin.

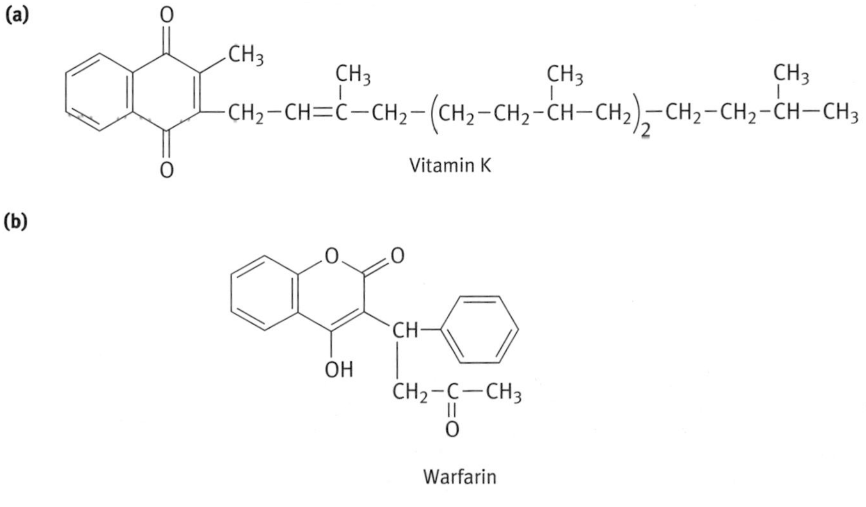

$$\begin{array}{c} | \\ CH_2 \\ | \\ CH_2 \\ | \\ COO^- \end{array} + CO_2 \longrightarrow \begin{array}{c} | \\ CH_2 \\ | \\ HC{-}COO^- \\ | \\ COO^- \end{array} \quad \begin{array}{l} \text{This group} \\ \text{efficiently} \\ \text{binds } Ca^{2+} \end{array}$$

A glutamic acid side chain of prothrombin

Carboxyglutamate

(a)

Vitamin K

(b)

Warfarin

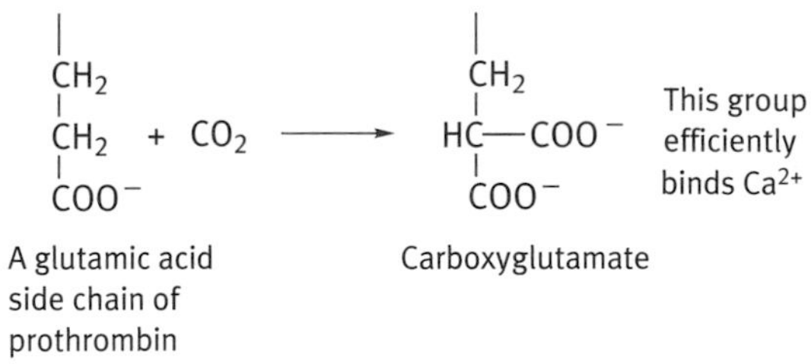

Fig. 17.4 Comparison of the structures of **(a)** vitamin K and **(b)** warfarin. Vitamin K is needed for blood clotting. Warfarin antagonizes vitamin K and prevents blood clotting.

The carboxyglutamate is needed to bind Ca^{2+}, which is essential in the prothrombin $\rightarrow$ thrombin activation process. The same modification also applies to other factors in the cascade.

Protection against ingested foreign chemicals

Large foreign molecules are dealt with by the immune system (Chapter 25). Small foreign molecules (which are not indicative of invasion by a living pathogen) are coped with by different systems.

Human beings ingest large numbers of foreign chemicals, collectively referred to as **xenobiotics** (*xeno* meaning foreign). These include pharmaceuticals, pesticides, herbicides, and industrial chemicals as well as complex structures such as the terpenes, alkaloids, and tannins of plants. Many of these are relatively insoluble in water, but soluble in fats, and they therefore tend to partition into the hydrocarbon layer of membranes and the fat globules of fat cells rather than being excreted in the aqueous urine. Unless they are rendered more polar and therefore more water-soluble, they will accumulate in the body with deleterious consequences. To facilitate their excretion, foreign chemicals are metabolized in several ways, but the **cytochrome P450** system (usually referred to simply as P450) in the liver is of central importance. A typical reaction of this P450 system is to add a hydroxyl group to an aliphatic or aromatic grouping. Other, different, enzymes add various highly polar groups, of which glucuronate is the major one, thus facilitating excretion in the aqueous urine.

We now need to look a little more closely at these processes. First the P450 system.

Cytochrome P450

P450 is a heme–protein complex as are the respiratory chain cytochromes (page 121). The name comes from P for pigment and 450 from the absorption maximum of the complex formed with carbon monoxide. (CO is not involved in the reaction—it just happens to give a complex with a spectrum that makes measurement of the amount of P450 easy). The enzyme is anchored into the smooth endoplasmic reticulum facing the cytoplasm. A foreign compound, AH, is attacked according to the reaction

$$AH + O_2 + NADPH + H^+ \rightarrow A-OH + H_2O + NADP^+.$$

It is called a **monooxygenase reaction** because it uses only one atom of oxygen from each O_2 molecule. NADPH is used to reduce the other oxygen atom to water. It is also called a **mixed-function oxygenase** because it both hydroxylates AH and reduces O to H_2O. The electrons from NADPH are transferred to the Fe^{3+} in the heme of P450 by a P450 reductase enzyme also present in the smooth ER membrane. You have already met this type of enzyme—phenylalanine hydroxylase on page 202. (Note

the difference between oxidation and oxygenation. Oxidation involves the removal of electrons; oxygenation involves the addition of an oxygen atom from O_2 to the molecule attacked. Oxygenation of hemoglobin is different again, for there a molecule of O_2 is loosely attached—see page 381.)

The amazing thing about the P450 system is the large number of different compounds that it attacks, including many that living organisms could not have significantly encountered before the advent of the modern chemical industry. One might speculate on the reason for this apparently anticipatory ability of evolution for, without it, it is difficult to see how we could have survived the vast variety of newly produced man-made chemicals to which we are now exposed. The somewhat weird collection of compounds such as terpenes, alkaloids, etc., found in plants may have been developed as protection for the plants against attacks (for example, grazing) by animals. The latter, therefore, it may be speculated, evolved a detoxifying system that could cope with almost anything present in plants and other food sources. So versatile is this system that it also copes with the chemicals newly devised by humans. The basis of this versatility is that a P450 enzyme has a wide specificity—it attacks a variety of related structures—and different P450 enzymes exist with different but overlapping specificities.

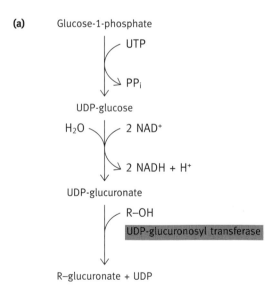

Fig 17.5 (a) The glucuronidation system. **(b)** Structure of a glucuronide.

Secondary modification—addition of a polar group to products of the P450 attack

Also present in the smooth ER membrane is a **glucuronidation system**. This transfers a glucuronate group from glucuronyl-UDP to the hydroxyl group generated on a foreign chemical by P450 (Fig. 17.5). UDP-glucuronate is produced by oxidation of UDP-glucose (page 81).

The response of the liver to the ingestion of foreign chemicals

When the liver is exposed to a foreign chemical such as phenobarbital, there is a massive response in which the smooth ER proliferates greatly. At the same time the P450 and glucuronidation systems are induced—the drug switches on the genes for these and the amount in the liver increases many fold. When all the drug is metabolized, the liver reverts to normal. The massive response underlines the threat that the ingestion of such compounds presents. The response has clinical relevance, for the effectiveness of many pharmaceuticals is limited by the rate of their metabolic destruction and hence by the level of the detoxifying enzyme systems, and these, as stated, increase on exposure to the drug. The P450 system is also involved in the disposal of natural compounds such as heme and steroid hormones. Other P450 systems, for example, in adrenal gland mitochondria, are involved in the synthesis of steroid hormones from cholesterol. An ironic twist to the story is that the oxidation of some substances by the P450 increases their carcinogenic effect. Benzpyrene in tobacco smoke is converted by P450 in the liver to a reactive species that can covalently react with DNA and is a carcinogenic mutagen.

Multidrug resistance

Another form of protection of cells against toxic chemicals is the reduction of their accumulation inside cells. Many cells, including those of human tissues, express a glycoprotein in their cell membranes that is an ATP-driven multidrug transporter that transports drugs out of the cell. A remarkable range of chemicals are transported, amongst them several of the anticancer drugs used in chemotherapy. This has caused much interest in the phenomenon but a much wider range of chemicals reacts with the transporter. These include many pharmacological agents and cytotoxic chemicals. Since steroids are also transported and the transporter protein is prevalent in steroid-secreting adrenal cortical cells, the system may have this as a primary biological role. The molecules transported have no chemical similarities but all are amphipathic compounds preferentially soluble in lipids.

Protection of the body against its own proteases

We have seen (Chapter 4) how the digestive system manages to escape the actions of its own proteolytic enzymes. However, proteases exist elsewhere in the body. A particularly important one, in the present context, is the **elastase of neutrophils**. Neutrophils are phagocytic white cells attracted to the site of infection or irritation. When activated at such sites, they secrete elastase which clears away connective tissue from the site. Elastin is the elasticity-conferring connective tissue (page 36) from which elastase gets it name. In the lung, air passages lead to minute pockets, the alveoli, which have a very large surface area needed for the diffusion of gases between blood and air. Neutrophils present in alveoli liberate elastase but this is prevented from destroying the lung structure by α_1-**antitrypsin** (α_1-**antiproteinase**), a protein that is produced by the liver and secreted into the blood. The α_1-antitrypsin inhibits several proteases, including trypsin as the name implies, but is especially effective on elastase. It inhibits by combining tightly with the enzyme and blocking its catalytic site.

α_1-Antitrypsin in adequate levels in the blood is essential for the protection of the lung. The molecule diffuses from the blood into the alveoli. If, due to a genetic defect, the level of α_1-antitrypsin is sub-normal, neutrophil elastase can destroy alveoli resulting in much larger pockets in the lung structure and consequent reduction of surface area available for gaseous exchange. The result is emphysema, a symptom of which is extreme shortness of breath.

Smokers are prone to emphysema for two reasons. The irritants in smoke are sufficient to cause neutrophil attraction to the lungs, with a consequent increased release of elastase. Secondly, oxidizing agents in the smoke destroy α_1-antitrypsin; they oxidize the sulfur atom of a crucial methionine side chain to a sulfoxide group (S $\rightarrow$ S=O) This is sufficient to prevent the α_1-antitrypsin from inactivating elastase and results in proteolysis of lung tissue, resulting in emphysema. Other antiproteases exist but α_1-antitrypsin is the most important one.

Protection against reactive oxygen species

As described in Chapter 7 (page 95), oxygen is an ideal electron sink for the energy-generating electron transport system. Its position on the redox scale means that, in energy terms, electrons from NADH have a long way to 'fall', meaning that the negative free-energy change of the overall oxidation of NADH to produce water is large. Again, as explained earlier (page 123), O_2 accepts four electrons and four protons to give H_2O the final product of the electron transport chain,

$$O_2 + 4e^- + 4H^+ \rightarrow 2H_2O.$$

The beauty of the system is that O_2 is relatively unreactive and therefore itself does no chemical damage and the product, H_2O, is entirely benign. There is, however, a darker side to the story, for oxygen has the potential to be exceedingly dangerous in the body. During evolution, the switch from energy generation by anaerobic metabolism to the use of oxygen as the electron sink was one of the most important events, but it was in some ways equivalent to jumping on to the back of a tiger for a ride. The danger occurs when a single electron is acquired by the O_2 molecule to give the **superoxide anion**, an extremely reactive corrosive chemical agent,

$$O_2 + e^- \rightarrow O_2^-.$$

The unpaired electron acquires a partner by attacking a covalent bond of another molecule.

Superoxide is formed in the body in several ways. In the electron transport chain, the final enzyme that donates electrons to oxygen, cytochrome oxidase, does not release partially reduced oxygen intermediates in any significant amounts—it ensures that an O_2 molecule receives all four electrons resulting in H_2O formation. However, inevitably, components of the electron transport chain (page 121) may 'leak' a small proportion of electrons to oxygen, one at a time, resulting in O_2^- formation. Moreover, mutations in mitochondrial DNA may block electron transport pathways and deflect electrons into superoxide formation (for example by direct ubiquinone oxidation by oxygen). Mitochondria have few or no DNA repair systems, so that mutations accumulate in them (see page 267).

In addition, there are other oxidation reactions in the body that produce small amounts of dangerous oxygen species such as H_2O_2. These include some oxidases that directly oxidize metabolites using O_2 (quite distinct from the respiratory pathways we have described). Spontaneous oxidation of hemoglobin (Hb) to methemoglobin (Fe^{3+} form) is another source; as a rare event, the oxygen in oxyhemoglobin (HbO_2) instead of leaving as O_2 and leaving behind hemoglobin in the Fe^{2+} form, leaves as O_2^- with the formation of methemoglobin, the Fe^{3+} form. It is also known that ionizing radiation causes superoxide formation.

When phagocytes ingest a bacterial cell there is a rapid increase in oxygen consumption; this is used to oxidize NADPH via a mechanism that deliberately generates superoxide anions. These are shed into the vacuole and converted to H_2O_2 (see below), which helps to destroy the contained bacterial cell. Excess neutrophils attracted to irritated joints may lead to superoxide release and contribute to arthritic damage.

Although the superoxide anion is present in minute amounts, it can set up chain reactions of chemical destruction in the body. In this, the superoxide anion attacks and destroys a covalent bond of some cell constituent, but in doing so generates a new free radical species from the attacked molecule that, in turn, attacks yet another molecule of cell constituent producing yet another free radical, and so on. The destruction initiated by the superoxide anion is thus a self-perpetuating chain of reactions.

The biological injuries caused by this process are not finally established but it has been suggested that superoxide damage contributes to ageing, cataract formation, the pathology of heart attacks, and other problems.

Basically, there are three protective strategies—one chemical and two enzymic.

Mopping up oxygen free radicals with vitamins C and E

The chemical strategy is to dampen or quench the chain reactions, initiated by superoxide, by using **antioxidants**. The requirements for such a quenching reagent is that it should itself be attacked by the superoxide anion but generate a radical insufficiently reactive to perpetuate the chain reaction. The main biological dampening agents are **ascorbic acid** (vitamin C) and **α-tocopherol** (vitamin E). The former is water-soluble, the latter lipid-soluble, and so, between them, they protect in both phases of the cell. These two vitamins are not the only antioxidants—some normal metabolites such as uric acid are effective; β-carotene is also an antioxidant. **Bilirubin**, produced from heme breakdown by heme oxygenase (page 378), is also an effective antioxidant. It is interesting that heme oxygenase, which catalyses the first step in bilirubin production from heme, is induced by ionizing radiation (which generates oxygen free radicals) and other agents causing oxidative stress. Whether these effects are direct protective mechanisms against oxidative stress by bilirubin generation remains to be proven.

Enzymic destruction of superoxide by superoxide dismutase

Probably all animal tissues contain the enzyme **superoxide dismutase** and, most appropriately, it occurs in mitochondria. It also occurs in lysosomes and peroxisomes (Chapter 16) as well as in extracellular fluids such as lymph, plasma, and synovial fluids. Superoxide dismutase catalyses the reaction

$$2O_2^- + 2H^+ \rightarrow H_2O_2 + O_2.$$

The hydrogen peroxide is destroyed by **catalase**,

$$2H_2O_2 \rightarrow 2H_2O + O_2.$$

Some oxidases also generate H_2O_2 directly. Flavoprotein oxidases, with FAD as their prosthetic group, generally catalyse reactions of the type

$$AH_2 + O_2 \rightarrow A + H_2O_2.$$

Xanthine oxidase, involved in purine metabolism (Chapter

18), is of this type. H_2O_2 is potentially dangerous because, in the presence of metal ions such as Fe^{2+}, it can generate the highly reactive hydroxyl radical ($OH\bullet$), not to be confused with the hydroxyl anion (OH^-). (H_2O_2 is the result of two electrons being added to O_2, when three electrons are added, $OH\bullet$ and OH^- are formed.)

$$H_2O_2 + Fe^{2+} \rightarrow Fe^{3+} + OH\bullet + OH^-.$$

The hydroxyl radical can attack DNA and other biological molecules. The two protective enzymes, catalase and superoxide dismutase, are therefore important. Another enzyme that destroys H_2O_2 is glutathione peroxidase, described in the next section. Since brain has little catalase, this latter enzyme possibly is the main one for protection against H_2O_2 in that organ.

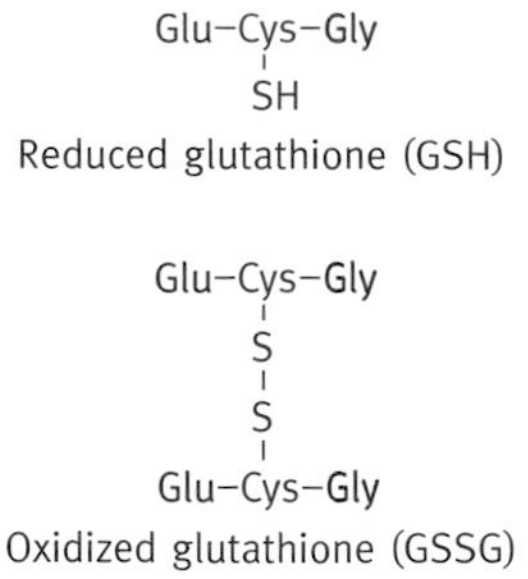

The glutathione peroxidase–glutathione reductase strategy

Glutathione is a thiol tripeptide (γ-glutamyl-cysteinyl-glycine), which is abbreviated to GSH (Fig. 17.6). It is found in most cells where, because of its free thiol group, it functions as a reducing agent, for example, to keep proteins with essential cysteine groups in the reduced state. This reaction with proteins is non-enzymic, but another protective action of GSH is to inactivate peroxides via the action of **glutathione peroxidase (GSSG)** ,

$$H_2O_2 + 2GSH \leftrightarrow GS\!-\!SG + 2H_2O.$$

Organic peroxides ($R\!-\!O\!-\!OH$) are also destroyed in this way. The GSSG is subsequently reduced by NADPH, the reaction being catalysed by **glutathione reductase**,

$$GSSG + NADPH + H^+ \rightarrow 2GSH + NADP^+.$$

Red blood cells depend for their integrity on GSH, which reduces any ferrihemoglobin (methemoglobin) to the ferrous

```
        Glu—Cys—Gly
              |
              SH
    Reduced glutathione (GSH)

        Glu—Cys—Gly
              |
              S
              |
              S
              |
        Glu—Cys—Gly
    Oxidized glutathione (GSSG)
```

Fig. 17.6 Structure of reduced glutathione (GSH) and of the oxidized form (GSSG). Glu, Cys, and Gly are abbreviations for glutamate, cysteine, and glycine, respectively, using the three-letter system. (A single-letter system is used for extensive amino acid sequences; see Table 22.2.)

form, as well as destroying peroxides. This explains why red blood cells have the pentose phosphate pathway (Fig. 13.1) to supply the NADPH needed for the reduction of GSSG. Usually, patients with defective glucose-6-phosphate dehydrogenase, the first enzyme in the pentose phosphate pathway, have enough enzymic activity for normal function. However, when extra stress is placed on the cell, for example, by the accumulation of peroxides due to the action of the antimalarial drug pamaquine, the supply of NADPH can no longer be maintained. The integrity of the cell membrane is impaired and hemolysis results in such patients from taking the drug.

In this chapter we have concentrated on protective mechanisms in animals. It is becoming increasingly evident that plants too have their protective devices—perhaps more varied because they have no immune system and cannot run away or flick off predators. A random example is the possession of an enzyme, in some plants, of chitinase, used to attack chitin in the cuticle of insects.

Further reading

Blood clotting

Scully, M. F. (1992). The biochemistry of blood clotting: the digestion of a liquid to form a solid. *Essays in Biochemistry*, 27, 17–36.
Very readable review of clotting, limiting of the process and medical aspects.

Cytochrome P450

Coon, M. J., Ding, X., Pernecky, S. J., and Vaz, A. D. N. (1992). Cytochrome P450; progress and predictions. *FASEB J.*, 6 669–73.
A crisp overview of different aspects of this important field.

Eastabrook, R. W. (1996). The remarkable P450s: a historical review of these versatile hemoprotein catalysts. *FASEB J.*, 10 202–4.
Summarizes the metabolic roles of P450s in broad terms and

outlines the current research interests in the field. Points out that there are over 2000 research papers per year on the subject.

Free radicals

Babior, B. M. (1987). The respiratory burst oxidase. *Trends Biochem. Sci.*, 12, 241–3.
Reviews the deliberate production of superoxide by phagocytes.

Rusting, R. L. (1992). Why do we age? *Sci. Amer.*, 267(6), 130–41.
Contains a section on the evidence that free radicals may be a significant factor in ageing.

Harris, E. (1992). Regulation of antioxidant enzymes. *FASEB J.*, 6, 2675–83.
A review of how the level of enzymes destroying oxygen free radicals are adjusted to needs.

Rose, R. C. and Bode, A. M. (1993). Biology of free radical scavengers: an evaluation of ascorbate. *FASEB J.*, 7, 1135–42. Clearly discusses the properties required for free radical scavengers and looks at ascorbate in detail.

Fridovich, I. (1994). Superoxide radical and superoxide dismutase. *Ann. Rev. Biochem.*, **64**, 97–112.

Concise review of the field.

Uddin, S. and Ahmad, S. (1995). Dietary antioxidants protection against oxidative stress. *Biochem. Education*, 23, 2–7. A useful review. The point is made that free iron and copper ions can catalyse OH· formation and this is presumably why they are always tightly bound to proteins.

Problems for Chapter 17

1 Blood clotting involves cascades of enzyme activations. What is the rationale for such a long process?

2 Explain how thrombin triggers the formation of a blood clot.

3 The spontaneous polymerization of fibrin monomers forms a 'soft clot'. How is this converted into a more stable structure?

4 Why is vitamin K needed for blood clotting?

5 What is the function of cytochrome P450?

6 Why should NADPH be involved in an oxygenation reaction?

7 What is the function of glucuronyl-UDP in disposing of water-insoluble compounds?

8 What is multidrug resistance?

9 Why does smoking cause emphysema?

10 What is superoxide?

11 What mechanisms exist for guarding against the deleterious effects of superoxide?

Chapter summary

Chapter 18

. . . .

Nucleotide metabolism

Several of the preceding chapters have been mainly concerned with energy production from food. In this, ATP occupies the central position, but GTP, CTP, and UTP are also involved in aspects of food metabolism. However, the involvement of these nucleotides described so far has all been concerned with the phosphoryl groups of the molecules. The nature of the bases, whether A, G, C, or U, has been important only for recognition by the appropriate enzymes but otherwise has not been directly relevant to the metabolic processes. For this reason we have not previously given information about the bases themselves.

We are now about to start, in the next chapter, a new main area of biochemistry in which information transfer is a main purpose—we refer to nucleic acids and protein synthesis and for this the structures of the bases of nucleotides become important. In this chapter, the structures, synthesis, and metabolism of nucleotides are dealt with. This is an essential prerequisite for the subsequent chapters.

. . . .

Structure and nomenclature of nucleotides

The term 'nucleotide' originates from the name of nucleic acids, originally found in nuclei, and which are polymers of nucleotides. A **nucleotide** has the general structure

phosphate—sugar—base.

A **nucleoside** has the structure

sugar—base.

Thus, AMP and the corresponding nucleoside, adenosine, have the structures shown.

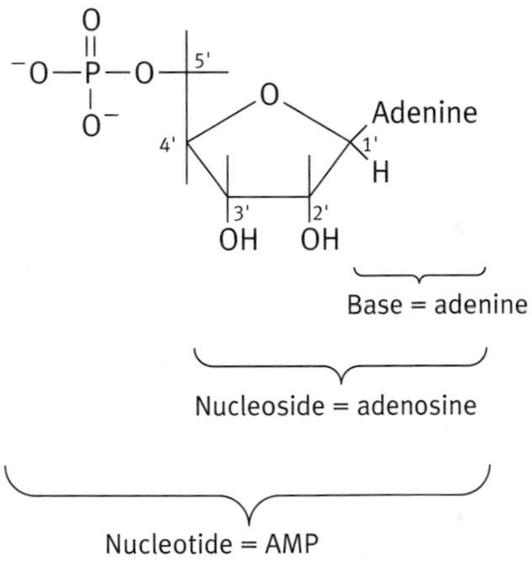

Strictly speaking, the AMP shown here should be written as $5'$ AMP. The prime (') indicates that the number refers to the position on the ribose sugar ring, to which the phosphate is attached, rather than to the numbering of atoms in the adenine ring. However, it is a common practice to assume that the phosphate is $5'$ unless specified, since this is the most common position. Thus $5'$ AMP is often called AMP, whereas if the phosphate is on the carbon atom 3 of the ribose, this is always specified as $3'$ AMP (a compound you'll meet later).

The sugar component of nucleotides

The sugar component of a nucleotide is always a pentose, most often ribose, or $2'$-deoxyribose, which are always in the D-configuration, never the L-form.

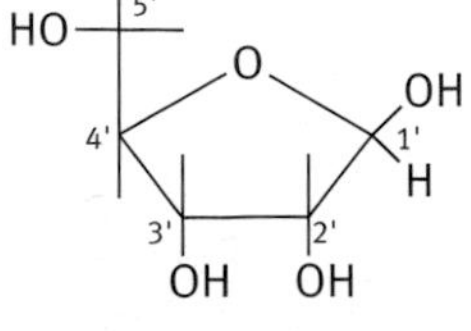

D-Ribose 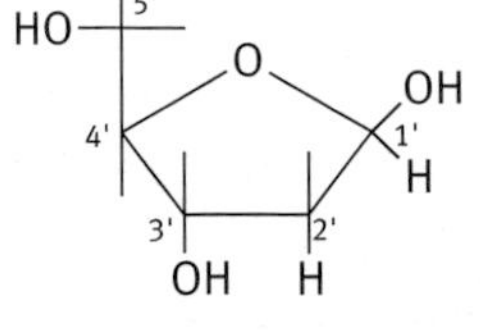   D-2'-deoxyribose

In **RNA** the sugar is always ribose (hence the name, **ribonucleic acid**) and in **DNA**, deoxyribose (hence, **deoxyribonucleic acid**). A nucleotide containing ribose is a ribonucleotide but this is not usually specified; unless otherwise stated, a named nucleotide such as AMP is taken to be a ribonucleotide. A deoxyribonucleotide *is* always specified; for example, deoxyadenosine monophosphate or dAMP, etc. (with the one occasional exception mentioned below).

The base component of nucleotides

Nomenclature

We are primarily concerned with five different bases—adenine, guanine, cytosine, uracil, and thymine, all often abbreviated to their initial letter.

A, G, C, and U are found in RNA;
A, G, C, and T are found in DNA.

The ribonucleotides are AMP, GMP, CMP, and UMP, but older and still used terms are adenylic, guanylic, cytidylic, and uridylic acids, respectively (or adenylate, guanylate, cytidylate, and uridylate for the ionized forms at physiological pH). The deoxyribonucleotides are dAMP, dGMP, dCMP, and dTMP. The latter often is called TMP, or thymidylate, without the d-prefix because T is found only in deoxynucleotides (with rare exceptions of no significance here).

When the intention is to indicate a nucleotide without specifying the base, the abbreviations NMP or 5′-NMP are often used, or dNMP or 5′-dNMP for deoxynucleotides.

Deoxy UMP exists only as an intermediate in the formation of dTMP; it does not occur in DNA (except as a result of chemical damage to the DNA, when it is promptly removed—see page 264).

The corresponding ribonucleosides (base–sugar) are, respectively, adenosine, guanosine, cytidine, uridine, and d-adenosine, etc. for the deoxyribose compounds. The names for the nucleosides, cytidine and uridine, sound like those of free purine bases (cf. adenine and guanine), while the name of the free base, cytosine, sounds like that of a nucleoside (cf. adenosine), so be careful here. Other so-called 'minor' bases exist and are found in transfer RNA (page 294). Hypoxanthine is one of these. It is also the first base produced by the pathway of purine biosynthesis as hypoxanthine ribotide or inosine monophosphate (see below); hypoxanthine riboside is called inosine.

Structure of the bases

The first point is that:

A and G are **purines**;
C, U, and T are **pyrimidines**.

These names originate from their being derivatives of the parent compounds, purine and pyrimidine, respectively (neither of

which occur in nature).

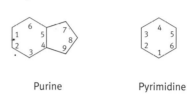

Purine   Pyrimidine

These rings in future structures will be represented by the simplified forms below.

Purine   Pyrimidine

The structures of the nucleotide bases are represented in Fig. 18.1.

Of especial importance, note that *T is simply a methylated U*. It will be useful to fix in your mind that T is essentially the same as U except that it is 'tagged' by a methyl group. T is found only in DNA; U only in RNA. The significance of this will be apparent later (page 264).

Attachment of the bases in nucleotides

The bases are attached to the sugar moieties of nucleotides at the N-9 position of purines and the N-1 position of pyrimidines. The glycosidic bond is in the β-configuration, that is, it is above

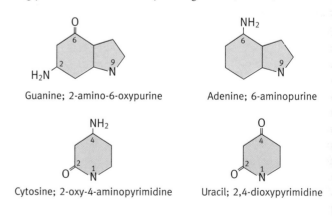

Guanine; 2-amino-6-oxypurine   Adenine; 6-aminopurine

Cytosine; 2-oxy-4-aminopyrimidine   Uracil; 2,4-dioxypyrimidine

Thymine; 2,4-dioxy-5-methylpyrimidine

Fig. 18.1 Diagrammatic representation of structures of purine and pyrimidine bases found in nucleic acids. (Other minor bases are found in tRNA, described in Chapter 22.)

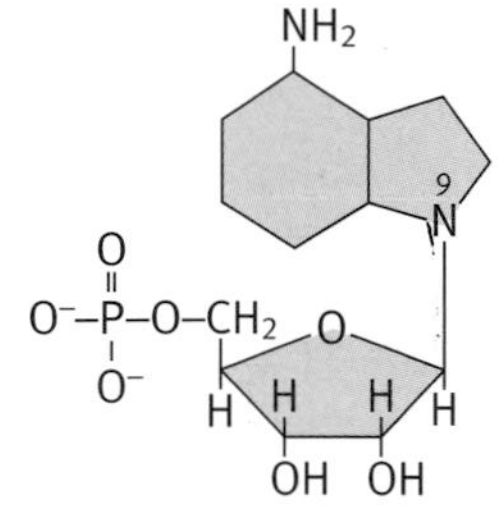

Adenosine monophosphate (AMP)

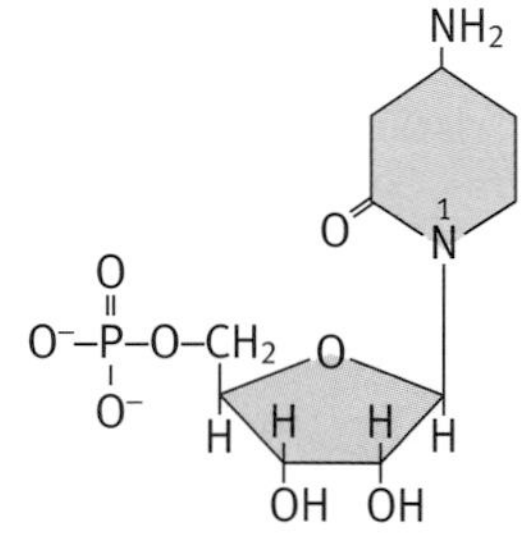

Cytidine monophosphate (CMP)

Fig. 18.2 Structures of a purine and a pyrimidine nucleotide.

the plane of the sugar ring. The structures of AMP and CMP are given in Fig. 18.2.

Synthesis of purine and pyrimidine nucleotides

Purine nucleotides

Most cells can synthesize purine bases *de novo* from smaller precursor molecules. In the *de novo* synthesis of purine nucleotides, bases are *not* synthesized in the free form but rather the purine ring is assembled piece-by-piece with all the intermediates attached to ribose-5-phosphate so that, by the time a purine ring is assembled, it is already a nucleotide. This refers only to the *de novo* synthesis of purines because free purine bases released by degradation of nucleotides *are* utilized for nucleotide synthesis by the separate **salvage pathway** to be described later. The mechanism of **ribotidation** (the addition of ribose-5-phosphate) is the same in both pathways, as well as in pyrimidine nucleotide synthesis (but the compound that is ribotidated differs in each case, as described below). Which brings us to PRPP, the metabolite that is involved in all ribotidation.

PRPP–the ribotidation agent

PRPP is **5-phosphoribosyl-l-pyrophosphate**. It is formed from ribose-5-phosphate (produced by the pentose phosphate pathway, page 181) by the transfer of a pyrophosphate group from ATP by the enzyme **PRPP synthetase**.

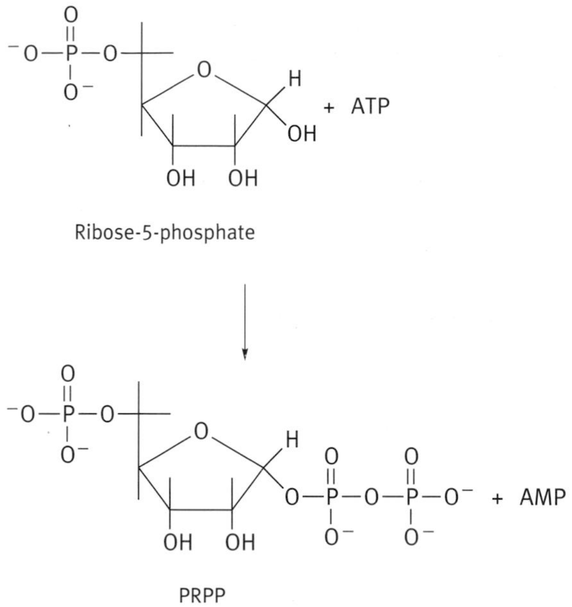

Ribose-5-phosphate

PRPP

The PRPP is an 'activated' form of ribose-5-phosphate; appropriate enzymes can donate the latter to a base forming a nucleotide, and splitting out —(P)—(P). Hydrolysis of the latter to $2P_i$ drives the reaction thermodymically.

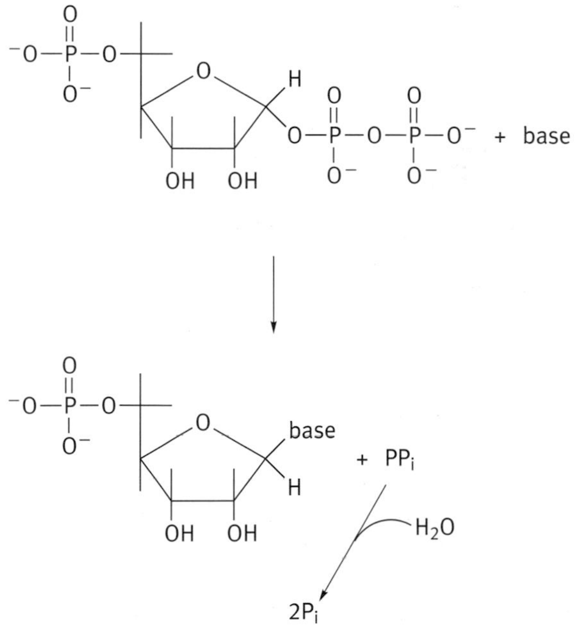

In this reaction the configuration at carbon atom 1 is inverted so that the base is in the required β position. Note that in the *de novo* pathway (see reaction 1 of this pathway in Fig. 18.4) the base is simply —NH$_2$, derived from glutamine and the product, 5-phosphoribosylamine; in the purine *salvage* pathway it is a purine (see page 229). We usually associate the term nucleotide with a purine or pyrimidine base but it can be applied to any base attached in the appropriate manner to the sugar phosphate.

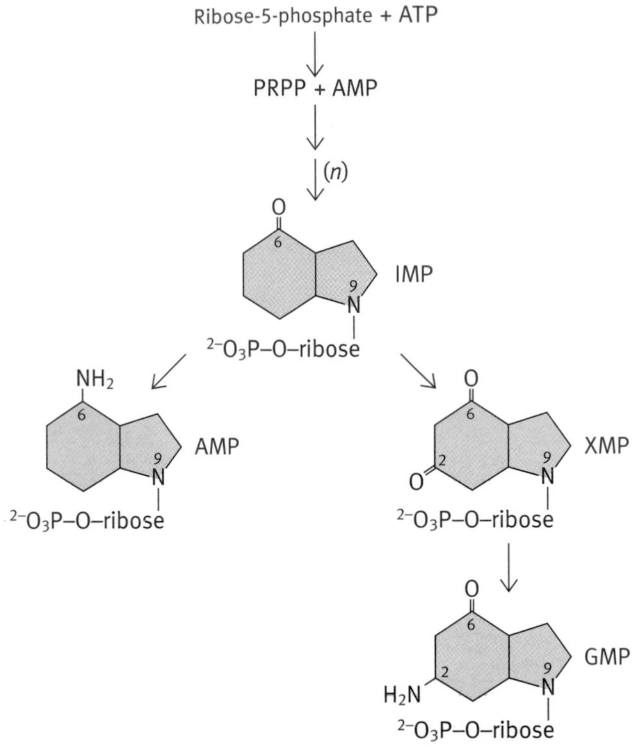

Ribose-5-phosphate + ATP

$\downarrow$

PRPP + AMP

$\downarrow$

$\downarrow (n)$

IMP

$^{2-}O_3P-O-ribose$

AMP

$^{2-}O_3P-O-ribose$

XMP

$^{2-}O_3P-O-ribose$

GMP

H_2N

$^{2-}O_3P-O-ribose$

Fig. 18.3 Diagram of the purine *de novo* pathway of GMP, AMP, and XMP synthesis. The base in IMP or inosine monophosphate is hypoxanthine. The base in XMP or xanthosine monophosphate is xanthine. The complete pathway of AMP and GMP synthesis can be seen in Figs 18.4 and 18.5.

To return from this general point to the purine *de novo* pathway in particular, after the formation of 5-phosphoribosylamine, there follows a series of nine reactions resulting in the assembly of the first purine nucleotide in which hypoxanthine is the base (see Fig. 18.3). This nucleotide is IMP or inosinic acid.

IMP is a branch-point since its hypoxanthine base is converted either to adenine or guanine yielding AMP and GMP, respectively. The pathway is summarized in Fig. 18.3, but you should also look at all of the reactions of the pathway as set out in Figs 18.4 and 18.5, which it is suggested do not need to be learned in detail. We will refer to some reactions of specific interest. You will see that six molecules of ATP are consumed in the synthesis of one purine nucleotide molecule. The ATP utilization refers only to —Ⓟ groups; there is no loss of the adenine nucleotide of ATP so that the pathway results in a net synthesis of AMP. We do not want to go into all of the reactions shown in Fig. 18.4, but the reactions numbered 3 and 9 of this pathway have a general importance and we need to divert to deal with these in some detail. This concerns a type of reaction not dealt with in this book before—**one-carbon transfer.**

The one-carbon transfer reaction in purine nucleotide synthesis

Reactions 3 and 9 of the pathway involve the addition of a formyl (HCO—) group to intermediates in the pathway (see Fig. 18.4). The donor molecule in both cases is N^{10}-formyltetrahydrofolate, a molecule not mentioned before in this book. **Tetrahydrofolate (FH$_4$)** is the carrier in the cell of formyl groups. It is a coenzyme derived from the vitamin folic acid (F) or pteroylglutamic acid (we suggest that you need to memorize only the relevant part of this and related structures below, not the whole molecules).

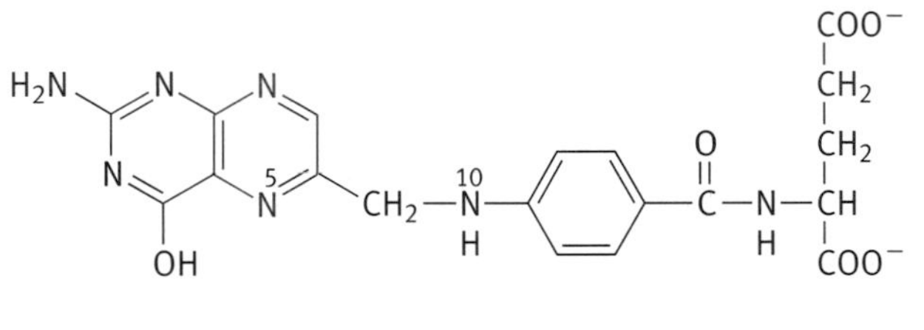

Folic acid (pteroylglutamic acid)

The vitamin (F) is reduced to FH$_4$ by NADPH in two stages.

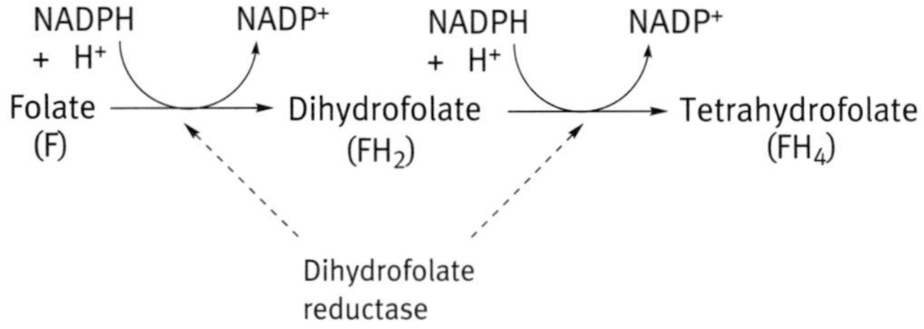

NADPH + H$^+$ → NADP$^+$ NADPH + H$^+$ → NADP$^+$

Folate (F) → Dihydrofolate (FH$_2$) → Tetrahydrofolate (FH$_4$)

Dihydrofolate reductase

The structures of FH$_2$ and FH$_4$ are given below.

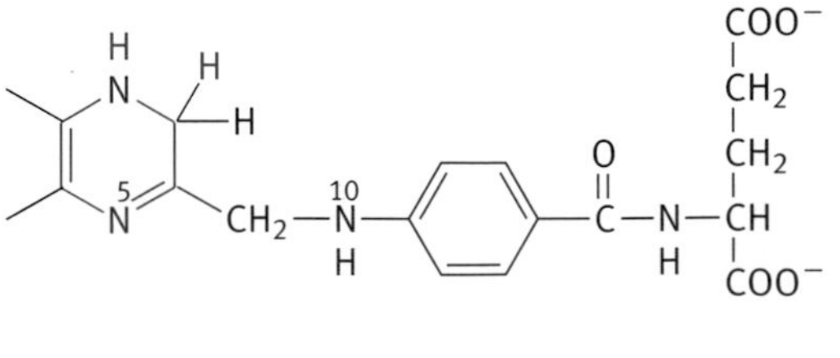

Dihydrofolate (FH$_2$)

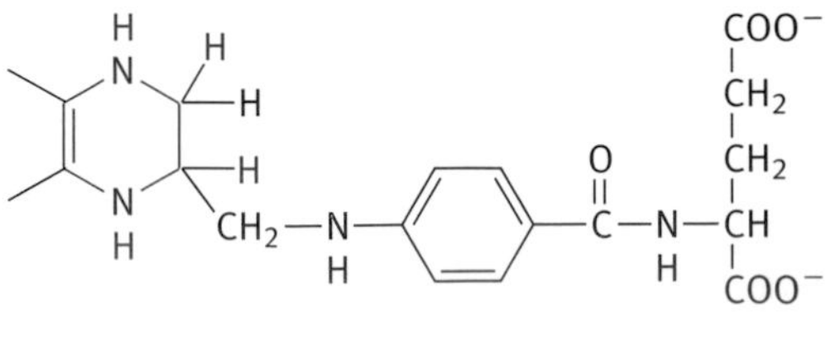

Tetrahydrofolate (FH$_4$)

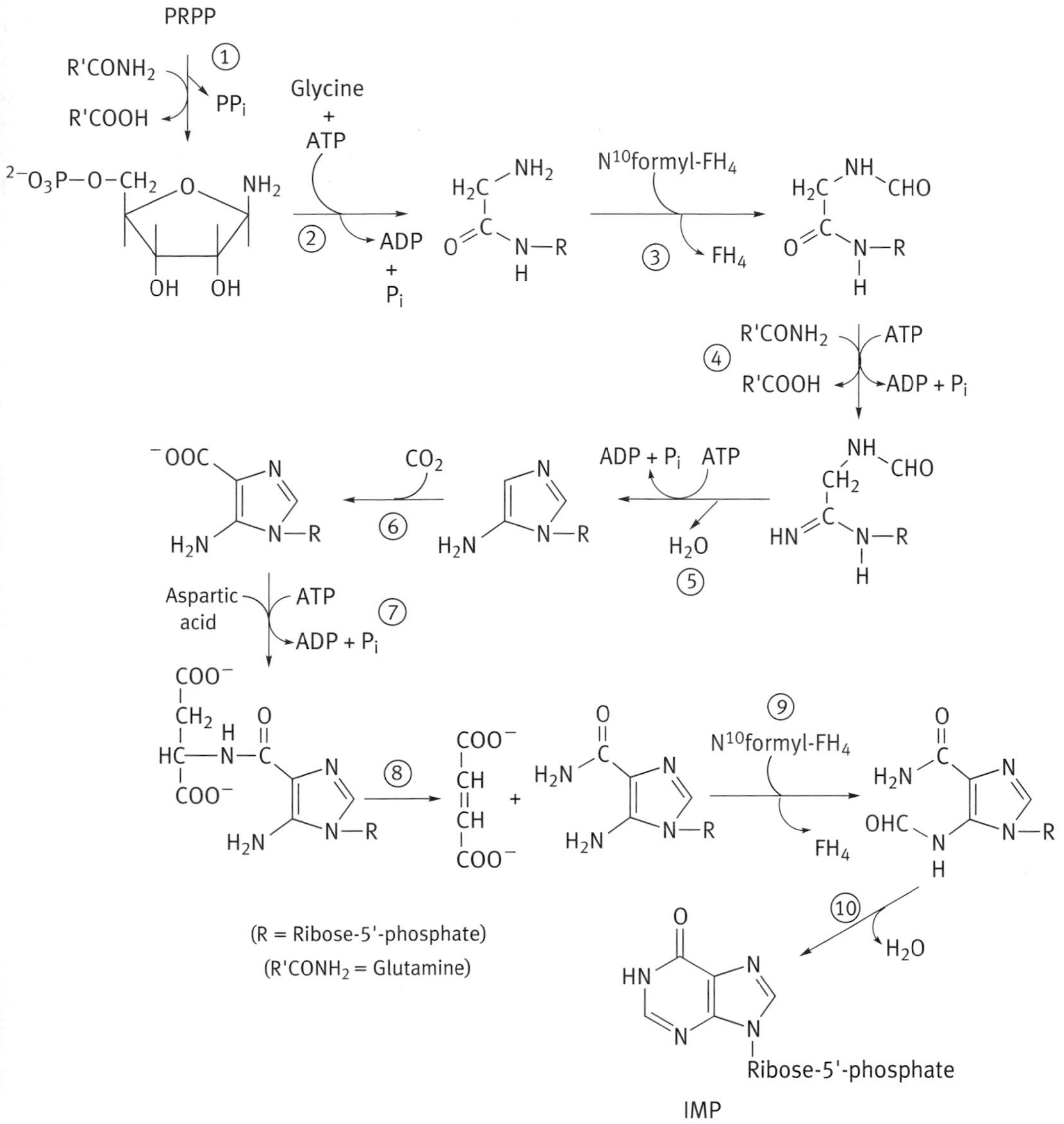

PRPP

R'CONH₂ ①

R'COOH ← PP$_i$

Glycine + ATP

② → ADP + P$_i$

$^{2-}O_3P-O-CH_2$ O NH₂

OH OH

N^{10}formyl-FH₄

③ ↘ FH₄

H₂C—NH₂

O=C—N—R
         H

H₂C—NH—CHO

O=C—N—R
         H

R'CONH₂ ↘ ATP

④

R'COOH ← ADP + P$_i$

NH—CHO
CH₂

HN=C—N—R
       H

ADP + P$_i$ ATP

← H₂O

⑤

⁻OOC N

       ‖

H₂N N—R

CO₂

⑥

N

‖

H₂N N—R

Aspartic acid ↘ ATP

⑦ → ADP + P$_i$

COO⁻
CH₂
HC—N—C N
     H ‖
COO⁻ H₂N N—R

⑧ →

COO⁻
CH
‖
CH
COO⁻

+

O
‖
H₂N—C N
         ‖
H₂N N—R

N^{10}formyl-FH₄

⑨ ↘ FH₄

O
‖
H₂N—C N
         ‖
OHC—N N—R
       H

⑩ ↘ H₂O

(R = Ribose-5'-phosphate)
(R'CONH₂ = Glutamine)

O
‖
HN N

N N
    Ribose-5'-phosphate

IMP

Fig. 18.4 Pathway for the *de novo* synthesis of the purine ring from PRPP to inosinic acid, given for reference purposes. (The circled reaction numbers are referred to in the text.) Blue indicates the structural change resulting from the latest reaction.

Have a look at the structure of FH₄ and notice that the N-5 and N-10 atoms are placed such that a single carbon atom can neatly bridge the gap between them. For our present purposes we can therefore represent FH₄ as

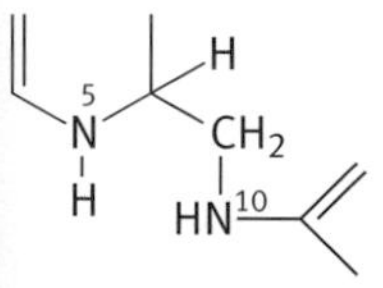

H
|
N⁵—CH₂
| HN¹⁰
H

and N^{10}-formyl FH₄ as

⁵NH   CH₂
H O=C—N¹⁰
        H

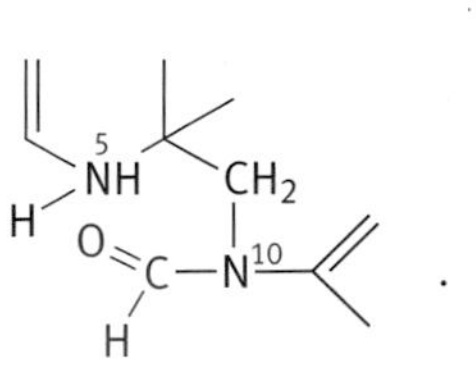

N^{10}-formyltetrahydrofolate
(formyl FH₄)

This is the donor of the formyl group in reactions 3 and 9 of the purine biosynthesis pathway; specific formyl transferase enzymes catalyse the reactions.

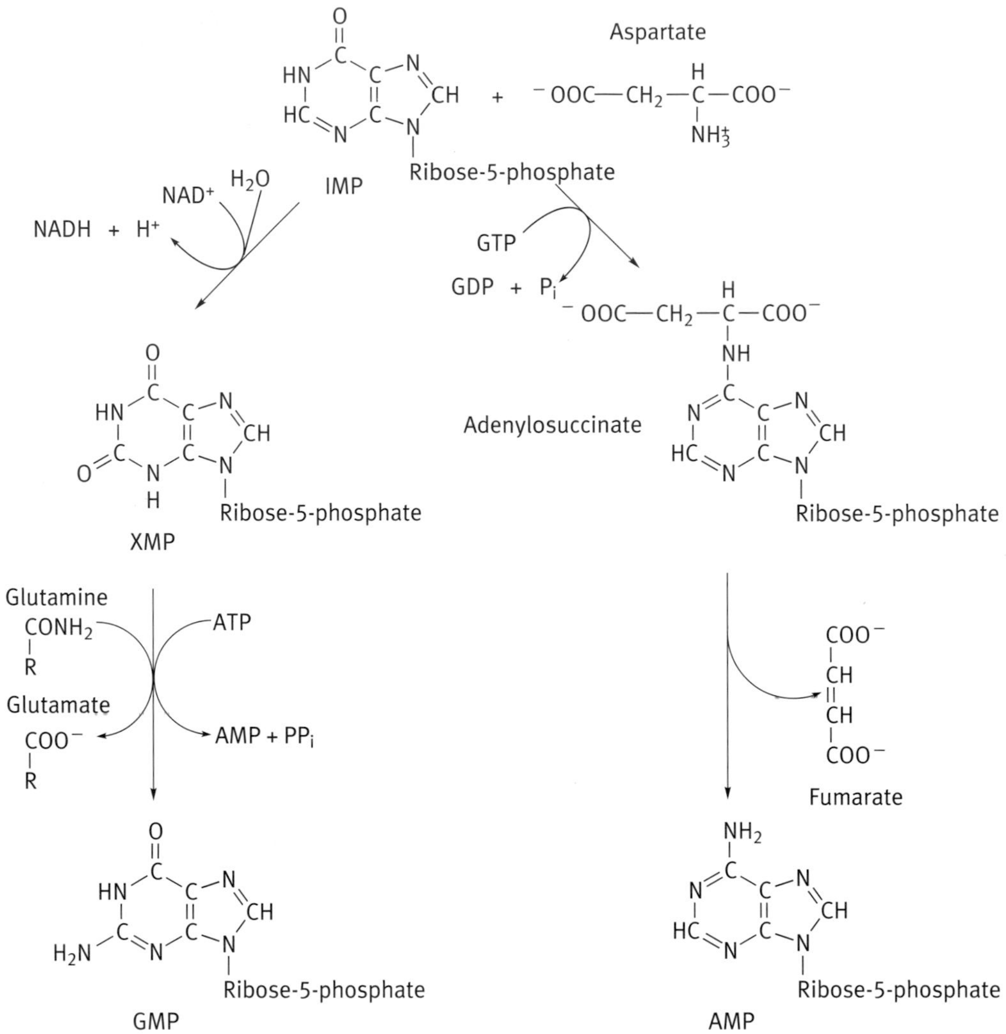

Fig. 18.5 Pathways for the synthesis of GMP and AMP from inosinic acid (IMP). The colour shows the change resulting from each reaction.

Where does the formyl group in formyl FH_4 come from? The answer is the amino acid serine (which is readily synthesized from the glycolytic intermediate 3-phosphoglycerate).

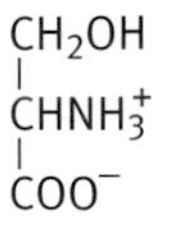

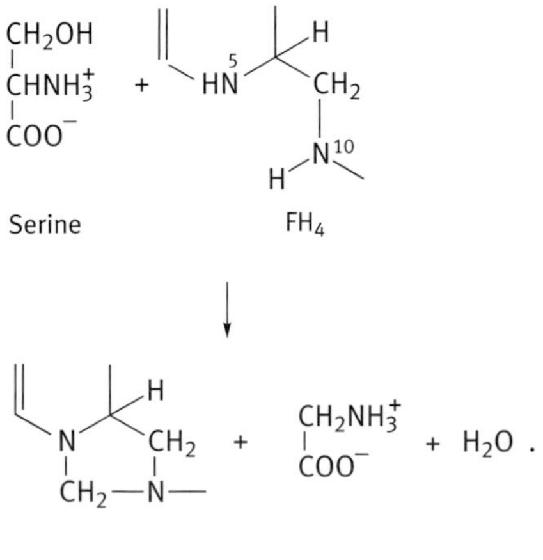

An enzyme, **serine hydroxymethylase**, transfers the hydroxymethyl group ($—CH_2OH$) to FH_4 leaving glycine and forming N^5, N^{10}-methylene FH_4.

The product methylene FH_4 is not quite what we want for formylation because the $—CH_2—$ group is more reduced than a formyl group. It is therefore oxidized by an $NADP^+$-requiring enzyme forming the methenyl derivative, which is hydrolysed to N^{10}-formyl FH_4, the formyl group donor.

AMP + ATP $\longrightarrow$ 2ADP   (adenylate kinase)

GMP + ATP $\longrightarrow$ GDP + ADP   (guanylate kinase)

GDP + ATP $\longrightarrow$ GTP + ADP   (nucleoside diphosphate kinase)
or   or
ADP   ATP

N^5,N^{10}-methylene FH_4 →(NADP$^+$, NADPH +H$^+$)→ N^5,N^{10}-methenyl FH_4

(→ H_2O →) N^{10}-formyl FH_4

Donates formyl groups in the purine nucleotide biosynthesis pathway

How are ATP and GTP produced from AMP and GMP?

Most of the synthetic reactions of the cell involve nucleoside triphosphates. As you will see later, these are needed for nucleic acid synthesis. It is a simple but especially important concept that enzymes (kinases) exist in the cell to transfer $—\text{P}$ groups between nucleotides *at the high-energy level*. There is little free-energy change involved so that $—\text{P}$ groups can be shuffled around from nucleotide to nucleotide with ease. The main source of $—\text{P}$ is, of course, ATP, for remember that the energy generating metabolism constantly regenerates ATP from ADP and P_i. Newly formed AMP and GMP are phosphorylated by kinase enzymes as shown.

The purine salvage pathway

We have emphasized that the *de novo* synthesis of purines does not involve free purine bases—purine nucleotides are produced. However, as already indicated, there is a separate route of purine nucleotide synthesis in which **free bases** are converted to nucleotides by reaction with PRPP. The free bases originate from degradation of nucleotides—they are salvaged and hence the name of the pathway. Two enzymes are involved—these are phosphoribosyltransferases, one of which forms nucleotides from adenine and the other from hypoxanthine or guanine. The latter enzyme, known as **HGPRT** (for **hypoxanthine–guanine phosphoribosyltransferase**), catalyses the reaction

Guanine or Hypoxanthine $+ \text{PRPP} \rightarrow$ GMP or IMP $+ \text{PP}_i$.

The enzyme salvaging adenine may be of lesser importance than that dealing with guanine and hypoxanthine in humans, for the main routes of nucleotide breakdown produce the free bases hypoxanthine (from AMP) and guanine (from GMP) as shown in Fig. 18.6.

What is the physiological role of the purine salvage pathway?

Since purines are energetically 'expensive' to make, a mechanism for re-utilizing free purine bases is economical since it can reduce the amount of *de novo* synthesis a cell has to carry out. Moreover, certain cells such as erythrocytes have no *de novo* purine synthesis pathway and must rely on the salvage pathway.

The physiological importance of purine salvage is underlined by the genetic disease of infants called the Lesch–Nyhan

Fig. 18.6 Production of free purine bases hypoxanthine and guanine by nucleotide breakdown. Patients lacking adenosine deaminase in lymphocytes have an immune deficiency that formerly could be treated only by keeping the affected child in a sterile plastic bubble. The disease was the first to be successfully treated by gene therapy in which the normal gene for adenosine deaminase was inserted *in vitro* into bone marrow stem cells and returned to the patient. (See page 333.)

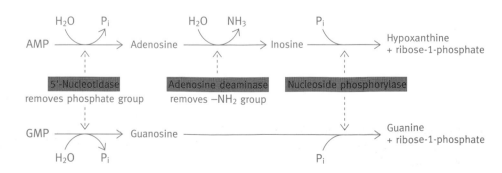

syndrome in which the enzyme HGPRT is missing. This results in neurological problems including mental retardation and self-mutilation. Brain possesses the *de novo* pathway only at low levels, so purine nucleotide synthesis is very sensitive to the salvage defect. Lack of the salvage reaction leads to a hepatic *overproduction* of purine nucleotides by the *de novo* pathway in these patients because the level of PRPP rises (due to lack of utilization by the salvage reaction) and stimulates the *de novo* pathway. This explains why in these patients excessive uric acid production occurs as in gout (see below), which may result in kidney failure caused by the urate crystals. The connection between the biochemical defect and the neurological symptoms is not clear in the Lesch–Nyhan patients. While uric acid overproduction is treatable with allopurinol (see below), this does not relieve the neurological problems. Nor do patients with gout develop the neurological symptoms.

The sources of free purine bases for salvage are probably several. Although the diet contains purines in the form of nucleic acids, there is evidence that most of them are destroyed by epithelial cells of the intestine and not absorbed. By contrast, purines injected into the bloodstream are utilized by cells. The liver is a major site of purine synthesis and some evidence exists that it releases the bases into the blood for use by other cells, such as reticulocytes, that do not have the complete *de novo* pathway. It is also probable that the salvage pathway recycles, within cells, purine bases released from breakdown of nucleic acids. Lysosomal destruction of cellular components containing nucleic acids (see page 210) would presumably release free bases. Although, as discussed further below, there is no doubt as to the importance of the purine salvage pathway, information on the traffic of free bases in the body is not complete.

The recycling of preformed purine bases has the obvious advantage of energy-saving provided, of course, that the *de novo* pathway synthesis is correspondingly reduced. This is achieved in two ways: (1) salvage reduces the level of PRPP and hence of the pathway; and (2) the AMP and GMP produced by salvage exert feedback inhibition on the pathway (see below).

Formation of uric acid from purines

Nucleotide degradation leads to the production of free hypoxanthine and guanine. Part of this is salvaged back to nucleotides but part is oxidized to produce **uric acid** (Fig. 18.7). The enzyme, xanthine oxidase, that produces uric acid is present mainly in the liver and intestinal mucosa. Gout is due to a raised level of urate in the blood, leading to the deposition of crystals in tissues. Although gout is traditionally associated with rich living, the main source of uric acid is probably excess *de novo* production of purine nucleotides due, in some patients, to a high level of PRPP synthetase activity. Also, as described above, deficiency of the HGPRT enzyme leads to the overproduction of purine nucleotides. The drug **allopurinol** used in the treatment of gout, mimics the structure of hypoxanthine. It is converted to

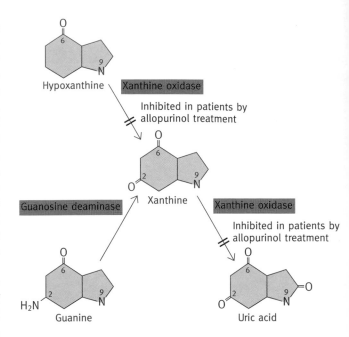

Fig. 18.7 Conversion of hypoxanthine and guanine to uric acid. XO, xanthine oxidase. The drug allopurinol is closely related in structure to hypoxanthine. It is converted to alloxanthine, which now inhibits xanthine oxidase. The conversion is carried out by xanthine oxidase itself.

a potent xanthine oxidase inhibitor, alloxanthine, by the xanthine oxidase itself, a process known as **suicide inhibition**. These structures are shown below. This inhibition results in xanthine and hypoxanthine formation rather than that of uric acid (see Fig. 18.7). These products are more water-soluble than uric acid and more readily excreted, thus preventing the deposition of insoluble uric acid crystals in tissues that results in the clinical symptoms of gout.

Allopurinol

Alloxanthine

Enol form of hypoxanthine

Control of purine nucleotide synthesis

As with all metabolic pathways, there must be regulation or chemical anarchy would prevail. The *de novo* pathway is a classical example of **allosteric feedback control** (see page 162). The first step of a pathway is the logical place for control. In the *de novo* pathway this is the PRPP synthetase. This enzyme is negatively controlled by AMP, ADP, GMP, and GDP. The next enzyme, which catalyses the first *committed* step to synthesis of purine nucleotides (reaction 2, Fig. 18.4), is inhibited also, as shown in Fig. 18.8. However, this isn't quite the end of the story, because the *de novo* pathway produces IMP, and then the IMP goes in two directions—to AMP and GMP. These latter feedback-control their own production, as shown in Fig. 18.8. Both negative and positive regulatory loops serve to ensure a balanced production of ATP and GTP since both are required for nucleic acid synthesis (Chapter 19).

Synthesis of pyrimidine nucleotides

Most cells of the body synthesize pyrimidine nucleotides *de novo* but, unlike bacteria, mammals do not appear to have significant pyrimidine salvage pathways for free bases, analogous to that for purines. The nucleoside, thymidine, however, is readily phosphorylated to TMP by thymidine kinase and, in that sense, salvage of this nucleoside does occur.

The **pyrimidine pathway** is summarized in Fig. 18.9 and given in full in Fig. 18.10 for reference purposes. It starts with aspartic acid and produces a ring structure compound, orotic acid. Orotic acid is converted to the corresponding nucleotide by the PRPP reaction and this is converted to UMP. UTP is produced by kinase enzymes much as in the purine pathway. CTP is produced by amination of UTP.

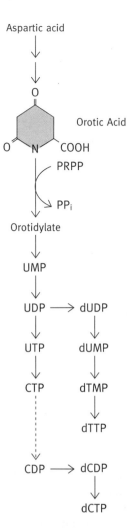

Fig. 18.9 Summary of pyrimidine nucleotide synthesis. (It is assumed here that CTP is converted to CDP.) The formation of deoxynucleotides is described later in this chapter. The complete pathway is given in Fig. 18.10.

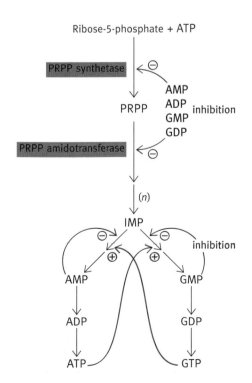

Fig. 18.8 Simplified scheme of the control of the purine nucleotide biosynthesis pathway. The positive feedback of ATP and GTP is via energy supply to the reactions.

How are deoxyribonucleotides formed?

For DNA synthesis dATP, dGTP, dCTP, and dTTP are required (Chapter 19). The reduction of ribonucleotides to deoxy compounds occurs at the diphosphate level with NADPH as the reductant (the electrons being transported to the reductase by a complex pathway, not described here).

ADP		dADP
GDP		dGDP
CDP	Ribonucleotide reductase	dCDP + H_2O
UDP	NADPH + H⁺ NADP⁺	dUDP

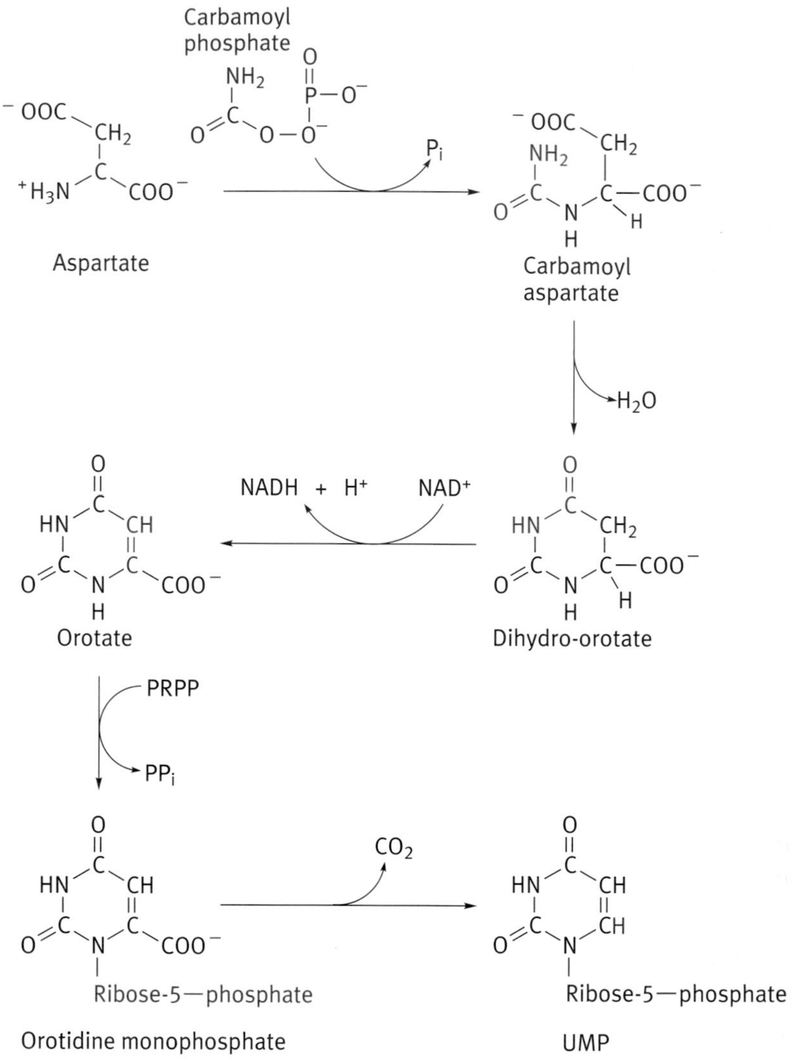

Fig. 18.10 Pathway for the *de novo* synthesis of pyrimidine nucleotides. Blue indicates the structural change resulting from the latest reaction.

The resultant dADP, dGDP, dCDP, and dUDP are converted to the triphosphates by phosphoryl transfer from ATP. However, dUTP is not used for DNA synthesis since DNA, you will recall, has thymine (the methylated uracil) as one of its four bases, but never U. The dUTP is converted to dTTP. This is done in three steps: first, dUTP is hydrolysed to dUMP,

$$dUTP + H_2O \rightarrow dUMP + PP_i.$$

The dUMP is converted to dTMP and then to dTTP by phosphoryl transfer from ATP. An appropriate system of allosteric feedback controls exists to keep the production of the four deoxynucleotide triphosphates in balance.

Conversion of dUMP to dTMP

The methylation of dUMP is of especial interest. The enzyme involved is called **thymidylate synthase** and it utilizes the coenzyme N^5, N^{10}-methylene FH_4. Recall (page 229) that, in purine synthesis, the methylene group of the latter is oxidized to produce a formyl group. In thymidylate synthesis, the methylene group is transferred and, at the same time, *reduced* to the methyl group of thymine; the reaction is catalysed by thymidylate synthase. The reducing equivalents for the reduction come from FH_4 itself, leaving it as FH_2 (note how versatile this coenzyme is). In the scheme below, only the relevant part of N^5, N^{10}-methylene FH_4 is shown (see page 227).

dUMP N^5,N^{10}-methylene FH$_4$

dTMP   Dihydrofolate (FH$_2$)

The FH$_2$ produced in this reaction is reconverted to FH$_4$ by dihydrofolate reductase. The FH$_4$ can be reconverted to methylene FH$_4$ by reaction with serine (page 228).

NADPH + H$^+$ NADP$^+$ serine   glycine

FH$_2$ ⟶ FH$_4$ ⟶ N^5,N^{10}-methylene FH$_4$

For thymidylate synthesis to continue, FH$_2$ must be reduced to FH$_4$ by dihydrofolate reductase. The antileukemic drugs **methotrexate** (amethopterin) and **aminopterin** (two of the so-called 'antifolates') inhibit FH$_2$ reductase by mimicking the folate structure.

The structure of methotrexate (amethopterin) is shown here for interest (you do not need to memorize it); aminopterin is similar but lacks the N^{10}-methyl group.

Methotrexate (Amethopterin)

Cancer cells, like those of leukemia, require rapid dTMP production to synthesize DNA and are thus selectively inhibited. The scheme is outlined in Fig. 18.11.

Tetrahydrofolate, vitamin B$_{12}$, and pernicious anemia

N^5, N^{10}-Methylene FH$_4$ in addition to the above reaction, can be reduced to N^5methyl FH$_4$ which supplies the methyl group to convert homocysteine to methionine (see page 203), a reaction catalysed by **methionine synthetase**. The latter requires coenzyme B$_{12}$ as cofactor (see page 135); it is believed that in the absence of this, as in the disease pernicious anemia, the available supplies of FH$_4$ become trapped as methyl FH$_4$ and are unavailable for reactions involved in purine nucleotide biosynthesis and other reactions. The **methyl trap hypothesis** would account for the fact that vitamin B$_{12}$ deficiency produces a megaloblastic anemia identical to that seen in folate deficiency. The neurological symptoms of pernicious anemia may be related to methylmalonic acidosis, referred to earlier (page 135).

Although several vitamin B-related enzyme reactions occur in bacteria only the two, methylmalonyl-CoA mutase (page 135) and methionine synthetase, are known in humans. Lack of vitamin B$_{12}$ in pernicious anaemia is due to the absence of a gastric glycoprotein, the intrinsic factor, required for absorption of the vitamin in the intestine, rather than to a dietary deficiency.

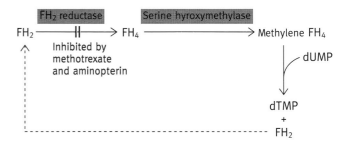

Fig. 18.11 Site of action of the anticancer agent, methotrexate. The relationship of the methotrexate structure to folic acid can be seen by comparing the structure on page 226.

Part 4

Information storage and utilization

Chapter summary

Previous page: Electron micrograph of messenger RNA molecules being transcribed from DNA. The central thread is DNA and the side branches RNA in the act of being synthesized. Transcription in this picture starts at the right and progresses along the DNA with the mRNA molecules becoming longer until completed and released (latter step not shown).

Photograph: Professor O L Miller/Science Photo Library

Chapter 19

DNA—its structure and arrangement in cells

In this chapter we will deal with the structure of DNA and what a gene is. We will then, in the following chapter, describe DNA synthesis, followed in Chapter 21 with the way in which DNA directs protein synthesis. The central point is that the function of DNA is to carry information needed for the synthesis of proteins whose amino acid sequences have been devised by millions of years of evolution to carry out the myriad processes on which life depends (page 23).

What are nucleic acids?

DNA was first isolated from cell nuclei; it is an acid because of its phosphate groups—hence the term **nucleic acid**. It contains a sugar, 2-deoxy-D-ribose and therefore is called deoxyribonucleic acid, or DNA for short. There is another acid of similar structure found in cells in which the sugar is D-ribose. It is therefore called ribonucleic acid or RNA for short (this is dealt with in Chapter 21). The two sugars are shown below. 2-Deoxy-D-ribose lacks the oxygen on the carbon 2 position; it is usually simply referred to as **deoxyribose**.

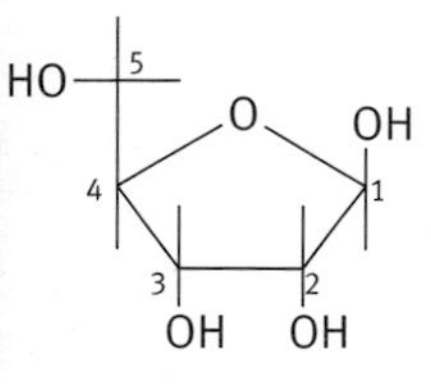

D-Ribose 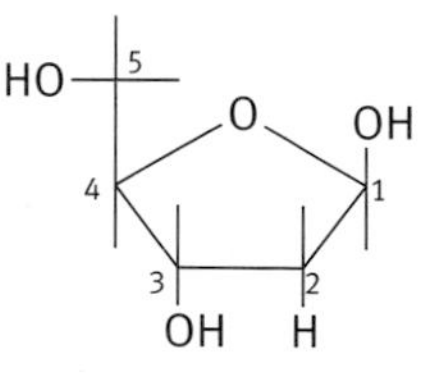   2-Deoxy-D-ribose

In eukaryote cells, the bulk of the DNA is confined inside the nuclear membrane, that in mitochondria and chloroplasts being the remainder, whereas most of the RNA is found in the cytoplasm although, with the exception of that made in mitochondria and chloroplasts, it originates in the nucleus. Although we have already discussed nucleotides in Chapter 18

which dealt with their synthesis, we will repeat some of the material here both for convenience and because of its importance. With that introduction let us turn to DNA.

The primary structure of DNA

DNA is a polynucleotide. A nucleotide has the structure

Phosphate–Sugar–Base.

The structure of a deoxyribonucleotide is shown in the non-ionized form for simplicity.

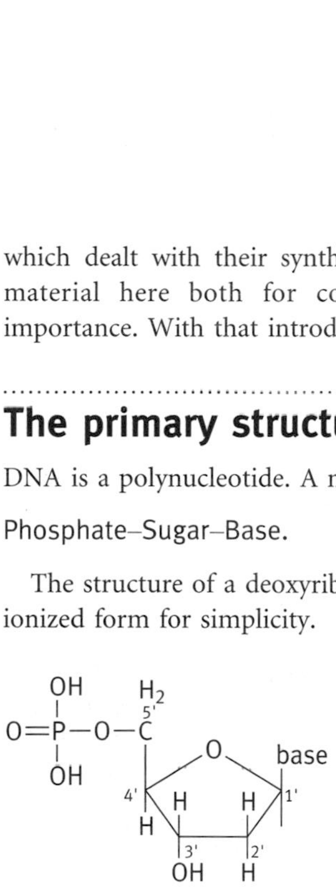

The sugar, as stated above, is 2-deoxy-D-ribose (deoxyribose). To specify a position in the deoxyribose moiety, a prime (′) is added to distinguish it from the numbering of the base ring atoms. Thus the sugar carbon atoms are $1'$, $2'$, $3'$, $4'$, and $5'$ (pronounced 'five prime', etc.) and indicated outside the ring. The sugar is in the furanose, five-membered ring form. The nomenclature of nucleotides is described on page 223.

What are the bases in DNA?

Unlike the situation in proteins where you have 20 different amino acids, in DNA there are only four different bases, but their structures are just a bit more difficult to remember than those of amino acids. The bases are adenine, guanine, cytosine, and thymine—abbreviated to A, G, C, and T. A and G are purines; C and T pyrimidines. The numbering of atoms in the bases is given inside the ring structures.

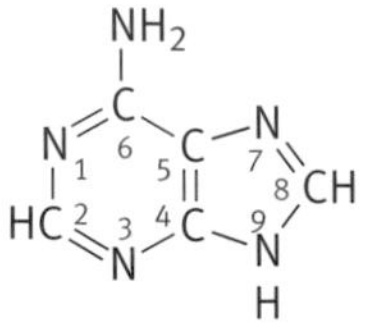

Adenine (A)

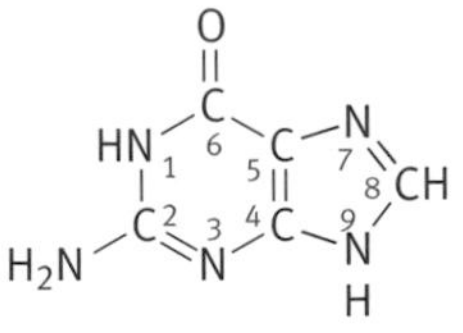

Guanine (G)

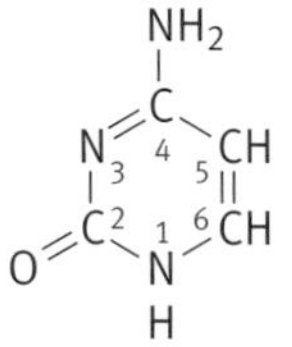

Cytosine (C)

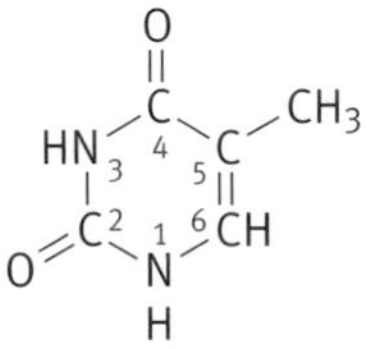

Thymine (T)

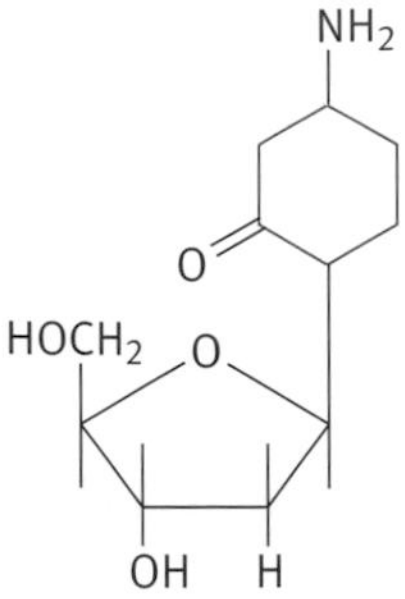

Deoxycytidine

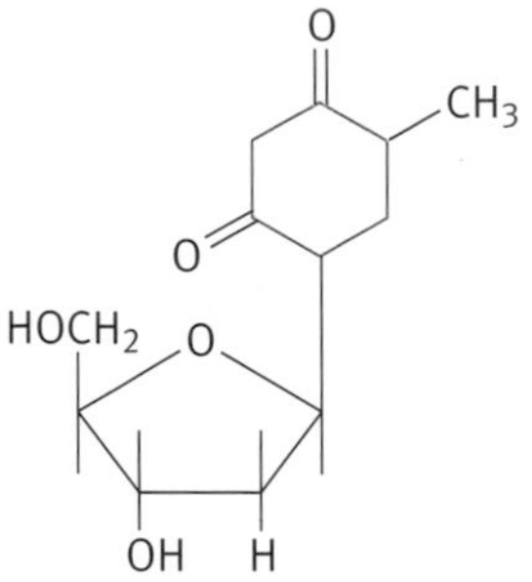

Deoxythymidine
(or, simply, thymidine)

Note the methyl group of thymine. In subsequent structures we will represent these diagrammatically.

Tautomeric forms of the bases can exist (keto–enol and amino–imino) but in DNA the bases are essentially always in the form shown with —C=O and —NH$_2$ groups the predominant form at neutral pH rather than C—OH and C=NH groups.

Attachment of the bases to deoxyribose

The glycosidic link is between carbon atom 1 of the sugar and nitrogen atoms at positions 9 and 1, respectively, of the purine and pyrimidine rings. The linkage is β (that is, above the plane of the ring).

The structure, base–sugar, is called a **nucleoside**; if the sugar is deoxyribose it is a **deoxyribonucleoside**. All the deoxyribonucleosides have specific names. Where the base is adenine it is deoxyadenosine; the guanine derivative is deoxyguanosine. Deoxycytidine and deoxythymidine are the deoxynucleosides of cytosine and thymine respectively. The structures are shown below.

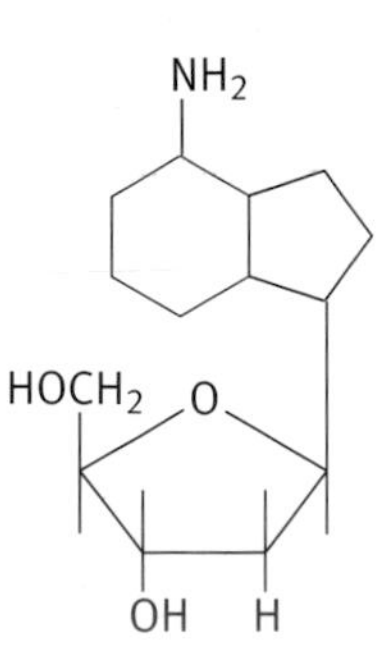

Deoxyadenosine

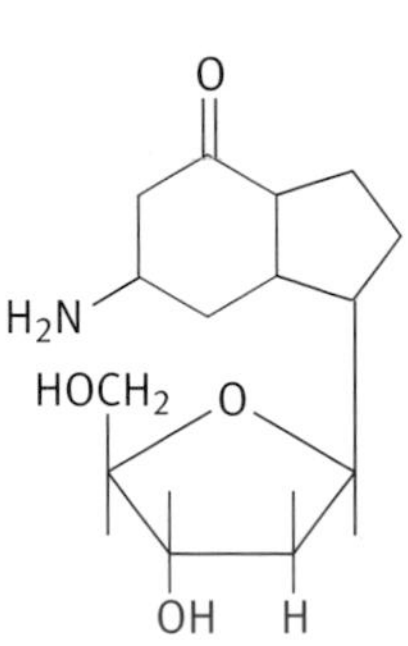

Deoxyguanosine

What are the physical properties of the polynucleotide components?

The nucleotides in DNA are the 5′ phosphate compounds, dAMP, dGMP, dCMP, and dTMP. As pointed out (page 224), the latter may be referred to as TMP since the ribose analogue is rarely encountered.

The phosphoric acid —OH groups are both ionized at physiological pH since one of the —OH groups has a pK_a of around 2 and a second of around 7. This means they are highly hydrophilic. DNA is strongly negatively charged. The sugar with its hydroxyl group is also strongly hydrophilic. The bases are different—and this is important—in that they are almost water-insoluble. Their flat faces are essentially hydrophobic but at the edge of each there is hydrogen-bonding potentiality.

Structure of the polynucleotide of DNA

Suppose we have two such nucleotides and notionally eliminate a water molecule as shown; the resultant structure is a **dinucleotide** (the nonionized forms are shown here for clarity).

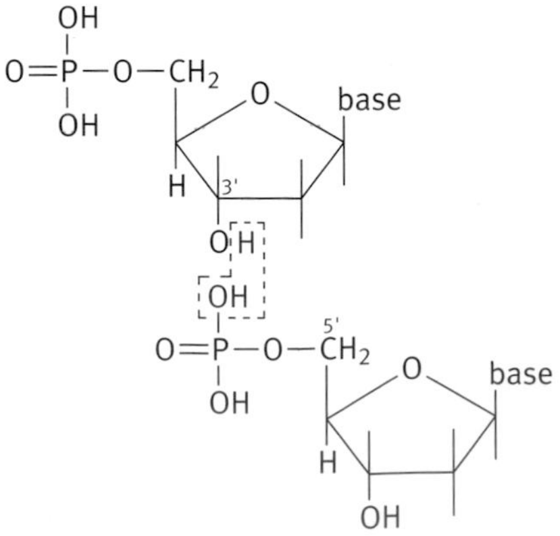

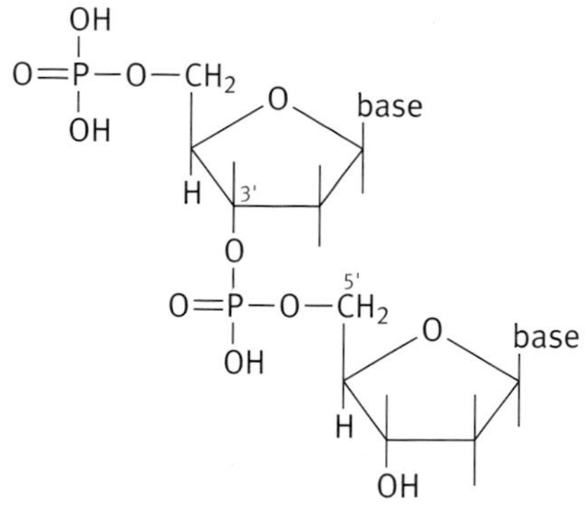

A dinucleotide

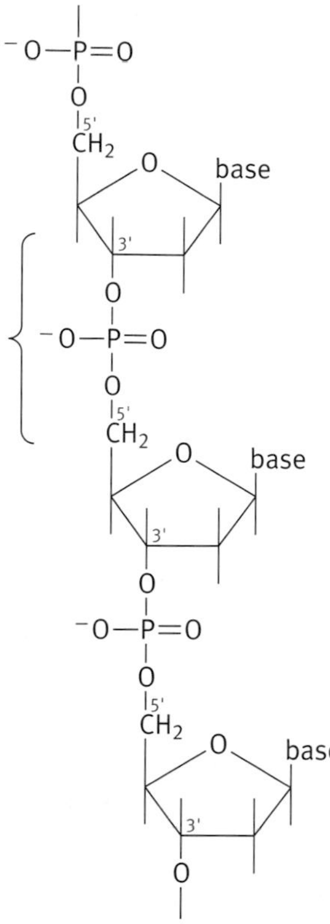

3'-5' phosphodiester link

The formation of a dinucleotide involves a large positive $\Delta G^{0'}$ value, so be clear that synthesis cannot occur by the direct condensation of two nucleotides. In a mononucleotide, the phosphate group is a primary phosphate ester by which we mean that there is only a single ester bond. In the dinucleotide, a **phosphodiester link** is formed—it is **diesterified**, the phosphate being linked to two groups.

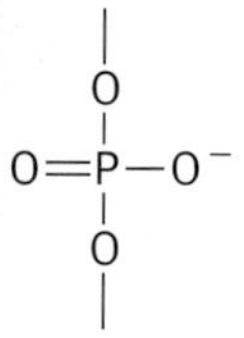

In the dinucleotide shown above it is a 3',5'-phosphodiester— the phosphate bridging between the 3'-OH of one nucleotide to the 5'-OH of the next. Nucleotides can be added in the same way indefinitely, giving a polynucleotide. DNA is in its primary structure a polynucleotide of immense length. In dealing with proteins you will recall that there is a **polypeptide backbone** (page 24) with amino acid residues attached. A polynucleotide has a backbone of alternating sugar–phosphate–sugar groups with a base attached to each sugar residue on the 1' position; coded information is carried in the sequence of the bases. DNA therefore has the primary structure

Backbone section	Informational or coding section of the structure
2'-deoxyribose	—base
phosphate	
2'-deoxyribose	—base
phosphate	
2'-deoxyribose	—base

or, in structural terms,

Why deoxyribose? Why not ribose?

The cell has several nucleotides (for example, AMP) with ribose as the sugar component, and the nucleic acid, RNA, has ribose as its sugar. In evolution it is probable that ribonucleotides predated deoxyribonucleotides and RNA predated DNA. The cell nevertheless goes to considerable energetic expense to convert ribonucleotides to deoxyribonucleotides required for DNA synthesis. One reason for this at least seems logical. DNA is the repository of genetic information gathered over countless millions of years and it is stored in chemical form in DNA molecules. This raises what might seem at first sight to be an insuperable obstacle to life—namely, that chemical molecules always have some degree of instability—they spontaneously break down, while DNA molecules have to remain largely unchanged, and certainly intact, for untold numbers of generations. The presence of the 2'-OH group of ribose makes a ribopolynucleotide less stable than the corresponding deoxyribose molecule—that is, DNA is more stable than RNA. This is because the 2'-OH group is suitably placed for a nucleophilic attack in the presence of OH$^-$ ions on the phosphorus atom, thus causing breakage of the phosphodiester link by forming a 2', 3' cyclic phosphate as shown; in DNA, lacking the 2'-OH group, this does not happen.

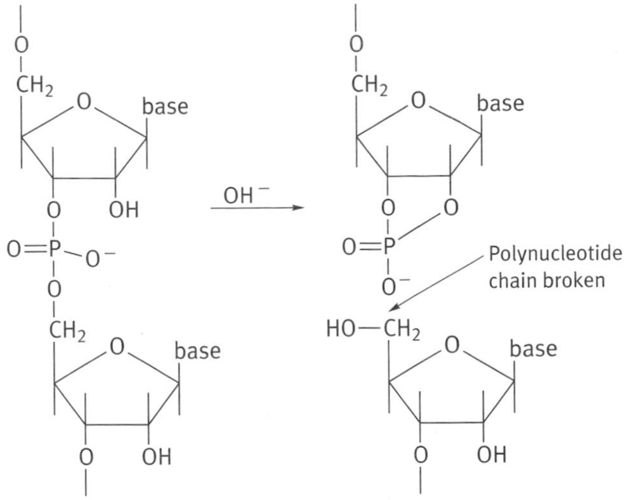

The OH^- ion facilitates the reaction because it can generate a $2'$-O^-, from the $2'$-OH group, which attacks the phosphorus atom and converts the phosphodiester group into a $2'$, $3'$-cyclic nucleotide, thus breaking the polynucleotide chain. Hydrolysis of the cyclic nucleotide produces a mixture of $2'$ and $3'$ nucleotides at the breakpoint.

The difference in stability is illustrated by the fact that dilute NaOH will completely destroy RNA at room temperature while DNA is unaffected. This is not the only chemical stability problem to be coped with—this will form the subject of a later separate section on the repair of DNA (page 263). DNA is therefore a more stable repository of genetic information than is RNA. The fact that some viruses can get away with RNA for this role does not contradict this concept as is explained later.

The DNA double helix

There will be few who have not heard of the double helix. DNA almost always exists as a double strand—only in a few viruses is it not double-stranded. In other words, you have two polynucleotide molecules paired together. What holds them together? The answer is **complementary base pairing**.

'Complementary' refers to base complementarity. It means that A and T in DNA chains are complementary in shape so that, when they are opposite one another in the two chains, they automatically form two hydrogen bonds between them. (We remind you that hydrogen bonds are short-range ones so that precise positioning of the pairing atoms is essential.) G and C are also complementary in shape and they can form three hydrogen bonds between them. Other combinations are *not* complementary so that G will not pair in the same way with A or T, etc. Only A–T and G–C pairing takes place in DNA, this being known as Watson–Crick base pairing. It is stressed that this is a completely automatic process requiring no catalysis. Because hydrogen bonds are weak, they are easily broken, for example,

by thermal energy (page 13). Mild heat will cause unpairing; cool the substance and re-association occurs.

The geometry of base pairing is shown in Fig. 19.1. Note that the base pairs always include one purine (larger molecule) and one pyrimidine (smaller size) so that the base pairs are essentially the same size. The reality of base pairing as a spontaneous process is shown by the phenomenon of **hybridization**. If a molecule of DNA is cut up into thousands of short double-stranded pieces, each say, about 20 to a few hundred nucleotides long and then the mixture is heated to an appropriate temperature (about 95°C), the two strands of each piece of DNA will separate—referred to as **DNA melting**. This is because heat disrupts the hydrogen bonding between them, resulting in thousands of single-stranded DNA pieces each of different base sequence. However, if the solution is cooled, the pieces will slowly re-associate with their original partners (Fig. 19.2). This hybridization technique, sometimes called **annealing**, lies at the heart of gene molecular biology (see Chapter 24). There is a thermodynamic driving force for hybridization of the pieces, since the formation of hydrogen bonds and associated weak forces releases energy. The most stable state in which free energy is minimized is that in which the bases are paired since this gives the maximum number of hydrogen bonds. This experimental illustration of hybridization shows that the G–C, A–T base pairing occurs spontaneously and is a real observable phenomenon.

Because of base complementarity, if you analyse the different base contents of different DNAs, the amount of G equals that of C and that of A equals T. Different DNAs vary in their percentages of $[A+T]$, and of $[G+C]$ because their base compositions are different, reflecting their different genetic information. For the first approximation then a stretch of DNA might be represented as shown below, the long solid line

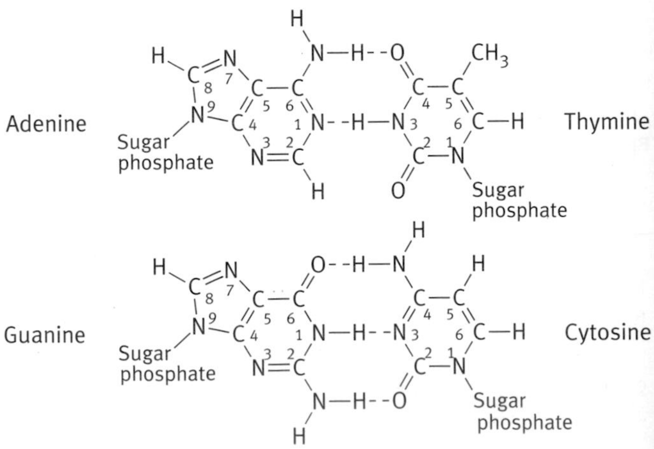

Fig. 19.1 Hydrogen bonding in the Watson–Crick base pairs.

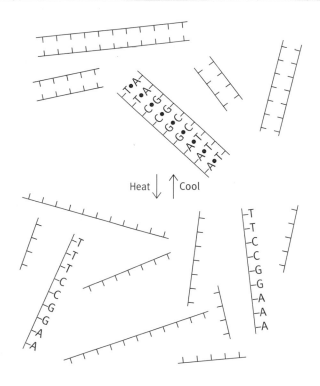

Fig. 19.2 Diagram to illustrate spontaneous hybridization of pieces of complementary DNA. Base sequences are shown on a single piece of DNA to illustrate the fact that hybridization depends on them.

representing the sugar–phosphate backbones and the attached bases interacting by hydrogen bonding.

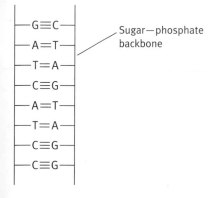

Note that, in a piece of DNA rich in G + C, the two strands will be more strongly held together than in a piece rich in A + T.

However, weak forces have a profound influence on the conformation of large molecules and this is no exception—the above structure, with ladder-like straight strands—is not 'allowed' under normal solution conditions. One important reason is shown in the diagram below.

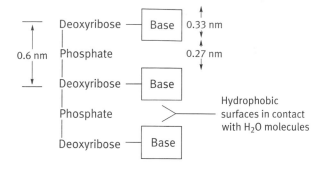

The length of the phosphodiester link is 0.6 nm (1 nanometer (nm) $= 10^{-9}$m) while the bases are about 0.33 nm thick, so that in the straight ladder-like structure there would be a gap between them. The faces of bases are, as mentioned, hydrophobic, and in the 'straight' structure above, they would be exposed to H_2O molecules, an unstable situation. What can be done to bring the hydrophobic bases together to exclude H_2O from their faces? It can be achieved by collapsing them together by sloping the phosphodiester link; or, perhaps more accurately, hydrophobic forces cause the bases to collapse together, as shown in Fig. 19.3(a).

The base pairs still lie flat, stacked on top of each other—a phenomenon known as **base stacking**. The hydrogen-bonding face at the edge is still exposed so that it can bond to its partner strand. However, the 'skewed ladder' structure shown is not the form in which DNA occurs, for such a structure has stereochemically unacceptable features. Instead, each of the two DNA chains forms a spiral, with the bases inside and the hydrophilic sugar and phosphate groups outside (Fig. 19.3(b), (c)).

The arrangement still permits base-pair stacking and the exclusion of water from between them, but the stacking cannot be exactly vertical. Instead, successive base pairs must rotate slightly relative to one another, as illustrated in Fig. 19.3(d), such that approximately 10 base pairs are required to rotate through one complete turn.

The helices are right-handed—as you move along a strand or a groove you continually turn clockwise; alternatively, imagine you are driving in a screw, holding the screwdriver in your right hand. The turning motion gives the direction of twist. The structure of the double helix is such that there is a major and a minor groove (see Fig. 19.3(b)). Although the grooves are shown in the illustrations, it is not necessarily easy to picture the three-dimensional arrangement from these. If you have access to a physical model of DNA it would be helpful to walk around it and follow the two grooves, for they are important. Any particular base pair can be viewed from both the major and the minor grooves but only their edges are visible. The major grooves provides easier access for proteins to 'recognize' (by which we mean attach to) the base pair edges. A given base pair 'looks' quite different when viewed from the two grooves (see

(a)

Sugar–phosphate backbone of DNA

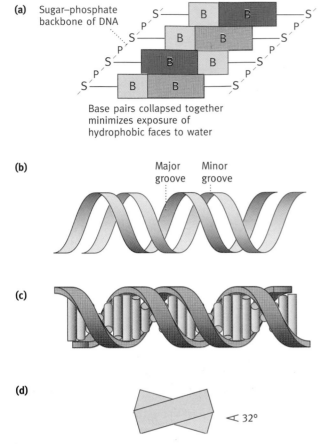

Base pairs collapsed together minimizes exposure of hydrophobic faces to water

(b)

Major groove   Minor groove

(c)

(d)

◁ 32°

Fig. 19.3 (a) Diagram showing how a skewed arrangement of the ladder collapses the bases together. **(b)** Outline of the backbone arrangements in the DNA double helix. **(c)** As (b), but showing the base pairs in the centre of the helix. (Note that each coloured band is a base *pair*.) **(d)** Diagram of two successive base pairs in a double helix showing the twist imposed by the double helix. A more realistic model corresponding to (c) is shown in Fig. 19.5.

Fig. 19.4), the significance of which will become apparent in Chapter 21 where gene regulation is discussed.

The DNA conformation described above is known as the **B form** (Fig. 19.5) and is the normal form that exists in cells. However, DNA *can* adopt different configurations in special circumstances. When dehydrated, the double helix is more squat in shape and the bases are tilted; this is known as the A form. It may exist in spores. Another form is known as Z (because the polynucleotide backbone zigzags); in this the double helix is left-handed (cf. right-handed in B DNA). It has been observed to occur in short synthetic DNA molecules with alternating purine and pyrimidine bases provided the solution is of high ionic strength. Whether either of the A or Z forms has biological significance is not known, but the possibility of DNA adopting modified configurations in localized sections of the chromosome is not excluded.

An important property of the double helix structure is that it can bend. A molecule of DNA may be a million times longer than the widest dimension of a cell or nucleus; to pack it in, flexibility is clearly needed.

It needs to be mentioned that, while DNA is almost always in the double helix form, single-stranded DNA does occur, for example, in certain bacterial viruses. In such situations the molecule takes up a complex internally folded structure to satisfy thermodynamic considerations. The main thing to note is that in such cases the life cycle involves a double helix form of DNA (Chapter 23) so that the basic principles of genetic information with complementary base pairing are the same in all cases. The existence of single-stranded forms is in this sense a specialized idiosyncrasy—not a fundamental difference.

DNA chains are antiparallel; what does this mean?

By **antiparallel** we mean that the two chains of a double helix have opposite polarity—they run in opposite directions. It may not be immediately clear what is meant by the polarity or direction of a DNA strand. It is worth spending a little time on this so that you are totally comfortable with the concept, because a lot of biochemistry requires an understanding of it. Two antiparallel strands of DNA are illustrated in Fig. 19.6.

The first point is that any single linear strand of DNA has (obviously) two ends. One end has a 5'-OH group on the sugar nucleotide that is *not* connected to another nucleotide. (It may have a phosphate on it.) This is the 5' end. The other end has a 3'-OH group that is *not* connected to another nucleotide (though it also may have a phosphate group esterified to it). This is the 3' end. At one end of a piece of double helix there is always one 5' end and one 3' end. You can never have two 5' or two 3' ends together. This is what antiparallel means. If a piece of DNA is circular, there are no free ends but there is still inherent polarity in the individual strands, as you can see from the sugar moieties.

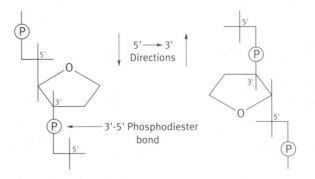

You can read off the direction by looking at the 5' and 3' positions. Thus the structure on the left runs 5' → 3' down the page and that on the right runs 5' → 3' up the page (by convention, the 5' → 3' direction is always specified).

Incidentally, in case it seems contradictory, as you move down a DNA strand in the 5' → 3' direction the actual phosphodiester bonds that you traverse are 3' → 5' (see above diagram). *Remember that a 5' → 3' direction in a DNA strand*

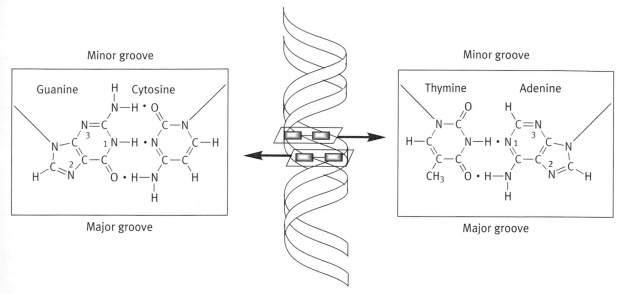

Fig. 19.4 Diagram illustrating that the edges of a given base pair in DNA look different when viewed from the major and minor grooves. DNA binding proteins designed to recognize specific sequences of base pairs in DNA can identify (bind to) the characteristic chemical groupings of different base pairs without unwinding the DNA. The importance of this is discussed in Chapter 21 which deals with gene control.

means that you are processing (in a linear molecule) from a terminal 5'-OH group towards a terminal 3'-OH group or, if you prefer it put it another way, you are travelling in the direction from the 5' group of the deoxyribose towards the 3' group of the same sugar as illustrated in the diagram above. This will all become familiar to you, as will its importance, as you progress through the next two chapters. The antiparallel arrangement fits the stereochemical requirements of the double helix structure.

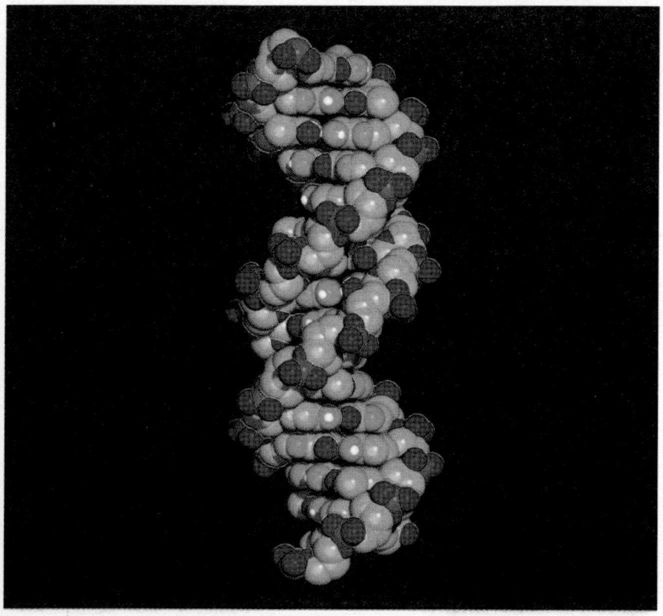

There are conventions in writing down the base sequence of DNA and it is important that you understand these. It is usual to represent a polynucleotide structure simply by a string of

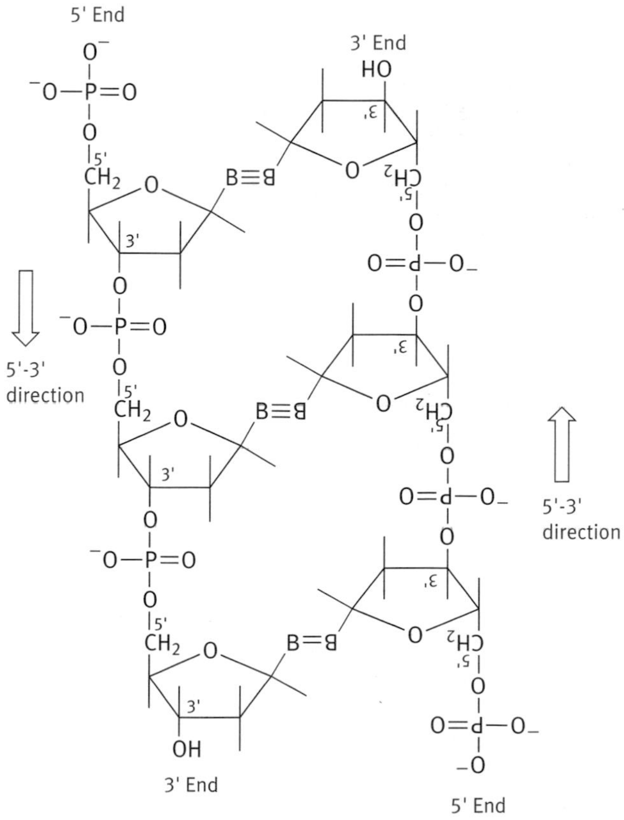

Fig. 19.5 A model of B DNA. Space-filling atomic model of a DNA segment with two major grooves and one minor groove.

Fig. 19.6 Two antiparallel strands of DNA. B, Base.

letters representing the bases of component nucleotides, but sometimes the phosphodiester link is indicated by the letter p inserted between the bases (for example, CpApTpGp, etc.). Suppose we have a piece of double-stranded DNA whose base sequence is

5′ CATGTA 3′
3′ GTACAT 5′.

Sometimes it is useful to write both strand sequences, but usually it is not necessary to write both sequences since, given one, the complementary sequence is automatically specified. So you will find that the structure of a gene is often given as a single base sequence, despite there being two strands. There is a convention that a single base sequence is written with the 5′ end to the left and there is no need therefore to specify the 5′ and 3′ ends of a sequence. Thus, if the structure illustrated above is part of a gene, it would be written as CATGTA.

How large are DNA molecules?

The variation can be enormous but the main thing is that even the shortest is a very long molecule in terms of the number of base pairs and the longest is gigantic. The DNA of a virus may be as small as a few thousand base pairs in length; *E. coli* DNA has about four million base pairs (it is circular in form). The human genome (all the chromosomes taken together) totals one to two metres of DNA in length and contains 6 billion base pairs. There are about 10^{13} cells in a human being and, if you add up the total length of DNA in a person, it comes out to an astronomical length, of the order of a diameter of the solar system.

Clearly, the more information that has to be recorded the greater the amount of DNA needed for this. You would have imagined, therefore, that the more complex the organism, the greater would be the DNA content of its cells. Surprisingly, it doesn't always work out like that. A broad bean cell has more DNA than a human cell—among vertebrates, amphibia have the most. In some comparisons, the DNA contents of organisms do correspond to perceived complexity. The apparent anomalies in DNA content are related to the fact that a good deal of DNA is not in the form of functional genes, as described later.

How is the DNA packed into a nucleus?

There are many cases in biochemistry where a problem facing life is so staggering that the solution boggles the imagination to a point where it would be reasonable to wonder if it could possibly work except for the fact that it obviously does. The packing of DNA is one of these.

In a human cell, there are 46 chromosomes giving, as already mentioned, a total length of DNA per cell of 1–2 metres, packed into a nucleus millions of times smaller in diameter. A very elaborate packing procedure indeed is needed. DNA in eukaryotic cells exists as **chromatin**—a DNA–protein complex.

The main proteins are **histones**; these are small basic proteins rich in arginine and lysine, giving them positive charges that could form ionic bonds with the negative charges on the phosphate groups on the outside of the double helix of DNA. The amino acid sequences of eukaryote histones are highly conserved throughout evolution. Indeed, one of the histones differs only in two amino acids in the entire molecules found in peas and cows and the changes are very conservative (valine for isoleucine, lysine for arginine). This extreme conservation presumably means that this protein must precisely combine with a structure, or structures, equally invariant. Possibly the histone has to participate in a complex structure with multiple essential contacts and any mutation to be tolerated must not abolish any of them.

We will start with the four histones called H2A, H2B, H3, and H4. These form an octamer protein complex called a **nucleosome core** around each of which the DNA wraps two turns (about 146 base pairs) with an intervening stretch linking successive nucleosomes (Fig. 19.7(a), (b)). If, in the test tube, chromatin is treated wth a microbial enzyme capable of hydrolysing DNA into nucleotides, the intervening DNA links between nucleosomes are hydrolysed but the 146 or so base pairs wrapped around the nucleosome core are protected from the enzyme—the latter cannot readily get at the DNA to hydrolyse it. If the remaining DNA is then extracted, it is found to contain these protected segments. This was the clue that led to the discovery of nucleosomes. This arrangement condenses (packages) the DNA somewhat, but nowhere near enough. The nucleosomes are about 10 nm in diameter and form the 10 nm fibre shown in Fig. 19.7(b). A fifth histone protein known as H1 is involved; it is not part of the nucleosome itself but its role is not certain. This 10 nm is now condensed further to form a fibre, known as the 30 nm fibre, illustrated in Fig. 19.7(c). The nucleosome packing in this fibre is often depicted as a solenoid arrangement of the nucleosomes, but this is but one of several models; another model has a zigzag arrangement. Since the exact structure has not yet been determined, the diagram of the 30 nm fibre has been left without detail of packing. An electron micrograph of such a fibre is shown in Fig. 19.8. The condensation of the DNA achieved so far is about 100-fold. The nucleosome 30 nm fibres now form long loops that are attached to a central chromosomal protein scaffolding (Fig. 19.7(d)). This looped structure forms yet more densely packed structures, not yet fully understood, involving folding and/or coiling and achieving a 10 000-fold packing of the original DNA.

We thus have several levels of organization: (1) winding around nucleosomes; (2) nucleosomes packed into a 30 nm fibre about 3–5 nucleosomes thick; (3) the fibres form loops, thousands of nucleosomes long, attached to a central scaffolding; and (4) the loops forming yet other coils and/or folds (see Fig. 19.7(a)–(d)).

As will be described in later chapters, to be functional all this packed DNA has to be accessible to enzymes that read its

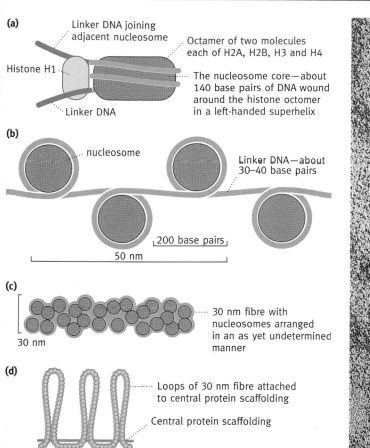

(a)
Linker DNA joining adjacent nucleosome
Octamer of two molecules each of H2A, H2B, H3 and H4
Histone H1
The nucleosome core—about 140 base pairs of DNA wound around the histone octomer in a left-handed superhelix
Linker DNA

(b)
nucleosome
Linker DNA—about 30–40 base pairs
200 base pairs
50 nm

(c)
30 nm fibre with nucleosomes arranged in an as yet undetermined manner
30 nm

(d)
Loops of 30 nm fibre attached to central protein scaffolding
Central protein scaffolding

Fig. 19.7 Order of chromatin packing in eukaryotes. **(a)** Diagram of a nucleosome. **(b)** Beads on a string form. **(c)** A 30 nm fibre of chromatin (see Fig. 19.8 for an electron micrograph of a 30 nm fibre). **(d)** Loops of the 30 nm fibre are attached to a central protein scaffold in a 360° array. It is believed that these loops are yet further condensed, perhaps by supercoiling and ultimately into the extremely compact metaphase chromosome. The latter condensation stage is not illustrated.

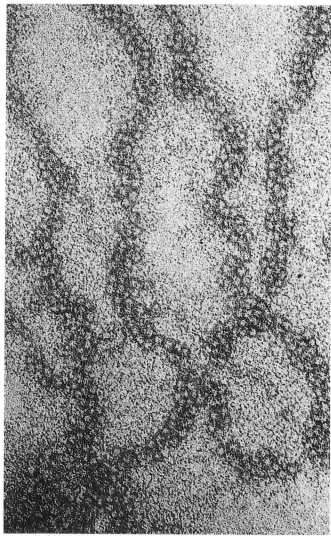

Fig. 19.8 Electron micrograph of a 30 nm fibre of chromatin.

informational content and to enzymes that replicate the entire DNA.

In an *E. coli* cell, the DNA is a circle of double helix. Its length is about 1000 times that of the cell. Prokaryotes have no nuclear membrane and there is no well-established structure corresponding to the eukaryote nucleosome organization. Nonetheless, proteins reminiscent of histones are found in *E. coli* that are postulated to play a somewhat analogous role. However, the proteins bind much less firmly to the DNA than do eukaryote histones so that nucleosome-like complexes have not been demonstrated. Looped DNA structures held in position by proteins are believed to occur in *E. coli*, but details in this area are uncertain at present.

How does the described structure of DNA correlate with the compact eukaryote chromosomes visible in the light microscope?

Chromosomes in stained dividing eurkaryote cells appear in the light microscope as compact solid structures such as those illustrated in Fig. 19.9.

The compact structures in dividing cells are **metaphase chromosomes** in which the DNA is in its most condensed state for the purpose of enabling the replicated chromosomes to be translocated into each daughter cell, a process described later (page 402). It is 'packaged DNA' for delivery to new cells. The **centromere** consists of a highly condensed section of DNA comprising repeated sequences. The DNA in this section holds together the pair of chromatids and, on its two outward

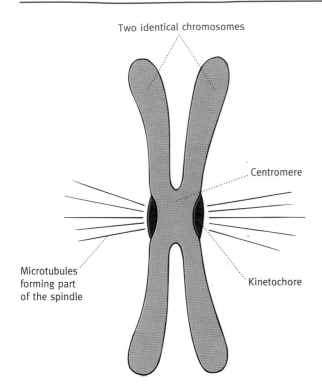

Two identical chromosomes

Centromere

Microtubules forming part of the spindle

Kinetochore

Fig. 19.9 Chromosome at metaphase consisting of two daughter chromatids. See Fig. 29.7 for mitotic phases.

facing sides, the **kinetochore** assembly of proteins is attached. The microtubules of the mitotic spindle attach at the kinetochore to cause the daughter chromosomes to separate. (Microtubules are described in Chapter 29.) Once cell division is achieved, the chromosome becomes unpacked into the **interphase chromosome**. The DNA of a metaphase chromosome is functionally inert—it is the unpacked form that is functional (but still much more packed than a simple DNA thread). The degree of 'unpacking' of DNA in an eukaryote cell is important. To use information in DNA the molecule must be accessible to enzymes and other proteins and not concealed in tightly condensed structures. Certain parts of eukaryote chromosomes contain DNA believed not to be involved in gene structure—that is, DNA that does not carry coded information for the synthesis of proteins (see below). The DNA in these regions even in interphase (the phase between cell division events) chromosomes remain condensed (called (**heterochromatin**) while the functional regions are less condensed (called **euchromatin**). The function of heterochromatin is not understood (with the exception of the centromere).

Proteins other than histones exist in chromatin, these 'nonhistone' proteins appear to be involved in the formation of the long loops of DNA attached to a central protein scaffolding. The essential message in all of this is that chromatin structure is complex; it varies in its degree of packing and the degree of packing of the DNA in chromatin is relevant to gene expression.

The term **gene expression** refers to protein production directed by a gene, but there are exceptions to this statement in that there are genes that code for specific RNA molecules, as described later.

What is a gene in molecular terms?

A **gene** is the unit of heredity. In molecular terms it has no independent existence—it is simply a stretch of DNA, part of a huge molecule, and carries coded information for the sequence of amino acids of one polypeptide chain (but see also page 274 for a more detailed description of a gene). For a protein with a single polypeptide chain, a gene therefore codes for that protein. The information resides in the base sequence of the DNA—there is no physical or molecular discontinuity between one gene and the next—only coded information. For a more familiar analogy a chromosome can be compared with an entire magnetic tape, a gene with a piece of music recorded on it, and the base sequence with the magnetic signals. The above describes what might be called the 'standard' typical protein-coding gene, which is the overwhelmingly predominant type.

Some variations on the 'standard' gene

Sometimes there are multiple copies of the same gene—this is true of some histone proteins. This situation can arise from gene duplication during evolution. Multiple copies of genes can provide for more rapid production of a given protein; there are multiple copies of histone genes that occur in clusters in some organisms—hundreds of repeating copies in the case of the sea urchin. In evolutionary terms, this also permits modification of one of the copies to an alternative function. Pseudogenes also exist—essentially copies of functional genes but ones that are not expressed or that make a nonfunctional protein or protein fragment.

A group of genes exist that do not code for proteins. These are the ribosomal RNA genes and transfer RNA genes. They code for RNA molecules that are essential for the synthesis of proteins (see Chapter 22).

The usual 'standard' gene is a stretch of DNA fixed in position in the chromosome. DNA sequences are not necessarily fixed in position; chromosomes sometimes have elements that are mobile—they move from one place in the DNA to another. They are called **transposons** or jumping genes (page 266). Chromosomes can also acquire new resident genes from retroviruses (page 320).

Gene composition may not always be completely fixed; under certain conditions a DNA section containing a given gene may be amplified so that there is a vast number of copies of the gene and a vast, corresponding increase in the production of the protein coded for by that gene—as much as 1000-fold. This has been observed experimentally in cultured mouse cancer cells treated with the drug methotrexate. Methotrexate is an 'antifolate' used in treatment of leukemia and is described on page 233.

Repetitive DNA

Not all, or even most, of the DNA of, say, a human cell is in the form of genes. For example, there exist in human chromosomes segments of DNA, a few hundred bases long, repeated hundreds or thousands of times and scattered throughout the chromosome. The repeat sequences are not always exactly the same in base sequence but are nearly so. These are known as **Alu sequences** (named after the abbreviated name of an enzyme that can hydrolyse them). No established functions for them are known.

Also a type of repetitive DNA, called **satellite DNA**, occurs particularly in the centromere regions of mouse chromosomes; this is part of the heterochromatin referred to earlier. The 'satellite' term may be misleading in that it is not a satellite in any sense in the cell; when eukaryote DNA is isolated by normal means it breaks into pieces, because of its gigantic size. When the mixture of pieces is analysed by a density gradient method,

most of the DNA runs in one band but the repetitive DNA described here has a slightly different density and runs as a 'satellite' (separate from) the main band. It consists of short DNA sequences repeated over and over in tandem arrangement (next to each other in the DNA molecule). Each section is about 10^5 bases long. Other repetitive DNA, by contrast, is dispersed all over the chromosome, as discussed above.

Where are we now?

To avoid losing sight of the main thread amongst all the necessary detail, we have earlier dealt with protein structure and started the longish trail to understanding how proteins are synthesized. We next must deal with DNA synthesis for two reasons—it comes logically after DNA structure and will help in understanding the next steps towards protein synthesis. So, in the next chapter we move on to DNA synthesis. After that we will move on to the way information in the genes is used to direct the synthesis of proteins.

Further reading

Ptashne, M. (1987). *A genetic switch: gene control and phage λ*. Cell Press and Blackwell Scientific Publications.
A classic book full of insights into DNA structure and function, as well as phage molecular biology.

Callandine, C. R. and Drew, H. R. (1992). *Understanding DNA: the molecule and how it works*. Academic Press.
A very clear discussion of DNA structure, supercoiling and DNA organization in the cell. Written in an interesting style.

Rennie, J. (1993). DNAs new twists. *Sci. Amer.*, **266**(3), 88–96
A most interesting account of the 'unorthodox' genes—jumping

genes, fragile chromosomes, expanded genes, edited mRNA, non-standard genetic code in mitochondria and chloroplasts.

Chambers, D. A., Reid, K. B. M., and Cohen, R. L. (1994). DNA: the double helix and the biomedical revolution at 40 years. *FASEB J.*, **8**, 1219–26.
Reviews a meeting to mark the 40th anniversary of the double helix. Biochemical nostalgia, but which summarizes the landmarks in the whole area and looks to the future. Strongly recommended for all students (and staff too!).

Problems for Chapter 19

1 Write down the structure of a dinucleotide.

2 Ribonucleic acid (RNA) almost certainly evolved before deoxyribonucleic acid. Why do you think DNA evolved?

3 The flat faces of the bases of DNA are hydrophobic. Explain the structural repercussions of this fact on the structure of double-stranded DNA.

4 What is the main form of double-stranded helical DNA called? Is it a right- or left-handed helix? Approximately how many base pairs are there in a stretch of DNA that completes one rotation of the helix?

5 Explain what is meant by DNA chains in a double helix being antiparallel.

6 Explain in everyday language what is meant by a $5' \rightarrow 3'$ direction in a linear DNA molecule.

7 If you see a DNA structure simply written as CATAGCCG, what exactly does this means in terms of a double-stranded structure and the polarity of the two chains? Explain your answer.

8 **(a)** What is a nucleosome?
   (b) What led to their existence being first suspected?

9 What are Alu sequences?

Chapter summary

Chapter 20

· ·

DNA synthesis and repair

In this chapter we will deal mainly with the *E. coli* cell, though the eukaryote situation is described where information is available. DNA synthesis is a very complex process, and much more is known of the mechanism in *E. coli* than in eukaryote cells. Sufficient is known of the latter to be sure that the processes are basically the same in both, even if not in absolute detail.

Every time a cell divides, its entire content of DNA must be duplicated or, as is more usually stated, the chromosome(s) must be replicated, so that a complete complement of DNA can be given to each daughter cell. A human cell has about 6 billion base pairs in its total DNA. The magnitude of the task of faithfully replicating these needs no emphasis. Even a single incorrect base in a gene may cause a protein to be produced with impaired function.

· ·

Overall principle of DNA replication

We will go into the question of *how* DNA is synthesized in due course, but for the moment let us look at it at a general level.

A chromosome is double-stranded DNA. Its replication is described as **semiconservative** in that the two original strands, called parental strands, are separated and each acts as a template for synthesizing a new strand; each new double helix has one old and one new strand. This was established in the classic experiment shown in Fig. 20.1.

The basis of the replication is that of complementarity in that a G will base pair with C, and A with T, so that a base on the parental strand automatically specifies which base is to be incorporated into the new strand as its partner. How this is specified to the synthesis machinery is described later in this chapter. Since this copying process depends on Watson–Crick hydrogen bonding of base pairs, it follows that strand separation is essential to unpair the bases and make them available for base pairing with incoming nucleotides. The *E. coli* chromosome is circular. It contains about four million base pairs. The strands are initially separated at one particular point called the **origin of**

replication and *two* **replication forks**, moving in opposite directions, synthesize DNA at the rate of about 500 base pair copies per second, with separation of parental DNA and synthesis of new DNA occurring at the same time (Fig. 20.2). The two forks meet at the opposite side of the circle. As a result of the replication, at the end there are two newly completed interlinked rings, a seemingly impossible situation. The prediction that this would occur was previously regarded as militating against the model of DNA replication. A remarkable insight of Crick was that evolution would have produced a mechanism to separate the circles. The topoisomerase type II described later (page 254) does precisely this.

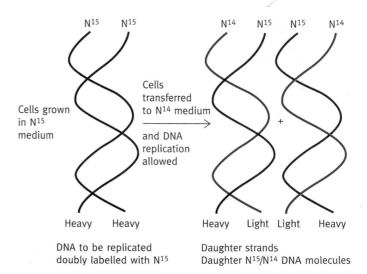

Fig. 20.1 Demonstration of semiconservative DNA replication by Meselson and Stahl. The DNA of cells was labelled by growing them in a medium in which the nitrogen source was N^{15}, so that both strands of DNA were 'heavy'. They were then transferred to N^{14} medium so that all subsequent DNA chains synthesized would be 'light'. The density gradient analysis indicated that, one generation after the transfer, each DNA molecule contained one 'heavy' and one 'light' strand. This is known as semiconservative replication. Continuation of the experiment for further generations confirmed the result. The red strands are newly synthesized.

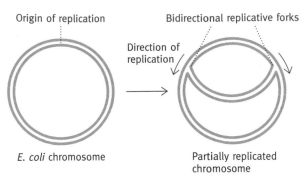

Fig. 20.2 Bidirectional replication of the *E. coli* chromosome. Parental strands are blue; newly synthesized strands are red.

Eukaryote chromosomes are linear. In eukaryotes, a replication fork synthesizes DNA at about an average 50 base pairs copied per second—too slow by a huge margin for a single replicon (explained below) to synthesize the vast lengths of DNA in a chromosome in the time allowed for it in cell division. To cope with this, there are hundreds of origins of replication along the chromosome from which replicative forks work in both directions (Fig. 20.3); thus, the process is analogous to the *E. coli* situation.

Control of initiation of DNA replication in *E. coli*

Before cell division occurs in *E. coli* there must be a complete duplication of the chromosome; cell division follows about 20 minutes later. Exactly how cell division and DNA replication are coordinated is not understood. Protein synthesis and a critical enlargement of the cell are required. As already seen (Fig. 20.2), in *E. coli* there is a single point of origin of DNA synthesis called

*ori*C at which replication commences bidirectionally. The entire chromosome is called a single **replicon**, the latter term referring to a stretch of DNA whose replication is under the control of a single origin of replication.

The origin of replication has a specific base sequence, very rich in A–T pairs, presumably to facilitate strand separation. (We remind you that A–T pairs have two hydrogen bonds and G–C pairs three and, therefore, the former are less tightly bound together.) At the time of initiation, a protein referred to as DnaA binds in multiple copies to this region and causes strand separation. This permits the main unwinding enzyme (helicase), which works at each replicative fork, to attach and begin progressive unwinding of the strands in both directions. The **helicase** (or DnaB, the protein coded for by the gene *dnaB*) is referred to later when we describe the mechanism of DNA synthesis.

Initiation of DNA replication in eukaryotes

Eukaryotic cells have a more complex cell cycle (Fig. 20.4) than that of *E. coli* cells, in that DNA synthesis is confined to a definite period of time called the S (for synthesis) phase. Different cells vary a great deal but, in cultured animal cells, the S phase takes about 8 hours out of the total cycle of 24 hours. Before the S phase is the G_1 phase, (G for gap in DNA synthesis). To proceed to cell division, a mammalian cell requires a mitogenic signal from outside itself. This takes the form of protein signalling molecules or **growth factors** that attach to surface receptors on cells and transmit the signal to grow to the interior of the cell. The latter process (that of cell signalling) is of immense importance and is the subject of Chapter 26. As explained, an animal chromosome may have

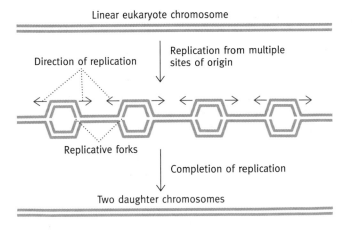

Fig. 20.3 Diagram of multiple bidirectional replicative forks in a eukaryotic chromosome.

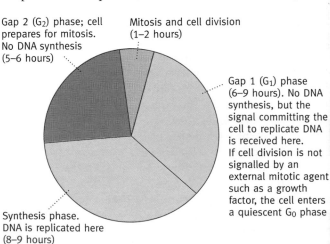

Fig. 20.4 The eukaryote cell cycle. The duration of the cell cycle varies greatly between different cell types. The times given here are for a rapidly dividing mammalian cell in culture (24 hours to complete the cycle).

hundreds of replicative origins from which bidirectional forks duplicate a replicon or stretch of DNA. What controls initiation of replication is not understood; the multiple replicons do not necessarily fire off at the same time. It is vital that each replicon fires off only once for each cell division.

Unwinding the DNA double helix and supercoiling

DNA strand separation by helicase presents topological (physical) problems, and to explain these we must deal with the subject of DNA supercoiling.

Since duplex DNA has two strands in the form of right-handed helices, it has an inherent degree of twist, there being one turn of the helix per approximately 10 base pairs. An isolated short piece of linear DNA that is completely free to rotate on its own long axis automatically adopts this strain-free configuration, known as the **relaxed state**. Suppose instead that you clamped one end of the duplex so that it was not free to rotate and you gave an extra twist to the other end, so that the coil of the double helix is tightened somewhat—the number of turns for a given stretch of DNA is increased; that is, the number of base pairs per turn is decreased. It is now **positively supercoiled** or **overwound**. If you twisted in the opposite direction, the coil would be opened up—the number of turns per unit stretch would be reduced, or the number of bases per turn increased. The DNA would be **negatively supercoiled** or **underwound**. Both the underwound and overwound states are under tension and one way of accommodating the strain is for the DNA double helix to coil upon itself forming a **coiled coil** or **supercoil** (Fig. 20.5). You can illustrate supercoiling very easily with a piece of double-stranded rope. Have someone hold one

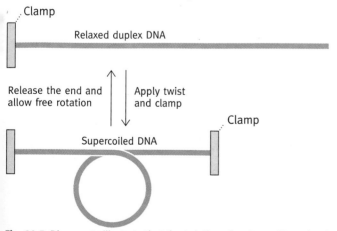

Fig. 20.5 Diagram to illustrate that the twisting of a piece of DNA that is not free to rotate induces supercoiling to accommodate the twisting strain. Cellular DNA is effectively clamped and is not free to rotate. If, somehow, free rotation is allowed, the supercoil will relax. If the applied twist is in the direction of unwinding, the supercoil will be negative; it will be positive if the applied twist is in the opposite direction (the difference between the two is not illustrated).

end or clamp it somehow, so that the rope cannot rotate freely, and twist the rope on its axis. Coils will form to take up the twisting strain. If you release the end of the supercoiled rope it will immediately spin back to the relaxed state.

What has this to do with DNA replication? DNA in the cell is not free to rotate on its own long axis; in *E. coli* the closed circle chromosome effectively 'clamps' the DNA. In eukaryotes, the DNA is of such vast length, arranged in fixed loops (see Fig. 19.7) and attached to protein structures, that once again free rotation is impossible. But, separation of DNA strands demands that the duplex rotates. This causes overwinding—it generates positive supercoils ahead of the replicative fork and, as the helix tightens, further separation is resisted. If unrelieved, the tension would bring strand separation and DNA replication to a halt.

A very simple experiment will convince you of this. If you take a short piece of double-stranded rope and pull the ends apart, the rope will spin rapidly, thus preventing the accumulation of positive supercoils. The rope strands will easily separate completely. Now take a long piece of the same rope coiled on the floor, or have someone hold one end of a reasonably long piece, so that it cannot freely rotate and try to pull the strands apart. Positive supercoils will rapidly snarl up the separating fork and prevent any further separation. This would be the situation in DNA replication in the cell if something weren't done about it.

It follows that, for DNA synthesis to proceed, the positive supercoils ahead of the replicative fork must be relieved and this, of necessity, involves the transient breakage of the polynucleotide chain.

How are positive supercoils removed ahead of the replicative fork?

A group of enzymes, known as **topoisomerases**, catalyse the process. They act on the DNA and isomerize or change its topology. There are two classes of topoisomerases, called types I and II. We will deal with the principles of their mechanisms first, and then explain their roles in DNA replication.

In type I, the enzyme breaks one strand of a supercoiled double helix, which permits the whole duplex to rotate on the single phosphodiester bond of the partner strand, effectively introducing a swivel into the DNA. After rotation has occurred, the enzyme reseals the duplex (see Fig. 20.6). It is important to note that the enzyme *does not hydrolyse the phosphodiester bond it attacks*—it simply transfers the bond from the deoxyribose-3′OH to the —OH of one of its own tyrosine side chains. Since no energy loss is involved in the cutting of the chain, the process is freely reversible. Note that this enzyme does not use ATP. A supercoiled DNA molecule is in a state of tension—it is at a higher energy level than the relaxed state whether the super-coiling is positive or negative. A topoisomerase type I can only relax supercoiled DNA—it cannot insert supercoiling. As it happens, the particular topoisomerase I of *E. coli* can relax only negatively supercoiled DNA so it cannot solve the unwinding

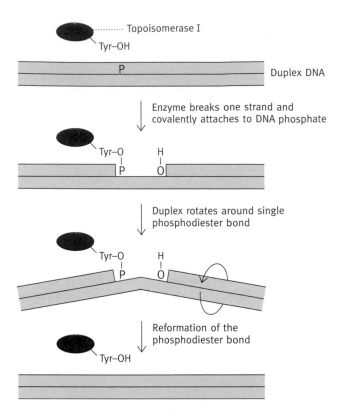

Fig. 20.6 The reaction catalysed by type I topoisomerases. The Tyr-OH represents a tyrosine residue of the enzyme.

problem discussed so far. But, as will become evident shortly, it has an important role.

The relaxation of the positive supercoiling ahead of the replicative fork in *E. coli* is done in a different way—by the active insertion of negative supercoils, which 'neutralize' as it were the positive supercoils. This is done by topoisomerase type II, called gyrase, described below.

A type II topoisomerase breaks two strands of the DNA double helix transferring the bonds to itself, making the breakage of the polynucleotide chains a freely reversible process. The enzyme physically transfers the DNA duplex of a coil through the double gap, ATP being required for this piece of work, possibly through its hydrolysis effecting a conformational change in the protein. In *E. coli*, the topoisomerase II is called **gyrase**; it introduces negative supercoils in the DNA. The mechanism of this is illustrated in Fig. 20.7. In this figure, the insertion of negative supercoiling into a simple relaxed circle of DNA (Fig. 20.7(a)) is given for simplicity, but the same principle applies to the DNA ahead of the replicative fork in *E. coli* where positive supercoiling has occurred as a result of the strand separation. As stated, insertion of negative supercoiling in this region is equivalent to relaxation of the positive supercoiling. The gyrase binds to the DNA such that the latter is wrapped around the protein. In doing so it creates overwinding (positive supercoiling) in the local region associated with the

protein (Fig. 20.7(b)). However, since no covalent bonds have yet been broken, there can have been no net change to the structure as a whole and, therefore, the local positive supercoiling caused by attachment to the protein must be compensated for by a negative supercoiling elsewhere in the molecule of DNA, thus keeping the net change in the supercoiling to zero. The enzyme now breaks both strands of the DNA (Fig. 20.7(c)) and, by passing the double strand from underneath to the upper side, followed by resealing of both breaks, converts the positive node to a negative one (Fig. 20.7(d)). In short, negative supercoiling has been inserted. The physical movement of the DNA through the gap is driven by

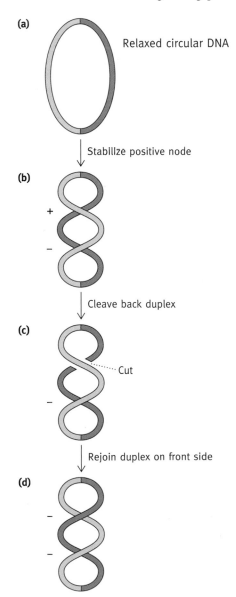

Fig. 20.7 Diagram of the reaction catalysed by *E. coli* gyrase to insert negative supercoils into circular DNA. This action can convert positively supercoiled DNA into negatively supercoiled DNA. Steps **(a)–(d)** are referred to in the text.

ATP hydrolysis. The negative supercoiled state is also under tension and, therefore, the *insertion* of negative supercoils is an energy-requiring reaction. As already stated the bonds broken are not hydrolysed but simply transferred temporarily to the enzyme itself without loss of free energy, the process of bond breaking is therefore freely reversible.

What are the biological implications of topoisomerases?

It is clear that is has been essential during evolution for living cells to cope with DNA topological problems and, as described, types I and II topoisomerases are the two mechanisms for doing so. However, once again, evolutionary tinkering has occurred and in various organisms the precise properties of the enzymes differ somewhat; at least 11 variants have been identified in different organisms.

In *E. coli*, the topoisomerase I relaxes only negative supercoils, not positive ones, while gyrase (type II) actively inserts negative supercoils and, since this will relax positive supercoils, permits DNA synthesis to proceed.

In eukaryotes, the topoisomerase I can relax both positive and negative supercoils; the type II topoisomerase can relax positive supercoils but cannot introduce negative supercoils (see Table 20.1). Thus, in prokaryotic and eukaryotic DNA replication, the potential snarl-up of strand separation through the accumulation of positive supercoils is averted.

However, *DNA supercoiling is important in more than DNA replication*. When DNA is carefully isolated from cells it is found to be negatively supercoiled. In relaxed DNA, the double helix has one turn per 10.5 base pairs; in cellular DNA it has about one turn per 12 base pairs—it is underwound. The degree of supercoiling is roughly comparable in the DNA of different cells, which suggests that it is of importance and that its generation is

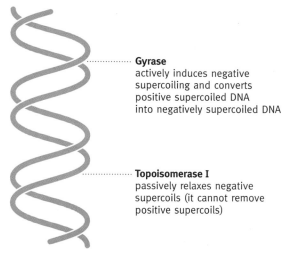

Gyrase
actively induces negative supercoiling and converts positive supercoiled DNA into negatively supercoiled DNA

Topoisomerase I
passively relaxes negative supercoils (it cannot remove positive supercoils)

Fig. 20.8 Diagram of the balancing act of gyrase and topoisomerase I in *E. coli* to achieve the correct degree of negative supercoiling required for DNA replication and transcription. Gyrase is ATP-driven. The amounts of the two enzymes synthesized (and hence the balance between them) is thought to be controlled by the degree of supercoiling of the gene promoters.

controlled. It is known that negative supercoiling is essential in *E. coli*, both for replication and transcription (Chapter 21). The reason for this is that, in such a state, DNA strand separation occurs more readily than in the relaxed or positively supercoiled state. It has also been suggested that protein binding to negatively supercoiled sections of DNA could relieve the strained conformation and, in doing so, increase the release of energy due to the binding. This would facilitate the attachment of such proteins. As will become evident in the next chapter, DNA binding proteins are of supreme importance in gene function.

To summarize the *E. coli* situation, the required negative supercoiling of DNA is introduced by gyrase (topoisomerase II) and the *E. coli* topoisomerase type I relaxes negative, but not positive coils. The balance between the two opposing activities (gyrase and topoisomerase I) maintains the correct degree of supercoiling (Fig. 20.8); the correct balance is essential to the life of the cell and the synthesis of the two enzymes is controlled to achieve this, the degree and sign of supercoiling in the DNA of the respective genes being a controlling factor. The essential nature of gyrase to bacteria is underlined by the fact that the antibiotic **nalidixic acid** that inhibits their multiplication does so by inactivating gyrase. Since eukaryotes do not possess this enzyme (see below) nalidixic acid can be used clinically to treat certain infections in man.

It is particularly interesting that, in a thermophilic bacterium living at extremely high temperatures, a 'reverse gyrase' has been found. A possible reason is that positively supercoiled DNA is less likely to undergo the strand separation that occurs at high temperatures.

Table 20.1   Summary of *E. coli* and eukaryote topoisomerases, types and action

Type*	Action on DNA	Effect on supercoiling
Topoisomerase I		
E. coli	Cuts one DNA strand	Relaxes negative supercoils
Eukaryotes	Cuts one DNA strand	Relaxes positive and negative supercoils
Topoisomerase II (gyrase)		
E. coli	Cuts two DNA strands; ATP-dependent	Relaxes positive supercoils; inserts negative supercoils; separates interlinked circles
Eukaryotes	Cuts two DNA strands; ATP-dependent	Relaxes positive supercoils but cannot insert negative supercoils

* Note that variants of these occur in other organisms.

Eukaryote DNA, like that of prokaryotes, is underwound or negatively supercoiled in the cell. However, unlike the situation in prokaryotes, no eukaryote topoisomerase is known that can actively insert negative supercoils into DNA. How then is this achieved?

When chromatin is assembled, the DNA winds around nucleosomes in such a manner that, in the local region in contact with the protein, it is in an underwound state. (This is achieved by a left-handed coil; it can be difficult to grasp and retain such topological details in three dimensions and, since it is not essential in the present context, we omit description of the handedness of such coils.) Since this nucleosome winding does not involve any bond breakage and since the chromosomal DNA cannot freely rotate, it follows that there cannot have been any *net* change in the supercoiling of the DNA. Therefore, the local negative supercoiling at the nucleosome must be compensated for by positive supercoiling elsewhere so that the net change in the structure is zero (Fig. 20.9(b)). The eukaryote topoisomerases I or II now relax the positively supercoiled section, thus achieving the insertion of a negative supercoil (Fig. 29.9(c)). A prokaryote type of gyrase is thus not needed; the fact that prokaryotes do not have the nucleosome structures correlates with the need for their own type of gyrase.

Thus far, we have helicase unwinding the strands and topoisomerases allowing the separation to proceed right along the chromosome by relaxing positive supercoiling induced by the turning force applied to the double helix during unwinding. This is not the end of the unwinding mechanism, for **SSB** is involved.

What is SSB?

The answer is DNA *s*ingle-*s*trand *b*inding protein, which has a high affinity for single-stranded DNA but with no base sequence specificity—it binds anywhere on the separated strands. The energy release in binding helps to drive strand separation to completion; the single-stranded form is prevented from re-annealing. When the strands function as templates for new DNA synthesis, it has to be swept out of the way by the replicating machinery.

The situation so far

So far we have dealt with the rather broad aspects of DNA replication—its semiconservative nature based on Watson–Crick base pairing, with the biological aspects of the cell cycle, with the initiation of replication, and with the mechanism of unwinding. It says a lot for the complexity of DNA replication that, after all this, we have only got as far as separating the DNA strands and nothing so far has been said about how new DNA is actually synthesized in the replication fork. We now want to turn to this; the simplest part by far is the chemistry by which nucleotides are linked together to form the new DNA chain. The enzyme(s) that catalyse this are called **DNA polymerases** for they polymerize nucleotides into DNA.

The basic enzymic reaction catalysed by DNA polymerases

First let us consider the basic chemical reaction. A series of facts first.

- There are three DNA polymerases in *E. coli* called Pol I, II, and III—named in order of their discovery.
- The DNA synthesis occurring in the replicative fork, is catalysed by Pol III or its eukaryote equivalents, but Pol I also plays an essential role in DNA replication as well as in repair. Less is known of Pol II, but it is believed to be associated with certain types of DNA repair.
- The substrates for DNA polymerases are the four deoxyribonucleoside triphosphates dATP, dCTP, dGTP, and dTTP. These are synthesized in the cell as described earlier (page 231). The regulatory mechanisms in their synthetic pathways ensure that they are produced in adequate amounts and in coordinated concentrations.
- The polymerase must have a DNA template strand to copy. 'Copy' is used in the complementary sense—a G on the template strand is 'copied' into a C in the new strand, and likewise A into T, C into G, T into A.
- A most important fact to fix in your mind: *a DNA polymerase can only elongate (add to) a pre-existing strand called a primer*. This **primer** may only be 2–10 nucleotides long but without it nothing happens. **DNA polymerases cannot start a chain—they cannot join together two free nucleotides.**

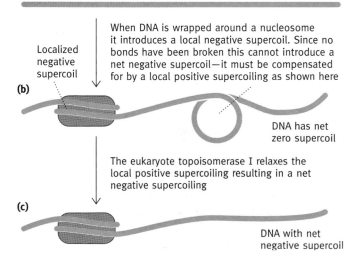

(a) DNA of chromosome (zero supercoil); note that this strand is not free to rotate

Localized negative supercoil

When DNA is wrapped around a nucleosome it introduces a local negative supercoil. Since no bonds have been broken this cannot introduce a net negative supercoil—it must be compensated for by a local positive supercoiling as shown here

(b)

DNA has net zero supercoil

The eukaryote topoisomerase I relaxes the local positive supercoiling resulting in a net negative supercoiling

(c)

DNA with net negative supercoil

Fig. 20.9 A mechanism by which eukaryote DNA becomes negatively supercoiled despite the absence of any enzyme capable of actively inserting negative supercoils such as the prokaryote gyrase. Steps **(a)–(c)** are referred to in the text.

- As illustrated in Fig. 20.10, the polymerase attaches a nucleotide to the 3′ free OH group of the end of the primer strand, liberating inorganic pyrophosphate. Hydrolysis of the latter increases the negative $\Delta G^{0'}$ value for the synthesis thus helping to drive the reaction (page 12). Incorporation of a nucleotide into the new strand of DNA involves the formation of hydrogen bonds with its template partner with the liberation of energy (page 14), thus adding to the thermodynamic drive of the process.

- Which of the four deoxyribonucleoside triphosphates is accepted by the DNA polymerase is determined by the base on the parental strand being copied.

- *DNA synthesis always proceeds in the 5′ → 3′ direction with respect to the growing strand.* Be sure that you know what this

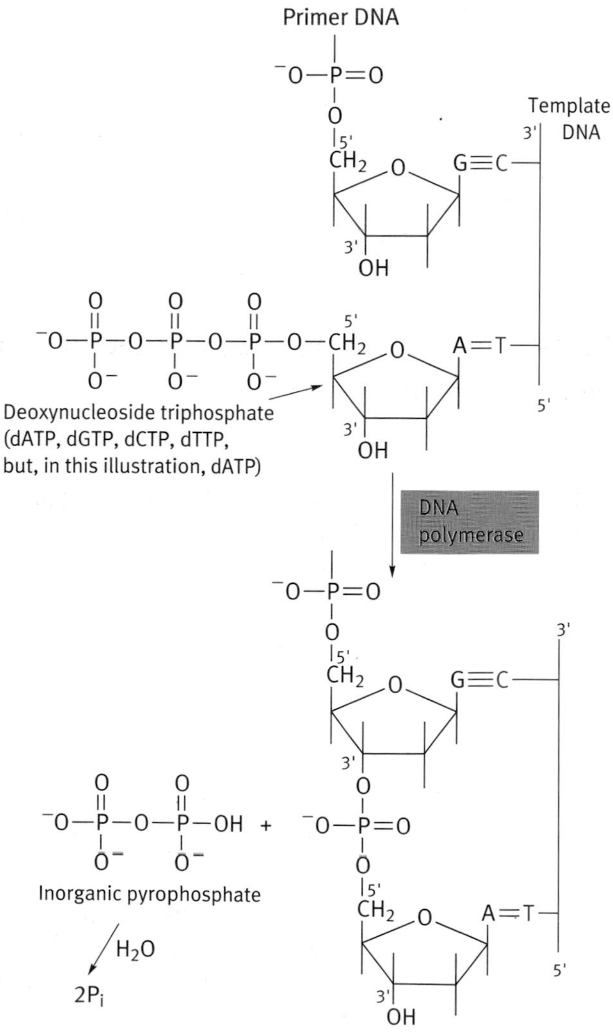

Fig. 20.10 The reaction catalysed by DNA polymerase. The diagram shows the addition of an adenine deoxynucleotide from dATP to the 3′ end of the primer DNA strand, the base selected for addition being determined by the base on the template strand. Note that the synthesis is in the 5′ → 3′ direction—the chain is being lengthened in the 5′ → 3′ direction.

means—that the growing DNA chain is being elongated in the 5′ → 3′ direction—a nucleotide is added to the free 3′-OH of the preceding terminal nucleotide. At the risk of overemphasizing the point, for it is important, when we talk of synthesis being in the 5′ → 3′ direction we always refer to the direction of elongation—the polarity of the *new* strand. We are *not* referring to the template strand, which has the opposite polarity. The polarity of DNA strands has been explained in the previous chapter (page 244).

Problems and more problems in DNA synthesis

How does a new strand get started?

As is now fixed in your mind, DNA polymerases cannot initiate new chains and yet, at each origin of replication, new chains must be initiated.

The solution to the question in the above heading is rather surprising in that DNA chains are initiated by RNA, which is slightly unfortunate because we don't want to deal with RNA and its synthesis until the next chapter. However, for the moment, RNA has the same structure as single-stranded DNA except that the sugar is ribose and the base thymine (T) is replaced by uracil (U). RNA is synthesized by RNA polymerases by essentially the same basic chemical mechanism as outlined above for DNA except that ATP, CTP, GTP, and UTP, are used. But, in the present context, the vital difference is the RNA polymerases *can initiate new chains*. They can take two nucleotides and link them (a template being required here also); DNA polymerase cannot do this.

When a small piece of RNA primer (perhaps five nucleotides) has been synthesized by a special RNA polymerase called **primase**, DNA polymerase takes over and extends the chain.

We now come to yet another problem due to the antiparallel nature of DNA.

The polarity problem in DNA replication

It might be useful to turn back again to page 244 to make sure you understand DNA polarity. Consider these facts. DNA is replicated at each replicative fork which steadily progresses along the chromosome. In *E. coli*, there are two DNA polymerase III molecules involved in each fork, one for each strand, the two enzymes being linked together into a single asymmetric holoenzyme dimer. As shown in Fig. 20.11, the polymerase molecules that are replicating the two strands must physically move in the same direction (that is, up the page as it were). However, one parent strand runs 5′ → 3′ in the reverse direction of fork movement and the other runs 5′ → 3′, the opposite way, in the direction of fork movement. Since DNA polymerase can synthesize only in the 5′ → 3′ direction, the template strand must run 5′ → 3′ in the *opposite* direction to that

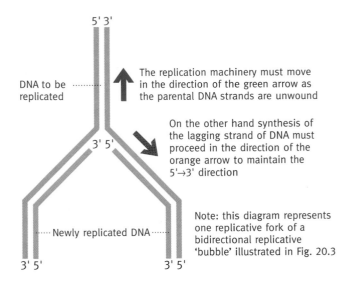

Fig. 20.11 The polarity problem in DNA replication.

of synthesis. This is fine for the synthesis of one new strand (the left-hand one in Fig. 20.11), but what about the other? It runs $5' \rightarrow 3'$ in the direction of fork movement and hence DNA synthesis must run the opposite way. A seemingly impossible problem is illustrated in Fig. 20.11 in which the polymerase on the right-hand strand *must* physically move up the page, as it were, but synthesize DNA in the direction of down the page. The left-hand strand, with no problems, is called the **leading strand** and the other the **lagging strand**.

First, how can the lagging strand initiate? The leading strand has no problem. Primase lays down a single primer at the origin of initiation and the DNA polymerase proceeds from there, but the same will not suffice for the lagging strand. The solution is that, as the DNA unwinds, there is repeated initiation by primase (an RNA polymerase), followed by many short stretches of DNA synthesis. The net result is illustrated in Fig 20.12 but do not be worried about how this is achieved, for it still looks to be an impossible situation.

The short stretches of DNA attached to RNA primers on the lagging strand are called **Okazaki fragments** after their discoverer. This still leaves the original problem of how a DNA polymerase can synthesize DNA backwards while moving forwards—a physical or topological problem. It also leaves the lagging strand not as a single piece of DNA but a series of disconnected short pieces attached to RNA primers that must be made into uninterrupted DNA. Let us deal with the physical problem first. To do this we must first discuss the replicative machinery at the replication fork.

Enzyme complex at the replicative fork in *E. coli*

The functional complex of proteins and protein subunits at the replication form is illustrated in Fig. 20.13. The key enzymes are the **helicase** to unwind the double helix, **primase** to synthesize

RNA primers at intervals on the lagging strand, and the extremely complex **Pol III**. The helicase unwinding activity is ATP-driven and moves along a DNA strand (probably by allosteric conformational changes) and, in doing so, separates the two strands of the double helix. Finally, **SSB** attachment stabilizes the single strand. As indicated, in *E. coli*, there are two connected molecules of Pol III in the replicative fork, one synthesizing the leading strand the other the lagging. They have the same core enzyme but the holoenzyme dimer is asymmetric with extra subunits present on the lagging strand side. Pol III has high processivity—once it locks on to a DNA strand it keeps on for thousands of bases without dissociating, which is appropriate for the leading strand. It is held on to the template strand by an annular 'clamp' (the β subunit) made of two halves. The annulus forms a sliding ring behind the Pol III and attaches to the enzyme, thus preventing the latter from falling off the template DNA strand. The circular clamp has a hole big enough for the DNA to slide through it easily. A complex of proteins assembles the clamp wherever there is an RNA primer and the polymerase becomes attached to it. The structure of the clamp is illustrated in a ribbon diagram in Fig. 20.14. On the lagging strand the enzyme (also with the circular clamp) stops when it meets the previous primer laid down by the primase. It is not known whether the Pol III assembly detaches and reassembles at the next primer to synthesize the next Okazaki fragment or what happens here.

Now for the physical or topological problem of how the Okazaki fragments are synthesized in the $5' \rightarrow 3'$ direction without the replicative machinery moving backwards. The basic principle is extremely simple. If the lagging template strand is looped, then for a short distance it is oriented with the same polarity as the leading strand template. The replicative machinery can therefore proceed in the direction of the fork and synthesize both new strands.

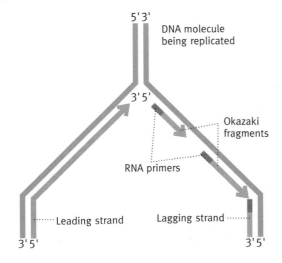

Fig. 20.12 Diagram of a replicative fork. The leading strand is synthesized continuously, while the lagging strand is synthesized as a series of short (Okazaki) fragments.

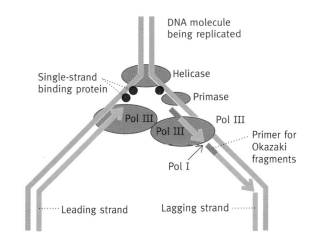

Fig. 20.13 Simplified diagram of the enzyme components of the *E. coli* replicative fork.

While the principle is simple, the mechanical problem of how the loop system can move along the entire length of the parental strand and permit the synthesis of Okazaki fragments is not simple. It requires that the loop be reformed and enlarged at regular intervals, and new RNA primers must be laid down.

A model put forward by Kornberg, called the loop model, is shown in Fig. 20.15. As the loop enlarges and the replicative machinery moves forward, the new DNA strand will meet the 5′-RNA end of the previous Okazaki fragment. At this point the polymerase must disengage and the loop fall away and a new

one started. Now to the other problem—how the Okazaki fragments are tidied up into continuous DNA.

What happens to the Okazaki fragments?

In *E. coli*, when the Pol III reaches the RNA primer of the preceding Okazaki fragment, it disengages from the DNA, leaving a nick at the DNA/RNA junction. This is where DNA polymerase I (Pol I) comes in. Pol I is an astonishing enzyme with three separate catalytic activities on the same molecule.

If we look at the problem, as illustrated in Fig. 20.12, the separate pieces of DNA, the Okazaki fragments, must be converted into a continuous DNA molecule. Each piece starts with RNA which must be removed, replaced with DNA, and the separate DNA pieces joined up. The Pol I attaches to the nicks, or breaks, between successive Okazaki fragments, and adds nucleotides to the 3′-OH of the preceding fragment, moving in the 5′ → 3′ direction; as with Pol III (or any DNA synthesis), nucleotide additions are always in the 5′ → 3′ direction. Since, as Pol I moves, it encounters the RNA of the next Okazaki fragment, the nucleotides of this are chopped out. Thus, as it were, the front end of Pol I chops nucleotides out, and a site further back adds nucleotides to fill the gap with DNA. The 'front' activity is a 5′ → 3′ **exonuclease** activity—'exo' because it works on the end of the molecule, 'nuclease' because it hydrolyses nucleic acids, and 5′ → 3′ because it nibbles away at the 5′ end of the RNA and moves in the direction of the 3′ end of the molecule. Note that Pol III does *not* have a 5′ → 3′

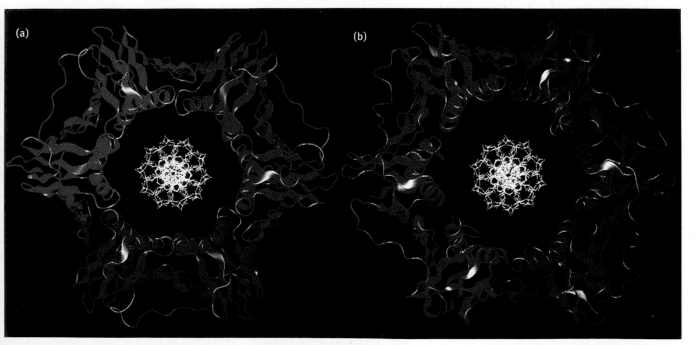

Fig. 20.14 Ribbon representations of the yeast and *E. coli* sliding 'clamps'. **(a)** The yeast clamp that confers processivity on DNA polymerase δ is a trimer (see page 264). **(b)** The *E. coli* clamp that attaches to DNA polymerase III is a dimer. The individual subunits within each ring are distinguished by different colours. Strands of β sheet are shown as flat ribbons and α helices as spirals. A model of B DNA is placed in the centre of each structure to show that the rings can encircle duplex DNA.

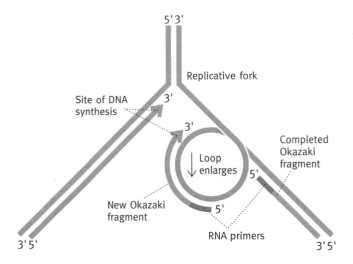

Fig. 20.15 The 'loop' model for Okazaki fragment synthesis. (Red arrowheads indicate DNA synthesis.) This model requires the loop to fall away when the new Okazaki fragment meets the old one. A new loop has then to be made. Both strands can be synthesized in this model in the required 5′ → 3′ direction as the replication machinery moves in the direction of the fork. It is not possible to specify the precise mechanical details of the looping mechanism.

exonuclease activity like Pol I. Thus, Pol III cannot chop out the RNA primer when it meets the preceding Okazaki fragment. It disengages from the DNA at this point and hands over the job to Pol I.

That is still not the end of the matter. At the 'rear' end of the Pol I molecule there is a 3′ → 5′ exonuclease as well as the polymerase activity. This exonuclease hydrolyses off the nucleotide at the 3′ end of the DNA chain that Pol I has just synthesized and it chops in the direction of the 5′ end—exactly the reverse of the 'front-end' 5′ → 3′ exonuclease. Hence, it is a 3′ → 5′ exonuclease. Put in another way, it is chopping backwards.

This is almost beyond reasonability—an enzyme whose 'rear end' both chops out DNA and synthesizes DNA. But there are other vital facts. *The 3′ → 5′ exonuclease hydrolyses the terminal nucleotide of the newly formed DNA strand only if it is unpaired*

with its corresponding base on the template strand and the DNA polymerase moiety will add nucleotides only if the preceding one (Fig. 20.16) *is properly base paired.* Thus, the enzyme checks the preceding base for correctness and cuts it out if unpaired, and replaces it. This is called **proofreading**—it checks the fidelity of what it has synthesized.

The DNA polymerase I has (unlike Pol III) low processivity— it does not hold on to the DNA template strand firmly and detaches relatively soon after the RNA has been replaced. It does not have the annular clamp to hold it on to the DNA. This is essential for otherwise it would go on replacing long stretches of the newly synthesized DNA. When it detaches, a nick is left in the chain, which is healed by a separate enzyme, called **DNA ligase**.

This enzyme synthesizes a phosphodiester bond between the 3′-OH of one DNA fragment and the 5′-phosphate of the next, a process requiring energy. In some prokaryotes and eukaryotes, ATP supplies this. The mechanism is that the enzyme (E) accepts the AMP-group of ATP, liberating pyrophosphate, and transfers AMP to the 5′-phosphate of the DNA. Finally the DNA-AMP reacts with the DNA-3′-OH, releasing AMP and sealing the break.

$$E + ATP \rightarrow E\text{-}AMP + PP_i;$$
$$E\text{-}AMP + \textcircled{P} - 5'DNA \rightarrow E + AMP - \textcircled{P} - 5'DNA;$$
$$DNA3' - OH + AMP - \textcircled{P} - 5'DNA \rightarrow$$
$$DNA3' - O - \textcircled{P} - 5'DNA + AMP.$$

The AMP is linked to the enzyme and to the DNA via its 5′-phosphate group.

In *E. coli*, instead of ATP, NAD^+ is used. This is a most unusual role for NAD^+ which you have met only as an electron carrier. However, NAD^+ can donate an AMP group just like ATP and, for some reason, *E. coli* uses this route. What happens then, in summary, is the following.

The DNA Pol I binds at the attachment site shown in Fig. 20.16—at the nick. The polymerase adds DNA nucleotides to the 3′ end of the left-hand fragment in the diagram, checking each addition, and moves in the 5′ → 3′ direction. The RNA,

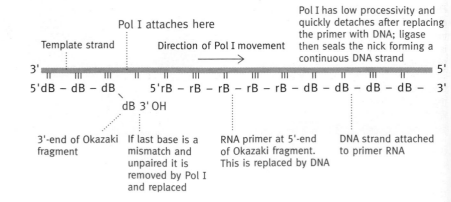

Fig. 20.16 Diagram of polymerase I actions in processing Okazaki fragments. dB, Deoxynucleotide; rB, ribonucleotide.

and some DNA, is nibbled away and the gap replaced by DNA. The Pol I detaches and a ligase joins the two fragments of DNA together. Thus, a series of Okazaki fragments becomes a continuous new DNA strand.

Proofreading by polymerase III

The base sequence of the newly formed DNA must be reproduced correctly, for even a single incorrect base could cause a lethal mutation. But, with such vast numbers of bases, mistakes are unavoidable.

When Pol III encounters, say, a G in the template strand, it is highly selective in that in the vast majority of cases only dCTP is accepted into the active site of the enzyme and added to the nascent DNA chain and so on with the other bases. The mechanism by which Pol III selects the correct nucleotide for the template base is most likely to be due to the fact that all Watson–Crick base pairs have the same geometrical shape, which is different from that of any other base pairs and which is precisely recognized by the enzyme. (Have a quick look at the illustration of this in Fig. 20.17.) Hydrogen bonding of correct pairs and base stacking also help the selection process since bond formation involves energy release.

There is a limit to the accuracy of such mechanisms; although Watson–Crick pairing is the predominant situation, in any collection of, for example, dATP molecules, a few will be in the imino form ($=NH$ instead of $-NH_2$) and this *can* base pair with C. The net result is that Pol III *can* make an error of DNA copying of about one base in 10^5–10^6. This would give much too high a rate of mutation. In *E. coli* the actual final error rate in DNA replication is much lower, about 10^{-10}.

To achieve a satisfactorily small error rate, Pol III also has a proofreading activity. It has a $3' \rightarrow 5'$ ('rear end') exonuclease activity that removes the last incorporated base if it is improperly paired, much as does Pol I. The major events involved in *E. coli* replication (as well as eukaryote replication, to be described shortly) are illustrated in Fig. 20.18.

Methyl-directed mismatch repair

Despite the proofreading mechanisms described above to achieve fidelity of DNA replication, in a system involving such vast numbers of nucleotide additions mistakes inevitably occur. Another backstop mechanism operates in *E. coli* to correct the

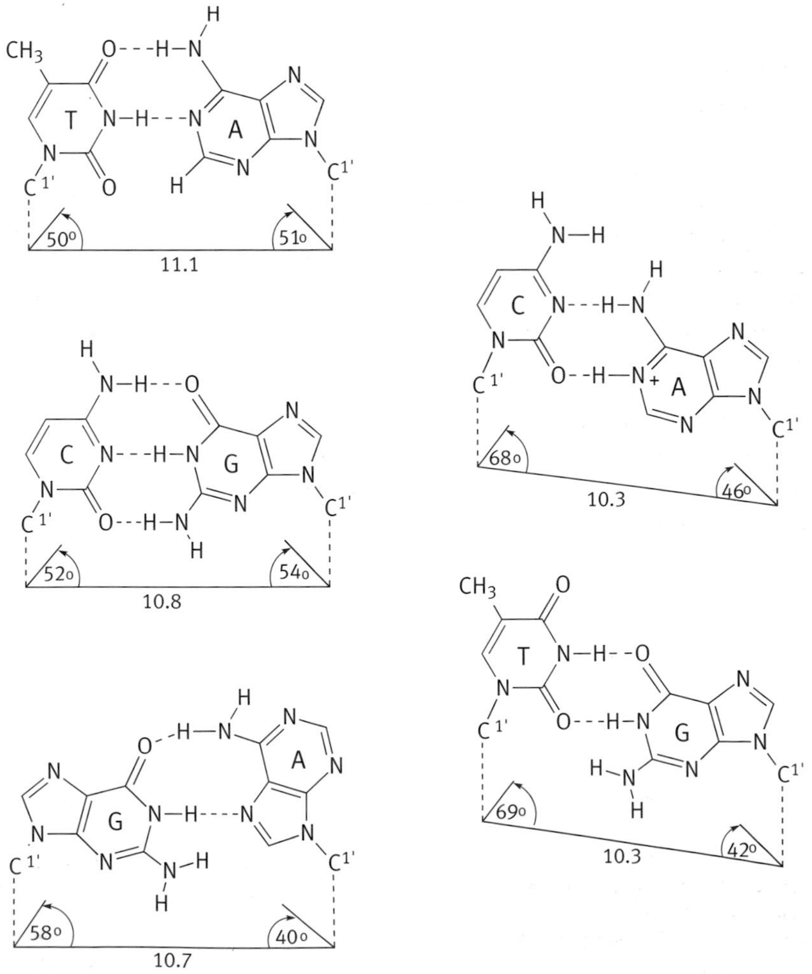

Fig. 20.17 Geometric characteristics of Watson–Crick and mismatched base pairs. The figure is based on X-ray crystallography of duplex B DNA oligonucleotides. The striking geometrical identity of the Watson–Crick pairs is not matched by the A–C and G–T wobble pairs or by the G (*anti*)–A (*syn*) pair. The term 'wobble pair' is explained on page 295. Distances (in nm) are those between the deoxyribose C–1 atoms of each pair, and angles are those of the glycosidic bonds.

Eukaryotes

Topoisomerase II
Relaxes positive
supercoils but cannot
insert negative ones

Topoisomerase I
Relaxes both
negative and
positive supercoils

Helicase—as in *E. coli*

SSB—as in *E. coli*

Polymerase δ on
leading strand;
Polymerase α on
lagging strand
contains **Primase**
activity

Polymerase γ in
mitochondrial
DNA
replicacation

E. coli

Gyrase (Topoisomerase II)
Relaxes positive supercoils
by insertion of negative ones

Helicase Unwinds double
helix at replicative fork,
causing positive supercoiling
ahead of fork

SSB
Single-strand binding protein

Primase—part of the
primasome. Synthesizes
primers for Okazaki fragment
synthesis

Polymerase III—one
molecule for each strand

DNA Polymerase I
Excises RNA primers from
Okazaki fragments and fills
in with DNA

DNA Ligase Joins nick in
DNA where Pol I lets go

Fig. 20.18 Diagram of events involved in DNA replication in *E. coli* and
in eukaryotes. (Initiation of replication and assemblage of the replicative
complex at the fork involves other proteins.)

mistakes that do occur in the base sequence of newly synthesized
DNA. If a mismatch has escaped the polymerase proofreading
correction, the error will cause some distortion in the duplex
chain illustrated diagrammatically below.

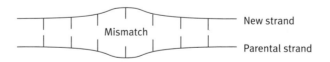

The error will be in the new strand and the repair system has to
recognize this and nick it as the first step in error correction.
The important point is that, in such a mismatch, the base on the
parental (template strand), by definition, is correct and it is the
complementary base on the new strand that is incorrect. The
repair system must discriminate between the two strands, for if
it replaced the template strand base it would confirm the
mutation. It must remove the base from the new strand and

correct it. How is this strand discrimination made? In *E. coli*,
wherever there is a GATC sequence in the DNA, the adenine of
this sequence is methylated by an enzyme in the cytoplasm
known as the Dam methylase. This does not affect base pairing
or DNA structure. It takes some time after its synthesis for the
new strand to be methylated and so, for this brief period, it is
unmethylated. Thus the parental strand is methylated but the
just-synthesized new strand is not. The repair system is little less

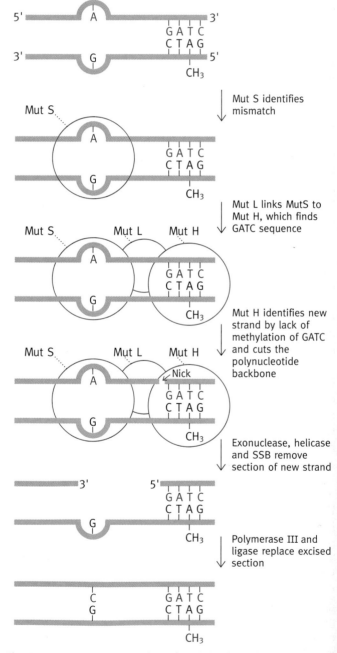

Fig. 20.19 Methyl-directed pathway for mismatch repair. If the GATC
sequence is distant from the error, bending of the DNA could bring the
two into proximity.

complex than that for DNA synthesis; first the protein Mut S recognizes the mismatch distortion in the helix; Mut H, linked to Mut S by Mut L, finds an *unmethylated* GATC on one of the duplex strands that identifies it as being newly synthesized and nicks it at this site (Fig. 20.19). This may be thousands of bases from the error. Then helicase, and SSB, and an exonuclease cooperate to remove the entire section from the nick to beyond the error and polymerase III replaces it with DNA. DNA ligase completes the repair by sealing the nick. To correct one base, thousands of nucleotides may be replaced (Fig. 20.19). This error correction system increases fidelity of replication as much as a thousandfold. There is evidence that mismatch repair occurs in eukaryotes. In humans, proteins corresponding to Mut S and Mut L proteins of *E. coli* are known. Mutations in the genes for these are associated with increased risk of cancer.

Repair of DNA damage in *E. coli*

The mechanisms described above ensure that DNA is replicated with the almost incredible degree of accuracy needed to ensure continuity of life. However, there is still a major problem. The base sequence of DNA must be essentially immortal, for much of it carries information that must remain unchanged for vast time spans. But DNA is an ordinary molecule in chemical stability terms, and chemical changes are occurring all the time and at a rate that would result in large numbers of mutations per day, in each cell, if there weren't constant repair. Some of the damaging changes are spontaneous. The glycosidic link that binds the purine and, to a lesser extent, pyrimidine bases to the deoxyribose moieties is fairly unstable and depurination and depyrimidation occur—large numbers of purines (and a lesser number of pyrimidines) break off the DNA every day in a human cell. In addition, cytosine and adenine chemically deaminate to become uracil and hypoxanthine, respectively (see page 224 and Fig. 18.7 for structures). Also, DNA is subject to 'insults' by a variety of agents. Ionizing radiation, oxygen free radicals (page 216), carcinogens, UV light, all can alter the DNA and thereby cause mutations.

An important general principle is that it will be relatively infrequent for damage to happen to *both* strands of a duplex DNA molecule at exactly the same place in both chains (though it does happen). When only one strand is affected at a given place, there is always the other strand to act as template and 'direct' the repair of the damaged part. A variety of repair systems exist in cells to repair DNA damage for, of all things, this is the area where maintaining the integrity of a molecule is of paramount importance. When other molecules such as proteins are damaged they are destroyed, but DNA must be repaired at all costs. There are different types of repair systems.

- **Direct repair.** Exposure of DNA to ultraviolet light can result in the covalent linking of two adjacent thymine bases (on the same strand), forming a T dimer.

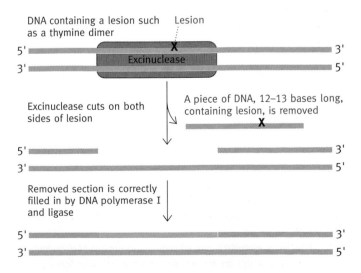

T dimer

In *E. coli*, the abnormal bonds are cleaved by a light-activated mechanism that restores the two thymine moieties to their original form. The system is widespread and probably of importance in plants. Alkyl groups on bases (which are formed by some mutagenic agents) may be directly removed by a 'suicide enzyme' that accepts the alkyl group and in so doing destroys its own action. It is more of a specific protein reagent than an enzyme.

- **Nucleotide excision repair.** Lesions that distort the double helix such as a T dimer can also be repaired by the excision of a short stretch of nucleotides, including the lesion, followed by its correct replacement, the opposite strand serving as the template for this. In *E. coli*, an unusual endonuclease (called *excinuclease*, or the *uv*rABC complex after the three genes coding for the enzyme) cuts the DNA on both sides of the lesion and removes a single-stranded section of 12–13 nucleotides (Fig. 20.20). DNA polymerase I attaches to the nick and adds nucleotides to the 3′ end of the nicked chain; ligase heals the nick. The system depends on it being possible to recognize which strand of the DNA is faulty.

- **Base excision repair and AP site repair.** Deamination converts cytosine to uracil and adenine to hypoxanthine. DNA glycosylases recognize the abnormal bases and hydrolyse them off, leaving AP (apurine or apyrimidine) sites in which the deoxyribose has no base attached to it

Fig. 20.20 The pathway of nucleotide excision repair in *E. coli*.

(Fig. 20.21). AP sites can also be formed spontaneously, since the purine–deoxyribose link especially is somewhat unstable. Repair of AP sites involves nicking of the polynucleotide chain adjacent to the lesion followed by replacement of the section containing the latter by DNA polymerase I and sealing by ligase.

The need to remove uracil formed from cytosine probably explains why DNA has T, instead of U. Remember that T is, in essence, simply a U tagged for identification purposes with a methyl group (page 224). If DNA *normally* contained U, it would be impossible to distinguish between a U that should be there and an 'improper' U, formed by deamination of C.

Cytosine   Uracil

Using T in DNA, instead of U, solves the problem. (As described in the next chapter, U can be used in RNA, because RNA has a relatively short lifetime, is much smaller, and errors do not have the same long-term consequences. Hence RNA is not repaired.)

There is rather a nice correlation; in *E. coli* cells there can be lots of nicks in DNA requiring Pol I to attach to and carry out repair. However, there are only a small number of replicative forks requiring Pol III. In an *E. coli* cell there are only about 10–20 molecules of Pol III but many more molecules of Pol I.

The machinery in the eukaryote replicative fork

The general principles of DNA replication in *E. coli* apply to eukaryotes, but there are differences in detail, especially that different polymerases synthesize the leading and lagging strands, unlike the Pol III dimer in *E. coli*.

In place of the *E. coli* polymerases I, II, and III, there are five DNA polymerases in eukaryotes, which are given Greek letter designations. Polymerases α and δ synthesize nuclear DNA (cf. Pol III in *E. coli*). Pol α has primase associated with it, but no demonstrated $3' \rightarrow 5'$ 'rear end' exonuclease. How fidelity is achieved is an open question. Polymerase δ has a $3' \rightarrow 5'$ exonuclease activity. It seems that polymerases α and δ form the replicative assembly in eukaryotes—δ for the leading strand. The δ polymerase of yeast has associated with it an annular sliding clamp, known as the PCNA (for proliferating cell nuclear antigen), as does the *E. coli* enzyme, conferring processivity on it. In Fig. 20.14 the structures of *E. coli* and yeast 'clamps' are shown in ribbon diagrams. Despite their almost identical

appearance, the two have very little amino acid sequence similarity—the *E. coli* clamp is a dimer while the yeast one is a trimer, suggesting they arose by independent evolutionary designs.

Two of the other polymerases (ε and β) in eukaryotes have repair function. The last, polymerase (γ) is for mitochondrial DNA replication.

The problem of replicating the ends of eukaryote chromosomes

Eukaryote chromosomes are linear and this poses a problem not encountered in the replication of circular chromosomes.

Consider the replication of the chromosome shown in Fig. 20.22 as a very short one for diagrammatic convenience. Let us

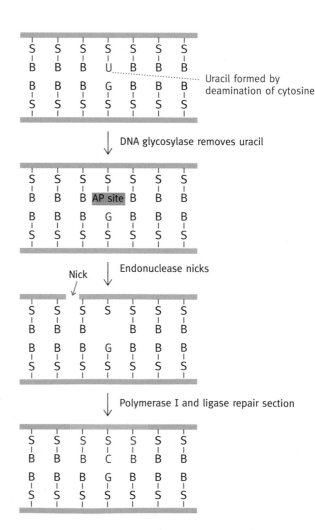

Fig. 20.21 Illustration of AP site formation and repair. In the example given, the site is created by removal of a uracil by a glycosylase, but sites are also formed by spontaneous hydrolysis of purine bases (and, to a lesser extent, of pyrimidine bases) from the nucleotide. S, sugar; B, base.

assume, again for simplicity, that it is replicated (in a bidirectional manner, don't forget) from a single initiation site in its centre (but remembering that a eukaryote chromosome has many such sites in actual fact). The 5′ end of each strand is fully replicated by leading strand synthesis. This is not true of the 3′ ends, which are not fully replicated by lagging strand synthesis, because the synthesis of the end Okazaki fragments requires RNA primers to be laid down as shown on the 3′ ends of the template strands. When the primers are removed it leaves these ends unreplicated and no mechanism exists by which it could be replicated by the DNA synthesizing machinery that we have described so far. To fill in the missing parts would require RNA primers but there are no templates against which they could be laid down. This means that, with cell division, chromosomes would become progressively shorter; a more potentially disastrous situation could hardly be imagined. The mechanism of DNA synthesis means that incomplete replication of the double-stranded DNA cannot be avoided. The solution adopted is that, at the ends of eukaryote chromosomes stretches of special DNA are attached in which the DNA has no informational content but is sacrificial DNA, called **telomeric DNA** (the ends of the chromosomes containing it are called

telomeres). The ends will still not be replicated by the DNA synthesis machinery so far described, but it no longer matters for the 'real' chromosome is fully replicated with the primer at the 3′ end of the template strand being laid down against telomeric DNA. In rapidly dividing cells, moreover, the telomere is added to so that the chromosome is never at risk.

How is telomeric DNA synthesized?

A telomere consists of repeating short stretches of bases—it varies between species. In humans there are hundreds of repeats of the TTAGGG sequence. The enzyme **telomerase** adds these sequences one after the other to the 3′ end of pre-existing telomeric DNA and so extends the latter (all chromosomes have telomeres to start with). Telomerase is a type of enzyme you have not met in this book before. It has two remarkable features: (1) it uses RNA as the template for DNA synthesis—it is a **reverse transcriptase** (see page 319 for a fuller explanation); and (2) it carries its own template in its structure. This RNA carries the sequence complementary to the TTAGGG sequence (in humans). The RNA template hybridizes to the end of a repeating unit (Fig. 20.23), which positions the template bases for the repeating unit correctly, and a unit is then added, one base at a time. When this is done the enzyme moves along and hybridizes to the end of the new repeating unit and thus the telomere is constructed in a discontinuous manner. The complementary strand to the telomeric extension, added as described, is synthesized so that the telomere is double-stranded, but there is some uncertainty as to how this occurs; a DNA polymerase is presumably involved.

The necessity for telomeres has been demonstrated by the use of **YACs** or **yeast artificial chromosomes**. These contain the three types of DNA essential for chromosome replication—centromeres, sites of origin, and telomeres. It was shown that YACs are correctly maintained for generations when inserted into yeast cells but that, when they lack telomeric ends, they disappear in time from the cells.

There is not the same problem in prokaryotes, for in a circular chromosome there is always a DNA template available for priming. The existence of telomeres in eukaryotes has profound potential significance. Telomerase has been found in germ cells, which are continuously rapidly dividing, and it might be expected that continual lengthening of the telomeres would be necessary to compensate for the continuing shortening. However, in somatic cells where cell division is slower, additions to the telomeres would be less appropriate and, indeed, telomerase is believed not to be present in such cells. If this is so, the implication could be that somatic cells receive their initial 'ration' of telomeric DNA to suffice for the lifetime of that cell and its progeny. This leads to the speculation that the telomeres of most cells become shorter with age and this could be an important factor in determining the possible lifespan of species. Somatic cells can be transformed to the rapidly dividing state if

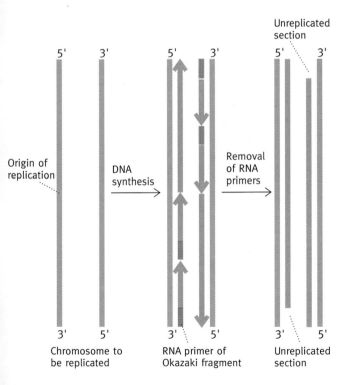

Fig. 20.22 Diagram illustrating the shortening of linear chromosomes by replication. For diagrammatic convenience the bidirectional replication of a very short piece of DNA is represented. It should be noted that primer removal from Okazaki fragments is a continuous process—it is represented here, for clarity, as occurring as a separate event. A typical chromosome will have multiple origins of replication. The red lines represent new DNA synthesis; the green lines the RNA primers of Okazaki fragments.

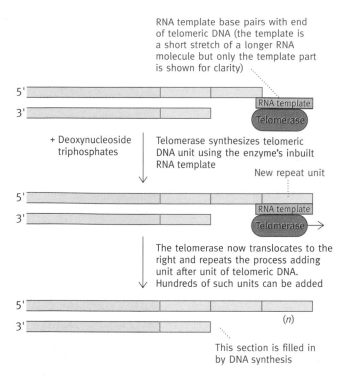

RNA template base pairs with end of telomeric DNA (the template is a short stretch of a longer RNA molecule but only the template part is shown for clarity)

+ Deoxynucleoside triphosphates

Telomerase synthesizes telomeric DNA unit using the enzyme's inbuilt RNA template

New repeat unit

The telomerase now translocates to the right and repeats the process adding unit after unit of telomeric DNA. Hundreds of such units can be added

This section is filled in by DNA synthesis

Fig. 20.23 Mechanism by which telomerase synthesizes telomeric DNA. The blue lines indicate the chromosomal or informational DNA (the 'real' chromosome) and the red lines the pre-existing telomeric DNA at one end of a chromosome. The telomerase has an inbuilt short RNA molecule that contains the sequence complementary to the repeating unit characteristic of the species. The enzyme becomes positioned with the RNA pairing with the terminal bases of the pre-existing telomere and adds one repeating unit of TTAGGG (in the case of humans) one base at a time. Synthesis is, as always, in the $5' \rightarrow 3'$ direction. The enzyme moves so that the RNA template is now paired with the end bases of the new repeating unit and a further unit is added, and so on. The newly synthesized telomeric DNA acts as the template for filling in the opposite strand, probably involving a DNA polymerase, so that the telomere is double-stranded.

they become cancerous; such cells in effect become immortal—there is no limit to their ability to divide in culture. It is interesting that telomerase has been reported to be present in cancer cells.

DNA damage repair in eukaryotes

The dealkylating 'suicide enzymes' described for *E. coli* occur in human cells. Excision repair also occurs in eukaryotes. As an example, in humans with the genetic disease xeroderma pigmentosum, normal excision repair of pyrimidine dimers does not occur. A thymine dimer is formed by covalent joining of two adjacent thymine bases (on the same strand) as a result of UV light on skin. In patients with this disease, exposure to sunlight causes cancerous skin lesions. Excision repair is known to be of general importance in humans. Genes for proteins corresponding to Mut S and Mut L of *E. coli* (page 262) have

been found; mutations in these genes have been found to be associated with the development of cancers.

Eukaryote repair systems play a major role in maintaining integrity of the genome. Whether ageing in humans is associated with decreased repair and a consequent increase in mutations is an interesting question that has attracted research effort. See also the section on mitochondrial mutations below.

Transposons or jumping genes

There is a phenomenon that is different from anything described so far in this chapter, in which a piece of DNA jumps from one place in the chromosome to another place in the same or different chromosome. The existence of **jumping genes** was first detected by McClintock in genetic studies on maize; she concluded that gene control elements moved from one place to another in the genome and influenced the expression of genes, giving rise to phenotypic variations. This work was largely ignored for about 30 years until studies in *E. coli* confirmed that genes do, in fact, move around, and a Nobel Prize was awarded to McClintock.

The simplest type of transposon in *E. coli* is called an **insertion sequence (IS)**. It consists of a stretch of DNA with a gene for transposase, an enzyme that catalyses the transposition event. From one point of view, an IS might be regarded as the ultimate in futility—a gene that does nothing but code for its own movement from place to place; it could be regarded as a parasite. An IS can jump to any place in a chromosome—there are no specific insertion sites for this class of mobile gene. If a transposon is inserted into another gene the latter is inactivated because the gene sequence is thereby disrupted, but the frequency of transposition is, however, low. In addition to the gene for the transposase, the insertion sequence has at the ends **inverted terminal repeats**—these consist of short stretches of bases oriented as shown in Fig. 20.24. The insertion process results in a short sequence of the recipient DNA being replicated to give non-inverted repeats.

There are many variations on the simple case given above. One variation is that, instead of the transposon itself transferring to a new location, it is replicated and the copy inserted elsewhere, leaving the original one where it was. The replication involves a reverse transcriptase—an enzyme that copies RNA into DNA (described in more detail on page 319). The creation of multiple identical sequences in the genome in this way could provide the opportunity for homologous recombination. It is possible that chromosomal rearrangement caused by mobile genes is of significance in the evolutionary process.

There are viruses that integrate into the chromosome of the host cell—HIV is one and lambda bacteriophage another. The former integrates anywhere; the latter only at a specific site in the *E. coli* chromosome. We will leave a discussion of these to Chapter 23 which deals with viruses.

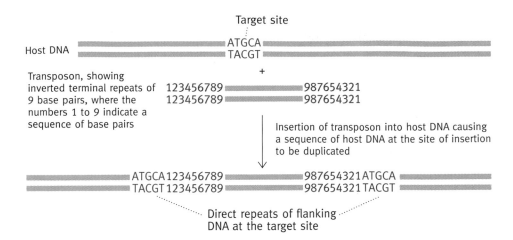

Fig. 20.24 Diagram of the insertion of a transposon (IS type) into the host chromosome.

Is the mechanism described above the only way in which DNA is synthesized?

As is so common, evolution has exploited many avenues to achieve the same ends. Some viruses, for example, use a protein instead of RNA for DNA chain synthesis priming. However, as already indicated, retroviruses have quite a fundamentally different mechanism in that the RNA genome is copied into DNA. This will be described in Chapter 23 on viruses. It is particularly interesting that mitochondria have their own DNA, which must be replicated for mitochondrial multiplication. There are no DNA repair systems in mitochondria and this presumably reflects the facts that the chromosome is very much smaller with fewer chances for errors and that, since there can be large numbers of mitochondria per cell, a small proportion of faulty ones has less significance. The absence of repair, however, means that mutations in mitochondrial DNA may accumulate with time and possibly be related to the process of ageing (see page 218). A large number of genetic diseases are now known to be caused by mitochondrial mutations.

Further reading

Books

Kornberg, A. and Baker, T. A. (1992). *DNA replication.* 2nd edn. W. H. Freeman.

Lewin, B. (1994). *Genes V.* Oxford University Press.

Singer, M. and Berg, P. (1991). *Genes and genomes: a changing perspective.* University Science Books.

DNA polymerase

Kelman, Z. and O'Donnell, M. (1995). DNA polymerase III holoenzyme: structure and function of a chromosomal replicating machine. *Ann. Rev. Biochem.,* **64**, 171–200.
A beautiful review of this complex enzyme, its subunit structure, sliding clamps, roles in leading and lagging strand synthesis. Very concise.

Replication

Coverley, D. and Laskey, R. A. (1994). Regulation of eukaryotic DNA replication. *Ann. Rev. Biochem.,* **63**, 745–76.
Describes the origins of replication, including mechanism and cell cycle control of DNA replication. Advanced but readable account.

Li, J. J. (1995). Once and once only. *Current Biology,* 5, 472–5.
Discusses the essential link between the eukaryote cell cycle and initiation of sites of origin of DNA replication.

Wyman, C. and Botcham, M. (1995). A familiar ring to DNA polymerase processivity. *Current Biology,* 5, 334–7.
Reviews the sliding clamps in DNA replication. Beautifully illustrated.

Telomeres

Greider, C. W. and Blackburn, E. H. (1996). Telomeres, telomerase and cancer. *Sci. Amer.,* **274**(2), 80–5.
Discusses the problem of DNA end-replication, DNA shortening and how telomerase protects chromosomal end segments. Possible relevance to ageing and cancer discussed.

Repair

Echols, H. and Goodman, M. F. (1991). Fidelity mechanisms in DNA replication. *Ann. Rev. Biochem.,* **60**, 477–511.
The article is an interesting advanced review on base insertion selectivity by DNA polymerases, proofreading by DNA polymerase, and mutations generated by damage to DNA.

Demple, B. and Karran, P. (1983). Death of an enzyme: suicide repair of DNA. *Trends Biochem. Sci.,* **8**, 137–9.
Reviews dealkylation DNA repair enzymes.

Mitochondrial DNA abnormalities

Hammans, S. R. (1994). Mitochondrial DNA and disease. *Essays in Biochemistry*, **28**, 99–112.

A timely discussion of diseases arising from abnormalities in mitochondrial DNA.

Problems for Chapter 20

1   What is a replicon?

2   In separating the strands of parental DNA during replication, what topological problem occurs?

3   How is the problem referred to in the preceding question solved, both in *E. coli* and eukaryotes?

4   By means of diagrams, explain the actions of topoisomerases I and II.

5   Eukaryotes have no topoisomerase capable of inserting negative supercoiling into DNA and yet eukaryote DNA is negatively supercoiled. Explain how this is brought about.

6   What are the substrates for DNA synthesis?

7   Why is TTP used in DNA synthesis—why not UTP as in RNA?

8   **(a)** Can a DNA chain be synthesized entirely from the four triphosphate substrates? Explain your answer.
(b) In which direction does DNA synthesis proceed? Explain your answer so as to be totally unambiguous.

9   What are the thermodynamic forces driving DNA synthesis?

10   *E. coli* DNA polymerase III is highly processive; explain what this means, how it is achieved, and why it should be so.

11   *E. coli* polymerase I is a complex enzyme. Describe its different activities and explain their roles in DNA synthesis.

12   Discuss the mechanism by which DNA polymerase III of *E. coli* achieves a high standard of fidelity in DNA synthesis.

13   The proofreading activity of *E. coli* polymerase III is important, but insufficient to give a sufficiently high fidelity rate. If an improperly paired nucleotide is incorporated giving a mismatch, it has to be replaced. This demands that the repair system recognizes which of the two bases in the mismatch is wrong. How is this done and how is the problem fixed? Does this mechanism exist in humans?

14   Explain what a thymine dimer is, how it is formed, and how it is repaired.

15   Explain how eukaryote chromosomes become shortened at each round of replication.

16   Explain how the DNA shortening problem in replication is coped with.

Chapter 21

Gene transcription—the first step in the mechanism by which genes direct protein synthesis

The information in DNA, encoded in the sequence of the four bases, is used to direct the assemblage of the 20 amino acids in the correct sequence so as to produce the protein for which a given gene is responsible. In case you wonder how four bases can code for 20 amino acids, and to anticipate the next chapter, the system is that a sequence of three bases codes for an amino acid. Since with four bases $4 \times 4 \times 4$ triplets are possible, there is no shortage of coding ability. A gene does not participate directly in protein synthesis; in eukaryotes the DNA is enclosed inside the nuclear membrane while the protein-synthesizing machinery is outside in the cytoplasm and the two never meet. How then does such a gene direct protein synthesis? It does so by sending out copies of its coded information to the cytoplasm. (In *E. coli* the copy is immediately in contact with the cytoplasm.) Since the information is in a sequence of bases it follows that the copy must also be a nucleic acid, but in this case it is RNA rather than DNA. There are several types of RNA and this one, for obvious reasons, is called **messenger RNA** or **mRNA** for short, or it is often referred to as 'messenger'.

We had better, therefore, now describe this mRNA.

Messenger RNA

The structure of RNA

RNA stands for ribonucleic acid. It is a polynucleotide essentially the same as DNA but with these differences:

- The sugar is ribose, not the deoxyribose of DNA. It has an —OH in the $2'$ position.

D-Ribose   2'-Deoxy-D-ribose

- mRNA is single-stranded, not a duplex of two molecules as is DNA. From this, you will infer that mRNA is a copy of only one of the two strands of the DNA of a gene, a point elaborated upon later.
- Its four bases are A, C, G, and U. There is no T. As we described earlier (page 224), U and T are identical in base-pairing properties—they both pair with A. You should regard T as U that has been tagged with a methyl group without changing its properties (see page 264 for the probable reason for the methyl group).

Were it not for these differences the structure of a single strand of DNA, shown on page 241, could be that of single-stranded RNA. There are the same $3' \rightarrow 5'$ phosphodiester bonds between successive nucleotides. There is a $5'$ end lacking a nucleotide substitution on the $5'$-OH and the $3'$ end lacks a nucleotide attachment on the $3'$-OH. The $2'$-OH group makes the molecule more unstable than DNA (page 241). In dilute alkaline solution at room temperature, RNA is destroyed while DNA is unaffected.

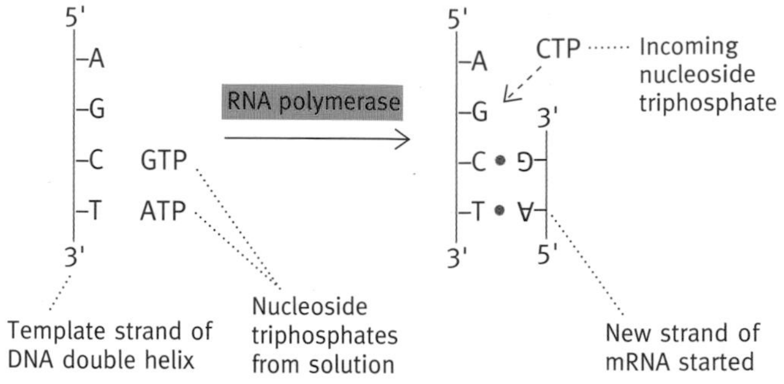

Fig. 21.1 Copying mRNA from a DNA template strand. The non-template strand is not shown. Note that the separation of the two strands is transitory. A bubble of DNA strand-separation moves along the DNA as the polymerase progresses along it.

How is mRNA synthesized?

The building block reactants for RNA synthesis are ATP, CTP, GTP, and UTP, which are produced in all cells (Chapter 18). mRNA is synthesized from these by a single enzyme in *E. coli*, DNA-dependent RNA polymerase, commonly referred to simply as **RNA polymerase**. In eukaryotes three such polymerases exist. The synthesis requires that the duplex DNA strands are separated so as to provide a single-stranded template for directing the sequence of nucleotides to be assembled into mRNA. The two strands are transitorily separated at the site of mRNA synthesis, and then come together again after the polymerase has passed. In effect, a separation 'bubble' moves along the DNA. The basic process of synthesis is much the same as in DNA synthesis in that the base of the incoming ribonucleotide is complementary to the base on the DNA template (Fig. 21.1).

The RNA polymerase works its way along the template, joining together the nucleotides in the correct order as determined by the DNA template. RNA synthesis is always in the $5' \rightarrow 3'$ direction (as is the case also in DNA synthesis). That is, new nucleotides are added to the 3'-OH and so the chain elongates in the $5' \rightarrow 3'$ direction. The template is antiparallel, running in the opposite $(3' \rightarrow 5')$ direction. The chemical reaction catalysed by the polymerase involves the transfer of the α-phosphoryl group (that is, the first one attached to the ribose) of the nucleotide triphosphates to the 3'-OH of the preceding nucleotide, splitting off inorganic pyrophosphate. The latter is hydrolysed to two P_i molecules, thus making the reaction, shown in Fig. 21.2, strongly exergonic.

An important point is that RNA polymerase *can* initiate new chains—it does not need a primer; it can synthesize the entire mRNA molecule from the four nucleoside triphosphates, provided a DNA template is there. This is quite different, we remind you, from the situation in DNA synthesis where primer is always required for DNA polymerase activity (page 256).

Some general properties of mRNA

In a typical chromosome there are thousands of different genes. An mRNA molecule is coded for by a single gene (or, in prokaryotes, often by a small group of genes) and, therefore, large numbers of different mRNA molecules are formed in the cell. It will be clear that, while the DNA molecule is vast in length, an mRNA molecule is minute in comparison. In the cytoplasm the mRNAs direct the synthesis of proteins for which their respective genes code.

DNA is immortal in cellular terms, but mRNA is ephemeral with a half-life of perhaps 20 minutes to several hours in

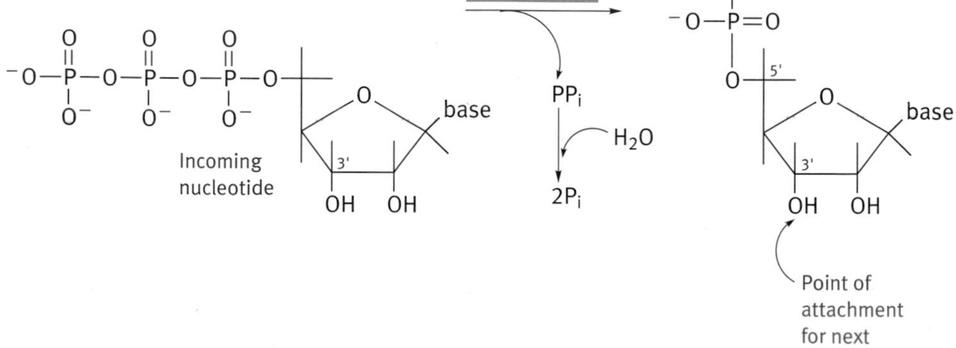

Fig. 21.2 The reaction catalysed by RNA polymerase.

eukaryotes and about 2 minutes in bacteria. Thus, for expression of a gene (the term 'expression' means that the protein coded for is actually being synthesized), a continuous stream of mRNA molecules must be produced from that gene. The gene, as it were, 'stamps out' copy after copy, RNA polymerase being the stamping machinery. This might seem wasteful but it gives the important benefit of permitting control of the expression of individual genes. Destruction of mRNA is the main 'off switch' in protein synthesis once a gene ceases to produce mRNA (how the gene is controlled is a major topic discussed later in this chapter).

The genes of an *E. coli* cell are in contact with the cytoplasm since there is no nuclear membrane. In eukaryotes, mRNA always corresponds to single genes (except in certain viruses) but in prokaryotes it may carry the coded instructions for the synthesis of several proteins, all joined together in a single RNA molecule. This is called **polycistronic mRNA** after the genetic term, **cistron**, which effectively refers to the coding for one polypeptide. Such clustered genes giving rise to a single mRNA results in the coordinated expression of several genes whose proteins function as a unit, such as the enzymes of a metabolic pathway. There are other very significant differences between mRNA production in prokaryotes and eukaryotes that will be described later.

Some essential terminology

The flow of **information** in gene expression is

　　(transcription)　　　　(translation)

DNA　− − − →　mRNA　− − − →　protein.

(Please be careful to note that the broken arrows represent *information* flow, not chemical conversions. DNA cannot be converted into RNA nor RNA into protein.)

The 'language' in DNA and RNA is the same—it consists of the base sequences. In copying DNA into RNA there is transcription of the information. Hence mRNA production is called gene transcription, or more often, simply **transcription**, and the DNA is said to be transcribed. The RNA molecules produced are called transcripts. The 'language' of the protein is different—it consists of the amino acid sequence whose structures and chemistry are quite different from those of nucleic acid. The synthesis of protein, directed by mRNA, is therefore called **translation**. If you copy this page in English you are transcribing it. If you copy it into Greek you are translating it.

We have so far talked of mRNA synthesis as 'copying' the DNA. The template DNA strand is 'copied' only in the complementary sense as you have seen from its method of synthesis, which is dependent on Watson–Crick base pairing of incoming ribonucleotides to the template bases. A in the template becomes U in the copy and so on. The terminology in this area can be confusing, especially if you consult different

texts. The DNA strand that acts as the template for mRNA synthesis is, not surprisingly, called the **template strand**, the other one the **nontemplate strand**. There's nothing ambiguous about that, but other terms are commonly used—'coding' and 'noncoding' strands and 'sense' and 'nonsense' strands, and the terms have quite different and contradictory usages in different texts. There is no 'authorized' version, but it is important to know what is being implied by the terms used. In this book we will base nomenclature on the mRNA. This has the information for the sequence of amino acids in a protein; it therefore carries the sense or message of the gene to the translational machinery. The base sequence of the mRNA is the same (apart from the T → U switch) as that of the nontemplate DNA strand and, *therefore, we will call the nontemplate strand, the coding or sense strand. The template strand therefore is defined here as the nonsense or noncoding strand* (Fig. 21.3). *In viruses the template strand is often called the minus strand (−) and the nontemplate the plus strand (+).*

A note on where we go from here

So far we have dealt with gene expression only in general terms in that we have referred to the gene only as supplying template DNA. However, there are many important and interesting questions such as how the genes that are to be expressed are selected out of the huge collection available in a cell. In transcribing a gene, which is part of a huge chromosome, how does the RNA polymerase 'know' where to start copying the DNA and where to stop? How is the rate of gene expression controlled? A protein coded for by one gene may be produced in large amounts and another in tiny amounts or not at all; and some genes may be expressed at one time and not at other times. How are these situations achieved? To answer these questions we now must look at the structures of the actual genes, by which, of course, we mean the base sequences of genes.

In this book we have, so far, largely concentrated on the biochemistry of animal cells for simplicity and also because the differences between eukaryotes and prokaryotes tend to be more at the level of detail in many areas of biochemistry rather than involving different fundamental principles. However, when it comes to gene expression, the differences are sufficiently great to warrant separate treatments. We will deal with *E. coli* first.

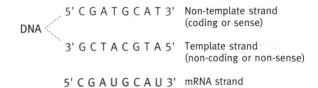

Fig. 21.3 Relationship of transcribed mRNA to template and nontemplate strands of DNA terminologies. There is no consensus in texts about the terminologies, but the system used here refers everything to the mRNA as carrying the sense of the information (see text). In viruses a frequently used terminology is: template, minus (−) strand; nontemplate, plus (+) strand.

Gene transcription in *E. coli*

What is specified by the term 'gene' in prokaryotes?

In Chapter 19 (page 248) we described what a gene is in molecular terms—namely, a section of a large DNA molecule. We now have to refine this description for reasons that will become clear.

A gene functions, as already mentioned, only by acting as a template for the transcription of RNA molecules whose base sequence corresponds to the nontemplate DNA strand. A gene is a specific section of DNA that is transcribed into RNA. There is a handful of genes whose function is to code for RNA molecules that are not messengers—these are the ribosomal and transfer RNAs whose role in protein synthesis is described in the next chapter. The rest of the vast array of genes produce mRNA molecules whose base sequences code for the amino acid sequences of specific proteins. How mRNA functions is the story of the next chapter. However, an mRNA molecule has sections at each end that are not translated into protein. The 5' **untranslated region** (UTR) contains encoded signals necessary for initiation of translation and the 3' untranslated region signals for its termination (again described in the next chapter), so that a gene includes sections of DNA that are transcribed into these regions as well as that for the protein itself.

This does not complete the list of DNA regions associated with the gene, for, in addition, there is a region of DNA adjacent to the 5' end of the gene, called the **promoter**, that is essential for transcription of the gene, but is not itself transcribed into RNA. At the opposite (3') end is a **terminator region** necessary, as the name implies, for termination of transcription and that also is not itself transcribed into RNA. A typical gene is illustrated in Fig. 21.4. There is a 'start' site where transcription begins; it provides the template for the first nucleotide of the mRNA. The first template nucleotide is given the number +1 and the nucleotide 5' to this −1 and so on. The start site is illustrated by an arrow (→) that indicates the direction of transcription. Nucleotides 5' to this are referred to as 'upstream'

and 3' to this as 'downstream'. You will notice that Fig. 21.4 includes one end of the gene being labelled 5'. This brings us to the next question.

What do we mean by the 5' end of a gene?

A typical gene has two strands of DNA of opposite polarity, and **duplex DNA** has therefore no intrinsic polarity. How then can we refer to one end of a gene being 5'? The 5' end of a gene is the end containing the promoter. Since RNA synthesis always proceeds in the $5' \rightarrow 3'$ direction, the 5' end of a gene corresponds with the 3' end of the template strand (but note that there are no physical ends—the DNA strand continues to the next gene).

A point to note is that, when we illustrate a *single* gene on a chromosome in a diagram, the 5' end is usually placed to the left. However, the template strand on one gene may not be the same DNA strand as the template strand of the next gene; strand usage can switch from one gene to the next. If the opposite strand is used as template on a second adjacent gene, then transcription would occur in the opposite direction, since synthesis occurs in the $5' \rightarrow 3'$ direction.

Phases of gene transcription

There are three phases—initiation, elongation, and termination. Initiation is by far the most complex.

Initiation of transcription in *E. coli*

In the promoters there are short stretches of bases that are called 'boxes' or 'elements'. These are accepted terms but the stretches of bases are not in any sense boxes nor do they have anything to do with atomic elements. In a typical *E. coli* promoter there are two boxes—the **Pribnow box** (named after its discoverer) centred at nucleotide − 10 and the other box centred at − 35. (The numbering is explained above.) The consensus sequences of the boxes are shown in Fig. 21.5. A **consensus sequence** for a box is obtained by determining the sequence of, for example, the Pribnow box in a number of different genes, for they are often not exactly the same. You then look at the first nucleotide position in the box and count up which base is most often used by the different genes and so on for all the other positions. In fact, the consensus sequence itself might never actually occur in any gene but the variation from it will be small. Single sequences are always given in the $5' \rightarrow 3'$ direction but, of course, in a gene there is the second DNA strand. Thus, although we say the

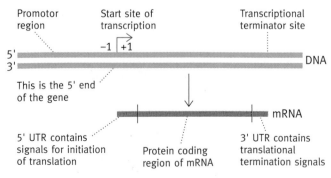

Fig. 21.4 Geography of a prokaryotic gene and its mRNA. Note that the 5' end of a gene is referring to the nontemplate strand or sense strand. In an adjacent gene, the other strand could be the nontemplate strand with transcription occurring from right to left.

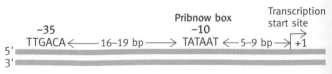

Fig 21.5 Consensus sequences of *E. coli* promoter elements.

Pribnow box has the sequence TATAAT, it really is

5′ – – TATAAT – – 3′
3′ – – ATATTA – – 5′

It is the *double strand* that is recognized by the proteins that control transcription (see below).

Correct initiation of transcription is obviously important. Synthesis of an mRNA must commence at the correct nucleotide on the template. The question is how the RNA polymerase is positioned in the correct place to start transcribing a gene. The − 35 and Pribnow boxes are the signals for this. DNA-dependent RNA polymerase of *E. coli* is a large protein complex. In the 'core' enzyme there are four subunits, and it has an affinity for any stretch of DNA to which it attaches at random, but it cannot recognize the correct initiation site. When it is joined by another protein from the cytoplasm, the sigma protein (σ) or **sigma factor**, the resultant **holoenzyme** loses much of its affinity for random DNA but binds tightly to the − 35 and Pribnow boxes and initiation of transcription can start. (We remind you that, although the DNA bases are Watson–Crick paired in the centre of the duplex, their edges are still 'visible' in the DNA grooves and can be recognized by proteins designed to do so; see page 243 and Fig 19.4.) This aligns the enzyme in the correct starting position, and the correct orientation (that is, pointing in the right direction). It can now separate the DNA strands and initiate. A 'bubble' of separated DNA strands is formed (about one and a half turns of the helix in length), thus making the template strand bases available for pairing with incoming bases of NTPs. The enzyme now synthesizes the first few phosphodiester bonds from nucleoside triphosphates and initiation is thus achieved. At this point the sigma protein flies off (to be used again) and the polymerase, now released, moves down the gene synthesizing mRNA at the rate of about 40 nucleotides per second until it reaches the terminator. It is not known what causes the sigma protein to fly off.

Untwisting the DNA

Gene transcription requires strand separation of the DNA double helix, but only in a transitory way, the strand re-annealing once the section has been transcribed. It is presumably no coincidence that the Pribnow box with its high A and T content and therefore weak hydrogen bonding is where initial strand separation occurs. DNA unwinds ahead of the polymerase and re-winds behind it. Negative supercoiling (underwinding) of the DNA facilitates the process (see page 255). Thus a temporary unwound 'bubble' passes along the gene with the polymerase. The newly synthesized mRNA is paired with the template strand for a short distance but then detaches. Figure 21.6 indicates how the RNA polymerase moves along the DNA double helix.

Termination of transcription

At the end of some transcribed prokaryote genes are sequences

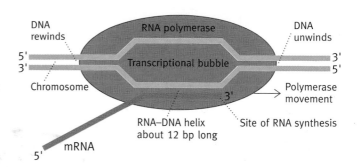

Fig. 21.6 Diagrammatic representation of DNA transcription by *E. coli* RNA polymerase. The polymerase unwinds a stretch of DNA about 17 base pairs in length forming a transcriptional bubble that progresses along the DNA. The DNA has to unwind ahead of the polymerase and rewind behind it. The newly formed RNA forms a RNA–DNA double helix about 12 base pairs long.

that result in the transcribed RNA having a **stem loop structure**. This needs to be explained. Although mRNA is single-stranded, it still has the thermodynamic obligation to maximize base pairing within itself. The newly synthesized mRNA is attached to the DNA template strand by base pairing. Near the end of the gene, the sequence of bases in the mRNA produced is such that the stem loop structure shown in Fig. 21.7 forms because the base sequence here permits formation of G–C pairs, a stable structure because of the triple bonding between G and C. Immediately following the stem loop structure in the mRNA transcript is a string of U residues, giving weak bonding of the RNA to DNA. This probably is to facilitate detachment of the mRNA and termination of transcription. The stem loop structure would prevent binding of the mRNA to the template at this point, since, if its bases are preferentially internally paired, they cannot pair with the template DNA. The string of Us would facilitate final detachment of the mRNA because of the weak A–U hydrogen bonding. Whatever the precise mechanism, this hairpin structure followed by several U residues gives precise termination.

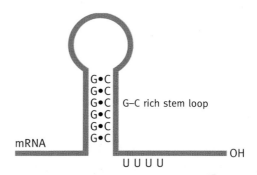

mRNA detachment from DNA template
facilitated by weak A–U pairing to the latter

Fig. 21.7 Diagram of the stem loop structure of an RNA transcript involved in the Rho-independent termination of gene transcription (see text).

There is a second method of termination of transcription in many prokaryote genes. This requires the assistance of a protein called Rho for the detachment of mRNA from the DNA–RNA hybrid. The **Rho factor** attaches to the newly transcribed mRNA and moves along it behind the RNA polymerase. At the termination site, the polymerase pauses, possibly because of a difficult·to separate G–C rich section of the DNA, and allows the Rho factor to catch up with the polymerase. The Rho factor has an unwinding (helicase) activity specific for unwinding the RNA–DNA duplex formed by transcription and so, having caught up with the polymerase, it detaches the RNA transcript and terminates transcription.

The mRNA directs protein synthesis; in fact, in *E. coli* it starts to do this before the full mRNA molecule is completed because the transcription occurs in contact with the cytoplasm.

However, we are not finished with *E. coli* gene transcription for there is still the question of gene control.

The rate of gene transcription initiation in prokaryotes

Genes that are **constitutively expressed** are those that are 'switched on' all the time and whose rates of expression are not selectively modulated. Such genes code for enzymes and other proteins that are needed at all times and in amounts that do not vary from time to time. However, amongst these constitutive proteins some will be required in much larger amounts than others. The major influence on the rate of gene expression in bacteria is the rate of mRNA production and this is largely determined by the frequency of initiation of transcription of a given gene. This varies because genes have promoters of different 'strengths'. A 'strong' promoter will initiate transcription frequently and thus cause many mRNA transcripts of the gene to be made and hence a lot of the specific protein. A 'weak' promoter has the reverse effect. The strength of a promoter is a function of the precise base sequence of the Pribnow and −35 boxes, the distance between them, and the nature of the bases in the +1 to +10 region. The greater the affinity of these regions for the polymerase, the stronger the promotion though this may not be the sole determinant.

Control of transcription by different sigma factors

A particularly neat method of controlling whole blocks of genes in prokaryotes is by using different sigma factors. Under certain conditions, the usual sigma protein is replaced by a different one which causes the RNA polymerase to initiate at a different set of genes. Such conditions include: (1) sporulating bacilli, in which a new sigma protein is produced after the signal of adverse conditions in the environment is received; this causes expression of a set of genes leading to sporulation; (2) a special σ factor that is produced in nitrogen starvation; (3) after a heat shock (a sudden rise in temperature), *E. coli* transitorily increases the synthesis, stability, and activity of a different sigma protein (σ^{32}) that normally is present at a nonfunctional level. This factor directs the transcription of genes for a set of 'heat shock proteins' that protect the cell against the consequences of the heat shock. We will refer to what some of these proteins do in the next chapter (page 306).

How are individual *E. coli* genes regulated in a variable fashion? The *lac* operon

Bacteria live in continually changing environments to which they must adapt for survival. It is relevant to the topic in hand to remember how fierce is the competition for survival amongst bacterial populations. A typical *E. coli* cell under optimal conditions will divide in about 20 minutes. If a strain lops 1 minute off that time, it will outgrow (wipe out) other strains very quickly. Wastage and inefficiency in biochemical processes is not tolerated in the face of such competition.

Production of enzymes consumes resources and energy, and it would not do if the *E. coli* cell produced enzymes that were unnecessary at the time. As mentioned already, some enzymes are constitutive—they are produced in all circumstances without any 'signal' being needed. Glucose-metabolizing enzymes come into this category for glucose is the commonest sugar and other sugars are shunted on to the glucose pathways. The cell assumes (as it were) that these enzymes will always be needed so the promoters of the genes coding for them have no 'on' and 'off' switches—they are always 'on'.

However, the *E. coli* cell may encounter other sugars—the disaccharide lactose present in milk is an example. If it is the sole source of carbon available, ability to utilize it would make survival possible. The enzyme needed to utilize this sugar is β-**galactosidase**, so called because lactose is a β-galactoside, and it must be hydrolysed to free galactose and glucose before it can be metabolized (Fig. 21.8). An additional transport protein, β-**galactoside permease**, is needed to transport the lactose into the cell. A third protein, **galactoside transacetylase**, is believed to be involved in protection of the cell against nonmetabolizable, potentially toxic β-galactosides that may be imported, though less is known of this. The three proteins are normally made in minute amounts (basal levels) because they are not required unless lactose is encountered. When lactose is encountered as the sole energy source, there is an almost instant burst of synthesis of the three proteins. The cell can then use the lactose as a carbon and energy source. However, if, in addition to lactose being present, there is also glucose, then production of the three enzymes would be wasteful since this merely leads to production of more glucose inside the cell when there is plenty of it available anyway. The cell therefore 'ignores' the lactose signal and does not produce the enzymes. The regulation, which we will now describe, is at the gene transcription initiation level.

Structure of the *E. coli* *lac* operon

First a few terms: production of a protein in response to a chemical signal is called **induction** of that protein and the

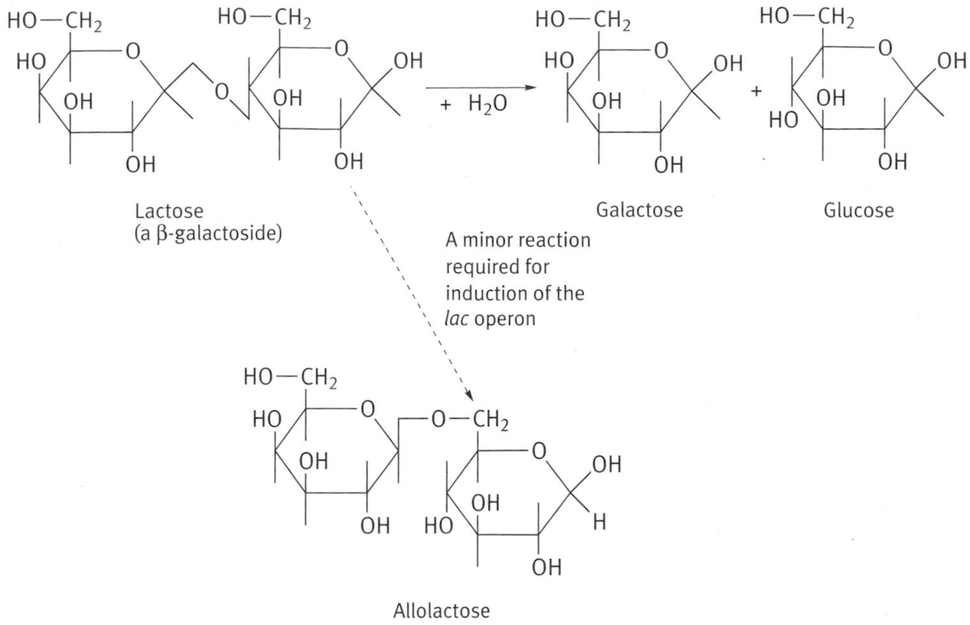

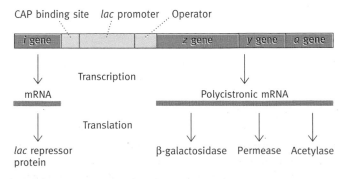

Fig. 21.8 The reactions catalysed by β-galactosidase. The term lactase is often applied to the enzyme catalysing this reaction in digestion. See text for explanation of allolactose formation.

responsible chemical, the inducer. Prevention of the production of a protein is called **repression**. As mentioned earlier, eukaryote genes are single units but many prokaryote genes are grouped together, the individual groups being under transcriptional control of a single promoter. The RNA polymerase transcribes through the entire group thus creating a polycistronic mRNA molecule with the coding instructions for several proteins.

Such a group of genes, with its single promoter control, is called an **operon** (the promoter being part of the operon). β-Galactosidase, lactose permease, and transacetylase genes belong to such an operon. The three genes are often referred to as *z, y,* and *a,* respectively. There is also an *i* gene (i for inducibility) that codes for a protein called the **lac repressor** and there is a stretch of DNA called the operator region to which the *lac* repressor protein can bind. Finally, there is a stretch of DNA to which a cyclic AMP receptor protein (**CAP** can bind). The latter stands for **catabolite gene activator protein**. It is given this general name because it is involved in the induction of other enzymes involved in catabolism of substrates. The approximate arrangement of these elements in the operon is shown Fig. 21.9. We can now systematically go through the control mechanism noting the following points. The *lac* promoter *on its own* is a weak one so that the RNA polymerase does not readily bind to it and initiate transcription. Without extra help the *lac* operon is not transcribed except at a low basal level. The extra help in polymerase binding is given by the attachment of the protein, CAP, to the adjacent site. Binding of CAP causes the double helix to bend at the site of attachment. When CAP is attached to the DNA, the promoter is a strong one. However, CAP does not attach unless cAMP is bound to it—it is an allosteric protein. *cAMP is produced by the cell only when glucose levels are low.*

When glucose is at a high level, cAMP is scarce in the *E. coli* cell, CAP does not bind, the RNA polymerase therefore does not bind effectively, and *lac* operon transcription is minimal. This situation is illustrated in Fig. 21.10(a).

Does this mean that the *lac* operon is always transcribed when glucose is scarce? The answer is no. As explained, there is no point in doing so unless lactose is present. In the absence of lactose the *lac* repressor protein is attached to the operator and blocks the RNA polymerase from transcribing the genes (Fig. 21.10(b)). The *lac* repressor is also an allosteric protein. In the absence of lactose in the environment, it has strong affinity for the operator, but, if lactose is present, a small amount is able to enter the cell via the basal level of permease, and it is hydrolysed, during the course of which a small amount is converted to the lactose isomer, allolactose (see Fig. 21.8). This binds to the repressor protein, causing an allosteric change in the protein,

Fig. 21.9 Diagram of the *lac* operon. Note that the *i* gene is an independent gene that codes for the *lac* repressor protein. Similarly, there is a completely independent gene producing the catabolite gene activator protein (CAP) to which cAMP can bind.

Situation (a). High glucose; no cAMP; no lactose; no transcription of *lac* operon

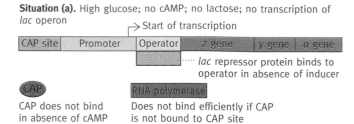

CAP does not bind in absence of cAMP

Does not bind efficiently if CAP is not bound to CAP site

Situation (b). Low glucose; high cAMP; CAP–cAMP complex binds CAP site; RNA polymerase can now bind to promoter; no lactose; repressor protein blocks operator; no transcription

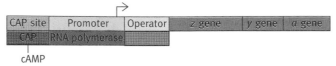

Situation (c). Low glucose; high cAMP; lactose present; repressor protein–allolactose complex detaches from operator; transcription of operon proceeds

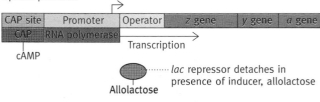

Fig. 21.10 Expression of the *lac* operon. **(a)** In the presence of high glucose there is no cAMP to cause CAP to bind and this binding is necessary for the attachment of RNA polymerase to the promoter. **(b)** With low glucose but no lactose, although CAP binds and assists the RNA polymerase to bind, transcription still does not occur because the *lac* repressor is bound to the operator blocking polymerase movement. In **(c)** the inducing allolactose binds to the repressor, causing its release from the operator, and transcription can proceed. (In the presence of lactose, a small amount of a lactose isomer, allolactose, is produced, which is the actual inducer—see text.) CAP, Catabolite gene activator protein.

which then dissociates from the operator and unblocks the operon.

Why is allolactose the inducer rather than lactose? A possible (and speculative) answer is that allolactose is produced as a minor side-product of the β-galactosidase reaction and is formed in sufficient amount to induce only when significant quantities of lactose are present. This would prevent traces of lactose firing off the synthetic process when the amount of sugar is insufficient to warrant this. Having lactose as the inducer but with a low affinity for the repressor protein would not achieve the same thing since, presumably, induced β-galactosidase would keep intracellular levels of lactose very low irrespective of the external concentration. To generate the inducer, allolactose, by the very process that destroys lactose would be an attractive mechanism. Additionally, when lactose was exhausted, allolactose would be destroyed by β-galactosidase, for which it is a substrate, and thus a mechanism for terminating the induction would be provided. The system would be constantly testing the rate of inflow of lactose to decide whether the induction should continue.

With the unblocking of the operator, the RNA polymerase is now free to move down the operon, producing the polycistronic mRNA (Fig. 21.10(c)); this is translated into production of the three enzymes.

The *lac* operon was the first understood example of prokaryote operon control, but it applies to many pathways. In *E. coli* the **tryptophan operon** (**trp operon**), which contains the five structural genes needed to produce three enzymes involved in tryptophan synthesis, is similarly controlled by a *trp* repressor protein. In this case, tryptophan bound to the latter causes it to block the operator site. This prevents most of the mRNA transcription but a second mechanism in the presence of high levels of tryptophan stops transcription by those RNA polymerase molecules that do manage to get through the repressor block. This second mechanism, called **attenuation**, depends on the fact that, in prokaryotes, translation of mRNA starts as soon as a small stretch of a messenger molecule has been synthesized, and therefore the polymerase is closely followed by the ribosome doing the translation. (The latter is dealt with in the next chapter so, for the moment, the **ribosome** is a large complex that moves along the messenger adding amino acid residues.) The operon first codes for a peptide only 14 amino acids in length; it is called the **leader peptide** but note that this is quite different from the leader peptide attached to certain proteins as described in the next chapter. It is destroyed immediately after its production. The attenuation leader peptide contains two tryptophan residues next to one another. If there is plenty of tryptophan, the ribosome adds these and moves along the mRNA; if tryptophan is scarce, the ribosome halts in the attenuation region. The crucial point is that, in the absence of a ribosome in this position (that is, when adequate tryptophan levels enable the ribosome to translate right through), the partially completed mRNA molecule internally base pairs to form a termination hairpin structure—the transcription is aborted; if a ribosome is in this region (due to it being stalled in the absence of tryptophan), it interferes with the base pairing such that the polymerase is allowed to complete its transcription of the operon. Thus the level of transcription is adjusted according to the level of tryptophan in the cell; the lower this is, the higher the proportion of polymerase molecules that are allowed to produce a complete mRNA molecule.

Attenuation is known to control six operons concerned with specific amino acid biosynthesis. (Only the *trp* operon amongst these has the repressor control as well.) All have the same ingenious method for assessing the level of the particular amino acid in the cell, and this is made sensitive by the occurrence of multiple residues of the particular amino acid in the leader peptide; in the case of the histidine operon there are seven histidine residues.

One of the features of the *lac* operon control (and that of other operons) in bacteria is that, once the facts are known, the mechanism is readily understandable. It all works with military precision—CAP binds and enables the polymerase to bind, a repressor protein binds and blocks transcription, and the inducer causes it to become unblocked. It is the sort of 'mechanical' control that one might have thought of oneself. Gene transcriptional control in eukaryotes is more complex as you will see later and is not remotely like anything one could have planned oneself. But first we need to look at the process of eukaryote gene transcription, our next topic.

Gene transcription and its control in eukaryotes

The basic processes involved in eukaryote mRNA production

The 'nuts and bolts' of the chemistry of RNA synthesis are the same as in prokaryotes. Thus Fig. 21.1 in which RNA is assembled from the four ribonucleoside triphosphates under the direction of a DNA template applies here. The DNA-dependent RNA polymerase that transcribes genes coding for proteins in eukaryotes is called **RNA polymerase II**, a very large multi-subunit enzyme. (We will leave RNA polymerases I and III to the next chapter where we describe RNA-producing genes from which RNA molecules, which are not messengers, are transcribed.)

The initiation of transcription of eukaryote genes is complex and is inseparable from the subject of gene control. So, for the moment, please just accept that initiation occurs and this will allow us to deal with mRNA production in eukaryotes (which is not exactly the same as in prokaryotes). We will then return to initiation in the control section.

Termination of transcription in eukaryotes

It is not understood how termination by RNA polymerase II occurs in eukaryote genes; there is nothing known that corresponds to the prokaryote termination signals. In fact it looks (erroneously, no doubt) as if during evolution the design of the process was somewhat fouled up. The RNA polymerase II, when it reaches the $3'$ end of the gene, synthesizes a sequence AAUAAA (coded for by the template strand) but then goes on to transcribe well beyond this and then somehow terminates. A separate enzyme cuts the RNA transcript a short distance beyond the above sequence and then another enzyme, not dependent on a template, adds adenine nucleotides (using ATP as their source)—as many as 200 to form a polyA tail. Histone mRNA does not have this tail so it can't be essential for translation. Evidence has been obtained that the polyA tail may be involved in the stability of the mRNA molecule in the cell (see page 288).

Capping the RNA transcribed by RNA polymerase II

The RNA of the eukaryote gene primary transcript immediately undergoes a modification at its $5'$ end, called 'capping', as soon as RNA synthesis is initiated. At the $5'$ end of the RNA there is a triphosphate group, since the first nucleotide triphosphate incorporated into the RNA simply accepts a nucleotide on its $3'$-OH group, leaving the initial triphosphate group unchanged. The terminal phosphate of this is removed and a GMP residue is added from GTP (Fig. 21.11). The $5'$–$5'$ triphosphate linkage (see Fig. 21.11) is very unusual in nature. The G is then methylated in the N-7 position and also at the $2'$-OH of the

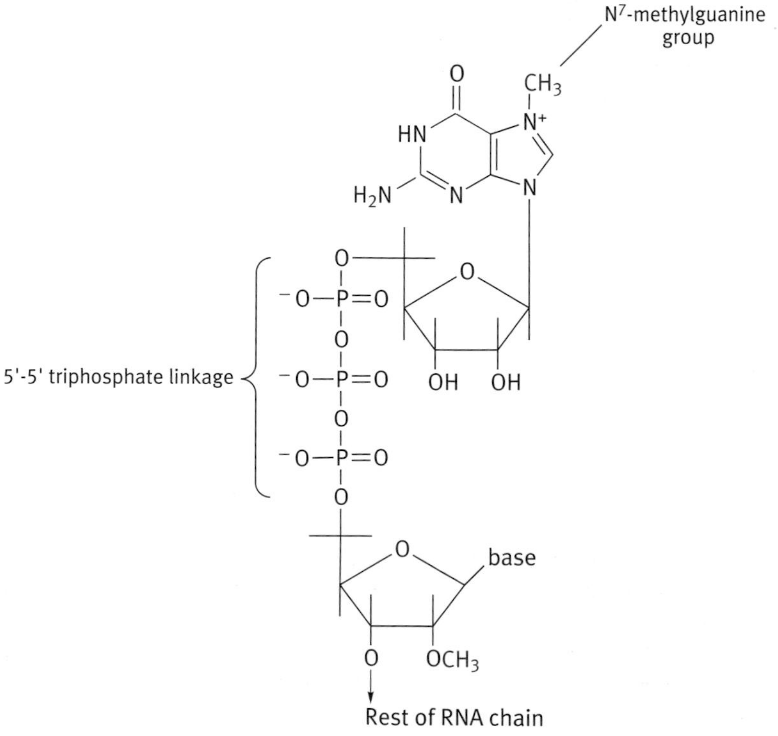

Fig. 21.11 Structure of the 5′-cap in eukaryotic mRNA. The terminal nucleoside triphosphate of the primary RNA transcript is converted to a diphosphate followed by a reaction with GTP in which pyrophosphate is eliminated. This is followed by methylation reactions. (A third methyl group may be added to the 2′-OH of the next nucleotide of the primary transcript.) The capped primary transcript is then processed to mRNA—see text for details.

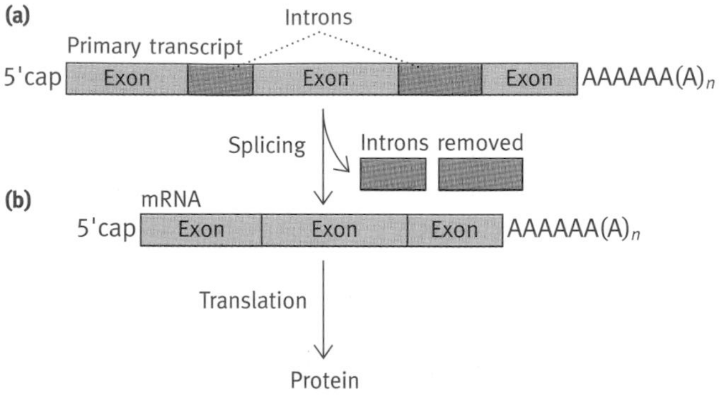

(a)

Introns

Primary transcript

5'cap | Exon | Exon | Exon | AAAAAA(A)$_n$

Splicing   Introns removed

(b)   mRNA

5'cap | Exon | Exon | Exon | AAAAAA(A)$_n$

Translation

Protein

Fig. 21.12 Primary polymerase II transcript of a eukaryote gene showing **(a)** introns after capping and addition of polyA tail. **(b)** Excision of introns to form the mature mRNA is called splicing.

second, and sometimes of the third, nucleotide. Because the GTP cap is formed on the first nucleotide of the RNA, the start site of a gene (the $+1$ nucleotide) is often referred to as the **cap site**. The cap is believed to protect the end of the mRNA from exonuclease attack and it is involved in initiation of translation as described in the next chapter.

These are not the only differences between mRNA production in prokaryotes and eukaryotes because most eukaryote genes are split genes.

What are split genes?

mRNA in both prokaryotes and eukaryotes is proportionate in length to the size of the protein it codes for (or, in polycistronic mRNAs, to the several proteins coded for), apart that is from the 5′ and 3′ untranslated regions, which are relatively short anyway. In bacteria this is true of the *primary* RNA transcript but that from most eukaryote genes is very much longer—perhaps even 10 times longer than would be expected on this basis. This is because the RNA coding for protein in the mRNA is split up into several parts, linked together by intervening stretches of RNA that do not code for amino acid sequences. In the DNA from which such RNA molecules are transcribed, the sections coding for the intervening sequences, with no protein-coding content, are called **introns** and the coding stretches are called **exons**. There can be two to 50 or so introns in a human gene (Fig. 21.12(a)) and introns can vary in length from about 50 to 20 000 base pairs. Exons are usually less than 1000 base pairs in length. The primary transcript is processed to eliminate the introns and link together the exons into one mRNA molecule. This is known as **mRNA splicing** (Fig. 21.12(b)). Although introns in general have been presumed to have no 'information' content, it is known that some contain regulatory sequences such as enhancers (see below).

Mechanism of splicing

Removing the unwanted RNA introns of a primary transcript and joining up the exons into mRNA looks a formidable task

but the key to it is the **transesterification reaction**. In this, a phosphodiester bond is transferred to a different —OH group. There is no hydrolysis and no significant energy loss.

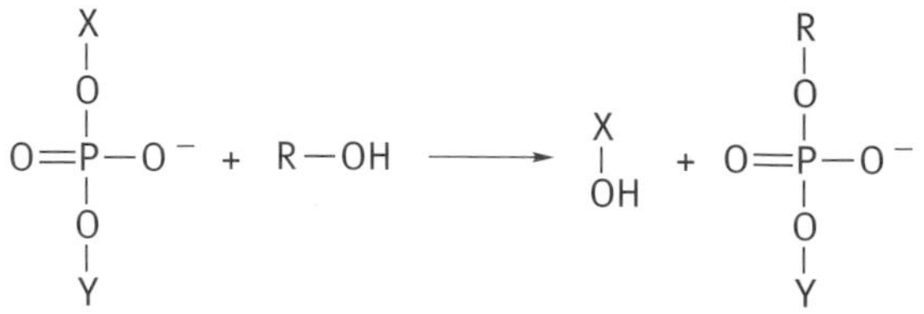

$$O{=}P{-}O^- + R{-}OH \longrightarrow \begin{array}{c} X \\ | \\ OH \end{array} + O{=}P{-}O^-$$

If X—Y is an RNA chain, it would be broken.

We turn now to RNA processing or splicing. The exon–intron junctions are 'labelled' by consensus sequences; all introns begin with GU and end with AG (though the identification consensus sequences are longer than these). The ROH of the above diagram is actually the 2′-OH of an adenine nucleotide in the intron chain (Fig. 21.13). The 2′-OH group attacks the 5′ phosphate of the G nucleotide at the splice site, forming a lariat structure. This breaks the chain at the 3′ end of exon 1, thus producing a free 3′-OH which attacks the 5′ end of exon 2, thus joining the two exons. As stated, since transesterification is involved, no energy input is required.

In most eukaryotes, the splicing reaction in the nucleus is catalysed by very complex protein-RNA bodies called **spliceosomes**. The contained RNA molecules are called **snRNAs**, meaning **small nuclear RNAs**. It is likely that the snRNAs hybridize by base pairing to splicing consensus sequences on the transcript RNA and position the intron ends for excision. A mutation in the splicing consensus sequence may result in

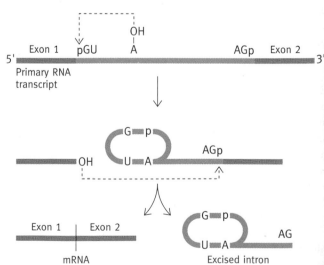

Fig. 21.13 Mechanism of mRNA splicing. Note that, for clarity, the process is shown in two stages; energy is not required for the process since transesterification reactions are involved.

improper splicing or failure to excise an intron. In the genetic disease *β* thalassemia, the *β* subunit of hemoglobin is not produced in normal amounts, because the G at the 5′ end of an intron is mutated to an A and, thus, primary transcripts are not properly processed to mRNA.

Against this complexity of the splicing mechanism, in the protozoan *Tetrahymena*, accurate splicing is done by the RNA transcript itself without any help from proteins. This discovery created shockwaves in biochemistry for here is an RNA molecule catalysing in itself specific chemical reactions. Until this point, only proteins were believed to accomplish such happenings. The discovery has opened up a new era in RNA biochemistry and this, incidentally, has strengthened theories that the first life forms were based on RNA chemistry with proteins coming later in evolution.

The primary polymerase II transcript of the protein coding eukaryote gene has been capped, trimmed, tailed, and spliced and it now emerges from the nucleus as mRNA.

What is the biological status of introns?

An important biological aspect of the existence of introns is that they may have facilitated evolution. At this point you might re-read page 33 on protein domains. As research on protein structure identified domains, and gene research identified exons, it was found that the latter often code for discrete protein domains though the correspondence is not always perfect. As stated earlier (page 33), new proteins are believed to have evolved by 'domain shuffling'—the concept is that the same domain is used repeatedly, for a partial function of, for example, an enzyme, combined with a variety of other domains so that a new family of proteins is assembled from pre-existing domains. The separation of the parts of a gene into exons coding for discrete protein domains would facilitate domain shuffling. The latter would provide a more rapid means of producing novel proteins by recombination events rather than by point mutations in DNA leading to single amino acid changes.

The role of introns in **exon shuffling** may need explanation. Exon shuffling will be the result of chromosome rearrangements. The DNA coding for a protein domain must be intact and the existence of introns on either side could facilitate successful transposition of the exon. If a crossover rearrangement occurred in the middle of a domain coding region, it would probably destroy the protein domain structure coded for and be a nonviable change since, as explained (page 33), a protein domain has a structure that could be expected to exist on its own as a discrete entity. Since introns have no protein coding function, gene crossovers could occur with less risk of hitting a domain and therefore lead to a potentially successful exon shuffle.

What is the origin of split genes?

There are two views on this question. One is that the prokaryote type, noninterrupted gene, is primitive and that introns were inserted later in evolution ('introns late' model). This gives no explanation of the origin of introns which, in the hypothesis, remains a complete mystery.

The alternative view is called the exon theory of genes, or more descriptively, the 'introns early' model, which postulates that introns are primitive. Introns could then represent the fused, untranscribed flanking regions of adjacent ancient minigenes from which it is postulated modern genes were constructed. The fact that RNA molecules can have the property of self-splicing (see above) removes the requirement for a complex splicing mechanism early in evolution. On this hypothesis, prokaryote uninterrupted genes are regarded as a later development to facilitate rapid cell division.

As stated, the introns early view has developed as a result of the often observed correspondence between exons and protein domains (page 34), which, in turn, led to the exon-shuffling concept in which new proteins are assembled from new combinations of domains.

However, it has recently been pointed out that, if introns had been inserted randomly into ancient noninterrupted genes, some parts of genes coding for protein domains might, by chance, have introns nicely placed on either side of them so that successful exon shuffling could occur. Assuming that such shuffling conferred an evolutionary advantage, then more modern genes arrived at by exon shuffling would automatically exhibit the correspondence between protein domains and exons. In this view, it is the shuffling that has produced the exon/domain correspondence, not the other way about, and the correspondence of exons with domains says nothing about the origin of introns.

To examine this, several ancient genes (ones coding for proteins conserved over long evolutionary times) were examined. It was found that there was no correspondence between the intron–exon structure and the structural features of the protein. The workers concluded that this does not support the introns early view and indicates that uninterrupted genes are primitive. (Note that the occurrence of exon shuffling is not questioned by either hypothesis; this is a different issue from that of the origins of introns that facilitated shuffling.) It has been pointed out that interrupted genes in prokaryotes would pose a difficult splicing problem in that mRNA is immediately translated by ribosomes that initiate at the 5′ end of the messenger before synthesis of the 3′ end has been completed.

Alternative splicing or two (or more) proteins for the price of one gene

There is one other well-established advantage that split genes confer—**alternative splicing**. In typical splicing, all the exons of the primary RNA transcript are linked together to form the mature mRNA leading to the formation of a single specific protein. However, there are many known cases where the splicing can occur in different patterns so that a particular group of exons forms one mRNA, and a different group from the same gene transcript forms another mRNA. This leads to different proteins. The mechanism may be employed to produce variant forms of a protein required in different tissues or at different times.

Mechanism of eukaryote gene transcription and its control

Although initiation of eukaryote gene transcription obviously precedes transcription and processing, etc., we have left discussion of the subject until now because it is inseparable from the control of gene activity and, moreover, it is a fairly complex subject in its own right.

Unpacking of the DNA for transcription

We have already dealt with the very great degree of packing of eukaryote DNA necessary to fit its huge length into the nucleus (page 246). Before (or during) transcription, the DNA structure is selectively loosened up at those parts of the DNA that are to be transcribed. We know that selective loosening up actually occurs from two lines of evidence. First, in insect salivary gland polytene chromosomes, the loosening can be directly seen in the light microscope. Since these chromosomes are unique and classically important, let us describe them.

Salivary gland cells of the larval form of the fruit fly, *Drosophila*, are very large and have a most unusual structure. The chromosomes are replicated about 1000 times without cell division occurring and the elongated chromosomes, still with much condensation packing, lie side by side precisely aligned lengthways. The packing arrangement along the DNA gives rise to bands, visible in the light microscope; in a single chromosome these would not be visible but, multiplied a thousand times, they become visible. During larval development successive banks of genes are expressed and it is seen that, during their transcription, the DNA of specific bands becomes loosened—forming what are called 'chromosome puffs' (Fig. 21.14).

The second means of knowing that transcription of eukaryotic genes is associated with loosening of the DNA packing is that transcriptionally active regions of chromatin become more susceptible in *in vitro* experiments to attack by added DNase—an enzyme that hydrolyses DNA. It is known that globin genes, for example, become hypersensitive to DNase attack only in chromatin from those cells that synthesize globin and only at the developmental time at which the genes are actively transcribed.

The gene control needs in differentiated eukaryotes

An *E. coli* cell has about 4000–5000 genes; in a human, the figure is more like 50 000–100 000. In an *E. coli* cell most, or all, of these genes are expressed during the division time of a given cell. It is true that the *lac* operon for example is active only in the special case of low glucose concentration and the presence of lactose but, in contrast to eukaryotes, in each *E. coli* cell, transcription of all genes is potentially needed.

In eukaryotes, the induction of certain genes by chemical inducers does occur—in liver, for example, phenobarbital causes a severalfold increase in enzymes metabolizing this compound in a few days. Repression also occurs—cholesterol represses the first enzyme of the pathway leading to its synthesis (page 147). Such controls are never as dramatic as the responses of, for example, the *lac* operon, but they are important.

However, in animals there is also a whole catalogue of hormones and other controlling agents that cause selective modulation of the synthesis of specific proteins via control of gene transcription. The complexity of controls is very great, as described in Chapter 26, and, indeed, a given gene may be subject to control by a whole variety of different signals. This is a level of control not found in *E. coli*. Next, there is need to express different genes in different tissues. Some genes, referred to as **housekeeping genes**, are expressed in all tissues, at all

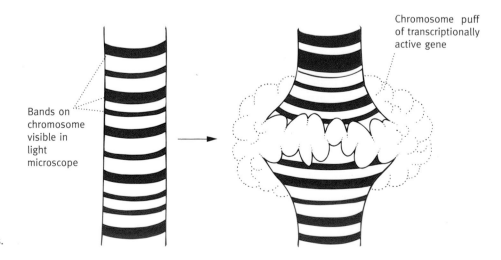

Chromosome puff of transcriptionally active gene

Bands on chromosome visible in light microscope

Fig. 21.14 Diagram of chromosome puffs at transcriptionally active chromatin on insect salivary gland polytene chromosomes.

times. Glycolysis enzymes, proteins needed for DNA synthesis and protein synthesis, etc. are needed by all cells and their genes (housekeeping class) are expressed in all cells except terminally differentiated cells such as mature erythrocytes. But other proteins are needed only in specific tissues. Immature red blood cells produce hemoglobin and this happens only in such cells. There are liver-specific proteins, muscle-specific proteins, kidney-specific proteins, and so on. The different cells all contain identical sets of genes, which means that the muscle cell, for example, expresses only certain genes, and never expresses liver-specific genes, and so on. Even this is not the end of the problems for, during embryonic development, certain genes are required at an early stage of the differentiation process and others later in the development of a specific tissue. The outcome of all this is that a eukaryote gene, at one end of the spectrum, might be constitutively expressed at a constant rate without further transcriptional controls or, at the other end of the spectrum, might be subject to multiple controls. The complexity is such that a gene might be expressed only in one or more tissues, but at rates affected by a whole battery of different hormones or other factors.

You can see the problem—in prokaryotes you can switch off a gene with a repressor or, in the case of, for example, the *lac* operon, switch it on by cAMP and lactose. These are readily understandable. But how do you exercise a whole group of controls, both positive and negative, on a eukaryote gene and how can the system allow for almost any permutation of controls to act on given genes?

The structure of the type II eukaryotic genes

We will discuss in this section only the transcription of the protein-coding genes, all of which use RNA polymerase II.

As is the case with prokaryote genes, the eukaryote version has a promoter 5' to the start site. The start site itself has a recognizable short sequence between nucleotides -3 and $+5$ called the initiator region (Fig. 21.15). About 80% of genes have a TATA box centred at about -25, whose sequence is reminiscent of the Pribnow box in the prokaryote promoter.

Further upstream, within about 100–200 or so base pairs of the start site, are elements or boxes—short specific DNA sequences that are recognized by specific proteins called **transcriptional factors**. Three common upstream elements are the CAAT (pronounced CAT) box, the GC box, and an eight-

base-pair octamer box; these are located at sites within the -100 to -200 region in different genes. However, in addition, there are other DNA elements or boxes that bind transcriptional factors that greatly activate transcription of the gene. These are called **enhancers**. The remarkable feature of these is that they can be located far away from the gene—even thousands of base pairs away, and their orientation in the DNA can be reversed without impairing their function.

This makes the definition of a eukaryote promoter less precise than that of a prokaryote gene. Is a distant enhancer part of the promoter? One useful working definition is that given in Lewin's textbook, *Genes V*—that the eukaryote promoter includes sequences of DNA that must be in a relatively fixed position with respect to the start site. The promoter therefore includes the initiator region, the TATA box, and upstream elements such as GC and CAAT boxes, but excludes the enhancers.

One of the most difficult aspects to accept is the apparent haphazard nature of eukaryote gene control. This refers to the remarkable and somewhat disturbing fact that no one element is essential for transcription and different genes have different combinations of elements. Twenty per cent of genes, including many housekeeping ones, lack TATA boxes, some lack other elements, some have multiple copies of an element. The variations found in three sample genes are illustrated in Fig. 21.16. Have a look at these. It is so very different from the comparatively orderly situation in prokaryotes.

The mechanism of eukaryotic gene transcriptional initiation
The basal initiation complex

In bacteria, the RNA polymerase recognizes the correct binding site on a promoter and binds directly to the DNA, helped in some cases by for example, CAP. This does not occur in eukaryotes. Instead a whole retinue of proteins, called **general transcription factors** and present in all cells, are needed to form the basal initiation complex on the DNA to which RNA polymerase attaches and which is needed for the transcription of all type II genes. In genes with TATA boxes, a protein universally present in eukaryote cells, called **TBP** (TATA binding protein), attaches to the TATA box. Eight or more other proteins called **TAFs (TBP-associated factors)** are associated with this, forming a complex known as **TFIID (transcriptional factor D for polymerase II)**. The TFIID is then joined by RNA polymerase II and other proteins to complete the basal initiation complex (Fig. 21.17). In TATA-less genes, the complex assembles at the initiator region possibly using another factor to achieve correct positioning. (In the case of genes transcribed by polymerases I and III, a different basal complex assembles and recruits the appropriate polymerase.)

However, as already indicated, there are different classes of genes in terms of their expression. Some are 'housekeeping' genes and expressed constitutively at all times. Some are tissue-specific and expressed only in certain cells. Some are inducible;

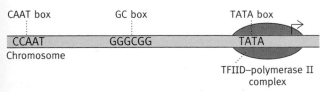

Fig. 21.15 Examples of upstream activating sequences in a eukaryotic gene promoter region.

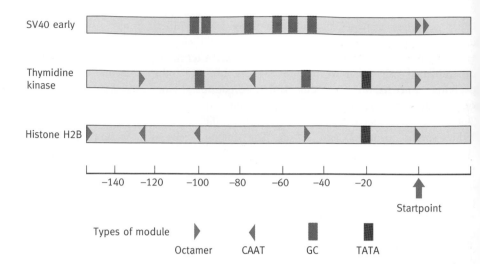

Fig. 21.16 The elements of three eukaryote gene promoters. Promoters contain different combinations of TATA boxes, CAAT boxes, GC boxes, and other elements.

for example, certain steroid hormones activate transcription of specific genes. Some are subject to negative control; some are subject to combinations of the above types of control. The big question now is how these different types of control are achieved. This is where we come to the specific transcriptional factors that combine with the upstream and enhancer elements.

What are the roles of transcriptional factors?

It appears that all of the genes in chromatin (which, you will recall (page 246), is DNA associated with nucleosomes and other proteins) in the absence of transcription factors are basically in a 'shut down' or repressed state because a nucleosome blocks the initiating region of each gene promoter. A gene cannot be transcribed until the nucleosome is displaced, thus allowing the basal initiation complex to assemble.

A variety of transcriptional factors may be involved in the activation of a given gene. First, there are the factors that attach to the upstream boxes such as GC and CAAT boxes. These are present in all cells. Constitutively expressed genes may require only these. How do transcriptional factors that bind to distant elements participate in a complex? It can occur by looping of the DNA. Each factor has at least two binding sites—one to attach

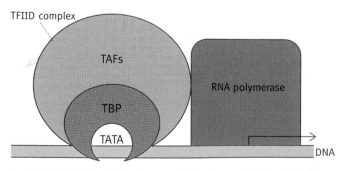

Fig. 21.17 Model for RNA polymerase II basal transcriptional initiation complex. The TFIID complex includes the TATA binding protein (TBP) and multiple TAFs (TBP-associated factors). The shapes, sizes, and positions of components are drawn arbitrarily.

to DNA and another to interact with the other proteins of the transcriptional complex as illustrated in Fig. 21.18.

How are genes selectively controlled? This is the function of elements that may be located anywhere in the DNA—even thousands of bases away—and still influence the transcription of the gene. As an example, there is a distant enhancer for the β-globin gene (involved in hemoglobin synthesis) whose presence in the DNA results in increased transcription by a factor of hundreds. Stimulatory factors may attach to enhancer elements.

The β-globin gene is expressed in red blood cells but not elsewhere, because the gene controls include a 'GATA' box; transcription is dependent on the presence of a transcriptional factor that attaches to this. It is present in red blood cells but not in other cells. Tissue-specific expression is achieved in this way.

How is the rate of transcription of an inducible gene controlled? An example is the activation of specific genes by steroid hormones. In target cells, there are **inactive transcriptional factors**—they cannot bind to their response elements in the DNA. When the hormone arrives, it activates the factor to bind to its response element and this in turn activates transcription. The mechanism of activation of the factor is described in Chapter 26 where we deal with cell signalling, for a major part of the latter consists of signals activating transcriptional factors in cells. Negative controls also exist. Figure 21.19 illustrates the situations just described.

The major question of *how* the interactions between the transcriptional factors and the initiating complex control initiation is unanswered. The essential point about eukaryote gene control is its enormous flexibility, for any number of factors can have a say in the rate of transcription of a gene provided there are appropriate binding sites for them.

The structure of DNA binding proteins

From what has been said in this chapter, it is clear that proteins that bind to specific sites on DNA play a central role in life. There are numerous repressors and transcriptional factors

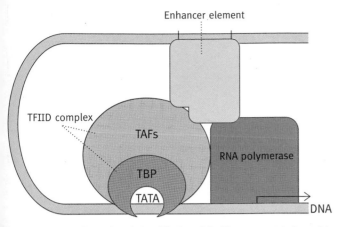

Fig. 21.18 Hypothetical and simplified model of how a protein bound to DNA at some distance away from the transcription start site can be involved in RNA polymerase II transcriptional initiation. The site(s) of interaction between the transcriptional factor, the basal initiation complex, and the RNA polymerase are drawn arbitrarily. Enhancer elements, perhaps thousands of base pairs away, can participate in the complex by the looping mechanism (see Fig. 21.19).

involved in differentiation, embryonic development, and gene control in general and all depend on their ability to recognize, and bind to, DNA.

Because of the vital importance of this, a large research effort has gone into elucidating exactly how such proteins bind to their appropriate sites on the DNA. It has emerged that most DNA binding proteins can be grouped into a small number of families on the basis of their structural characteristics and we will describe these shortly but, before this, a general overview may be useful. The main families of such proteins are the helix–turn–helix proteins, the homeodomain proteins, the leucine zipper proteins, and the zinc finger proteins.

The structures or motifs that give rise to these names are only small sections of the proteins that are characteristic of each family. When you therefore see a diagram of a helix–turn–helix protein, for example, it usually refers only to this small part of the total protein.

The DNA binding proteins bind to double-stranded DNA; site-specific binding involves weak bond formation between the amino acid side chains of the protein and the DNA bases, though additional stabilizing bonding to the sugar–phosphate–sugar backbone may occur. The contacts of the protein recognition motifs with the bases occur predominantly in the major groove of the double helix where the edges of the bases are exposed. In many cases, the contact is made by a recognition α helix that readily fits into the major groove.

A point of particular importance is that often there are two adjacent sites on the DNA necessary for specific attachment of a protein. These may be identical, formed by a palindromic arrangement of the bases (see below), or they may be dissimilar. These two situations are catered for by the protein being a dimer, made of two identical subunits (a **homodimer**) or

different subunits (a **heterodimer**), respectively. In the case of zinc finger proteins, there may be multiple DNA binding motifs on the same protein.

The essence of variable gene control is that the DNA binding proteins, such as transcriptional factors and repressors, often bind only when they are 'instructed' to do so and are detached at other times (or vice versa). We have already met situations in which transcriptional factor activation is achieved by the regulatory protein being allosterically modified by the binding of a ligand (see the *lac* operon control above). In eukaryotes, the 'instruction' usually comes from a chemical signal external to the cell, such as that provided in the form of hormones and growth factors. The mechanisms by which such signals are transmitted to activate transcriptional factors to bind to the DNA is a major topic, which will be dealt with in Chapter 26.

With that preamble, we will now deal with the different families of DNA binding proteins.

Helix–turn–helix proteins

This was the first type to be identified; it occurs commonly in prokaryotes and many examples exist. We will use the lambda repressor as an example; lambda is an *E. coli* bacteriophage (described on page 321). We will not go into the role of the repressor protein here for it is a very complex story, but it binds to a region of the lambda DNA and is involved in the decision of whether an infecting virus enters the lytic or lysogenic cycles (explained on page 321). For the present, its importance is that its mechanism of binding to DNA has been fully elucidated.

The helix–turn–helix (HTH) motif is, as explained, the small section of the protein that makes the binding contact to the operator site on the DNA. (It is not a protein domain for it could not exist as such in isolation; motif means a recognizable

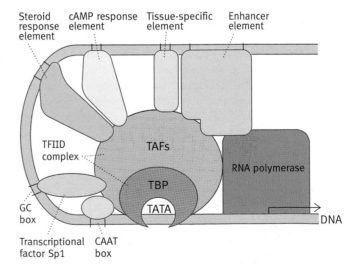

Fig. 21.19 Diagram to illustrate multiple controls on eukaryote gene transcription initiation. The sizes, positions, and interactions of proteins are arbitrarily drawn. The combination of controlling elements is similarly arbitrary. Enhancer elements may be thousands of bases distant.

structural feature.) The motif has two α helices linked by a β turn; one of the two (the recognition helix) sits in the major groove of DNA (Fig. 21.20(a)). The operator, or **DNA recognition site**, is an imperfect palindrome. A palindrome is a phrase or word that reads the same forwards or backwards. An example is 'Madam I'm Adam'—it is symmetrical about the letter 'I'. Or, as is usually stated, palindromic DNA binding sites have a **dyad symmetry** (dyad means two units treated as one, or a group of two). The structure of the operator (not to be remembered) is

```
5' TATCACCGCCAGTGGTA 3'
3' ATAGTGGCGGTCACCAT 5'
```

The two halves of the palindrome represent identical binding sites, although the coloured nucleotides show that the palindrome is not perfect. The repressor protein is a dimer, the recognition helices of the two HTH motifs fitting into them (Fig. 21.20(b)).

Leucine zipper proteins

These are found in many eukaryote transcriptional factors. Note that the name does *not*, in this case, refer to the DNA-recognition motif as in HTH proteins, but rather to the characteristic structure found in these proteins that causes dimerization of two subunits (which may or may not have identical recognition sites). The whole molecule is an α helix but, in the 'leucine zipper' motif, every seventh amino acid is leucine. Since there are 3.6 residues per turn, the leucines all appear on the same side of the α helix forming a hydrophobic face. Two such subunits attach by hydrophobic forces between the leucine side chains (Fig. 21.21(a)). The term 'zipper' is a misnomer resulting from the initial belief that the leucine residues interdigitated. The actual DNA-binding region of each monomer is a region rich in positively charged arginine and lysine residues. The dimer attaches to the DNA in a 'scissor' grip, the two arms being in adjacent major groove sections of the duplex (Fig. 21.21(b)).

Zinc finger proteins

In this DNA binding motif, a zinc atom attached to two histidine residues and two cysteine residues stabilizes a finger-like structure (Fig. 21.22(a)), one part of which is a recognition α helix that sits in the major groove of DNA (Fig. 21.22(b)). Transcriptional factors may have more than one such finger, making multiple contacts with DNA (Fig. 21.22(c)). One such gene regulatory protein has 30 zinc fingers. Not all zinc fingers necessarily bind specifically to DNA; some are nonspecific and help stabilize the specific binding. Zinc finger proteins are

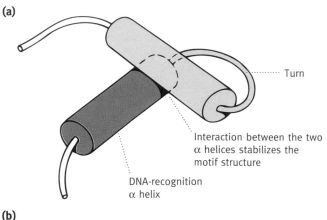

(a)

Turn

Interaction between the two α helices stabilizes the motif structure

DNA-recognition α helix

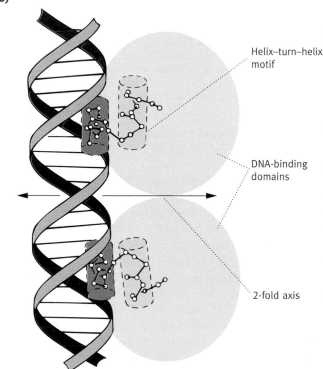

(b)

Helix–turn–helix motif

DNA-binding domains

2-fold axis

Fig. 21.20 (a) Diagram of the helix–turn–helix motif of a DNA-binding protein monomer. **(b)** Diagram of dimer helix–turn–helix protein binding to DNA in major grooves.

common in many different eukaryote gene regulatory proteins. Intracellular receptors to which steroid hormones bind (page 354) have variants on the zinc finger theme.

Homeodomain proteins and development

One of the biggest remaining unsolved areas of biology is the mechanism of **differentiation** or **embryological development**. A fertilized egg of a multicellular organism divides and its progeny differentiate into various tissues—epidermis, nerves, muscle, liver, and so on. Since every cell has the same collection of genes, differentiation involves selective expression of genes in different

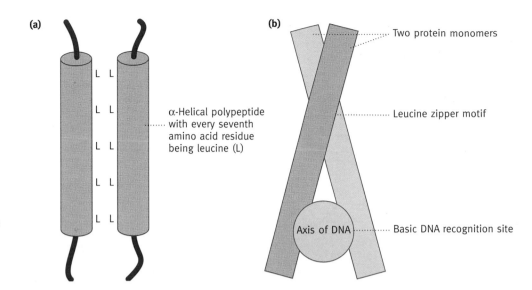

Fig. 21.21 (a) Diagram of the leucine zipper motif. The hydrophobic leucine residues are opposed, not intercalated as in a zipper. **(b)** Diagram of the leucine zipper protein attached to DNA, looking down the axis of DNA. The two arms of the protein lie in adjacent sections of the wide groove. The attachment sites are rich in basic amino acids.

tissues. Relatively little is known of how this occurs, but in the last few years certain genes of overriding importance to the proper embryological development have been identified, known as **homeotic genes**. Their importance in development has been established from the effects of mutations in these genes on the development of organisms such as the fruit fly, *Drosophila*. The important point in the present context is that homeotic **gene products**—that is, the proteins they code for—all have basically similar homeodomains, the section of DNA coding for the latter being known as a **homeobox**. Homeodomains are DNA-recognizing domains, 60 amino acid residues in length, reminiscent of the helix–turn–helix motif with a recognition helix lying in the major groove. It has been shown that homeotic gene-coded proteins bind at specific sites to DNA. It is believed from this that homeotic genes code for transcriptional factors. The target genes for these factors have not yet been identified but it is known that they, in some cases, are gene activators and in others are repressors. Homeodomains are highly conserved, being similar in structure in the insect *Drosophila* and in vertebrates, including humans.

There is the danger of some confusion in terminology in this area. The term 'box' is usually associated in transcription with a sequence of DNA that is recognized by a protein. Examples are the TATA and GC boxes, described earlier. The 'box' in homeobox genes, however, is used in the sense of referring to the sequence of bases that *codes* for the homeodomain of the homeobox protein. This is quite different from the previous usage of the term 'box', although both refer to specific stretches of DNA.

mRNA stability and the control of gene expression

Although regulation of gene transcription is of overriding importance in the control of gene expression, the stability of individual mRNAs is also of significance. In most situations the rate of synthesis of a protein is a reflection of the level of mRNA for

that protein except where translation of a messenger is controlled such as occurs in the synthesis of erythrocyte aminolevulinate synthase and of globin as described on pages 375 and 377, respectively. The level of an mRNA in a cell is a function of its synthesis and breakdown rates. At a given rate of mRNA production, a messenger with a long half-life will be present at a higher steady state level within the cell than a less stable one, resulting in a higher rate of synthesis of the cognate protein. Mechanisms that result in the alteration of the half-life of a given mRNA can thus provide a way of regulating gene expression.

Prokaryote mRNAs in general have an ephemeral existence with half-lives of about 2–3 minutes. Rapid messenger turnover permits rapid responses to changing circumstances. In mammals, the half-life of individual mRNAs ranges from about 10 minutes to 2 days in extreme cases. The mRNA for globin, which is regarded as a stable one, has a half-life of about 10 hours. Regulatory proteins tend to be coded for by short-lived mRNAs so that changes in the rate of transcription of their genes have a rapid effect on the rate of expression of the latter. Transcriptional factors usually have mRNAs with half-lives of less than 30 minutes. A given cell therefore contains mRNAs with widely different rates of degradation, and the stability of individual mRNAs may change within a cell as conditions change.

Despite the potential importance of mRNA stability, much less is known of the seemingly mundane process of mRNA breakdown than of mRNA synthesis and its control. Relatively little is known of the enzymes involved in the breakdown but some of the ways by which the lifetimes of mRNAs are determined have been identified.

Determinants of mRNA stability and their role in gene expression control

Almost all eukaryote mRNAs have a polyA tail added to their 3′ ends before they emerge from the nucleus (page 279 and Fig.

(a)

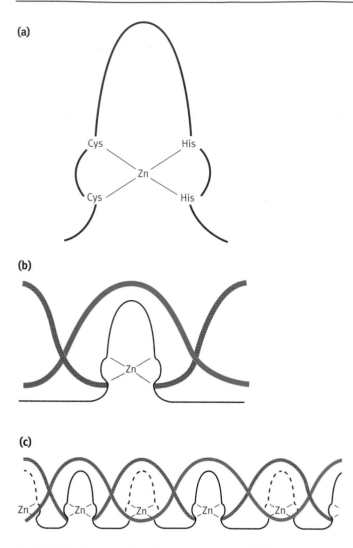

(b)

(c)

Fig. 21.22 (a) Diagram of the zinc finger structure. **(b)** Model of the interaction between the target region of DNA and a zinc finger protein. One side of the finger structure is in the form of an α helix that attaches to the major groove of DNA. **(c)** Multiple zinc finger motifs attached to DNA.

21.23(a)), histone mRNAs being exceptions. The polyA tail of eukaryote mRNAs is believed to offer protection against rapid breakdown of messengers in general. mRNA destruction often is preceded by deadenylation. A protein that binds to the polyA tail has been discovered and this may inhibit degradation of the mRNA from the 3′ end of the molecule but the mechanism by which polyA protects mRNAs is not well understood. It is believed to have effects on the transport and translation of mRNAs that may affect the breakdown of the latter.

Histone mRNAs lack a polyA tail and their stability is determined by a stem loop at the extreme 3′ end of the molecules (Fig. 21.23(b)). Synthesis of histones is required only during the S phase of the eukaryote cell cycle when DNA synthesis and nucleosome (page 246) assembly are occurring. Histone genes are transcribed during S phase but this ceases in

the G_2 phase when the level of histone mRNAs rapidly falls (see page 252 for the cell cycle phases). The latter is partly due to cessation of transcription but, in addition, the half-life of the mRNAs decreases from 40 to 10 minutes, a change dependent on the presence of the 3′ stem loop referred to above. The experimental transfer of this structural feature to globin mRNA produces a hybrid messenger, which after insertion into cultured cells, is destabilized at the end of S phase as if it were a histone mRNA. The destabilization of histone mRNAs requires the presence of free histone monomers, which accumulate as soon as DNA synthesis (and therefore nucleosome assembly) ceases at the end of S phase. Prompt switch-off of histone synthesis is necessary since histone monomers are toxic to cells. The fourfold reduction in the histone messenger half-life means that, after cessation of histone gene transcription, mRNA is exhausted in about 2 hours but, without destabilization, it would take almost 9 hours for this (Fig. 21.24). The reduction in half-life is thus essential, for otherwise histone synthesis could proceed throughout the G_2 phase but exactly how this is achieved remains to be discovered.

The synthesis of β-tubulin is also regulated by such a feedback mechanism modulating mRNA stability. In Chapter 29 we describe the role of this protein in aggregating to form microtubules (do not be concerned with what these are until you come to that chapter). The presence of free tubulin monomers destabilizes the mRNA for this protein. The mRNA has to be associated with active translation for the breakdown to occur since the partially synthesized peptide is itself somehow involved in the destabilization process. The mechanism by which the tubulin monomer causes mRNA breakdown is not understood but the regulatory rationale of the system is self-evident.

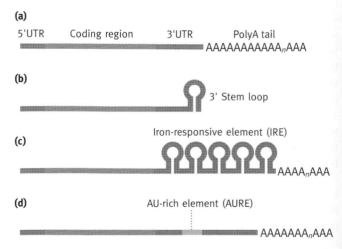

Fig. 21.23 Diagram of some structures present in the 3′ untranslated regions (UTR) of mammalian mRNAs that influence the half-lives of the molecules in the cell. **(a)** The polyA tail found in the majority of eukaryote mRNAs; **(b)** the 3′ stem loop found in histone mRNAs; **(c)** the iron-responsive element (IRE) of transferrin mRNA; **(d)** the AU-rich element (AURE) found in a large number of unstable eukaryote mRNAs.

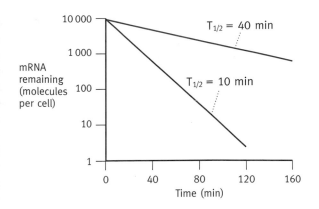

Fig. 21.24 The effect of a fourfold reduction in the half-life of an mRNA on the amount of the latter remaining in a cell at increasing times after the cessation of synthesis of that mRNA.

The synthesis of transferrin receptor protein is another case where mRNA stability regulates synthesis of the protein. The receptor is responsible for the transport of iron into cells (discussed on page 375). In the 3′ untranslated region of the mRNA is a group of five stem loops called an **iron-responsive element** (**IRE**) and shown in Fig. 21.23(c). In the absence of iron, an IRE binding protein attaches to the IRE and stabilizes the mRNA, thus increasing receptor synthesis and increasing import of iron. In iron abundance, the iron complexes with the protein, which then no longer binds to the IRE. It is believed that this exposes the region to attack by an endoribonuclease.

Short-lived mRNAs may contain 'instability' sequences that target the mRNAs containing them for rapid destruction within the cell. A structural feature of many unstable mRNAs is the **adenine/uracil-rich elements** (**AUREs**) found in the 3′-untranslated regions of the molecules (Fig. 21.23(d)). These elements vary from one messenger to another but contain AU sequences at least nine bases in length. They cause messenger destabilization possibly by virtue of AURE binding proteins. It is not known how AUREs function; their effect on mRNA half-life may be indirect such as by influencing translation but their presence is associated with a rapid deadenylation of the mRNA. A particularly interesting example is the mRNA derived from the c-*fos* gene, which codes for a leucine zipper class of transcriptional factor and contains an AURE in its 3′-untranslated region (the terminology of c-*fos*, v-*fos*, and oncogenicity is explained on pages 320–1). The corresponding v-*fos* gene is strongly oncogenic (cancer-producing); the mRNA derived from it lacks the AURE found in the c-*fos* mRNA and has a longer half-life than the latter within the cell. The normal c-*fos* gene is converted to an oncogene by deletion of the AURE region. (The relationship of regulatory proteins to oncogenicity will become clear when you deal with Chapter 26.)

The mechanisms by which mRNA stability is determined in the cell appear to be complex. In addition to those outlined above, it is known that several hormones modulate the stability of specific messengers. Control of mRNA breakdown is likely to assume greater importance in our understanding of the ways by which gene expression is regulated.

Gene transcription in mitochondria

In the earlier chapters on energy generation from food oxidation, the central role of mitochondria was described. **Mitochondria** are replicating organelles of eukaryote cells, with their own DNA and protein-synthesizing machinery (the same is true of plant chloroplasts). They divide to maintain the number appropriate to the cell—about 1000 in rat liver to 10^7 in a frog's oocyte. The DNA of mitochondria is usually a circular duplex—less than 20 000 nucleotides in mammals but there can be five or 10 copies in each mitochondrion.

The mitochondrion is far from self-contained for its DNA codes for only a small fraction of mitochondrial proteins. The rest are coded for by the nuclear DNA of the cell; the proteins are synthesized in the cytoplasm and transported into the mitochondria, a topic dealt with in the next chapter. In mitochondria *both* strands of the DNA are completely transcribed into single long transcripts. (The polymerase molecules doing this move in opposite directions since synthesis is always in a 5′ → 3′ direction.) The single primary transcripts are then processed, in a not totally understood way, into mRNAs, the tRNAs, and rRNAs (see below for explanation of the latter two). To produce more of the rRNA, many shorter transcripts encompassing the rRNA genes are made—again explained below.

As will be described in the next chapter, mitochondrial translation (protein synthesis) has prokaryotic characteristics as compared with the eukaryotic features of cytoplasmic translation. This has given rise to the belief that mitochondria evolved from engulfed prokaryote cells that became symbiotic (a similar theory holds for the chloroplasts in plant cells). The mitochondrial transcriptional system really does appear to be mixed up as to whether it is prokaryote or eukaryote. In mammalian mitochondria, the mRNAs are polyadenylated (eukaryote) but not capped (prokaryote), and there are no introns in the genes (prokaryote).

In trypanosome mitochondria, mRNAs are produced that then have to be **edited**. RNA transcripts have extra 'Us' inserted at specific places to produce the correct coding sequence for proteins. These Us are *not* coded for in the DNA. However, after this discovery and more shocking still, it was found that mRNA editing may occur even in nuclear transcripts. Thus a gene for as respectable a protein as mammalian apolipoprotein B (page 86), coded for by a nuclear gene, has two mRNAs produced from a single transcript. The unedited mRNA codes for apolipoprotein B_{100}. The smaller apolipoprotein B_{48} is coded for by an edited version in which a DNA coded C is converted to a U to form a translational stop codon (see next chapter). A simian virus (which is eukaryote in its gene characteristics) has an mRNA edited by insertion of two G residues not specified by the DNA. Whether there are fundamental reasons requiring these unsettling (to us) variations from the normal or whether it is another example of evolutionary tinkering and the policy of accepting anything that works, is unclear.

......

Genes that do not code for proteins

Exceptions to the rule that genes code for proteins are genes that code for special RNA molecules that are not messengers. In the next chapter you will learn that proteins are synthesized by small bodies called ribosomes and that the amino acids to be incorporated into proteins are carried on small RNA molecules called transfer RNA or tRNA, many distinct species of which exist. Ribosomes contain RNA molecules collectively called ribosomal RNA or rRNA. Prokaryote ribosomes have three different rRNA molecules in them and eukaryotes, four. Relatively vast amounts of rRNA are made, accounting for perhaps half of total transcription. Since mRNA is unstable while tRNA and rRNA are long-lived, the majority of RNA in a cell is in these forms. In *E. coli*, the three rRNAs plus a few tRNAs are transcribed together and the large precursor molecule is cut up into the appropriate pieces. The other tRNA molecules are also transcribed in larger precursor molecules which are processed into the tRNAs. In eukaryotes three of the four rRNAs are produced from the primary transcript of one gene and processed into the final rRNA molecules. Unlike prokaryotes there are three RNA polymerases. RNA polymerase II, which transcribes protein coding genes, has already been described. RNA polymerase I transcribes most of the rRNA and RNA polymerase III transcribes the tRNA and smaller rRNA. Since so much rRNA and tRNA is needed, multiple copies of their genes exist. In eukaryotes there may be hundreds or thousands of rRNA genes tandemly arranged head to tail—they exist in the nucleolar regions of the nucleus. The genes transcribed by polymerase III are unusual in that regulatory elements or boxes occur, not only in the DNA 5′ to the start site but also in the region that is transcribed, namely, downstream from the start site.

......

Further reading

Lewin, B. (1994). *Genes V*. Oxford University Press.

Gene expression

Lewin, B. (1994). Chromatin and gene expression: constant questions, but changing answers. *Cell*, **79**, 397–406.
A very readable general discussion of nucleosomes and their relevance to eukaryotic gene expression.

Introns and exons

Gilbert, W. (1987). The exon theory of genes, *Cold Spring Harbor Symp. Quant. Biol.*, **LII**, 901–5.
Describes the 'introns early' theory of assemblage of genes from minigenes.

Go, M. and Nosaka, M. (1987). Protein architecture and the origin of introns. *Cold Spring Harbor Symp. Quant. Biol.*, **LII**, 915–24.
Considers the two theories of the origins of introns.

Mattick, J. S. (1994). Introns—evolution and function. *Curr. Opin. Genet. Develop.*, **4**, 823–31.
Summarizes the introns early—introns late debate very clearly and also deals with evolutionary aspects of the topic in a most interesting way. A very readable review.

Protein modules or domains

Doolittle, R. F. (1995). The multiplicity of domains in proteins. *Ann. Rev. Bichem.*, **64**, 287–314.
A fascinating discussion of domains, domain shuffling, exons, and introns.

Splicing

Breitbart, R. E., Andreadis, A., and Nadal-Ginard, B. (1987). Alternative splicing: a ubiquitous mechanism for the generation of multiple protein isoforms from single genes. *Ann. Rev. Biochem.*, **56**, 467–95.
Fairly detailed but gives a good overview of the biological role.

Sharp, P. A. (1987). Splicing of messenger RNA precursors. *Science*, **235**, 766–71.
Concise summary.

Orgel, L. E. (1994). The origin of life on earth. *Sci. Amer.*, **271**(4), 52–61.
Growing evidence supports the idea that the emergence of catalytic RNA was a crucial early step.

RNA self-cleavage

Symons, R. H. (1989). Cell cleavage of RNA in the replication of small pathogens of plants and animals. *Trends Biochem. Sci.*, **14**, 445–50.
Covers viroids and the classical 'hammerhead' RNA self-cleaving structure.

Eukaryote transcription factors

Sheldon, M. and Reinberg, D. (1995). Tuning up transcription. *Current Biology*, **5**, 43–6.
Eukaryote transcription is a complex process and recent results identify multiple steps that need to be stimulated to activate transcription, one of which is a conformational change of transcription complex TFIID. Superb diagrams.

Tjian, R. (1995). Molecular machines that control genes. *Sci. Amer*, **272**(2), 38–45.
The activity of protein coding genes are tightly regulated by a dozen transcription factors, which can each include several proteins.

Zawel, L. and Reinberg, D. (1995). Common themes in assembly and function of eukaryotic transcription complexes. *Ann. Rev. Biochem.*, **64**, 533–61.
A detailed review of knowledge on this centrally important topic.

Surridge, C. (1996). The core curriculum. *Nature*, **380**, 287–8.
A news and views article which discusses the composition of the TFIID initiation complex in eukaryote gene transcription.

DNA-binding proteins

Brennan, R. G. and Matthews, B. W. (1989). Structural basis of DNA-protein recognition. *Trends in Biochem. Sci.*, **14**, 287–90.
Reviews the structures of proteins involved in the control of gene transcription and the way they interact with DNA.

DNA-binding proteins—zinc fingers

Klevit, R. E. (1991). Recognition of DNA by Cys_2, His_2 zinc fingers. *Science*, **253**, 1367 and 1393.
Crisp, concise summary.

Rhodes, D. and Klug, A. (1993). Zinc fingers. *Sci. Amer.*, **263**(2), 56–65.
Discusses structures, functions, and distribution of these transcriptional factors.

Klug, A. and Schwabe, J. W. R. (1995). Zinc fingers. *FASEB J.*, **9**, 597–604.
A complete account of the structure of this protein motif and its role in DNA binding.

DNA-binding proteins—leucine zippers

McKnight, S. L. (1991). Molecular zippers. *Sci. Amer.*, **264**(4), 32–9.
A very clear account of these transcriptional factors.

Ellenberger, T. E., Brandl, C. J., Struhl, K., and Harrison, S. C. (1992). The GCN4 basis region leucine zipper binds DNA as a dimer of uninterrupted α helices: crystal structure of the protein-DNA complex. *Cell*, **71**, 1223–37.
A research paper, but readable and worth looking at for the molecular illustrations.

DNA-binding proteins—histones

Grunstein, M. (1992). Histones as regulators of genes. *Sci. Amer.*, **267**(4), 40–7.
Discusses the concept that histones can both repress and facilitate activation of genes.
Lu, Q., Wallrath, L. L., and Elgin, S. R. (1994). Nucleosome positioning and gene regulation. *J. Cellular Biochem.*, **55**, 83–92.
Discusses how nucleosomes can act as repressors of eukaryote transcription.

mRNA editing

Hodges, R. and Scott, J. (1992). Apolipoprotein B mRNA editing; a new tier for the control of gene expression. *Trends Biochem. Sci*, **17**, 77–81.
An interesting account of how mRNA editing leads to the production of two forms of apolipoprotein B from one mRNA transcript.

Transcription in mitochondria

Clayton, D. A. (1984). Transcription of the mammalian mitochondrial genome. *Ann. Rev. Biochem.*, **53**, 573–94.
It differs from nuclear transcription. General review of the topic.

Problems for Chapter 21

1 In what ways does RNA synthesis differ from DNA synthesis?

2 By means of notes and diagrams, describe the components of an *E. coli* single gene and its associated flanking regions.

3 Describe the process of initiation of transcription of a gene in *E. coli*.

4 Describe two methods by which, in *E. coli*, gene transcription is terminated.

5 What factors determine, in a constitutive gene of *E. coli*, the strength of a promoter?

6 Describe how the *lac* operon is controlled.

7 Describe, in broad terms, the main ways in which the formation of mRNA in eukaryotes differs from that in prokaryotes.

8 By means of a diagram, explain the mechanism of splicing. What possible biological significance does the existence of introns have?

9 In what ways does eukaryote initiation of transcription differ from that in prokaryotes?

10 In terms of structural motifs, there are several families of transcriptional factors. What are these?

Chapter summary

Chapter 22

Protein synthesis, intracellular transport, and degradation

In the previous chapter we dealt with production of mRNA. In this chapter we deal with the way in which this mRNA directs the assembly of amino acids into the finished protein coded for by the gene from which the mRNA was transcribed. Most of the protein of a eukaryote cell is synthesized in the cytoplasm (mitochondria and chloroplasts form the exception) but specific proteins may be destined for the nucleus, the cell plasma membrane, mitochondria, or other organelles and a mechanism for delivery to their destinations is needed. Finally, some proteins are programmed for a short life, some for a long life. In this chapter we deal with the synthesis, delivery, and selective destruction of proteins.

The essential basis of the process of protein synthesis

First, we will look at the problem in broad terms. mRNA is a long molecule with four different bases in its component nucleotides. Proteins are synthesized from 20 different species of amino acids and the sequence of bases in the messenger specifies the sequence of amino acids in the protein. A one-base code (in which a single base represents an amino acid) could code for only four amino acids; a two-base code could code for 16, still not enough. Therefore a three-base code is the minimum requirement and this is the situation. Each amino acid is represented on the messenger RNA by a triplet of bases called a **codon**. With four bases we have 64 different triplets or codons $(4 \times 4 \times 4)$.

The assignment of codon triplets to amino acids is complete and the list constitutes what is now known as the **genetic code**. Thus UUU means phenylalanine to give a single example. The process by which protein is synthesized is called **translation** because the language of nucleic acid bases is translated into the language of protein amino acids.

With 64 different triplet codons available for the 20 amino acids, you might, at first sight, think that evolution would have picked out 20 to be used and ignored the rest. However, this would mean that chance mutations would frequently result in unusable triplets corresponding to no amino acid, which would thus render genes inoperative since the synthesis of the protein from its messenger would halt when an unusable triplet in the mRNA was encountered.

An alternative policy has been adopted. Three codons have been reserved as 'stop' signals that indicate to the protein synthesizing machinery that the protein is complete. These three codons (UAA, for example, is one) have no amino acids assigned to them in the genetic code. If, by chance, a mutation produces a stop signal in an mRNA coding region, the protein will not be produced from that gene. However, with only three stop triplets, the chance of this is much less than if there were 44 of them. (As happens so often, statements have to be slightly qualified for evolution is endlessly tinkering—in bacteria a 'stop'-generating mutation can be overcome by a suppressor mutation; we will not give details of this here.)

Of the remaining 61 codons, all code for amino acids, which means, of course, that an amino acid is likely to have several different codons, giving what is known as a degenerate code. The assignment of codons to amino acids, the genetic code, is shown in Table 22.1, given for reference. Only two amino acids, methionine and tryptophan, have single codons (AUG for methionine). The rest have more than one—leucine has six. Codon assignment is not random. Where several codons exist for one amino acid, they tend to be closely related. For example, those for isoleucine are AUU, AUC, and AUA which differ only in the third base. Not only that, but codons for similar amino acids tend to be similar. For example isoleucine and leucine are very similar aliphatic hydrophobic amino acids (page 25); their codons include CUU (leucine) and AUU (isoleucine), respectively. These facts have important genetic consequences, since

Table 22.1 The genetic code

5' base	Middle base				3' base
	U	C	A	G	
U	UUU Phe	UCU Ser	UAU Tyr	UGU Cys	U
	UUC Phe	UCC Ser	UAC Tyr	UGC Cys	C
	UUA Leu	UCA Ser	UAA Stop*	UGA Stop*	A
	UUG Leu	UCG Ser	UAG Stop*	UGG Trp	G
C	CUU Leu	CCU Pro	CAU His	CGU Arg	U
	CUC Leu	CCC Pro	CAC His	CGC Arg	C
	CUA Leu	CCA Pro	CAA Gln	CGA Arg	A
	CUG Leu	CCG Pro	CAG Gln	CGG Arg	G
A	AUU Ile	ACU Thr	AAU Asn	AGU Ser	U
	AUC Ile	ACC Thr	AAC Asn	AGC Ser	C
	AUA Ile	ACA Thr	AAA Lys	AGA Arg	A
	AUG Met†	ACG Thr	AAG Lys	AGG Arg	G
G	GUU Val	GCU Ala	GAU Asp	GGU Gly	U
	GUC Val	GCC Ala	GAC Asp	GGC Gly	C
	GUA Val	GCA Ala	GAA Glu	GGA Gly	A
	GUG Val	GCG Ala	GAG Glu	GGG Gly	G

*Stop codons have no amino acids assigned to them.
†The AUG codon is the initiation codon as well as that for other methionine residues.

many mutations involving single base changes have no effect on the protein synthesized (changing CUU to CUC still represents leucine) or else substitute a very similar amino acid (changing CUU to AUU substitutes isoleucine for leucine). Isoleucine and leucine are so similar in size and hydrophobic properties that the substitution may not impair the function of the protein. Thus the arrangement of the genetic code provides a 'genetic buffering' action whereby the effects of many single base change mutations on the proteins synthesized are minimized.

How are the codons translated?

It is important at this stage to be clear that protein synthesis occurs only on particles called ribosomes—what these are and how they function we'll come to shortly. But for now, and remembering firmly that *it all happens on ribosomes*, we can deal, *in principle*, with how codons on mRNA are translated into amino acids of proteins.

There is no physical or chemical resemblance or relationship between an amino acid and its codon that could lead to their association directly. It was predicted that there must be adaptor molecules to associate amino acids with particular codons and that, since the hydrogen bonding potential of codons was most likely to be of importance, the adaptor molecules were postulated to be small RNA molecules. Almost at the same time these were discovered—the **transfer RNA or tRNA molecules.**

Transfer RNA or tRNA

These are small RNA molecules. Diagrammatically they each have a clover-leaf structure (Fig. 22.1(a)). Internal base pairing

forms the stem loops. The really important parts (from our present viewpoint) are the three unpaired bases that form the anticodon (explained below) and the 3'-CCA flexible arm to which an amino acid can be attached.

As you will see shortly, two of these tRNA molecules at a time have to be positioned side by side on the ribosome with their anticodons paired with adjacent codons on the mRNA. In real life the tRNA molecules are folded up into quite a narrow shape, diagrammatically represented in Fig. 22.1(b). In Fig. 22.2 you can see a structural model of a tRNA.

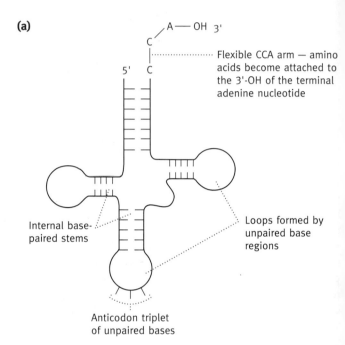

(a)

A — OH 3'
C
5' C

Flexible CCA arm — amino acids become attached to the 3'-OH of the terminal adenine nucleotide

Internal base-paired stems

Loops formed by unpaired base regions

Anticodon triplet of unpaired bases

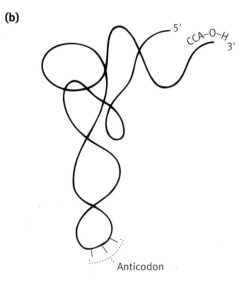

(b)

5'
CCA-O-H
3'

Anticodon

Fig. 22.1 (a) Diagram of the clover-leaf structure of transfer RNA. **(b)** Diagrammatic representation of the folded structure of tRNA molecules. See the molecular model in Fig. 22.2.

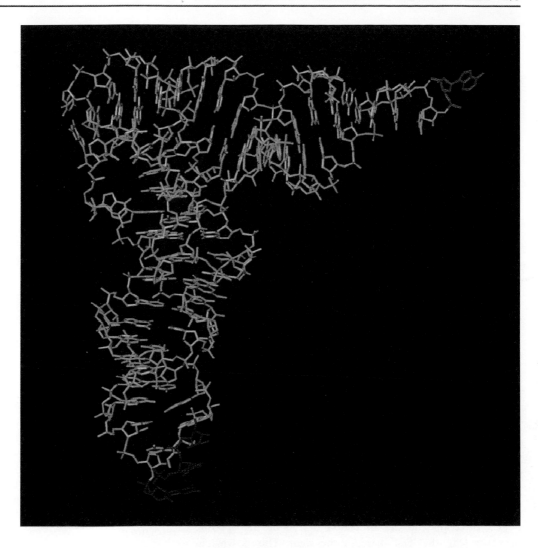

Fig. 22.2 Structural model of the three-dimensional structure of yeast tRNA^Phe. CCA terminus, yellow; anticodon, red.

The structures of tRNA molecules are a subject in themselves. They have modified and unusual bases in them but, at this stage, the essentials are the **anticodon** and the amino acid accepting site. The anticodon is a triplet of bases complementary to a codon (Fig. 22.3). Thus, if a codon on the mRNA is UUU (coding for phenylalanine), the anticodon corresponding to this on a tRNA molecule will be AAA. The vital point is that this particular tRNA molecule will accept (or have put on to it) *only* phenylalanine. We will come to how the amino acid is attached shortly. As stated, there are 61 codons, each of which represents an amino acid. To translate these, it might be expected that there would be 61 different tRNA molecules each with its own anticodon complementary to one codon and each accepting the one amino acid represented by its codon. In fact, there are fewer than 61 tRNA species—obviously, there must be at least one for each of the 20 amino acids but some tRNA molecules can recognize several codons—in this case each of the codons recognized by a single tRNA must, of course, represent the same amino acid. The arrangement means that the cell needs to make

fewer tRNA molecules. How is this achieved? The answer is 'wobble pairing'.

The wobble mechanism

In view of the importance of complementarity in DNA replication and transcription where Watson–Crick base pairing is absolutely sacrosanct, it may be somewhat disconcerting that, in codon–anticodon base pairing, the rules are bent a little. This applies *only* to the first base of the anticodon (with one exception, described later). The 'improper' pairing of this particular base is the result of flexibility in the adjacent structure in the tRNA (see Fig. 22.3), such that a U in this position will pair with A or G on the codon, and G with C or U. This is known as **wobble pairing** (see Fig. 20.17).

The meaning of the term 'first base of the anticodon' needs explanation. When an anticodon sequence is given *on its own*, the convention is followed of writing it in the 5′ → 3′ direction as is the case for any nucleotide sequence. In the anticodon GGC, G is the first (5′) base. However, codon–anticodon

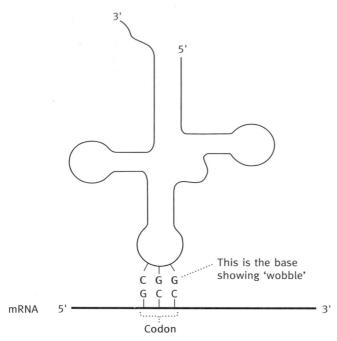

Fig. 22.3 Diagram of the base pairing of an anticodon of a tRNA molecule to an mRNA codon. To achieve antiparallel pairing the tRNA molecule is flipped over. This is why a tRNA structure is presented in one way in Fig. 22.1 (with 5′ by convention to the left) but in this paired form in the reverse way; the mRNA is written with the 5′ end to the left. (In the example shown, the tRNA can base pair with codons GCC and GCU, both of which code for alanine.)

interaction is antiparallel so that, if we want to show the same anticodon interacting with the codon, the latter is written in the 5′ → 3′ direction and the anticodon has to be the opposite direction with the 'first' base to the right. This is why, in diagrams, a tRNA molecule on its own (as in Fig. 22.1) is shown with the 5′ to the left but, when it is base paired on a codon, it is flipped over (as in Fig. 22.3). Thus, the anticodon GGC will base pair as

3′ CGG 5′ anticodon
5′ GCC 3′ codon.

The wobble mechanism permits the same anticodon to pair 'improperly' as

3′ CGG 5′ anticodon
5′ GCU 3′ codon.

Since GCC and GCU both code for the amino acid alanine, wobble pairing does not alter the amino acid sequence of the protein synthesized, but it enables a single tRNA to translate more than a single codon. In short, as stated, it permits the cell to synthesize fewer species of tRNA molecules.

Evolutionary tinkering has produced other ways of enabling more flexible codon–anticoding interactions without jeopardiz-

ing translational accuracy. One is to use the nonstandard base hypoxanthine (see Fig. 18.7 for its structure) in the anticodon for this will pair with C, U, or A in codons. (Hypoxanthine is often referred to as the inosine base because hypoxanthine riboside is called inosine.)

How are amino acids attached to tRNA molecules?

It will be clear to you that the system depends on a tRNA molecule having attached to it the particular amino acid specified by the codon which is complementary to the anticodon on that tRNA molecule. Thus a tRNA molecule with the anticodon AAA must be 'charged' only with phenylalanine since UUU is the complementary codon for that amino acid. If any other amino acid were to be attached to that tRNA, a mistake would be made in the synthesis of a protein molecule—phenylalanine will be replaced by the other amino acid. tRNA specific for phenylalanine is depicted as tRNA^{Phe} and so on for each of the 20 amino acids, using the first three letters of amino acid names as abbreviations. (Note that tRNA^{Phe} specifies only the tRNA; it does not mean that it has Phe attached. For the latter situation the term Phe-tRNA^{Phe} is used.) Enzymes that attach amino acids to tRNAs are called **aminoacyl-tRNA synthetases**. Each cell must have at least 20 different species of these enzymes which each can attach a specific amino acid to an appropriate tRNA. This means that each enzyme recognizes one or more specific tRNAs and the appropriate amino acid and joins them together. There is no general pattern for the way in which a synthetase recognizes the appropriate tRNA—the anticodon is recognized in some cases but, in others, a specific base or several bases elsewhere in the molecule are recognized.

The overall reaction, which involves ATP breakdown to supply energy, is

Amino acid + tRNA + ATP ↔ Aminoacyl-tRNA + PP$_i$ + AMP.

The process is sometimes referred to as **amino acid activation**.

The inorganic pyrophosphate is hydrolysed to 2P$_i$ and this drives the reaction to the right. However, there is more to this reaction. On the accurate selection by these enzymes of the correct amino acid depends the accuracy of translating mRNA into protein. After an amino acid is loaded on to a tRNA, the aminoacyl-tRNA enters into the protein synthesizing machinery but the latter has no means of checking that a particular tRNA is carrying the correct amino acid. It 'assumes', as it were, that it is correct so that it is very important that the enzyme attaching the amino acid does not have a high error rate. The active site of an enzyme is usually highly specific for its substrate but there are limits to the accuracy. It is relatively easy for an enzyme to distinguish between amino acids with markedly different characteristics, but much harder to do this between very similar amino acids such as valine and isoleucine. We remind you of their structures.

CH$_3$ CH$_3$
 CH
 |
CH—NH$_2$
 |
COOH

Valine

CH$_3$
 CH$_2$ CH$_3$
 CH
 |
CH—NH$_2$
 |
COOH

Isoleucine

The difference in binding energies due to a single —CH$_2$ group is not enough to give a sufficiently high degree of selectivity between isoleucine and valine, for without additional safeguards, the error rate of more than 1% would be much too high. The aminoacyl-tRNA synthetase specific for leucine would attach valine to tRNA^{Leu} at a rate that would result in an unacceptable rate of errors in mRNA translation, unless there were a corrective mechanism. A 'proofreading' mechanism is based on the fact that the overall reaction above occurs in two stages.

1. Amino acid + ATP ↔ Aminoacyl-AMP + PP$_i$.
2. Aminoacyl-AMP + tRNA ↔ Aminoacyl-tRNA + AMP.

The aminoacyl-AMP does not leave the enzyme. In the case of the isoleucine-specific enzyme, when tRNA^{Ile} attaches, a conformational change exposes an additional catalytic site that can destroy, by hydrolysis, valyl-AMP, while isoleucyl-AMP is not so attacked, possibly because it is too large to enter the site. The error rate of isoleucine attachment is reduced to about 1 in 60 000. Other aminoacyl synthetases do their 'proofreading' by hydrolysing incorrect aminoacyl-tRNAs. This is catalysed by a site on the same enzyme. The hydrolytic site recognizes tRNAs carrying a smaller amino acid than the correct substrate, or one of a similar size carrying a different type of group. For example, threonine is the same size as valine but the hydrolytic site of the valine-specific enzyme preferentially binds to the hydrophilic —OH group of threonine rather than to the hydrophobic —CH$_3$ of valine. The double selectivity—preferential loading of the correct amino acid on to the tRNA and preferential hydrolysis of aminoacyl-tRNAs carrying the incorrect amino acid—reduces the error rate to one in several thousand. Not all aminoacyl-tRNA synthetases have proofreading mechanisms. They are needed only where structurally similar amino acids occur. An occasional mistake in protein synthesis is not serious since huge numbers of protein molecules are made and all are ultimately destroyed (the faulty ones probably more quickly). Such mistakes do not have the same potential long-term consequence as errors in DNA synthesis but, all the same, a high degree of accuracy is still necessary.

The tRNA molecule has a terminal 3' trinucleotide sequence of CCA, the terminal A nucleotide having a free 3'-OH group on the ribose moiety. It is to the latter that the amino acid is attached by an ester bond (see Fig. 22.4). This CCA trinucleotide forms a flexible arm that can position the aminoacyl group on

(a)

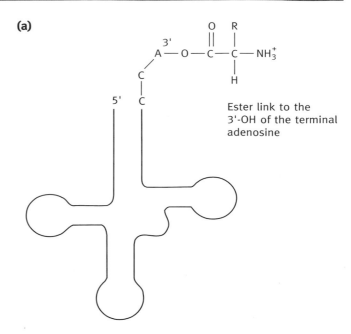

Ester link to the 3'-OH of the terminal adenosine

(b)

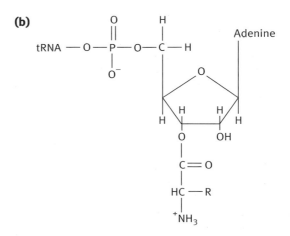

Fig. 22.4 (a) Diagram of tRNA molecule showing the CCA base sequence at the 3' end, where the amino acid is attached by ester link. **(b)** Structure of the terminal nucleotide with the attached amino acid. The aminoacyl group in solution migrates rapidly between the 2'- and 3'-OH of the ribose moiety. For convenience in the text we treat it as if it is on the 3'-OH.

the appropriate reactive site on the ribosome (see below). The ester formed by the aminoacyl group has essentially the same energy level as that of a peptide. Thus, there is no thermodynamic problem in transferring the aminoacyl-ester group to the —NH$_2$ of another aminoacyl group to form a peptide bond, as described below. In other words, the energy required for the formation of a peptide bond is inserted into the process by the aminoacyl-tRNA synthetase using ATP. However, this is jumping ahead for we now have to deal with how the amino acids on tRNA molecules are assembled into a polypeptide chain.

Ribosomes

Ribosomes are small particles present in cells in large numbers (except for terminally differentiated cells such as erythrocytes, which no longer synthesize proteins). Their name comes from their content of RNA or ribonucleic acid, which accounts for about 60% of their dry weight. A ribosome consists of two subunits—in *E. coli*, a large one containing two molecules of RNA and a small subunit with one RNA molecule. You have already met mRNA and tRNA; **ribosomal RNA** or **rRNA** is different again. Because of internal base pairing, rRNAs assume highly folded compact structures. Figure 22.5 is a diagram presented to give you some idea of the complex shape of one of the RNA molecules of an *E. coli* ribosome. This is the 16S rRNA of the small subunit. The large subunit has two different rRNAs, sedimenting at 23S and 5S (see below for S values). With it are associated many proteins forming solid particles—34 proteins in the case of the large subunit and 21 for the small.

When one gets to very large structures such as a ribosome, their sizes are measured in terms of the rate at which they sediment in an ultracentrifuge—expressed as **Svedberg units** or usually as **S values**. (An ultracentrifuge spins so fast that large molecules in solution move towards the bottom of the tube.) An *E. coli* ribosome is 70S, the subunits being 50S and 30S (the values are not simply additive, since S depends both on size and shape).

The overall *principle* of protein synthesis can be given very simply. The ribosome becomes attached near the 5′ end of mRNA and then moves down the mRNA towards the 3′ end, assembling the aminoacyl groups of charged tRNA molecules (that is, tRNA molecules with their amino acids attached), according to the sequence of codons, into a polypeptide chain. At the end, it meets a stop codon, at which point the protein is released, the ribosome detaches and dissociates into its subunits. Note that there are no 'special' ribosomes. In a given cell, any ribosome can use any mRNA just as a tape player can play any recorded tape; a qualification exists in that mitochondrial and chloroplast ribosomes are different from those in the cytoplasm (see page 304).

That is the principle—to understand the process we need to give more detail. Synthesis of a protein molecule can be divided into three phases—initiation, elongation, and termination.

Initiation of translation

It is of vital importance that a ribosome begins its translation of a mRNA at exactly the correct point—in other words, that it **initiates** correctly. mRNAs have 5′ and 3′ untranslated regions and the coding region lies in between. The ribosome must be able to recognize the *first* codon of the coding sequence and start translating there. Absolutely precise initiation is essential, for the correct translation of a mRNA depends on the ribosome being in the correct reading frame. This may need explanation. Suppose the **coding region** of an mRNA starts with the sequence

5′ AUGUUUAAACCCCUG · · · · · · · · 3′.

The first five amino acids are specified by the codons AUG, UUU, AAA, etc. There is nothing to indicate what constitutes a codon other than that the first three bases of the coding part of the mRNA encountered constitute codon 1, the next three, codon 2, etc. There are no commas or full stops between them. The message therefore depends on starting to read *exactly* at AUG. Suppose an error of one base is made and we start instead at base two. The codons then read would be

(A) UGU, UUA, AAC, CCC, UG · · · .

In other words, the codons translated would be totally different and the amino acids inserted, correspondingly, totally different. This error would be a **reading frameshift**; mutations deleting or adding one or two bases from an mRNA cause reading frameshifts and result in the amino acid sequence of the polypeptide synthesized after the frameshift being incorrect and the resultant protein being useless garbage instead of one designed by millions of years of evolution, the one specified by the gene. The general outline of protein synthesis and the components involved, apply to both prokaryotes and eukaryotes but there are significant differences in actual mechanisms between the two and we will deal with them separately. We will start with prokaryotes.

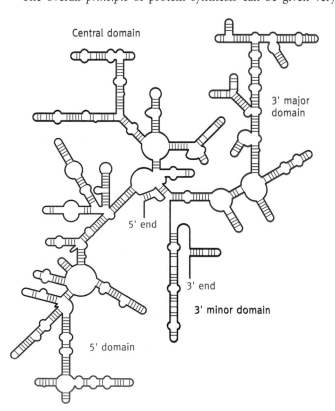

Central domain

3′ major domain

5′ end

3′ end

3′ minor domain

5′ domain

Fig. 22.5 Diagram of 16S rRNA.

Initiation of translation in *E. coli*

The entire mRNA is not translatable—there is a stretch of RNA 5′ to the translational start site; this is the ribosome positioning section. Initiation is *not* at the 5′ end of the mRNA and therefore the first codon must be identified in some other way.

The start site is the codon AUG (less frequently GUG); however, oddly enough, an AUG codon can occur *anywhere* in the mRNA since it codes for the amino acid methionine and, for a lengthy period, biochemists were forever wondering how the ribosome initiated only at the first AUG and not one of the others further down the mRNA. Why wasn't there a special codon always used for initiation? This confusing situation is resolved by the fact that there are two *different* tRNAs both specific for methionine—they have the same anticodon, but one tRNA is used exclusively for initiation and the other exclusively for adding methionine in the elongation process.

The tRNA that initiates translation can base pair with either of the codons AUG or GUG due to 'wobble'. Normally, wobble involves the 5′ base of the anticodon. In the tRNA used for initiation, it is the 3′ base of the anticodon that is the wobble base (left-hand as written when it is shown as bonded to its codon). This is illustrated in Fig. 22.6.

What then determines the different functions of the two methionine-specific tRNA species? The answer is that the *initiating* aminoacyl-tRNA species has structural features that are recognized by an **initiating protein** or **factor (IF2)** and delivered by this to the initiation complex. The aminoacyl-tRNAs involved in elongation following initiation, are recognized by a different cytoplasmic factor (described below), which delivers them to the ribosome. This factor does not bind to the initiating tRNA. There is another quirky difference in *E. coli*—the methionine, which becomes attached to the initiating tRNA, is formylated on its —NH$_2$ group by a transformylase using N^{10}-formyltetrahydrofolate (page 227) as formyl donor, so that prokaryote proteins, for reasons that are not clear, are synthesized with *N*-formylmethionine as the first unit. The formyl group, and frequently the methionine also, are removed before completion of the synthesis. The initiating tRNA is

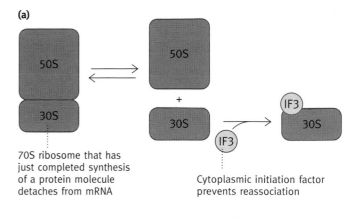

(a)

70S ribosome that has just completed synthesis of a protein molecule detaches from mRNA

Cytoplasmic initiation factor prevents reassociation

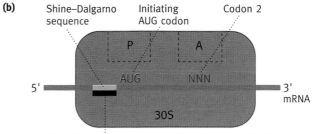

(b)

Shine–Dalgarno sequence   Initiating AUG codon   Codon 2

Part of 16S RNA sequence of the 30S subunit, with the complementary sequence to the Shine–Delgano sequence, this binding determining the precise site for the initiation codon AUG

Fig. 22.7 (a) Dissociation of 70S ribosome into 50S and 30S subunits. The initiating factor IF3 prevents 50S subunit association. In initiation of translation, the IF3 must be released before the 50S subunit can join. **(b)** The diagram showing the role of the Shine–Dalgarno sequence in positioning the 30S ribosome of *E. coli* on the mRNA at initiation. P, Peptidyl site; A, aminoacyl site. The P and A sites are only partial sites; when the 50S subunit joins (see text) the sites are completed.

Fig. 22.6 Diagram of the detail of the anticodon region, showing the 'wobble' position of: **(a)** general tRNAs involved in elongation; and **(b)** initiating tRNA$_f^{Met}$. The anticodon is coloured red. The wobble base is determined by the adjacent structure of the tRNA. The meaning of tRNA$_f^{Met}$ is explained in the text.

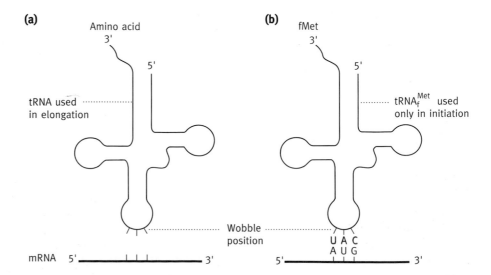

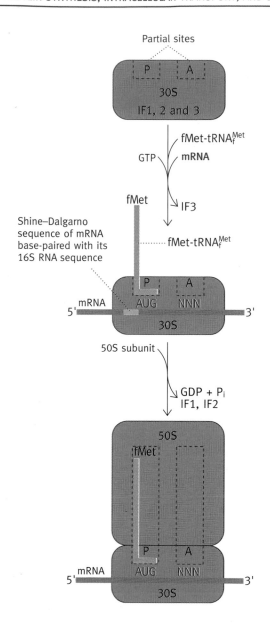

Fig. 22.8 Initiation of translation in *E. coli*. The initiating tRNA, tRNA$_f^{Met}$, is represented by the blue line, the anticodon being the horizontal short line. The fMet-tRNA$_f^{Met}$ is delivered to the 30S subunit by IF2. NNN represents any codon (N for any nucleotide). *Note*. The ribosome also has an exit site not shown in the diagram. This site will be discussed later.

usually called tRNA$_f^{Met}$ (f for formyl) and the charged version fMet-tRNA$_f^{Met}$ often shortened to fMet-tRNA$_f$. (You may find some differences in nomenclature in different texts.) This then clears up the mystery of how the correct methionine-specific tRNA is used for initiation and elongation—specific protein factors involved in initiation and elongation recognize only fMet-tRNA$_f^{Met}$ and Met-tRNA$_m^{Met}$, respectively.

In the cytoplasm there is a pool of 30S and 50S ribosomal subunits in equilibrium with 70S ribosomes. An initiating factor in the cytoplasm (IF3) binds to the 30S subunit and prevents

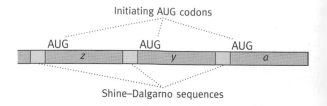

Fig. 22.9 Diagram of the structure of a polycistronic prokaryote mRNA. *z*, *y*, and *a* refer to coding regions of the *lac* mRNA (see Fig. 21.9).

premature reassociation with the 50S subunit at this stage (fig. 22.7(a)). Two other factors, IF1, IF2 are also needed. These protein factors are involved transitorily in initiation—they bind, participate in the process, and are released into cytoplasm to function again in a new round of initiation. All are found associated loosely with 30S subunits. The function of IF1 is not known; IF2 is necessary for fMet-tRNA$_f^{Met}$ to bind.

In the presence of mRNA, fMet-tRNA$_f^{Met}$, and GTP, a complex of these with a 30S subunit is formed and IF3 is released. As shown in Fig. 22.7(b), there is a sequence of bases on the mRNA, known as the Shine–Dalgarno sequence, that is complementary to a section of the 16S rRNA. Binding of the two correctly positions the mRNA on the small ribosomal subunit. The tRNA is positioned at the partial P site on the subunit with its anticodon paired with the AUG codon (Fig. 22.8). After IF3 release, this complex now associates with a 50S subunit. The event is accompanied by the hydrolysis of GTP and the release of GDP and P$_i$ and of IF1 and IF2 (Fig. 22.8). We now have a complete 70S ribosome positioned on the mRNA in the P site with the donor fMet-tRNA$_f$ anticodon base paired with the start AUG codon and the A site vacant, awaiting delivery of the second amino acid on its tRNA. Initiation is complete.

Some bacterial mRNA molecules are polycistronic; the *lac* mRNA is one such example (page 277). In this, there are three regions in the one mRNA molecule coding for three different proteins. In this situation, each coding region has a Shine–Dalgarno sequence adjacent to it so that each can be translationally initiated independently (Fig. 22.9).

Once initiation is achieved, elongation is the next step

Cytoplasmic elongation factors

There are two soluble protein **elongation factors** in the cytoplasm. Both are 'G' proteins—they bind a molecule of GTP and only in this state do they bind to ribosomes. Both are latent GTPases. On the ribosome they hydrolyse the bound GTP molecule to GDP and P$_i$ and release of the latter causes a conformational change in the proteins. In their GDP-bound state they detach from the ribosome; in the cytoplasm their GDP is exchanged for GTP and the factors are ready to participate in a new round of elongation.

The two factors are **EF-Tu** (**elongation factor, temperature unstable**) and **EF-G**. EF-Tu has the task of delivering the incoming aminoacyl-tRNA to the ribosome; EF-G that of moving the ribosome along the mRNA in the $5' \rightarrow 3'$ direction to the next codon, once an aminoacyl group has been added to the growing peptide.

It is useful to keep in mind this central concept of a pair of factors alternately hopping on to the ribosome in their GTP form, performing their tasks, and detaching in their GDP form to be recycled for subsequent rounds of elongation. With that introduction we can move to the details of peptide bond synthesis.

Mechanism of elongation

We start with the initiation complex illustrated in Fig. 22.10 (state a). In this complex, we have an fMet-tRNA$_f^{Met}$ in the P site and the A site is vacant. *Only* in the initiation process does the P site accept tRNA charged with an amino acid—in this case N-formylmethionine; *all* subsequent aminoacyl-tRNAs enter the A site. Aminoacyl-tRNAs (other than the initiating species) are complexed with the elongation factor EF-Tu, carrying a molecule of GTP bound to it. The EF-Tu-GTP-aminoacyl-tRNA complex binds to the ribosome such that the aminoacyl-tRNA occupies the A site with its anticodon positioned at the mRNA codon (Fig. 22.10 (state b)). It is especially important to note that EF-Tu-GTP does *not* bind to fMet-tRNA$_f$, the initiating tRNA—it binds only to aminoacyl-tRNAs involved in elongation. Initiation and elongation are thus kept quite separate. EF-Tu exists in high concentration in *E. coli*— sufficient to bind all of the aminoacyl-tRNA in the cell. On the ribosome, EF-Tu is a GTPase and the bound molecule of GTP is hydrolysed to GDP and P$_i$, the latter being released. This is the signal for the EF-Tu to detach from the aminoacyl-tRNA, presumably due to allosteric changes occurring, and the EF-Tu-GDP leaves the ribosome to be regenerated by a GDP–GTP exchange process in the cytoplasm, giving state (c) in Fig. 22.10. The aminoacyl groups on the two tRNA molecules on the P and A sites are in the vicinity of a ribosomal enzyme, **peptidyl transferase**, that transfers the fMet group of the tRNA in the P site to the free amino group of the aminoacyl-tRNA in the A site, producing a dipeptide attached to the tRNA (state (d) in Fig. 22.10). Although we have described the peptidyl transferase as an enzyme, it has been found that the activity remains with the ribosome structure even after 95% of the ribosomal protein has been extracted, raising the strong possibility that it is, in fact, a property of the ribosomal RNA.

The peptidyl transferase reaction is as follows

tRNA—O—CO—fMet + NH$_2$—CH—CO—O—tRNA $\longrightarrow$
(P site) | (A site)
                                R'

tRNA—OH + fMet—CO—NH—CH—CO—O—tRNA .
(P site) | (A site)
                                    R'

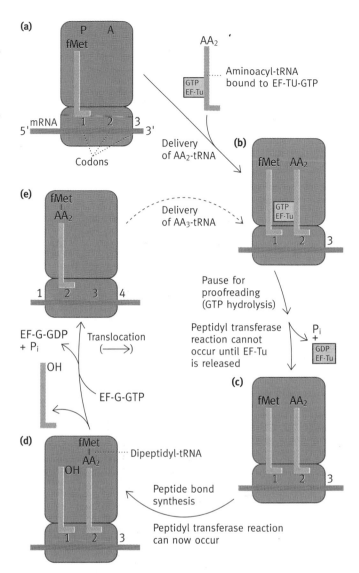

Fig. 22.10 Diagram of the elongation process in protein synthesis following initiation. tRNAs are shown as blue lines; AA, aminoacyl group. The positioning shown of the EF-Tu-GTP on the tRNA and on the ribosome are arbitrary. This diagram does not make evident why the ribosomal enzyme catalysing the peptide bond synthesis is called peptidyl transferase. However, if you do the next round of synthesis yourself (see text), you will see that in all subsequent rounds of synthesis it is a peptide that is transferred to the incoming aminoacyl tRNA—hence the name.

After this first peptide bond synthesis, we have the A site filled with a peptidyl tRNA and the P site containing an uncharged tRNA. The ribosome is now re-positioned one codon further along the mRNA, a process called **translocation**. The discharged tRNA is moved to a third exit site on the 50S subunit (not shown in Fig. 22.10) from which it escapes to be re-used, giving state (e). The movement of the ribosome relative to the mRNA requires EF-G, which is also known as **translocase**. GTP hydrolysis is required for the translocation. The binding of the EF-Tu-GTP and the EF-G occur alternately on the ribosome;

only one at a time can be bound. Thus peptide synthesis and translocation alternate one after the other.

In Fig. 22.10, the mechanism by which the peptidyl transferase reaction and translocation occur from stages (d) to (e) is not specified. More recent work has indicated that, during the process, the tRNAs each straddle two sites on the ribosome (see Fig. 22.11). The discharged tRNA (after losing its peptidyl group) straddles the P and E (exit) sites—the anticodon end of the molecule is still in the P site but the other end is in the E site. Similarly, the tRNA in the A site (now carrying the peptide) straddles the A and the P sites.

Two alternative models have been proposed to account for this straddling. Model I (Fig. 22.11) envisages that one end of the aminoacyl-tRNA swings over as shown; peptide transfer occurs, and the discharged tRNA swings one end to the E site as shown. Translocation straightens up the situation and we are back to that shown in Fig. 22.10 (state e), ready for the next round of elongation.

In the alternative model II, the large subunit moves but the small one remains stationary. This creates hybrid P/E and A/P binding sites as shown. After the peptidyl transferase reaction, the hybrid sites are occupied as shown. Translocation of the small subunit then produces the same situation as in model I, ready for the next amino acid addition.

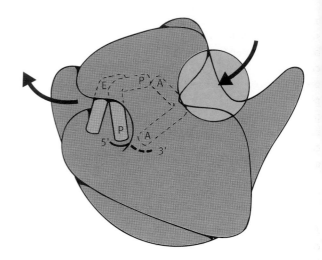

Fig. 22.12 Diagram of a ribosome with plausible locations for ribosome-bound tRNAs in the A/A, P/P, and E states (see Fig. 22.11 for the terminology). The shaded area at the right shows the approximate site of interaction of EF-Tu. The polarity of a fragment of mRNA containing the A- and P-site codons is shown. The arrows indicate the likely path of a tRNA as it transits the ribosome.

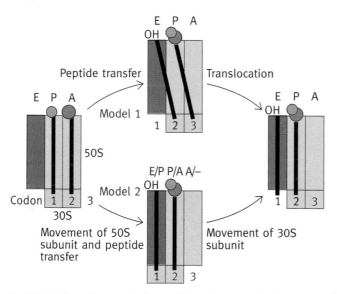

Fig. 22.11 Alternative models for peptide synthesis on the ribosome and translocation. Both reflect the observation that the tRNAs straddle the P/E and A/P sites on the ribosome. In model I, the AA-tRNA swings to make the straddle; in model II, hybrid sites are formed on the ribosome. Peptide synthesis is presumed to occur such that the peptide always remains in the same position relative to the large subunit as shown. In this illustration, synthesis of the first peptide bond has been chosen as the example so that the red circle represents fMet attached to tRNA on the P site which is aligned with mRNA codon number one; at the end of the round it is aligned with the second codon. The coloured rectangles represent binding sites: E, exit site; P, peptidyl site; A, acceptor site. E/P and A/P are hybrid sites formed by large subunit movement relative to the small subunit.

Both models have the important feature that *the nascent peptide remains in a fixed position relative to the large subunit*, as shown. This would eliminate the problem of how a tRNA physically moves with a relatively huge polypeptide attached to it. Model II would additionally explain why ribosomes always have two subunits. The ribosome is a very complex structure indeed. In Fig. 22.12 there is a more realistic diagram of what it is actually like with its tunnel-like structure.

The ribosome is now in the situation shown in Fig. 22.10(e). The P site is occupied by the dipeptidyl-tRNA while the A site, positioned at the third codon is vacant. The elongation of the polypeptide chain involves round after round of the same process. An aminoacyl-tRNA complexed to EF-Tu-GTP is delivered to the A site, the peptidyl group is transferred to it from the P site, the discharged tRNA moves to the exit site, the peptidyl-tRNA in the A site moves to the P site and the ribosome translocates one codon, and the whole process starts again. If the protein being synthesized is 200 amino acids long, the final round involves transfer of the 199 amino acid long peptide to the final amino acid on its tRNA giving a protein–tRNA complex. The dog is added to the tail, not the other way round. The amino terminal end is synthesized first; the last amino acid added forms the carboxyl terminus.

The fidelity of translation depends on the correct aminoacyl-tRNA as specified by the codon aligned with the A site being bound in that site in each round of elongation. This must be due to codon/anticodon interaction selecting the correct aminoacyl-tRNA but exactly how this selection is effected is not completely clear; the EF-Tu does not 'know' which aminoacyl tRNA has to react next and it would seem that it delivers them to the A site randomly. One hypothesis proposes that an incorrect aminoacyl-tRNA-EF-Tu-GTP complex diffuses away before it reacts since it will not be held in the A site as strongly by hydrogen bonding to the codon as a correct one would be. Peptide synthesis cannot occur until EF-Tu-GDP is released and this cannot happen until the GTP is hydrolysed. A slight delay in GTP hydrolysis is postulated to provide sufficient time for this 'proofreading' process to be effective. The possibility has been raised that the ribosome in some unspecified way participates in recognizing the correct aminoacyl-tRNA, but at present this is highly speculative and no mechanism for this has been postulated.

Termination of protein synthesis

At the end of the mRNA there is at least one of the three stop codons for which no tRNA exists; these are UAG, UAA, and UGA. When the ribosome reaches one of the stop codons, a specific cytoplasmic **protein release factor** causes release of the finished protein from the tRNA by altering the peptidyl transferase such that it hydrolyses the ester bond between the —COOH of the protein and the —OH of the 3′ terminal nucleotide (that is, it transfers the peptide to water). The ribosome detaches from the mRNA and dissociates into subunits ready for the next round of initiation. Two release factors are known, recognizing different stop codons.

What is a polysome?

It takes about 20 seconds for a ribosome to synthesize an average protein in *E. coli*, which adds about 15 amino acids per second. If only a single ribosome at a time moved along the mRNA molecule, the latter could direct the synthesis of the protein at the rate of one molecule per 20 seconds. However, as soon as an initiated ribosome has got underway and has moved along about 30 codons, another initiation can occur so that one ribosome after another hops on to the mRNA and they follow one another down the mRNA, each independently synthesizing a protein molecule—a typical case would be about five ribosomes per mRNA molecule but this varies with the length of the mRNA. The structure is called a polyribosome or more usually a **polysome**.

How does protein synthesis differ in eukaryotes?

The process is the same in the essentials, but important differences exist. Eukaryote ribosomes are larger (80S, with 60S and 40S subunits) and have extra rRNA molecules and proteins. Methionine is always the first amino acid but the methionyl-tRNA used for initiation is not formylated. The purpose of formylation in prokaryotes is not understood. However, as in prokaryotes, there is a special methionyl-tRNA for initiation distinct from that used in elongation. This does not mean that

all eukaryote proteins start with methionine for frequently this amino acid is removed from the polypeptide. A 40S initiation complex is assembled that is analogous to the 30S complex in prokaryotes and this then reacts with the large (60S) ribosomal subunit to give the final initiation complex. GTP is again hydrolysed in the process. However, the mechanism by which mRNA is positioned correctly so that the P site corresponds with the AUG initiation codon is completely different in eukaryotes. You will recall (page 279) that eukaryote mRNAs are 'capped' with a methylated guanine nucleotide at the 5′ end. There is no Shine–Dalgarno sequence in the eukaryote messenger RNA. Instead, a group of protein factors attach to the cap, which is joined by a 40S ribosomal subunit (Fig. 22.13).

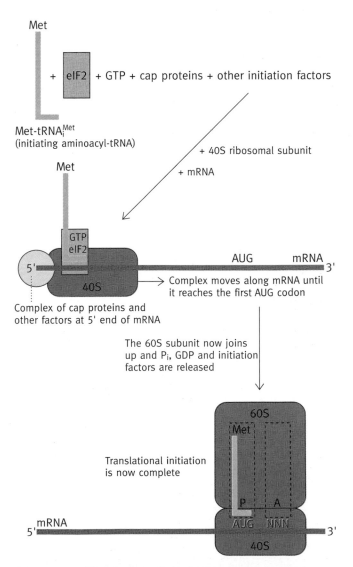

Fig. 22.13 Simplified diagram of initiation in eukaryotes. Note that several eukaryote initiation factors besides eIF2 are involved. tRNAᵢ^Met, initiating RNA. eIF2 is the eukaryotic initiation factor corresponding to IF2 in prokaryotes.

We thus have a 40S pre-initiation complex at the very 5′ end of the mRNA, some distance away from the initiating AUG codon. By an ATP-driven mechanism, the 40S subunit complex moves along the mRNA until it encounters the first AUG triplet and then the 60S subunit joins up to complete the initiation. GTP is hydrolysed in the process. This has one interesting repercussion. There can, with this scanning mechanism for selecting the AUG initiation codon, be only one initiating site per mRNA molecule. Eukaryote mRNAs are **monocistronic** (code for a single polypeptide) as contrasted with prokaryote mRNAs, which are often polycistronic with Shine–Dalgarno sequences provided for the initiation of translation of each cistron. The counterpart of EF-Tu in eukaryotes is called EF1α; that of EF-G is called EF2.

What is the situation in mitochondria?

Mitochondria contain DNA of their own and have their own protein synthesizing machinery. A widely held hypothesis is that mitochondria originated from prokaryote cells that became incorporated into eukaryote cells (page 289). The ribosomes of mitochondria are prokaryote-like and use fMet-tRNA$_f^{Met}$ for initiation. They have other features of interest such as a slightly different genetic code, and codon–anticodon interactions are simplified so that mammalian mitochondria can manage with only 22 tRNA species. The possibility of such simplification is related to the fact that mitochondria synthesize only a handful of different proteins. The mitochondrion is not autonomous. Most of their proteins are coded for by genes in the cell nucleus and are transported into the organelle (see below). Chloroplasts are also believed to have originated from incorporated (photosynthetic) prokaryotes.

Effects of antibiotics and toxins on protein synthesis

Antibiotics are the chemical missiles microorganisms throw at each other in the competition for survival. They attack highly important points in cellular processes and translation offers many such targets.

In prokaryotes, streptomycin affects initiation, kirromycin prevents EF-Tu release, and erythromycin and chloramphenicol inhibit translocase and peptidyl transferase respectively (this is true of mitochondrial ribosomes too). Fusidic acid inhibits translocation by blocking EF-G-GDP release.

In addition to these examples of antibiotic action, diphtheria toxin inhibits EF2, the eukaryote translocase (equals EF-G in bacteria). Ricin, the toxin of the castor bean, is an *N*-glycosidase that removes a single adenine base from one of the eukaryote ribosomal RNAs and inactivates the large subunit. One molecule of ricin can destroy a cell containing tens of thousands of ribosomes.

How does the polypeptide chain synthesized on the ribosome fold up?

Not so many years ago it was generally assumed that, once the polypeptide chain was assembled on the ribosome, that was the end of the matter, for the protein would then automatically fold up into the three-dimensional structure determined by the interactions of its amino acid residues. This, of course, was of fundamental importance because we know of only one type of information available to determine the structure of a protein and that is the base sequence of a gene which is translated into the amino acid sequence of a polypeptide. There is, moreover, experimental evidence for the concept. The enzyme ribonuclease was denatured (unfolded) by a hydrogen-bond-breaking agent such as urea and the reduction of disulfide bonds. On removal of the urea by dialysis and re-oxidation, a proportion of the enzyme refolded correctly and regained catalytic activity. There is, therefore, no doubt that the amino acid sequence of a protein determines its final native configuration. However, instead of the synthesis of the polypeptide chain being the end of the matter, we come instead into yet another area of biochemistry that almost boggles the mind.

In a polypeptide chain there are almost infinite possibilities of the amino acid residues associating with each other. A stretch of hydrophobic amino acids in one part of the chain might associate as it is formed with another such stretch, but the association may be completely 'improper' and not found in the native protein. It theoretically is possible that the polypeptide tries out every conceivable internal association pattern until it arrives at the minimal free energy of the native protein—a totally random method of folding. However, it has been calculated that such a process would require countless millions of years to find the folded form, while in the cell it all has to happen over in a time-scale of a minute or so.

While the ribonuclease *in vitro* refolding has been repeated with other proteins, these were of small molecular weight and usually consisted only of a single domain. Attempts to refold larger proteins, especially those with multiple domains, have been less successful. Since the majority of proteins are probably in the latter category there is clearly a problem of very great dimensions.

The mechanism of folding of proteins *in vivo*, by which we mean the pathways by which the unfolded newly synthesized polypeptide acquires its correct three-dimensional configuration, is not yet clearly defined. It is, however, believed that certain sections of a polypeptide may rapidly assume a secondary structure and that these somehow facilitate correct folding of the entire molecule. The concept therefore is of a modular folding in which folded modules rapidly form and these facilitate the folding of the rest of the molecule. While it thus appears that protein folding occurs in a series of steps, the details of the process are far from understood.

So far we have discussed protein folding as an entirely automatic unaided process. However, there are two classes of proteins that are involved in protein folding. One class is of conventional enzymes and the second is of molecular chaperones. We will now describe these in that order.

Enzymes involved in protein folding

The first is **protein disulfide isomerase** or **PDI**. This enzyme 'shuffles' —S—S— bonds in polypeptide chains. If an incorrect S—S— bond were to be formed, being covalent it would not break spontaneously and would fix the polypeptide in an incorrect configuration. PDI by breaking and reforming S—S bonds between different cysteine residues (page 30) permits the folding to correct itself. High concentrations of PDI are found inside the endoplasmic reticulum involved in folding proteins destined for secretion, many of which have disulfide bridges. (We will be describing this later in the chapter so don't worry about understanding that statement just yet.)

Another enzyme is **peptidyl proline isomerase** or **PPI**. Wherever a proline group occurs in peptide linkage the configuration can be *cis* or *trans*. The PPI plays the role of 'shuffling' proline residues between the configurations so as to permit the whole protein to assume the correct configuration. A quite remarkable class of proteins called **cyclophilins** has been discovered with proline isomerase activity. These bind the antibiotic immunosuppressant, **cyclosporin** (used in organ transplants). Just how proline isomerase correlates with the biological roles of cyclophilins is not clear at present.

Molecular chaperones and folding

There exists in all cells a family of proteins collectively known as **chaperones**. They are not enzymes, but rather proteins that have the ability to recognize, and bind to, partially folded proteins or, put in another way, partially unfolded proteins. The concept is that, as the polypeptide emerges from the ribosome, 'improper' associations may occur (hence the term 'chaperone' for agents preventing this) such as associations between hydrophobic sections. Such associations could prevent the formation of correct folding associations. Binding of chaperones to such sections is believed to stabilize the partially folded molecule until a stage is reached when correct associations can occur. This, of course, envisages that the chaperones have to dissociate from the polypeptide in the correct sequence and that, when this happens, the opportunity exists for correct folding. Although the details are unclear, chaperones play an important role in folding of polypeptide chains. The complexity of the system is indicated by the ATP requirement for the chaperones to dissociate from the polypeptide chain, as required by the folding process. Note that chaperones are concerned with the kinetics of folding—not with the nature of the final folded form, which is determined by the amino acid sequence.

Prion diseases and protein folding

A group of fatal neurological degenerative diseases that affect humans and animals exists. These include the Creutzfeldt–Jakob disease and kuru in humans. The latter, known as the laughing disease because of the facial grimaces it causes, was believed to be transmitted in certain New Guinea tribes by cannibalism. In sheep, the disease is known as scrapie because the animals scrape off their wool by rubbing against fence posts; in cows there is bovine spongiform encephalopathy (BSE) or mad cow disease. The diseases can be transmitted by consumption of infected tissue or, rarely, can be an inherited trait. The diseases were believed to be caused by 'slow viruses' because they are infectious and can take years to develop. However, all attempts to find nucleic acid in infectious material purified from brain failed and the evidence strongly eliminates the possibility of any infectious agent such as a virus being involved. Despite this, the infectious agents appear to replicate. Infection is associated with a protease-resistant form of a normal protein (prion protein or PrP) found in brain. The disease-producing unit is called a **prion** (for *pro*teinaceous *in*fectious particle). A protein that can infect mice with scrapie is called PrP^{sc}, while its *normal* counterpart is called PrP^c (c for constitutive). The two have the same polypeptide and are coded for by the same gene but their folded conformations are different, PrP^{sc} having a high β-sheet content and PrP^c having almost none. PrP^{sc},, unlike PrP^c, readily forms aggregates, which result in the amyloid plaques found in prion diseases. It is believed that the different conformation of the protein is what results in the disease. The question arises then as to how an improperly folded protein can be infectious and reproduce itself, since you will appreciate that none of the biochemical mechanisms for protein production that we have discussed allows a protein molecule to direct its own replication. There is, however, strong evidence that PrP^{sc} somehow causes PrP^c (the normal protein) to convert to the abnormal form. This has been demonstrated by incubating the two together whereupon the conversion was observed.

Models for the mechanism of this conversion include the intervention of a chaperone and an energy source for an unfolding of PrP^c and a refolding under the influence of a PrP^{sc} molecule. An alternative 'nucleation' model is that molecules of PrP^c are trapped by an aggregate of PrP^{sc} and conformational re-arrangement of PrP^c takes place. Because of experimental limitations, it has not been shown that in the demonstration of the *in vitro* protein conversion described above, the protein converted was in fact infectious.

The conversion of PrP^c to PrP^{sc} is a very rare event in the absence of infection so that spontaneous occurrence of the disease is very rare. It is believed that mutations in the gene for the normal PrP may increase the probability of this post translational conversion that results in the hereditary origin of some cases of the disease. Once some PrP^{sc} is formed, it would then trigger the autocatalytic formation of more.

Other roles for molecular chaperones

We have described chaperones here because of their role in protein folding. But it emerges that this is but one aspect of the wide role of molecular chaperones. As stated, chaperones have the ability to combine with unfolded or partially folded proteins—the polypeptide emerging from the ribosome is one such case.

However, proteins can be denatured, which means some degree of unfolding, by all sorts of abuses—heat is one; excessive oxidation, ionizing irradiation, and UV light are others. These are all stresses to which cells are subject and that can cause protein denaturation. For a long time it has been known that heat shock induces the synthesis of special proteins whose presumed purpose is to protect the cell. If *E. coli* is raised to 42°C, there is a burst of synthesis of **heat shock proteins** (page 276). It turns out that these are chaperones. It has been suggested that they should be called **stress proteins**.

The functions of chaperones, apart from assisting folding of newly synthesized proteins, include the untangling and stabilization of denatured proteins and the targeting of aged proteins for destruction by lysosomes (page 209). They are also required for protein translocation across mitochondrial membranes as described later in this chapter, for rearrangement of subunits in complexes, and for helping in the assembly of oligomeric protein complexes. A particularly useful definition of this class of proteins has been given by Hendrick and Hortl (reference given in the Further reading list): 'Presently we define a molecular chaperone as a protein that binds to and stabilizes an otherwise unstable conformer of another protein—and, by controlled binding and release of the substrate protein, facilitates its correct fate *in vivo*: be it folding, oligomeric assembly, transport to a particular subcellular compartment or controlled switching between active/inactive conformations.'

The latter point concerning switching refers to internal receptors for glucocorticoid hormones (see page 354 for discussion of such receptors).

It is inevitable that such diverse functions will require many different chaperones and already a good many are known. It is an area that is certain to grow in importance.

How are newly synthesized proteins delivered to their correct destinations?

There is only a single type of ribosome in a cell and they are all located in the cytoplasmic compartment of the cell (apart from those in mitochondria and chloroplasts, which are prokaryote-like). Which protein a given ribosome synthesizes at any one time is solely a function of the mRNA that it happens to be translating at the time, but different proteins, once synthesized, often have different destinations. Many are cytoplasmic proteins and, since they are synthesized in the cytoplasm, there is, in this case, no problem as regards their location in the cell. They are released from the ribosome and stay there.

However, there are many proteins located in the cell membrane (see page 47). How do they get there? Going further, consider a liver cell; it produces blood serum proteins in large amounts. These are large proteins and cell membranes obviously are designed not to leak proteins. How are blood proteins selectively released? The same applies to release of any extracellular proteins, for example, digestive enzymes or insulin from the pancreas. There are yet more problems. Most mitochondrial proteins are synthesized in the cytoplasm. How are these selectively transported into the mitochondria? Lysosomes and peroxisomes (Chapter 16) are membrane-bound vesicles full of enzymes but they cannot synthesize proteins. How are these transported into the correct vesicle?

We will now deal with the various types of protein translocation, starting with those proteins that are first transported into the endoplasmic reticulum or ER. The latter is something we haven't yet described.

What is the endoplastic reticulum or ER?

If you look at a section of a liver or pancreas cell in the electron microscope, there is a large amount of membrane *inside* the cell forming a huge reticulum enclosing the ER lumen. It is a complex network of interconnecting flattened bags. The cytoplasm is outside the ER membrane and the ER lumen is a totally separate compartment inside the ER membrane (see the electron micrograph, Fig. 3.15). In Fig. 22.14 you will see that part of the ER is studded with ribosomes, giving it a rough appearance in the electron microscope; part of the ER lacks these ribosomes. It is biochemically distinct and is called **smooth ER**. Proteins destined for other than the cytoplasm, nucleus, or mitochondria are first transported into the rough ER lumen, which leads to the next question.

How are proteins secreted through the ER membrane?

When a protein is to be secreted or has to end up in the external plasma membrane, inside lysosomes, or in the lumen of the ER itself, it has on its amino terminal end a **leader sequence** of about 25 ± 11 amino acids. The leader amino acid sequences of different proteins have a pattern, rather than a fixed sequence, as shown diagrammatically in Fig. 22.15. The **proteolytic cleavage**

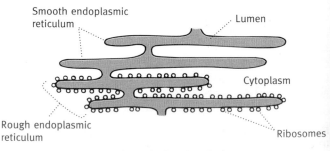

Figure 22.14 Diagram of the endoplasmic reticulum.

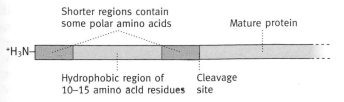

Fig. 22.15 Diagram of a typical leader sequence attached to the N-terminal end of a protein destined to be transported through the ER membrane. Such leader sequences show the same general pattern of amino acids but no specific amino acid sequence. Acidic amino acids are not present.

site is specified by a sequence pattern and specifies where the final mature protein will start.

Note that a *free* ribosome in the cytoplasm synthesizes the leader sequence of the protein first. This is immediately recognized by an **SRP** or **signal recognition particle** (an RNA–protein complex in the cytoplasm), which binds to the ribosome–nascent peptide complex and arrests further elongation of the polypeptide chain. The mechanism by which the 'arrested' ribosome transfers the nascent polypeptide through the ER membrane, as it is synthesized, is complex, but the essential features are now known.

In the ER membrane there are SRP receptors or **docking proteins** to which the cytoplasmic ribosome–SRP complex attaches. In a series of steps, during which a molecule of GTP is hydrolysed to GDP, the ribosome becomes attached to proteins in the membrane that form the **polypeptide translocating pore**, sometimes called a **translocon**, as shown in Fig. 22.16 and the SRP is released into the cytoplasm to be used again. The growing polypeptide is translocated through the protein pore in the membrane in a looped fashion with the signal peptide fixed to the pore (see Fig. 22.16). Associated with the pore is the **signal peptidase**, which hydrolyses off the leader sequence, and the polypeptide proceeds through into the lumen of the ER. It is not known what drives the translocation of the growing peptide. Folding of the polypeptide probably involves chaperones in the lumen.

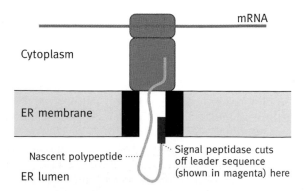

Fig. 22.16 Diagram showing the manner in which a polypeptide is transported across the ER membrane via the membrane 'translocon'. See text for the sequence of events leading up to this.

Glycosylation of proteins in the ER lumen

In Chapter 3 (page 48) we mentioned that some proteins, particularly membrane proteins and secreted proteins, have complex oligosaccharides added to them. The attachment points are either the amide $-NH_2$ of asparagine side groups (*N*-glycosylation) or the $-OH$ of serine and threonine residues (*O*-glycosylation). The additions are made in a series of steps. Inside the ER, *N*-glycosylation is carried out. The first step of this is particularly interesting in that a 'core' oligosaccharide of 14 sugar units is assembled in the cytoplasm and transported through the membrane attached to a long hydrophobic chain (which can exceed 100 carbon atoms in length) called **dolichol phosphate**. A transferase enzyme on the inside the ER membrane transfers the group to the nascent polypeptide as it enters the ER lumen. *O*-glycosylation occurs in the Golgi (see the next section).

What happens to the polypeptide translocated into the rough ER lumen?

The ER is a closed sac with no exits. Protein in the lumen of the rough ER is budded off into transport vesicles (from ribosome-free regions) and these deliver ER lumen contents to another organelle—the **Golgi apparatus** or **complex**—by fusing with its membrane. In some ways the Golgi complex is the most remarkable functional unit of the cell, almost to the point of improbability, so astonishing is the job it does.

The gross structure of the Golgi is simple—about half a dozen flattened membranous sacs without any entrances or exits. It receives proteins from the ER via the transport vesicles referred to above and it sends out proteins by budding off vesicles, as shown in Fig. 22.17. An EM of Golgi can be seen in Fig. 3.15.

Not all of the proteins arriving in the Golgi apparatus are for secretion; some are destined to be enclosed in lysosomes or peroxisomes; some, such as the protein disulfide isomerase (page 305) and glycosylating enzymes, are to be returned to the rough ER because they are swept along to the Golgi with all of the other proteins. The Golgi apparatus sorts out these proteins and sends them out in vesicles that deliver their contents to the correct place. So the really remarkable function of the Golgi is that of sorting proteins according to their destination and sending them out in appropriately addressed 'parcels'.

The Golgi packages proteins for secretion into vesicles that migrate towards the plasma membrane. There are two types of secretion. Some proteins are released continuously as produced, such as serum proteins which are released from the liver without stimulus. In this process, the vesicles fuse with the cell plasma membrane as they arrive there and release their contents by exocytosis. In the case of digestive enzymes from, for example, the pancreas, enzyme release is wanted only when there is food in the gut to digest. In this case the vesicles from the Golgi containing these enzymes are larger secretory vesicles. These store the enzymes until wanted, at which point a neuronal, or hormonal, stimulus causes a massive release by exocytosis of the

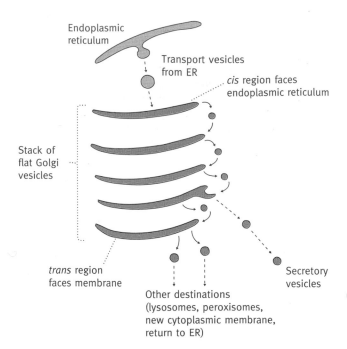

Endoplasmic reticulum

Transport vesicles from ER

cis region faces endoplasmic reticulum

Stack of flat Golgi vesicles

trans region faces membrane

Other destinations (lysosomes, peroxisomes, new cytoplasmic membrane, return to ER)

Secretory vesicles

Fig. 22.17 Diagrammatic representation of the central role of the Golgi apparatus in posttranslational sorting and targeting of proteins. In addition, newly synthesized membrane lipids are targeted to the appropriate destination by transport vesicles. Vesicle formation for transport between the Golgi cisternae involves a remarkable mechanism in which GTP–protein complexes bind to the membrane, which leads to binding of 'coatamer' protein molecules. Budding of a vesicle at the binding site leads to the formation of a coated vesicle. These uncoat at the target membrane as a result of GTP hydrolysis to GDP; uncoating leads to fusion of the vesicle and target membranes. Targeting molecules on the vesicle direct the vesicle to its destination. The 'coating' mechanism for vesicle formation appears to be a very widespread general mechanism.

secretory vesicles (the signal to discharge, and its mechanism, are described on page 366). Budding of transport vesicles from the Golgi sacs involves the coating of an area of membrane by coat protein molecules together with a G protein with attached GTP. GTP hydrolysis is the signal for uncoating and fusion of the vesicle with its target membrane. The vesicles may 'recognize' target membrane sites via special recognition proteins on vesicle and target membranes.

What are the 'destination signals' on proteins involved in Golgi complex sorting?

As described below, in the case of enzymes destined for inclusion in lysosomes, the signal is on the carbohydrate part of the glycoprotein. However, it seems that in most cases the sorting mechanism must include receptors that recognize certain structural features or specific sequences of the proteins

to be sorted. These are not yet elucidated fully but in one category they are known. In those proteins (such as protein disulfide isomerase) that have to be returned to the ER, the 'label' is a peptide sequence of four amino acids in the protein, Lys-Asp-Glu-Leu; in the single-letter abbreviation system for amino acids (Table 22.2), this comes out as KDEL.

Table 22.2   Single-letter symbols for amino acids

Alanine	A	Leucine	L
Arginine	R	Lysine	K
Asparagine	N	Methionine	M
Aspartic acid	D	Phenylalanine	F
Cysteine	C	Proline	P
Glutamine	Q	Serine	S
Glutamic acid	E	Threonine	T
Glycine	G	Tryptophan	W
Histidine	H	Tyrosine	Y
Isoleucine	I	Valine	V

Packaging of lysosomal proteins

So far we have dealt with proteins destined for secretion. As described in Chapter 16, **lysosomes** are small intracellular membrane-surrounded organelles containing a group of highly destructive hydrolytic enzymes. The enzymes are produced on the ER in the same fashion as for secreted proteins; they are *N*-glycosylated in the ER lumen.

In the Golgi apparatus an enzyme system recognizes lysosomal enzymes and the sugar mannose on their oligosaccharide attachments becomes phosphorylated. The Golgi membrane has internal receptors to which the mannose phosphate attaches and the region buds off vesicles containing lysosomal enzymes. These vesicles fuse with other 'sorting vesicles' with a low internal pH (due to proton pumps in their membranes). The low pH dissociates the mannose phosphate receptors from the glycoproteins and they are returned (again by vesicle budding) back to the Golgi while the enzymes are delivered to a lysosome. In the genetic disease called I-cell disease (I for inclusions of mucopolysaccharides) the mannose phosphorylating system is deficient, resulting in a group of lysosomal enzymes not being properly directed to lysosomes. This, in turn, results in failure to destroy mucopolysaccharides, which then accumulate in the lysosomes with serious clinical effects.

When this 'tagging' of a lysosomal enzyme by phosphorylating mannose residues was elucidated, there was an expectation that carbohydrate 'labelling' would be widespread. This appears not to be the case and, as already mentioned, in most cases some structural feature(s) of proteins are probably recognized by the Golgi sorting mechanism.

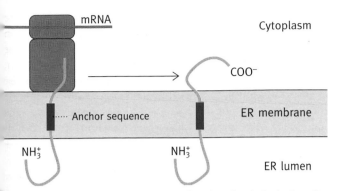

Fig. 22.18 Diagram illustrating the role of anchor signals in the insertion of integral membrane proteins into the ER membrane, with the polypeptide oriented with the *N*-terminal outside and the *C*-terminal inside. Membrane proteins involved in translocation are omitted for simplicity. Other arrangements of anchor signals can position the proteins in reverse orientation.

How are integral proteins of the cytoplasmic membrane targeted to their sites?

Integral membrane proteins are also synthesized on the rough ER, in fact, placed *in* the ER membrane, transported to the Golgi as vesicles, and thence to the plasma membrane also as vesicles. New membrane lipid synthesis also occurs in the ER and hence new membrane originates in this organelle.

An interesting question, of course, is why, in the case of a secretory protein, the polypeptide goes right through the rough ER membrane while an integral protein becomes anchored into the membrane. All the required information that determines whether and how a polypeptide becomes integrated into the membrane structure is in the amino acid sequence of the protein.

The simplest situation is where there is a 'stop-transfer' or **anchor sequence** in the polypeptide (Fig. 22.18). This fixes itself in the hydrophobic interior of the bilayer and halts any further transfer. The ribosome then completes the *C*-terminal part of the polypeptide resulting in a protein with its amino-terminal end outside of the ER and its carboxy terminal projecting into the ER lumen. When the piece of ER is transferred to become part of the cell membrane, the protein is in the same orientation, the amino-terminal projecting outside of the cell. Cell membrane proteins are also integrated with the opposite orientation and with the polypeptide criss-crossing the membrane several times. The latter involves multiple internal leader and stop-transfer signals.

Is all protein translocation cotranslational?

The message that we have given so far is that proteins pass through or into membranes while they are being synthesized. However, there is, in eukaryote cells, another completely separate type of protein translocation—into mitochondrial (or chloroplast in plant cells) compartments. The mitochondrion, as already mentioned, has chromosome and protein-synthesizing machinery of its own, but these account for only a small

proportion of the proteins, or subunits, found in the organelle. Several hundred different proteins, or subunits, are transported into mitochondria, constituting more than 95% of the total proteins present there. They are synthesized in the cytoplasmic compartment. You might guess from the ER situation that this happens on mitochondrial membrane-bound ribosomes, but this is not the case. They are synthesized on free ribosomes. Nuclear-coded mitochondrial proteins (that is, coded by genes in the cell's nucleus) are synthesized as pre-proteins and released into the cytoplasm but, as a polypeptide emerges from the ribosome, chaperones attach and keep it in an unfolded form. The released protein traverses the membrane in an extended, not folded, form, driven by the membrane charge potential. After traversing the membrane, other chaperones are required to fold the protein inside the mitochondrial matrix.

The mitochondrial protein leader sequence has about 12–70 amino acids. The actual sequences differ from one pre-protein to another, but a common structural feature is that they form an amphipathic α helix in which one side is positively charged and the other side is largely hydrophobic. This attaches to receptors in the outer membrane.

In the simplest case, an *N*-terminal leader sequence takes the protein into the matrix of the mitochondrion through a pore straddling the inner and outer membrane at points where the two membranes are in contact, and then the leader is cleaved off (Fig. 22.19). Where a protein is destined for the intermembrane

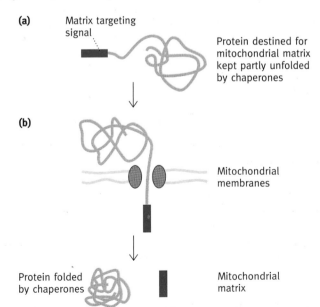

Fig. 22.19 (a) Diagram of a mitochondrial pre-protein synthesized in the cytoplasm but destined to be transported into the matrix. **(b)** The polypeptide traverses the mitochondrial membrane via a protein pore and the leader sequence is removed by a matrix signal peptidase. The initial attachment of the polypeptide to the mitochondrion is via a receptor protein in the membrane (not shown). In cases where the protein is destined for the intermembrane space, a second leader sequence directs the movement of the protein from the matrix across the inner mitochondrial membrane.

space, the pre-protein has two leader sequences. The first leads it into the matrix as already described, but removal of this leader exposes the second which leads the polypeptide out of the matrix, back through the inner mitochondrial membrane into the intermembrane space. There, the second leader sequence is removed.

The nucleus of a cell is surrounded by a double nuclear membrane. Proteins synthesized in the cytoplasm translocate into the nucleus via special pores. It has been shown in many cases that a short stretch of amino acids is the targeting sequence that marks proteins to pass through these pores though this does not appear to be needed for the small histone proteins. In the case of transcriptional factors (destined for the nucleus) the signal sequence is internal in the polypeptide chain.

A **pre-protein** should not be confused with a **pro-protein**. Insulin is secreted into the ER lumen as a single polypeptide protein, proinsulin. This is processed into two polypeptides linked by disulfide bridges by removal of a middle section of the proinsulin chain. The mRNA for insulin codes for a leader sequence to proinsulin giving rise to the term pre-proinsulin, though, since the pre-sequence (the leader) is removed on traversing the ER membrane, pre-proinsulin does not exist as such in the cell.

Degradation of proteins

We have now dealt with protein synthesis and protein translocation. This leaves a final topic to be described in this chapter, that of **protein degradation**.

The most important facts are: (1) that protein destruction occurs in all cells giving rise to protein 'turnover' as old protein molecules are replaced by new ones; and (2) the destruction of protein molecules in a cell is highly selective. Some proteins have a half-life of more than 20 hours (liver proteins average a few days); others 10 minutes or even 2 minutes.

The reasons for destroying proteins are several. Since perfect accuracy in translation is not likely to be possible, protein molecules containing incorrect amino acids are inevitably produced that may cause improper folding in the molecule. Proteins can also suffer chemical damage such as oxidation and have to be destroyed.

However, this 'garbage disposal' is far from the end of the story. Some proteins appear to be deliberately engineered to be degraded very rapidly—certain amino acid sequences correlate with rapid destruction of the protein containing them.

Why should the cell arrange to destroy perfectly good enzymes so rapidly? The answer in many cases may be in the interests of metabolic control. If an enzyme is destroyed quickly, then the level of that enzyme can be sensitively and continuously

controlled by the rate at which it is synthesized. Rate-controlling enzymes appear to be relatively short-lived compared with non-rate controlling enzymes. Structural proteins and hemoglobin are at the long-lived end of the spectrum.

One mechanism for the selective destruction of proteins involves **ubiquitin**, a small protein found universally in eukaryotes (but not in bacteria, so the name exaggerates somewhat). By an ATP-dependent reaction its —COOH terminal group becomes linked to the ε-NH$_2$ of the side chain of lysine residues of its target protein and, once attached, that protein is marked for destruction (Fig. 22.20). The basis of selection by ubiquitin of proteins destined for destruction appears to rest at least partly on the particular N-terminal amino acid of each protein. Chaperones also appear to play a role in targetting proteins for destruction.

Lysosomes may participate in the destruction of the longer-lived structural proteins (see Fig. 16.2). They are also responsible for the autolytic destruction of cells during development. The disappearance of the tadpole's tail is an example.

We know much less about protein degradation on the whole than about the far more complicated business of protein synthesis, which is something one could never have predicted. It is an area of vital importance. You might recall (see page 153) that protein breakdown of muscle supplies amino acids for gluconeogenesis during starvation—without which a human being could not survive starvation for much more than the 24 hours or so needed for exhaustion of glycogen reserves. Not surprisingly, protein degradation in the cell is a very active research field.

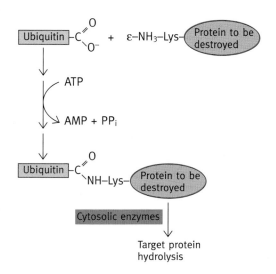

Fig. 22.20 Diagram of how ubiquitin tags proteins for degradation. The method of selection of proteins for destruction is not known though the terminal amino acid appears to influence protein half-life.

Further reading

Transfer RNA

Schimmel, P. and de Pouplana, L. R. (1995). Transfer RNA: from minihelix to genetic code. *Cell*, **81**, 983–6.
This review addresses the question of why particular nucleotide triplets correspond to specific amino acids. tRNA can be thought of as comprising of two informational domains: the acceptor-TψC minihelix encoding the operational RNA code for amino acids, and the anticodon-containing domain with the trinucleotides of the genetic code.

Ribosomal RNA

Noller, H. F. (1991). Ribosomal RNA and translation. *Ann. Rev. Biochem.*, **60**, 191–227.
A review covering the structure and function of rRNA with a useful section on interactions of antibiotics with rRNA. Removes any idea that rRNA is only an inert scaffolding for the ribosome. Very readable.

RNA and DNA molecules as therapeutic agents

Cohen, J. S. and Hogan, M. E. (1994). The new genetic medicines. *Sci. Amer.*, **271**(6), 50–5.
Sequences of DNA are being developed as drugs. These synthetic strands are called antisense and triplex agents; they can potentially attack viruses and cancers without harming healthy tissue.

Elongation factors and translational fidelity

Thompson, R. C. (1989). EF-Tu provides an internal kinetic standard for translational accuracy. *Trends Biochem. Sci.*, **13**, 91–3.
Describes the 'pause' mechanism for translational fidelity.

Weyland, A. and Parmeggiani, A. (1993). Towards a model for the interaction between elongation factor Tu and the ribosome. *Science*, **259**, 1311–14.
Discusses the discovery that two molecules of GTP are hydrolysed in the EF–Tu associated reactions, rather than one as hitherto believed. See also next reference.

Weyland, A. and Parmeggiani, A. (1994). Why do two EF-Tu molecules act in the elongation cycle of protein biosynthesis? *Trends Biochem. Sci.*, **19**, 188–93.
Discusses the evidence that two EF-Tu–GTP molecules are involved in the addition of each amino acid in protein synthesis and the possible role of this in translational fidelity.

Powers, T. and Noller, H. F. (1994). The 530 loop of 16S rRNA—a signal to EF–Tu? *Trends in Genetics*, **10**, 27–31.
A penetrating discussion of how fidelity in protein synthesis is achieved.

Translational control

Vassalli, J. D. and Stutz, A. (1995). Awakening dormant mRNAs. *Current Biology*, **5**, 476–9.
Discusses the role of polyA tail length in activating oocyte mRNAs for translation.

Chen, J.-J. and London, I. M. (1995). Regulation of protein synthesis by heme-regulated eIF-2α kinase. *Trends Biochem. Sci.*, **20**, 105–8.
A comprehensive review of the control of hemoglobin synthesis by this mechanism. (Hemoglobin synthesis and its control is dealt with in Chapter 27 of this book.)

Cyclophilins, peptidyl proline isomerase and protein folding

Schmid, F. X. (1995). Prolyl isomerases join the fold. *Current Biology*, **5**, 993–4.
Discusses the evidence that cyclophilins (which have proline isomerase activity) are involved in mitochondrial protein folding, though their link to immunosuppression is likely to involve other roles.

Molecular chaperones and protein folding

Ellis, R. J. and van der Vies, S. M.(1991). Molecular chaperones. *Ann. Rev. Biochem.*, **60**, 321–47.
The concept, a definition, and examples of their actions.

Ali, N. and Banu, N. (1991). Heat shock proteins: molecular chaperones. *Biochemical Education*, **19**, 166–72.
A comprehensive introduction to the family of proteins and their roles.

Craig, E. A. (1993). Chaperones: helpers along the pathway to protein folding. *Science*, **260**, 1902–3.
Very readable summary.

Agard, D. A. (1993). To fold or not to fold. *Science*, **260**, 1903–4.
A summary of the strategies which cells might use to prevent polypeptide aggregation, with specific illustrations of the employment of each of them.

Hendrick, J. P. and Hortl, F.-U. (1993). Molecular chaperone functions of heat-shock proteins. *Ann. Rev. Biochem.*, **62**, 349–84.
Comprehensive review of the many functions of chaperones.

Welch, W. J. (1993). How cells respond to stress. *Sci. Amer.*, **268**(5), 34–41.
Reviews heat shock proteins in their roles as chaperones and other functions.

Hortl, F.-U., Hlodon, R., and Langer, T. (1994). Molecular chaperones in protein folding: the art of avoiding sticky situations. *Trends Biochem. Sci.*, **19**, 20–5.
An excellent general review of protein folding.

Horwich, A. L. (1995). Resurrection or destruction? *Current Biology*, 5, 455–8.
Discusses how chaperones are involved both in rescuing proteins and directing them towards proteolytic destruction.

Glick, B. S. (1995). Can hsp70 proteins act as force-generating motors? *Cell*, 80, 11–14.
A minireview on the nucleotide-dependent configurational changes in one class of molecular chaperones, the hsp70 family, and the way they could act as translocation motors.

Hendrick, J. P. and Hortl, F.-U. (1995). The role of molecular chaperones in protein folding. *FASEB J*, 9, 1559–69.
Well illustrated account of their general role and role in transport of proteins across membranes.

Buchner, J. (1996). Supervising the fold: functional principles of molecular chaperones. *FASEB J.*, 10, 10–19.
A well illustrated survey of the field.

Prion diseases

Weissmann, C. (1995). Yielding under the strain. *Nature*, 375, 628–9.
A concise summary of the molecular basis of prion diseases.

Prusiner, S. B. (1995). The prion diseases. *Sci. Amer*, 272(1), 30–7.
Excellent general review of the field.

Thomas, P. J., Qu, B.-H., and Pedersen, P. L. (1995). Defective protein folding as a basis of human disease. *Trends Biochem. Sci.*, 20, 456–9.
Discusses the possibility that a large number of diseases, in addition to prion diseases, may be due to protein folding abnormalities.

Taubes, G. (1996). Misfolding the way to disease. *Science*, 271, 1493–5.
A research news item presents the general hypothesis that protein misfolding may cause amyloid diseases such as Alzheimers as well as the prion diseases. Introduces the concept that aggregation of proteins into insoluble complexes may be more prevalent and more important than hitherto suspected.

Protein targeting

Von Heijne, G. (1995). Membrane protein assembly: rules of the game. *BioEssays*, 17, 25–30.
Interesting discussion of how proteins can be 'stitched' into membranes with multiple transmembrane helices.

McNew, J. A. and Goodman, J. M. (1995). The targeting and assembly of peroxisomal proteins: some old rules do not apply. *Trends Biochem. Sci.*, 21, 54–8.
Reviews evidence that folded proteins may be imported into peroxisomes as such.

Schmid, S. L. and Damke, H. (1995). Coated vesicles: a diversity of form and function. *FASEB J.*, 9, 1445–53.
Reviews the different ways in which coated vesicles are budded off from the Golgi and other membranes.

Schekman, R. and Orci, L. (1996). Coat proteins and vesicle budding. *Science*, 271, 1526–33.
An excellent review of vesicle transport and their role in sorting of proteins.

Schatz, G. and Dobberstein, B. (1996). Common principles of protein translocation across membranes. *Science*, 271, 1519–25.
Particularly interesting in that general principles are sought for all protein transport across membranes.

Gorlich, D. and Mattaz, I. W. (1996). Nucleocytoplasmic transport. *Science*, 271, 1513–18.
A comprehensive, readable summary of the protein and RNA traffic between nucleus and cytoplasm via nuclear pores and the mechanisms involved. Includes the role of GTP hydrolysis in the process.

Protein glycosylation

Abeijon, C. and Hirschberg, C. B. (1992). Topography of glycosylation reactions in the endoplasmic reticulum. *Trends Biochem. Sci.*, 17, 32–6.
Reviews the different protein glycosylation reactions occurring in the endoplasmic reticulum.

Degradation of proteins

Mayer, R. J. and Doherty, F. J. (1992). Ubiquitin. *Essays in Biochemistry*, 27, 37–48.
The structure, mechanism of action, and roles of ubiquitin.

Ciechanover, A. (1994). The ubiquitin-proteasome proteolytic pathway. *Cell*, 79, 13–21.
This wide-ranging review highlights the role of the ubiquitin-proteasome system as a major, non-lysosomal, proteolytic and regulatory pathway for such cellular substrates as oncoproteins and transcriptional regulators, cyclins, cell surface receptor molecules and major histocompatability complex (MHC) class I-restricted antigens.

Jentsch, S. and Schlenker, S. (1995). Selective protein degradation: a journey's end within the proteasome. *Cell*, 82, 881–4.
DNA repair, cell cycle progression, signal transduction, transcription, and antigen presentation are all processes regulated by ubiquitin-mediated degradation.

Jentsch, S. (1996). When proteins receive deadly messages at birth. *Science*, 271, 955–6.
A newly discovered unusual mechanism in bacteria. If an mRNA lacks a stop codon (due to truncation), the synthesized polypeptide could jam the ribosome. A tagging method ensures that the polypeptide is hydrolysed and the ribosome freed.

Problems for Chapter 22

1 There are 64 codons available for 20 amino acids. Why do you think 61 codons are actually used to specify the 20 amino acids?

2 Despite the facts stated in question 1, there are fewer than 61 tRNA molecules. Explain why this is so.

3 In diagrams, when a tRNA molecule is shown base paired to a codon, the molecule is shown flipped over as compared with the same tRNA shown on its own. Why is this so?

4 At which points in protein synthesis do fidelity mechanisms operate?

5 Describe the participation and, where known, the role of GTP in protein synthesis.

6 Protection studies have indicated that, in *E. coli*, tRNA molecules (with their aminoacyl or peptidyl attachments) straddle A, P, and E sites on the ribosome. Explain why this occurs and the possible mechanisms.

7 The mechanism of initiation of translation in eukaryotes is not compatible with polycistronic mRNA. Explain why.

8 Explain the role of chaperones in protein synthesis.

9 What disease may be associated with improper protein folding?

10 Explain how proteins are secreted through the endoplasmic reticulum.

Chapter 23

....

Viruses and viroids

Quite apart from the biological and medical importance of viruses, they are important in biochemical research; they provide simple models for studying complex processes in gene function and also provide important research tools for many purposes, including crucial roles in recombinant DNA technology, described in Chapter 24.

If we ignore the special survival strategies such as the formation of spores in which chemical activities are minimized, living cells are always metabolically active, carrying out functions of ATP production, osmotic work, and synthesis of cellular materials, etc. The cell increases in size and divides. To do this it requires thousands of genes to specify the proteins required.

Viruses are quite different. They are much smaller, an electron microscope being needed to see them rather than the light microscope which is adequate for most bacteria. They have no metabolism—a virus particle or **virion**, as it is called, on its own does absolutely nothing. It generates no energy and catalyses no reactions; it is an inert, organized, complex of molecules that may often be crystallized. Nonetheless, they are reproduced when they infect living cells. Different viruses infect animal and plant cells and also bacteria (the latter ones are known as bacteriophages). Their strategy is to get their genetic material into cells and use the host cell machinery for their own replication.

A virus particle, therefore, is essentially a small amount of nucleic acid surrounded by one or more protective shells. Although the outer shell or shells contain many copies of protein molecules there are only a few different species of these in most viruses. The total number of genes may be as many as about a hundred for a large virus, such as vaccinia, or three or four for the smallest. The protein shell surrounds the genome and is called the **nucleocapsid**. In some viruses there is an additional lipid-bilayer membrane envelope into which are anchored the external viral protein molecules, exposed on the viral surface, that are involved in the first stages of the infective process, as described below.

....

The life cycle of a virus

A virus must gain entry to the cell that it infects and release its genetic material (RNA or DNA) into the cell. The viral genes must direct the synthesis of the viral-specific enzymes necessary for the reproduction of the virus and also those proteins required for the assembly of new virus particles. Multiple copies of the original genome have to be made, and new virions assembled from the synthesized components. Finally, the progeny virions must escape from the cell.

We will now describe the different stages in the life cycle of viruses. In doing so, we will first deal with different strategies used by different viruses in general terms but, following this, a number of specific viruses will be described in more detail, to illustrate some of the general principles.

A living cell is surrounded by a lipid bilayer membrane that is a barrier to virus penetration. However, there are ways past it. For example, cells have mechanisms for engulfing molecules; receptor-mediated endocytosis in animal cells is one such route (see page 210). Many animal cells are very active in this process and some viruses exploit this to gain entry themselves; they hitch a ride on a normal cellular import mechanism. The first event is that the virion binds to the cell surface, via a **coat protein** complementary to a specific receptor on the host cell surface. The latter requirement is the main restriction on which cells a given virus can infect. After attachment, the receptor–virus complex moves in the membrane to a depression in the latter which has a protein-coat on its internal surface, called **clathrin**, hence the term **coated pit** (see Fig. 23.1). As with normal receptor-mediated endocytosis, the pit invaginates more and more, until it engulfs the virion into a coated vesicle inside the cell. The clathrin returns to the cell membrane and the vesicle containing the virus fuses with an **endosome**, or cytoplasmic vesicle, leaving the virus inside the latter. In normal (nonviral) endocytosis, the endosome delivers its contents to vesicles from the Golgi sacs, containing an array of destructive

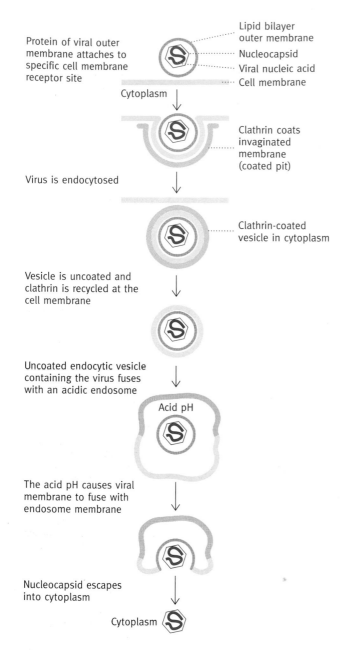

Protein of viral outer membrane attaches to specific cell membrane receptor site

Lipid bilayer outer membrane
Nucleocapsid
Viral nucleic acid
Cell membrane

Cytoplasm

Virus is endocytosed

Clathrin coats invaginated membrane (coated pit)

Clathrin-coated vesicle in cytoplasm

Vesicle is uncoated and clathrin is recycled at the cell membrane

Uncoated endocytic vesicle containing the virus fuses with an acidic endosome

Acid pH

The acid pH causes viral membrane to fuse with endosome membrane

Nucleocapsid escapes into cytoplasm

Cytoplasm

Fig. 23.1 The process of invasion of an animal cell by a membrane-enveloped virus using the receptor-mediated endocytosis route of entry. Receptor-mediated endocytosis is discussed further on page 210.

enzymes, forming lysosomes in which the engulfed particle is destroyed (Fig. 16.2). However, the endosome has an acidic internal pH due to a proton pump in its membrane. The low pH causes the virus to dissociate from the host cell membrane receptor (to which the virus initially attached). In the case of a membrane-enveloped virus, the lipid membrane envelope now fuses with the endosome membrane and the core of the virus escapes into the cytoplasm. There, it loses its protein coat and the contents are free in the cytoplasm. It is not so obvious how a

naked nucleocapsid (that is, without a lipid membrane) virus escapes from the endosome into the cytoplasm.

A second route, available only to certain of the viruses with a lipid membrane coat outside the nucleocapsid, is direct fusion with the external membrane of the host cell. In this, the virus again attaches to the cell surface by interaction of a viral surface protein with a specific protein on the cell surface. The two membranes fuse, a contiguous hole forms in both, and the virion contents enters the cell via this hole. HIV uses this route.

In bacterial viruses, or **bacteriophages**, a different route is followed. The bacterial cell has a rigid cell wall around it. Phage lambda has a tadpole-like shape (Fig. 23.2). The **head** is a capsule formed by protein molecules in which resides the viral DNA molecule. The phage attaches to the cell wall by its tail fibre and injects the DNA into the bacterial cell, almost like the action of a hypodermic syringe. It may be necessary to remind oneself that this is not a free-living cell, or organism, but a lifeless collection of molecules self-assembled into the phage structure.

Types of genetic material in different viruses

The genetic material of viruses may be double-stranded DNA, single-stranded DNA, double-stranded RNA, or single-stranded RNA.

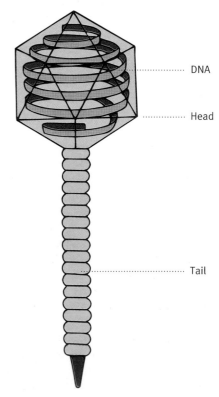

DNA

Head

Tail

Fig. 23.2 A λ particle. The λ chromosomome—some 50 000 base pairs of DNA—is wrapped around a protein core in the head.

So far in this book, double-stranded DNA has been taken to be the genetic material of organisms. We have pointed out earlier (page 241) that DNA is chemically more stable than RNA and double-strandedness means that single-strand damage can be corrected (page 263) using the other strand as template for the repair. The DNA replication machinery in the nucleus of cells includes proofreading, which reduces errors by a huge factor (see page 261)—without this, the rate of mutation would be unacceptably high. No such correction machinery has been developed in any RNA synthesis mechanism. Nevertheless, many viruses use RNA as their genetic material. Why then can some viruses get away with having RNA as genetic material? The reasons are probably: (1) that viral genomes are exceedingly small compared with the size of a cellular genome and therefore the chances of deleterious mutations during replication of viral RNA are smaller; (2) although errors in RNA replication are still higher than occur with DNA, giving rise to a rapid mutation rate in RNA viruses some of which will be lethal, this is tolerated because of the rapid multiplication rate and fierce selective pressure, and it becomes an advantage in that rapid mutation helps the virus to escape immunological attack by the animal host.

Different types of viral genetic material demand different biochemical strategies on entering the host cell.

Double-stranded viral DNA is transcribed by host RNA polymerase to produce mRNA in the way with which you are familiar. Vaccinia in this class is unusual in that it carries its own RNA polymerase in its virion, for a reason described later. The minus (template) strand of the double-stranded genome is copied to make mRNA. The mRNA directs the synthesis of virus-specific proteins. In the case of **single-stranded DNA viruses**, whether (+) or (−), double-stranded DNA is formed by replication using host cell enzymes, from which mRNA is transcribed (see below for strand definition).

Cells have no machinery for replicating viral RNA. **Double-stranded RNA viruses** carry in their structure molecules of an RNA-dependent RNA polymerase. Once in the cell, this enzyme transcribes mRNA from the viral RNA template. This is translated by the host cell machinery into viral-specific proteins necessary for replication of the virus.

Single-stranded RNA genomes in viruses occur in two forms. You will recall (page 273) that, in a cell, one of the two strands of DNA is called the sense (nontemplate) strand, whose base sequence is identical (apart from T becoming U) to the mRNA transcribed from the other (nonsense, or template) strand. In different single-stranded RNA viruses, the RNA strand might be equivalent either to a sense strand or a nonsense strand. These are called the 'plus' (+) and 'minus' (−) strands, respectively. Put in another way, *the RNA of a (+) strand RNA virus is itself a messenger RNA while that of a (−) strand virus is not.* This makes a difference because, if a (+) RNA strand enters the cell, since it *is* an mRNA, the protein synthesizing machinery of the cell can immediately translate it into proteins, including enzymes

necessary for viral reproduction. Such a virus requires only its RNA, coding for whatever proteins it needs, for reproduction of the virus to occur after infection. The situation with a (−) strand RNA virus is quite different. It is not an mRNA; the cell cannot translate it; the cell cannot replicate the RNA. An additional RNA-replicating enzyme (see below), carried by the virus, is needed to copy the (−) RNA into (+) RNA, which, as stated, is an mRNA.

Another class of (+) **strand RNA viruses** are the **retroviruses** of which HIV (human immunodeficiency virus), causing AIDS (acquired immune deficiency syndrome) is of major interest. The virion carries an enzyme that converts single-stranded RNA into double-stranded DNA (described later). The latter integrates into the host chromosome.

How are viruses released from cells?

Some viruses are released simply by lysis of the cell; polio is an example. In the case of bacteriophage λ, the rigid inert cell wall would prevent release of new phage particles. However, genes on the phage DNA code for an enzyme, **lysozyme**, which destroys the cell wall, resulting in cell lysis. The gene for this is

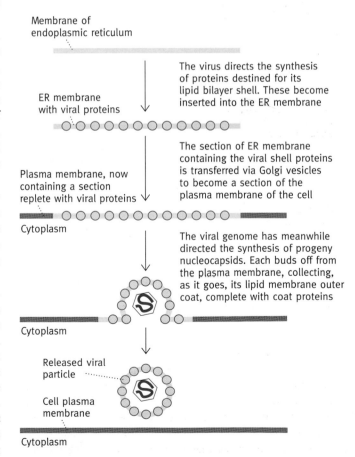

Fig. 23.3 Release of viral particles from cells by budding off from the cell surface.

transcribed only late in the lytic cycle of infection so that the cell is not destroyed before the new phage particles are produced. Despite the fact that bacteriophage λ is essentially just a small piece of DNA, 48 502 nucleotides in length with 63 genes, the latter are subject to an elaborate control system.

The release of animal viruses that are surrounded by a lipid bilayer membrane is more complex. The virus exploits a normal cellular function. You will recall (page 309) that the normal cell produces new cell membrane containing appropriate membrane proteins by inserting the proteins into the rough endoplasmic reticulum. The proteins are glycosylated and transported via Golgi vesicles to the cell membrane; the vesicles fuse with the latter so that we have new cell membrane containing the correct membrane proteins.

Some viruses with lipid membrane envelopes adapt this process to their own ends. The messenger for the envelope proteins codes for leader and anchor sequences (see page 309), and so positions the viral proteins in the rough ER whence they are delivered as described above to the cell membrane. There, the viral membrane proteins accumulate. A new **virus nucleocapsid** (that is, the viral genome plus a shell of capsid proteins) buds off from the membrane, collecting its bilayer envelope containing membrane proteins as it goes. Influenza and the AIDS viruses have this release method of budding off particles (Fig. 23.3).

Replication mechanisms of some selected viruses

The detailed strategies by which viruses cope with replication problems are diverse. Instead of summarizing many of these molecular strategies, which would be a very large topic, we will deal with a few viruses selected because they illustrate some of the general principles given above and because of their intrinsic interest.

Vaccinia

This is the virus causing cowpox and was the one used for smallpox vaccination. It is one of the largest and most complex viruses. It has the type of genome with which you are most familiar—double-stranded DNA. The latter is transcribed into mRNA leading to the synthesis of all of the protein components needed for the entire production of new virions (including enzymes needed for viral DNA replication) by the host ribosomal system.

In the case of most double-stranded DNA viruses, the viral DNA finds its way to the nucleus where host RNA polymerase transcribes the genes to produce mRNA. Vaccinia is unusual in that the entire replication occurs in the cytoplasm. There is no host RNA polymerase in the cytoplasm to transcribe the viral genes and the virus has to carry molecules of its own RNA polymerase in the virion particles, thus permitting mRNA production and the synthesis of all required proteins including, of course, more RNA polymerase.

Poliovirus

This is a **naked virion** (that is, it has only a nucleocapsid shell of coat protein but no membrane). It attaches to specific receptors found only on epithelial cells of humans and other primates. Since its single-strand (+) RNA *is a messenger RNA*, it is translated on entering the cytoplasm producing an **RNA replicase**, which synthesizes RNA using viral RNA as its template. The replicase copies the original (+) strands into (−) strands, which act as templates for more (+) strand synthesis. Polio virus RNA replicase is unique amongst RNA-synthesizing enzymes in that it cannot, on its own, initiate RNA chains. In this case, the primer is not an oligonucleotide but a protein to which it adds the first nucleotide and then elongates the chain.

Translation of the mRNA occurs via the production of a single large 'polyprotein', which is subsequently cleaved into separate proteins. This strategy of polyprotein production is, perhaps, surprising for, although many more coat protein molecules are needed for new virion production than are replicase molecules needed for replication, the process produces equal numbers of these.

Influenza virus

This is a (−) strand RNA virus. Its genome is divided into eight sections, each contained in a helical nucleocapsid and the whole surrounded by a lipid membrane into which are inserted molecules of two different glycoproteins (Fig. 23.4). One of these is **hemagglutinin**, present as a surface protein, so called because, if the virus is mixed with red blood cells, it causes the latter to agglutinate. The red blood cell has on its surface a carbohydrate attachment to one of its proteins (**glycophorin A**), terminating in sialic acid, otherwise known as **neuraminic acid** (page 44). When mixed with red blood cells, the hemagglutinin attaches to the neuraminic acid and crosslinks the cells into

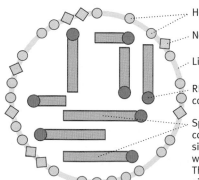

Hemagglutinin, a glycoprotein

Neuraminidase, a glycoprotein

Lipid bilayer outer membrane

RNA replicase, an enzyme complex carrried by the virus

Spiral nucleocapsids; these consist of segments of single-stranded RNA associated with nucleocapsid protein. The genome is divided into eight such segments

Fig. 23.4 Diagrammatic structure of the influenza virion. A layer of matrix protein underlying the outer membrane is not shown.

aggregates. (This is simply an *in vitro* effect.) The virus gains entry to its host cell by attaching to a receptor on the latter which, likewise, terminates in neuraminic acid. It enters via the endocytotic route.

The virion has several hundred hemagglutinin molecules on its surface. It also has an enzyme, **neuraminidase,** as another surface protein, which hydrolyses terminal neuraminic acid groups off glycoproteins. There have been various purposes suggested for this function, for, at first sight, it seems odd that a virus has a surface enzyme to destroy cellular receptors on which it depends for infection. One possibility is that its action liquefies mucins that contain sialic acid (page 32). This would both permit access to cell surfaces for infection and facilitate escape of new virus particles. Liquefaction of mucins might also help to distribute the virus more efficiently by the sneezing reaction. Finally, it may be that the escaping virus might become anchored to the host cell by neuraminic acid receptors, thus inhibiting its escape. The neuramidinase would guard against this by destroying the receptors. There is evidence that the enzyme is essential for efficient virus multiplication in an infection.

As stated, the virus contains its genome in eight separate (−) RNA strands, each as a helical nucleocapsid (Fig. 23.4). Replication of these requires an RNA replicase, since influenza (−) strand RNA is not a messenger. The virus carries in its nucleocapsids molecules of an RNA replicase that copies the (−) strand into (+) RNA (equals mRNA) and protein synthesis can then produce all the necessary viral proteins including more RNA replicase.

The immune system attack is principally against the hemagglutinin of the influenza virus, which is neutralized in this way, so that an individual experiencing the disease is immune to a second infection by the identical strain of virus. The viral hemagglutinin is constantly changing its amino acid composition as a result of mutations (remember, it is an RNA virus), so that the protection is *gradually* lost as this occurs. This is known as **antigenic drift**. Such mutation of individual amino acids does not produce major epidemics because the different antibody binding sites on the protein are not eliminated by gradual change, but only reduced in number, so that only partial loss of immunity occurs in the population and many infections are mild. However, if a totally new hemagglutinin is present in the virus, a world-wide virulent pandemic can result from this antigenic shift because there is no residual immunological protection against it. Such a situation may occur when two different strains of virus infect a cell. Because of the divided genome, reassortment of the RNA particles can occur when assembly of new virus nucleocapsids takes place. One such pandemic, in 1918, is believed to have resulted from reassortment between a human and bird strain of influenza, and is estimated to have caused 20 million deaths.

The role of structural determination of proteins in new approaches to disease therapy is illustrated by a current 'rational therapy' approach to influenza. Using crystalline influenza neuraminidase, the precise structure of the neuraminic acid binding site of the enzyme was determined by X-ray crystallography. From this, a synthetic analogue of neuraminic acid was devised that attaches with high affinity to the enzyme and blocks its normal function. Clinical trials of the substrate analogue against the virus are underway. The active site of an enzyme has limited opportunity to mutate and remain functional but this does not expose the active site to immunological attack for it is buried in a cleft not accessible to immune attack. It is, however, accessible to the neuraminic acid analogue. In this particular case, the analogue does not block the hemagglutinin site that binds to the host cell receptor and so would not inhibit infection. It depends on inhibiting the neuraminidase believed to be necessary for new virus release from cells.

Retroviruses

As already stated, there is a fourth class of single-stranded RNA viruses of immense current interest—the **retroviruses**. Retroviruses, of which that causing AIDS is one example, are (+) strand RNA viruses. However, they do not, on infection of a cell, follow the molecular strategy outlined above for polio virus. The retrovirus particle carries within itself a few molecules of an enzyme whose discovery caused initial disbelief, to be followed by the award of a Nobel Prize to its discoverers, Howard Temin and David Baltimore; it is called **reverse transcriptase**. Before its discovery it was, of course, known that DNA could direct RNA synthesis, but the reverse was not known. Viral reverse transcriptase is an amazingly versatile enzyme; when a retrovirus infects a cell, the viral reverse transcriptase copies the (+) RNA into DNA. As with all DNA synthesis, a primer is necessary. Retroviruses, remarkably, use a host cell tRNA molecule (page 294), which hybridizes to the start of the template strand, and it carries the appropriate tRNA in the virion particle. The RNA/DNA hybrid so formed is converted to single-strand DNA by RNA hydrolysis, an enzyme activity also present in the reverse transcriptase molecule. The single-stranded DNA is then copied by the same enzyme to form double-stranded DNA. The viral genome is now in the form of normal duplex DNA which becomes integrated or inserted into the host cell chromosome (Fig. 23.5). The double-stranded DNA copy of the viral genome (called **proviral DNA**) has at each end a base sequence called a **long terminal repeat (LTR)** (Fig. 23.6). An integrase enzyme, also carried in the virus, causes integration of the proviral DNA into the host chromosome where it is now, in essence, an extra set of genes carried by the cell. Once in, it is replicated along with host DNA for cell division. For the production of new retrovirus particles, the proviral genes (that is, viral genes in the host chromosomes) are transcribed into (+) RNA transcripts. Some of these become the genome of new retrovirus progeny; some are processed to mRNA to provide the proteins for virus particle assembly.

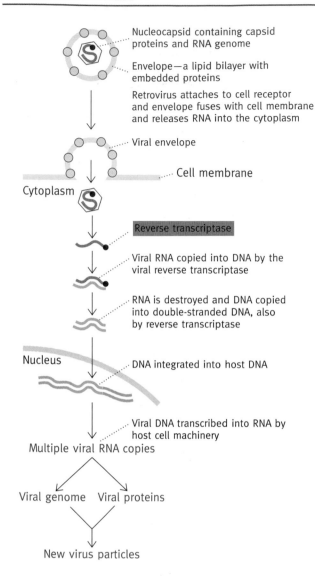

Fig. 23.5 Replication of a hypothetical retrovirus. See Fig. 23.3 for the release process.

The drug AZT (azidothymidine), whose structure is shown below, is used as therapy against the HIV (AIDS) virus; it works by inhibiting the viral reverse transcriptase. For this, it has to be converted into the triphosphate form which is an analogue of

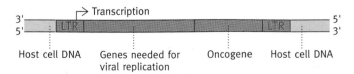

Fig. 23.6 Proviral genome of an oncogenic retrovirus incorporated into a host genome. LTR, Long terminal repeat. The oncogene may code for an abnormal protein involved in cell control or an abnormal amount of such a protein (see Fig. 26.20).

dTTP. AZT-triphosphate inhibits reverse transcriptase and also, when incorporated, terminates DNA chain extension because of the lack of a 3'-OH group. Host cell DNA polymerase is inhibited to a smaller extent.

Azidothymidine

Retroviruses have the potential to permit gene therapy—to place in a patient's cells that lack a normal essential gene a 'good' gene to remedy the deficiency. The principle is to cripple the reproductive capacity of the retrovirus by gene deletion and replace this with the RNA equivalent of the therapeutic gene, which becomes integrated in DNA form into the host chromosome. The method has the disadvantage that integration occurs randomly in the chromosome, which has potential hazards. Alternative methods for site-specific insertion of genes are being actively developed.

Cancer-inducing, or oncogenic, retroviruses

An **oncogenic retrovirus**, when integrated in its DNA form (Fig. 23.6) into the host chromosome, can cause tumour formation (*oncus* equals solid mass or tumour). The first tumour-causing virus discovered was the avian Rous sarcoma virus, which causes tumours in chicken muscle. This is due to a single gene in the retrovirus—an **oncogene**.

Since then, 18 or 20 oncogenes have been discovered in mammals. We described DNA hybridization earlier (page 242). In this technique, a piece of DNA made single-stranded by heating will unerringly find, and hydrogen-bond to, a complementary piece of DNA (that is, with a base sequence complementary in the Watson–Crick base pairing sense) amongst thousands. Techniques exist for detecting the hybridization complexes (see page 243). It was astonishingly discovered that, in each case, a cancer-producing retroviral oncogene had its closely similar counterpart in normal cells. This applies only to the exons of the normal gene since the viral RNA form of the gene lacks introns due to splicing (page 280 and see below). The viral oncogene is clearly derived, in the evolutionary sense, from its cellular counterpart. A single base change may be the only difference between the two. The **viral gene** is given the prefix **v** and the corresponding **cellular gene**, **c**. Genes are given curious

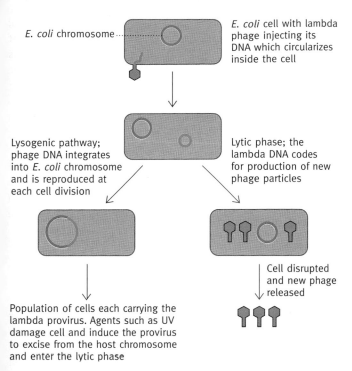

E. coli chromosome ·········

E. coli cell with lambda phage injecting its DNA which circularizes inside the cell

Lysogenic pathway; phage DNA integrates into *E. coli* chromosome and is reproduced at each cell division

Lytic phase; the lambda DNA codes for production of new phage particles

Cell disrupted and new phage released

Population of cells each carrying the lambda provirus. Agents such as UV damage cell and induce the provirus to excise from the host chromosome and enter the lytic phase

Fig. 23.7 Injection of bacteriophage λ DNA into the host bacterial cell, and the lytic and lysogenic pathways of replication.

different again. It has an elaborate set of genetic controls; on entering the cell, the DNA has two choices. It can immediately replicate and produce new virus particles to be released by a **phage-coded enzyme, lysozyme,** that disrupts the bacterial cell wall resulting in cell lysis. This is the **virulent** or **lytic route** (Fig. 23.7). The alternative is the **lysogenic route** in which the genome integrates into the *E. coli* chromosome where it can remain inert indefinitely being replicated with the host DNA. Thus large populations of otherwise normal *E. coli* cells can result, each carrying the phage DNA. A cell carrying lambda DNA in its chromosome is called a lysogenic cell and the incorporated

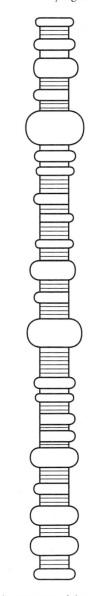

Fig. 23.8 Diagram of the structure of the coconut cadang-cadang viroid. The loops represent unpaired regions and the cross lines, paired bases. This viroid is the smallest known (246 nucleotides). It almost invariably kills coconut palms 5–10 years after infection. Kindly provided by Professor R. Symons, Department of Plant Science, Waite Institute, University of Adelaide.

abbreviations (see explanation of names on page 364) so we have names such as c-*myc* and v-*myc*, c-*ras* and v-*ras*, etc. The cellular gene is called a **protooncogene.** It turns out that protooncogenes code for proteins involved in important cellular control mechanisms (to be described in Chapter 26), abnormalities of which can lead to uncontrolled cell division. Some of them code for abnormal transcription factors (discussed on page 363). It would appear that retroviral oncogenes originated (in the evolutionary sense) through retroviruses accidentally picking up an mRNA molecule derived from a cellular control gene. The oncogenic viral form of the control gene may differ from the normal cellular counterpart due to mutation during, or subsequent to, the 'picking up' process. On re-insertion of this gene (by retroviral infection) into the host chromosome by the mechanism already described, the presence of the oncogene results in either an abnormal control protein, or abnormal amount of such a protein, in turn resulting in abnormal cellular control and cancer. This will become more understandable after you read Chapter 26, which deals with control pathways involving protooncogene protein products.

Bacteriophage lambda

This bacterial virus has been extensively studied and also used in molecular biology and in gene cloning (see Chapter 24 and Fig. 23.2). It is a double-stranded DNA phage but its life cycle is

phage DNA a **prophage**. If the DNA of an *E. coli* cell is damaged by UV light, or by a chemical or by ionizing radiation, so-called 'SOS' functions designed to institute repair are activated. However, in a lysogenic cell this results in the prophage being excised from the DNA by a mechanism too complex to describe here. The excised prophage enters the lytic cycle of replication. Bacteriophage lambda provides a beautiful example of the self-assembly process. If the component protein and DNA molecules of the virus are mixed together in the test tube, complete functional virus assembles, the DNA being packaged into the head. This constitutes a powerful tool in gene cloning.

What are viroids?

Viroids are the smallest infectious particles known—even smaller and simpler than viruses. They are very small, naked RNA molecules without any protein or other type of coat (see Fig. 23.8 for an example, which you do not need to memorize).

Infection of plant cells by them is dependent on mechanical damage of the host plant (to infect plants experimentally the viroid RNA is usually rubbed on the leaves with some abrasive material). The effect of viroid infection varies from some impairment of the health of the plant to certain death of, for example, palm trees. The really remarkable feature is that no protein coding genes have been identified in viroids. To detect a protein coding gene, you look for an **open reading frame**—that is, a sequence of codons that code for the long stretch of amino acids required for proteins. In viroids, although translational initiation sequences are found, the subsequent reading frames are so short (due to hitting stop codons—page 293) that protein production is excluded. It is not known how viroids produce disease nor how the viroid RNA is replicated.

Further reading

Watson, J. D. (1987). *The molecular biology of the gene.* 4th edn, Benjamin/Cummings.

Ptashne, M. (1987). *A genetic switch: gene control and phage λ.* Cell Press and Blackwell Scientific Publications.

Influenza virus

Laver, G. and Garman, E. (1994). Flu neuraminidase: the virus' Achilles heel? *Today's Life Science*, **6**(9), 40–7.
This gives an excellent account of the new rational therapy approach to combatting influenza, referred to briefly in the text of this book. Beautifully illustrated.

Hughson, F. M. (1995). Structural characterisation of viral fusion proteins. *Current Biology*, **5**, 265–74.
Deals with infection by influenza virus, and fusion of lipid bilayers. Describes how virus escapes from the endosome, with excellent diagram.

Retroviruses

Katz, R. A. and Skalka, A. M. (1994). The retroviral enzymes. *Ann. Rev. Biochem.*, **63**, 133–73.
Advanced and detailed review of the protease, reverse transcriptase, and integrase enzymes of these viruses, also with HIV inhibitors, and resistance to the latter.

Nowak, M. A. and McMichael, A. J. (1995). How HIV defeats the immune system. *Sci. Amer.*, **273**(2), 42–9.
Describes how continuous evolution of the human immunodeficiency virus in the body leads to the immune devastation that underlies AIDS—good graphics of the immune system.

Viroids

Symons, R. H. (1989). Self-cleavage of RNA in the replication of small pathogens of plants and animals. *Trends Biochem. Sci.*, **14**, 445–50.
Includes description of the self-cleaving 'hammerhead' RNA structure.

Problems for Chapter 23

1 Explain how a virus might gain access to the interior of a eukaryote cell.

2 Referring only to non-retroviruses, in what way must a (+) single-stranded RNA virus differ from a (−) single-stranded RNA virus?

3 Why does vaccinia carry with it a DNA-dependent RNA polymerase in its virion when the host cell already has this?

4 A retrovirus is a (+) single-stranded RNA virus. In what way does it differ from a non-retroviral (+) single-stranded RNA virus?

5 Explain how a virus with a lipid membrane envelope, such as influenza, acquires this envelope with its assemblage of integral proteins.

6   Influenza epidemics sweep the world regularly. Most are mild but occasionally, as in 1918, a highly lethal one occurs. Why is this so?

7   Why does influenza virus cause red blood cells to agglutinate when the two are simply mixed together?

8   Why is it somewhat surprising that the influenza virus has a surface neuraminidase activity? What might its function be?

9   Describe the 'choice' that bacteriophage lambda makes on infecting an *E. coli* cell.

10. **(a)** What is a viroid?

    (b) Given that viroids infect plants and are reproduced, one might expect it to have genes for protein production. Is this the case?

Chapter 24

Gene cloning, recombinant DNA technology, genetic engineering

The terms in the chapter title refer to the same area of molecular biology—they all involve DNA manipulation. Although the experimental aspects of biochemistry are outside the scope of this book, DNA manipulation techniques have caused such a revolution in biology and biochemistry in particular that a brief account is included here.

The technologies make it possible to isolate individual genes, to determine their base sequences, to manipulate the latter in any desired way, and to transfer genes from one species to another. The base sequence of the coding region also permits deduction, from the genetic code, of the amino acid sequence of the protein for which it codes (and is often the easiest way to determine the latter). Structural information is a prerequisite for elucidating how a gene functions and what are its control mechanisms. Much of the information in Chapter 21 on the control of gene activity has arisen from studies involving recombinant DNA technology. The isolated gene can be placed in a functional role in many different types of cells to produce the protein it codes for in whatever amount is required. The associated technologies can be used in medical diagnosis of genetic diseases. The general biological applications are almost unlimited.

What were the problems in isolating genes?

A major problem was the gigantic size of the DNA molecules of chromosomes. In a diploid eukaryote cell there are two copies of a given homozygous gene, one on each chromosome of a pair, amongst scores of thousands of other genes, which, in turn, form only a fraction of the total DNA of a cell. The only thing that distinguishes a gene from all the other DNA is the coded information of its base sequence. The task of isolating a gene resembled not so much finding a needle in a haystack, as finding a particular piece of hay in a haystack. It was not necessarily unduly pessimistic to wonder whether the subject had come up against an impenetrable barrier to further progress. Recombinant DNA technology changed all that.

The first step—cutting the DNA with restriction endonucleases

The first step in isolating a gene is to cut the cellular DNA into defined, suitable pieces, small enough to be handled. This, itself, seemed to be an impossible goal. The only enzymic method of cutting the phosphodiester bonds (page 239) of DNA was with DNases of the type found in the digestive enzymes of pancreatic juice. If DNA is incubated with such an enzyme to give partial hydrolysis, the molecule is fragmented into pieces, but the cutting is completely random.

This point emphasizes the revolutionary importance of the discovery of a different class of **endonucleases** or **DNases** in bacteria. (An endonuclease attacks internal bonds in the molecule, not the terminal bonds as occurs with exonucleases.) The new class of enzymes are usually referred to simply as **restriction enzymes**, the act of cutting with them, as restriction, and the DNA so treated as having been restricted. Such enzymes cut DNA, not randomly, but each recognizes a specific short sequence of bases so as to make a cut at a precise point in the sequence. Different bacteria have different restriction enzymes that recognize different base sequences in the DNA and therefore have different cutting sites. A large number of restriction enzymes cutting at specific base sequences are now known. To illustrate this point, an enzyme from *E. coli* cuts double-stranded DNA at the sequence

GAATTC
CTTAAG.

and that from *Bacillus amyloliquefaciens* cuts at the sequence

GGATCC
CCTAGG.

(There is no point in memorizing such sequences, but note that they have a twofold symmetry—that is, the partner strand read in the $5' \rightarrow 3'$ direction is identical in sequence.) The enzymes are named after the bacterium (and bacterial strain) of their origin and a Roman numeral where more than one enzyme occurs in that species. Thus the two enzymes mentioned above are called *Eco*RI, and *Bam*HI, respectively. *Eco*RI was the first to be isolated from *E. coli* strain R. Other restriction enzymes recognize four-, five-, and eight-base sequences. Restriction enzymes make it possible to cut DNA at precise base sequences with surgical neatness, producing defined fragments.

What is the biological function of restriction enzymes?

Bacterial cells have restriction enzymes to destroy invading foreign DNA. For example, a bacteriophage such as lambda (λ) phage injects its DNA into an *E. coli* cell (page 321). The cell's restriction enzyme, by cutting that DNA at the restriction sites, prevents a successful phage attack; it restricts the ability of foreign DNA to infect the cell. The question arises as to what prevents the enzyme from destroying the cell's own DNA. A hexamer base sequence such as that recognized by *Eco*RI must occur many times in the *E. coli* chromosome—statistically, every 4^6 (4096) base pairs. The cell guards against cutting its own DNA by adding a methyl group to one of the bases in all of the recognition sequences on each new DNA strand as soon as it is synthesized. This does not interfere with base pairing or gene function, but the restriction enzyme no longer recognizes the methylated sequence and the cell's own DNA is therefore immune from attack by that enzyme.

The biological role of restriction enzymes can be illustrated by a brief discussion of lambda phage. Lambda infects *E. coli* cells, but different strains, or isolates, of *E. coli* exist differing in their restriction enzymes. Thus, a λ phage that has replicated in a given strain of *E. coli* will reinfect that same strain with high efficiency, because the phage DNA is protected by methylation as is the *E. coli* host DNA. However, if this phage attempts to infect a different strain of *E. coli*, with a different restriction enzyme, the latter will rapidly attack the phage DNA. Whether, in the latter situation, infection is *ever* successful will be the outcome of a race between the cell's methylation system and the restriction enzyme, for, as soon as the lambda DNA enters the cell, it will be subject to both attacks. The protection against infection is very high, for the lambda DNA is sufficiently long to incorporate several restriction sites for typical restriction

enzymes. When phage DNA, unmethylated in these sites, enters the cell, methylation has to win the race in all cases for successful infection to result, since a single cut at any one of the sites destroys the infectivity of the phage DNA.

The isolation of specific genes to be used in various ways is of central importance and we will turn now to this technology.

Gene cloning, or how genes are isolated

In this section, we will describe two procedures used in recombinant DNA technology. One of these is to isolate a **genomic clone** from DNA; we take human DNA as the example though the methods apply to DNA from any source. A genomic clone provides a piece of DNA identical in base sequence to the corresponding stretch of DNA in the cell and often is designed to contain a specific gene. The second procedure is the isolation of a human **cDNA**, the 'c' meaning 'complementary'. It is the double-stranded DNA copy of a human mRNA. The genomic DNA and the cDNA differ in that the former contains introns. An important practical aspect of this is that, since introns are not spliced out of RNA transcripts (page 280) of genes in *E. coli* cloned eukaryote *genes* cannot direct protein synthesis in that organism. cDNA on the other hand is transcribed into an mRNA-like transcript by *E. coli* (provided that appropriate transcriptional signals are placed on the cDNA). The transcripts will direct protein synthesis in *E. coli*; this also requires that appropriate translational signals have been added to the cDNA. Sometimes the best way to isolate a gene is to first isolate the cDNA related to that gene, which can be used as a hybridization probe (see below). This is particularly true where mRNA preparations rich in one species of mRNA can be obtained, as might be the case for an inducible enzyme. The enriched mRNA preparation may be used as the hybridization probe for cDNA isolation (see below).

Genomic clones are required for studies on gene structure, while cDNA clones are required when, for example, the aim is to produce human insulin or other proteins in *E. coli* cells.

The two procedures, genomic cloning and cDNA cloning, have been selected because they serve to illustrate some general principles of gene cloning. The aim is not to give practical guides on the procedures, but to remove some of the mystery from gene manipulation. (A current laboratory manual on 'molecular cloning' comprises three volumes each of about 500 pages.)

What does 'cloning' mean?

A **clone** refers to multiple identical copies produced from a single origin. Cloning a gene or a cDNA gives us large numbers of copies of a piece of DNA. This is necessary since many of the techniques used require visualization or quantification of the DNA and this requires many copies of the molecule for detection to be possible.

The essence of cloning is, firstly, that a *single* piece of DNA (inserted into a replicating vector—see below), out of a mixture of thousands, produced, for example, by restricting total DNA, enters a bacterial cell and is replicated in that cell. Secondly, a single bacterial cell will grow into a colony containing large numbers of identical cells—for example, a single *E. coli* cell will produce $\sim 10^7$ cells in a single colony when grown on an agar plate overnight. Furthermore, there can exist multiple copies of the vector, each with its piece of inserted DNA, in each cell. Since this will happen to each piece of DNA in the restriction mixture, cloning provides a means of both separating and amplifying each piece of DNA, for it is easy to isolate separate *E. coli* cells by growing them into colonies on a nutritive plate. Provided you can pick out of the thousands of different colonies one that contains the piece of DNA you are interested in, it is possible to produce virtually unlimited numbers of copies of that piece by growing up pure cultures of the selected colony.

Isolation of a genomic clone of a human gene

Preparation of a human gene library

First a cloning vector must be chosen that can accept a piece of human DNA and then be replicated in *E. coli* cells. There are several choices and we will choose lambda bacteriophage (described on page 321). The centre portion of the DNA molecule of lambda can be replaced with a piece of human DNA, without impairing the ability of the **recombinant phage**, as it is called, to replicate in *E. coli*, as explained below. The lambda genes essential for its replication are contained in the arms on either side of the insert. Lambda phage can accept a foreign piece of DNA 15–20 kilobases (kb) long without impairing its replicative ability. This is a convenient size for most cases of genomic cloning.

How are recombinant molecules constructed?

The principle of this lies in 'sticky ends' of DNA pieces. Many restriction enzymes do not make 'straight through' cuts across both strands of a double-stranded DNA, but instead make a 'staggered' cut. *Eco*RI, for example, cuts like this.

—X—X—G ↓ A—A—T—T—C—X—X—

—X—X—C—T—T—A—A ↑ G—X—X—

       *Eco*RI ↓

—X—X—G   A—A—T—T—C—X—X—

—X—X—C—T—T—A—A   G—X—X—

              Sticky ends

The staggered cut produces cohesive or sticky ends. This is because the 'overhang' sequences automatically will base pair with each other, in this case A with T. If two pieces of DNA, both having identical sticky ends, are mixed, they will join together as indicated in Fig. 24.1. If human DNA and lambda phage DNA are restricted so that they have the same sticky ends, and the two sets of pieces are mixed, they will associate to form the recombinant lambda DNA molecules described above. (A ligation enzyme is used to seal the nicks in the chain by synthesizing a bond between the 3'-OH and the 5' phosphoryl group. Such a ligase is described on page 260.)

When these are added to the separate protein components of lambda phage, the lambda phage assemble and the recombinant phage DNA is automatically packaged into the lambda head to produce infective phage, a remarkable illustration of self-assembly. ('packaging kits' with the necessary phage components are commercially available.)

The packaging process requires that, for any DNA piece to be enclosed in an infective phage, it must have, at its two ends, lambda phage DNA and it must be of a correct length. Thus,

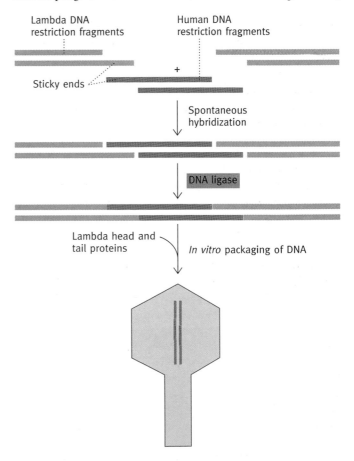

Fig. 24.1 Sequence of steps in producing in lambda bacteriophage a recombinant DNA library. Note that, although other associations of DNA pieces will occur due to sticky ends, only those containing a lambda left arm and right arm separated by a human DNA fragment of appropriate length (15–20 kb) will be packaged.

although the mixture of pieces with sticky ends will associate in all sorts of ways, only the desired type of recombinant molecule illustrated in the text above will be selected. (In preparing the human DNA fragments, steps are taken to ensure that pieces of appropriate size are obtained; we will not go into the details of this.)

The lambda phage, now carrying recombinant DNA, are used to infect *E. coli* cells by a procedure in which only a very small proportion of the cells become infected, to maximize the chance that only one phage will enter a given cell. It is essential also that the *E. coli* is a mutant strain lacking its restriction enzyme. The result of all this is that the entire fragmented human genome is collectively contained in a culture of *E. coli* cells, one piece per cell. The collection is known as a **genomic library**. The culture is spread on to a solid growth medium and incubated. The uninfected cells grow to form an opaque 'lawn' over the whole plate, but, wherever there is a phage-infected cell, a clear spot or **plaque** will appear. This is because the single phage multiples in the cell, the progeny are released to infect other, adjacent cells and so on, thus destroying the lawn cells surrounding the originally infected bacterium. (Release of phage involves lysis of the cell.) Each plaque represents multiple copies of phage originating from a single infective particle and each carrying the same copy of a piece of human DNA (Fig. 24.2).

If we can now identify which plaque contains the lambda carrying the desired gene, we can grow unlimited numbers of the phage in *E. coli* and isolate its DNA. How is the correct plaque identified?

Screening of the plaques for the desired human gene

To identify the phage carrying the wanted human gene DNA–DNA hybridization is most commonly used (see page 242). For this, a **hybridization probe** is needed, consisting of a short piece of DNA (perhaps 20 nucleotides long) whose base sequence is complementary to a known sequence in the gene in question (see below for how we get this). The method is to transfer phage, from each plaque, to a DNA-absorbent membrane by laying a disc of it on the surface of the culture plate (remembering to mark its orientation so that the position of individual plaques can be identified); phage from each plaque then sticks to it. The membrane is treated with alkali to disrupt the protein–DNA complexes of the phage head and release the DNA and also, importantly, to cause DNA strand separation. The DNA is then fixed to the membrane in single-stranded form by baking or UV light irradiation (which covalently crosslinks DNA to the membrane) and the ability of the membrane to fix further DNA is destroyed by flooding with nonspecific DNA. The hybridization probe (made radioactive by enzymic phosphorylation using radioactive ATP) is incubated with the membrane and allowed to hybridize to any complementary DNA on the membrane. Nonhybridized probe is washed away. The probe will stick only to complementary DNA—at the target gene.

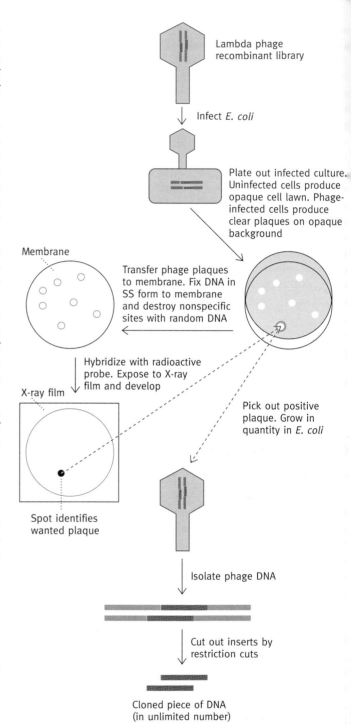

Fig. 24.2 Illustration of gene cloning using bacteriophage lambda phage as the cloning vehicle. See text for explanation.

Exposure of the membrane to X-ray film exposes the silver grains, so that, wherever the probe has hybridized, the developed film will show a black spot due to the radioactivity in the probe (Fig. 24.2). Once the plaque containing the human gene is identified, unlimited numbers of the phage within this plaque can be grown by infecting *E. coli*, and the human DNA insert released from the phage DNA, in this example, by appropriate restriction.

How is a suitable probe obtained, considering that the gene cannot be base-sequenced until we have isolated it? That is, how can we construct a probe of complementary sequence to identify the presence of the gene in a plaque? If the protein coded for by the gene has been isolated and all, or part, of its amino acid sequence determined, we can work backwards from the genetic code (Table 22.1) and determine what the base sequence of the gene coding for a particular stretch of amino acids in the protein must be. Well, not quite, because of the redundancy of the genetic code (page 293). Where an amino acid has several codons we cannot deduce which one was actually used by the cell to synthesize that protein. (Gene cloners are delighted if they find a stretch of methionines and tryptophans for these have only single codons.) Even where redundancy is unavoidable, there are methods for coping with the ambiguity, such as by synthesizing a mixture of probes. Machines exist that readily chemically synthesize short DNA probes of predetermined base sequence. A radioactive phosphate group added by an appropriate enzyme completes the probe. If such a probe is not available there are alternative screening possibilities, which we will not go into here.

We will now turn to the isolation of a cDNA clone.

Cloning a human cDNA

Preparation of a human cDNA library

The principle is that mRNA from human cells expressing the gene in question is isolated, copied into double-stranded DNA *in vitro*, and cloned. A significant advantage of this approach is that often a tissue can be selected that expresses the wanted gene in large amounts so that the relevant mRNA is abundant. It is especially favourable if the level of a given protein (and hence its mRNA) can be induced in a tissue, since this results in enrichment of the mRNA specific for the protein.

mRNA is only a small fraction of the total RNA of a cell and, in the case of human cells, has (with a few exceptions) polyA tails (page 279). If the RNA preparation is passed down a column of inert support carrying synthetic oligo-dT ligand (short, single-stranded 'DNA' with only T bases), the ribosomal and transfer RNA run through while mRNA binds due to A=T hydrogen bonding of the tails to the ligand; the mRNA can then be eluted using solutions of low ionic strength.

The isolated mRNA mixture is now copied in the test tube into DNA using the polymerase domain of viral reverse transcriptase (page 319). This gives an RNA–DNA duplex. The RNA is destroyed with NaOH or RNase treatment and the single-stranded DNA converted to double-stranded DNA by an exonuclease-free DNA polymerase I (page 259) derived from a bacteriophage that happens to be particularly suited to the task and is commercially available.

The procedure now is to clone the cDNA molecules. This may be done using lambda as the cloning vehicle, as already described for genomic libraries. The cDNA molecules are not sticky ended, but methods are available for enzymically adding appropriate sticky ends to such 'blunt-ended' molecules, as they are called, or alternative procedures for 'blunt-end ligation' are now used.

What can be done with the cloned DNA?

At this point, we have a purified lambda phage containing a desired genomic DNA insert or a wanted cDNA insert. We can very easily grow as much as we want of these in *E. coli* cells, isolate the progeny DNA, and cut out the DNA inserts by restriction.

Subsequent use of the latter, for the various purposes described below, almost always involves the DNA of interest being inserted into a different cloning vector, most frequently a **bacterial plasmid**. Lambda is convenient for the initial preparation of libraries because it can accept large pieces of DNA (around 15 kb) and efficiently infect cells so that a complete genomic 'library', or a complex cDNA library, can be obtained. However, when it comes to studying the cloned piece of genomic DNA or cDNA of interest, lambda has the disadvantage of possessing a large amount of its own DNA in the arms (about 30 kb), required for its replication but of no interest in the current context. Plasmids are much smaller and, experimentally, much more convenient to handle. They cannot accept pieces of foreign DNA as large as can lambda, but if, for example, the aim is to sequence a genomic clone originally isolated in lambda, it is cut up into pieces of appropriate size to be inserted into an engineered sequencing plasmid (described below) and the individual pieces sequenced. cDNA clones may be treated in the same way. However, cDNA clones are commonly small enough to be inserted into **expression vectors**—these include *E. coli* plasmids into which appropriate transcriptional and translational signals have been added. When these are infected into *E. coli* cells, the protein coded for by the cDNA may be produced in quantity. We must now describe what a bacterial plasmid is.

Bacterial plasmids

The *E. coli* cell has a single major circular chromosome carrying the few thousand genes that constitute most of the cell's genetic makeup. In addition, however, there are tiny separate minichromosomes or plasmids in the cytoplasm—circular DNA molecules carrying a handful of genes that usually have a protective role in the cell; typically they carry genes conferring antibiotic resistance on the cell by coding for an enzyme that, for example, destroys the antibiotic. Each plasmid has a replicating origin to provide for duplication of the plasmid in the cell.

Plasmids are not infectious in the sense that bacteriophage

are. If, however, *E. coli* cells are treated somewhat unkindly—e.g. exposed to CaCl$_2$ at 0°C and then the temperature suddenly raised to 42°C—they become 'competent' to take up plasmids.

Let us suppose we are dealing with a cloned cDNA molecule isolated as described above. First, a plasmid selected for the purpose in mind is cut with a restriction enzyme and, second, appropriate sticky ends are added to the cDNA molecules by procedures we need not describe. When the cut plasmids and the modified cDNA are mixed and ligated (that is, the nicks are enzymically sealed), **recombinant plasmids** are obtained (see Fig. 24.3). (A procedure that to avoid complexity, we will not describe is available to prevent the cut plasmid merely being recircularized to its original form and favours recombinant plasmids being produced.) *E. coli* cells are infected with a low multiplicity of the recombinant plasmids and grown into colonies on agar plates such that each colony arises from a single cell. Because of the low multiplicity of plasmid infection, only a very small proportion of the cells become infected, and the possibility of two plasmids being taken up by one cell is minimized. Since untransformed cells grossly outnumber transformed cells, it is necessary to have a quick method for selecting the latter.

One of the 'engineered' plasmids used for cloning is known as pBR322; this has, inserted into its DNA, genes for ampicillin and tetracycline resistance. Each of these genes has a different restriction site. Suppose that the foreign piece of DNA is inserted into the tetracycline gene; the latter is no longer capable of directing the synthesis of the protein conferring resistance. A cell carrying such a plasmid will be resistant to ampicillin but sensitive to tetracycline. A cell containing a plasmid without a DNA insert will be resistant to both; a cell without a plasmid will be sensitive to both. This gives a basis for selecting only those cells containing the wanted plasmid.

Determination of the base sequence of a cloned piece of DNA

The entire information in a gene lies in its base sequence and the determination of this is of central importance. There are two methods for DNA sequencing as the process is called. One is a direct chemical method—the **Maxam–Gilbert method**. The other is based on enzymic replication of DNA and this is called the **Sanger** or **dideoxy method**; this is the one most often used.

Outline of the dideoxy DNA-sequencing technique

Suppose that you have a cloned piece of DNA and want to sequence it. It is inserted into a specially engineered sequencing plasmid, which is multiplied by infecting *E. coli* cells with it. The plasmids are isolated so that you now have large numbers of recombinant plasmids containing your piece of DNA. The sequencing procedure (described below) requires that your piece of DNA to be sequenced is copied, *in vitro*, by DNA polymerase. This requires (apart from the four deoxynucleoside triphosphates (dNTPs)): (1) that the DNA to be copied is single-stranded; and (2) that a primer is hybridized to the start site because DNA polymerase cannot initiate new chains (page 256). The plasmid is rendered into single-stranded form by treatment with NaOH and heat. The sequencing plasmid is so engineered that at the 3′ end of each strand of your inserted cloned DNA piece is attached a known priming sequence to which a synthetic DNA primer will hybridize. A different primer site is used for the two strands enabling you to sequence one or other of the strands according to which primer you use in the copying process. Often, both strands are sequenced to confirm the results.

Incubation of the single-stranded DNA to be sequenced with primer, exonuclease-free polymerase I, dATP, dGTP, dCTP, and dTTP permits copying of the piece of DNA to proceed. One of the triphosphates (or the primer) is radioactive so that all of the new DNA copies are labelled.

The sequence determination depends on two things. First, an electrophoretic method to separate DNA molecules in which they migrate along a slab of acrylamide gel in an electrical field;

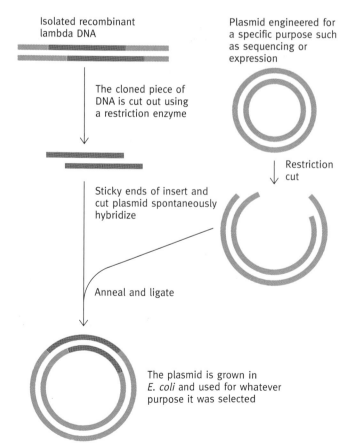

Isolated recombinant lambda DNA

The cloned piece of DNA is cut out using a restriction enzyme

Sticky ends of insert and cut plasmid spontaneously hybridize

Anneal and ligate

Plasmid engineered for a specific purpose such as sequencing or expression

Restriction cut

The plasmid is grown in *E. coli* and used for whatever purpose it was selected

Fig. 24.3 Transferring a lambda DNA insert into a plasmid. If it is desired to sequence a DNA insert that has been cloned in lambda or, for example, to express a eukaryote cDNA in *E. coli* to produce a specific protein, the selected DNA is transferred to a plasmid engineered for the particular purpose such as sequencing or expression.

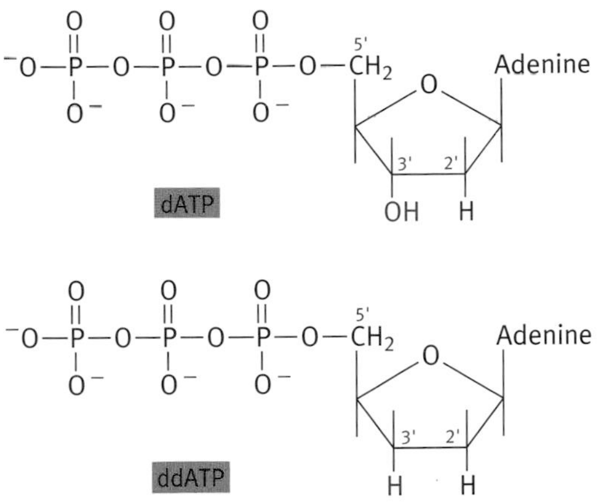

O O O
‖ ‖ ‖
⁻O—P—O—P—O—P—O—CH₂   Adenine
| | |
O⁻ O⁻ O⁻

dATP   OH   H

O O O
‖ ‖ ‖
⁻O—P—O—P—O—P—O—CH₂   Adenine
| | |
O⁻ O⁻ O⁻

ddATP   H   H

Fig. 24.4 The structures of deoxyATP (dATP) and dideoxyATP (ddATP). The absence of the 3′-OH group means that, when a ddNTP is added to a growing DNA chain, the chain is terminated.

the smaller the DNA chain, the faster it moves. Each addition of a nucleotide alters the migration so that chains form separate bands, each being one nucleotide different in length from the next. They are visualized by autoradiography, which involves exposing the gel to X-ray film sensitive to radioactivity. The second point concerns dideoxy derivatives of nucleoside triphosphates (ddNTPs). DNA polymerase adds a nucleoside to the 3′OH of a growing DNA chain. Dideoxy NTPs lack the 3′OH group (Fig. 24.4); they can still be added to a chain via their 5′P, but the chain is thereby terminated.

It is important to emphasize that, when we talk of a 'piece' of DNA being sequenced, multiple copies of that piece are involved in the experiments. Even a minute amount of DNA contains a large number of individual molecules. (Remember that one mol of any chemical compound contains the astronomic number of 6.03×10^{23} molecules.) Suppose that we have in the copying process all four deoxy NTPs plus a small amount of *one* dideoxy NTP—take dideoxy ATP as an example—every time addition of an A is specified, most of the new chains will have a normal adenine nucleotide added, but a fraction will, by chance, have a dideoxy form of the adenine nucleotide added, thus terminating those particular chains. (The fraction terminated will depend on the relative proportions of dATP and ddATP.) The terminated chains are sufficient in number to be detected as a separate band on an electrophoretic gel. The rest will go on being added to until another A is due to be added and the same will happen again.

How is all this interpreted as a base sequence?

Suppose the piece of DNA being sequenced has a sequence with Ts placed as shown.

3′—X—X—T—X—X—X—X—T—X—X—T—5′

If, during the copying dideoxy ATP is present (along with excess

deoxyATP) such that, at the addition of each A, a small proportion of the growing chains are terminated, the copying will produce the following population of chains (attached to the primer)

(a) 5′—X—X—ddA—3′
(b) 5′—X—X—A—X—X—X—X—ddA—3′
(c) 5′—X—X—A—X—X—X—X—A—X—X—ddA—3′.

On a sequencing gel these bands will be seen as the bands in Fig. 24.5 (left-hand column).

If a second incubation contains dideoxy TTP instead of the dideoxy ATP, an analogous set of chains terminating in T will be produced and, similarly, terminating in G and C if dideoxy CTP or GTP, respectively, are present in duplicate incubations. If all four incubations are carried out and the products are run side by side on a gel, the **sequencing ladder** shown in Fig. 24.5 is obtained, from which the base sequence of the piece of DNA can be simply read off. The sequence is read from the bottom of the gel upwards as the sequence of the copy chain (and therefore of the partner to the template strand) rather than of the template. This gives the sequence in the 5′ → 3′ direction since synthesis always proceeds in that direction.

About 200–300 bases can be determined in each sequencing ladder. If overlapping pieces from the original cDNA are obtained by using different restriction enzymes, the entire sequence can be determined. A part of an experimental sequencing ladder is shown in Fig. 24.6.

If the purpose of the sequencing of a cDNA is to determine the amino acid sequence of the protein coded for by the relevant gene, using the genetic code, it is necessary to determine the correct reading frame out of six possible on two strands. The

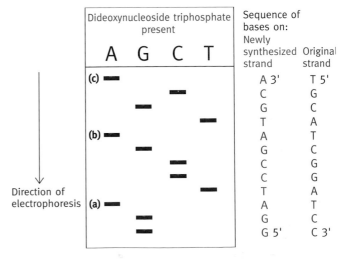

Fig. 24.5 Diagram of an autoradiograph sequencing gel. The sequence is read from the bottom to the top. **(a)**, **(b)**, and **(c)** are the bands produced in the presence of dideoxy ATP. See text for explanation. A photograph of a real sequencing ladder is shown in Fig. 24.6.

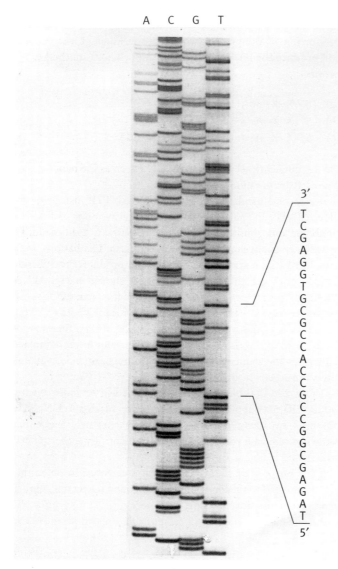

A C G T

```
3′
T
C
G
A
G
G
T
G
C
G
C
C
A
C
C
G
C
C
G
G
C
G
A
G
A
T
5′
```

Fig. 24.6 Photograph of an actual sequencing ladder. A, C, G, and T specify which dideoxynucleoside triphosphate was present in each incubation. Kindly provided by Dr. Chris Hahn, Department of Biochemistry, University of Adelaide.

correct frame is the one that does not prematurely run into stop codon-generating triplets.

The above procedure is employed in many laboratories. However, sequencing technology has advanced to permit automation. The principle of the new method is unchanged from that devised by Sanger, but fluorescent-labelled dideoxy nucleoside triphosphates are used, the fluorescence being of a different colour for each of the four. A single DNA synthesis incubation is used containing all four dideoxy compounds, plus of course the four deoxynucleoside triphosphates and the products analysed on a single electrophoresis track. The gel is automatically scanned for the different colours, and a computer prints out the base sequence.

The polymerase chain reaction (PCR) for amplifying a specific DNA segment

We mention this (out of an array of elegant techniques) because it constitutes a major recent landmark in DNA manipulation. In effect, it is a method of selecting a particular stretch of DNA on a long molecule and amplifying it by making vast numbers of copies. This means that a vanishingly small amount of DNA, perhaps from a few cells, is all that is needed for many purposes.

The principle is as simple as it is brilliant (Fig. 24.7). Take a piece of double-stranded DNA and separate the strands by

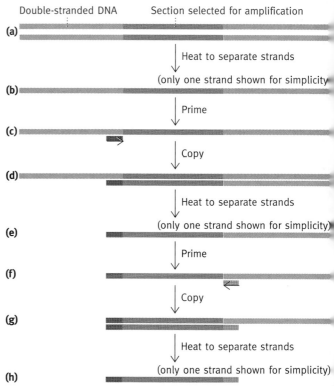

Double-stranded DNA   Section selected for amplification

(a)

    ↓ Heat to separate strands
    (only one strand shown for simplicity)

(b)

    ↓ Prime

(c)

    ↓ Copy

(d)

    ↓ Heat to separate strands
    (only one strand shown for simplicity)

(e)

    ↓ Prime

(f)

    ↓ Copy

(g)

    ↓ Heat to separate strands
    (only one strand shown for simplicity)

(h)

This is a copy of the wanted section of DNA; with successive reaction cycles it will be amplified exponentiially

Fig. 24.7 Diagram illustrating the basic principle underlying amplification of a DNA section by the polymerase chain reaction (PCR). **(a)** The green bars represent the section chosen for amplification by the use of appropriate primers (→ ←), each complementary to the 3′ end of the section in one strand to be amplified. **(b)** Heating separates the strands and, from this point, to keep the diagram to a manageable size, amplification of only one strand is shown; the process amplifies both strands. **(c)** Incubations contain four dNTPs, a heat-stable DNA polymerase, and primers. The primer appropriate for this strand binds. **(d)** DNA is replicated (in this case beyond the end of the desired piece). **(e)** Heating separates the strands. **(f)** The new strand is primed. **(g)** The new strand is replicated. **(h)** The separated wanted section. You can see that we now have reproduced our selected piece. With about 25 rounds of replication, heating, priming, and synthesis, the piece can be amplified a millionfold. Note that, every time priming and copying occurs, all of the original and new molecules are so primed and copied so that a detailed diagram of what happens becomes rather involved. The essential is that an enormous amplification of the selected piece occurs.

heating, and copy a selected portion of them into DNA; repeat the process over and over (copy the copies), and this gives an exponential increase in copy numbers. The section to be amplified on a longer piece of DNA is selected by using primers for the DNA synthesis reaction corresponding to the limits of the section to be amplified. Different primers are needed for the synthesis in the left and right direction. To obtain these, of course, one must know the base sequences of the DNA sample to which the primers will hybridize. At the end of each round of synthesis, the DNA duplexes formed by the copying process are separated by heat to permit further priming, and hence copying—automatic machines are available for achieving this and heat-stable DNA polymerases (from thermophilic bacteria) avoid the necessity of adding more enzyme after each round of copying and stand separation by heating.

The practical limit of the length of DNA that can be amplified in this way was about two kilobases, but recent techniques have enabled amplification of up to 35 kb. The DNA copies so obtained can be used for many purposes. The technique is important in forensic science and in diagnosis of fetal genetic abnormalities *in utero* (see below). It is also extensively used in biochemical research.

Applications of recombinant DNA technology

As stated already, a central role for gene cloning is to determine base sequences of genes, an essential prerequisite for studying gene function and control. The DNA base sequence is also a direct 'window' on evolutionary relationships. There are other uses of practical importance that will be mentioned only briefly here. These include the following (a far from exhaustive list).

Expression of the gene in bacterial and eukaryote hosts

As already briefly referred to, an isolated cDNA clone or a bacterial gene may be inserted into an 'engineered' expression vector. This may be a plasmid into which have been inserted the appropriate bacterial DNA promoter signals and translational initiation signals such that the inserted cDNA molecule in each plasmid is expressed—the cell synthesizes mRNA from the cDNA and translates this into the protein it codes for. Thus, an inserted prokaryote gene or eukaryote cDNA (which, remember, has no introns) can be efficiently transcribed and translated into protein. Expression vectors are available commercially.

Human proteins of therapeutic value, otherwise virtually unobtainable, may be produced in unlimited quantities in *E. coli* or other cells. Human insulin is produced in bacteria and there is an ever-increasing list of such examples. Apart from making available, in quantity, human proteins present in the body in minute amounts, production in *E. coli* eliminates the danger of a protein prepared from human sources being contaminated with a human infectious agent. Thus a human growth hormone is produced via a cDNA. An immunogenic protein of hepatitis B virus is produced in yeast to be used as a vaccine with no danger

of infective virus being present. In some cases, *E. coli* can be made to produce a foreign protein as a major percentage of the total cell protein—so much that it precipitates out as **inclusion bodies.** There is often a snag, however. Produced like this, the protein may not fold up properly or only a small proportion may do so (see page 304). In some cases, the inclusion bodies have been isolated and the protein partially correctly refolded *in vitro.* Bacteria do not glycosylate proteins (page 32), but, if appropriate animal cell systems are used as expression hosts, glycosylated proteins can be produced in quantity. Insect cells have been used for this.

Site-directed mutagenesis

This is a very powerful technique for analysing the functions of particular amino acid residues in a protein. Consider a case in which you have an enzyme for which the amino acid sequence is known, either from direct sequencing or by deduction from the base sequence of its gene or cDNA. You now wish to examine the role of a particular amino acid side chain in the function of that protein. If the bacterial gene or eukaryote cDNA is cloned into an expression vector such as a plasmid, the protein may be produced in *E. coli* in sufficient quantity for study. For example, in the case of an enzyme you could be interested in the rate of its catalytic activity or its specificity. **Site-directed mutagenesis** permits the specific replacement of one or more selected amino acids in the protein. This can be done by replacing the appropriate section of the isolated gene with a synthetic oligonucleotide whose base sequence is such that it codes for the altered amino acid. Methods for such procedures are well established. The altered catalytic activity of the mutant protein can then be studied.

Transgenesis

Genes can be inserted into animals or plants to give different phenotypic characteristics. This is referred to as **transgenesis.** Insertion of normal genes into patients to correct faulty genes (gene therapy) has the potential to cure genetic diseases. As vectors for the latter, retroviruses, genetically crippled to prevent their replication, might be used to insert genes into human chromosomes (see page 320) though the insertion point cannot be determined by this procedure and random insertion has potential hazards. Methods for gene insertion at specific points are being developed. Already the insertion of the gene for human adenosine deaminase into patients' white cells has cured immune deficiency in children (Fig. 18.6). In plants, foreign genes can be inserted into the chromosomes by using as cloning vector a naturally occurring (but suitably altered) plasmid (the Ti or tumour-inducing, plasmid) contained in the pathogenic soil bacterium, *Agrobacterium tumefaciens,* or the DNA molecules may be literally shot into plant cells, where they become functionally incorporated, with a gun-type instrument. In this way, crop plants are being engineered with specified phenotypic characteristics—for example, to be resistant to herbicides. The purpose of this is to control weeds by blanket spraying with the herbicide so that only the resistant crop plant survives. Transgenic animals can also be used as 'expression vectors'. For example, a wanted human protein can be produced by attaching the isolated human gene to the 'signalling' part of a milk protein resulting in secretion of that protein into the milk. The human protein may then be obtained in quantity in the milk; it has been used to produce human proteins in sheep milk.

Detection of genetic abnormalities by restriction analysis or Southern blotting

Where the structure of genes is known, hybridization probes can be made to detect gene abnormalities. The amount of DNA required is sufficiently small to permit examination of fetuses so information on potential diseases is available to permit decisions to be made on possible termination of a pregnancy (use of the polymerase chain reaction amplification means that only minute amounts of DNA need be obtained for examination). For example, in muscular dystrophy, the gene abnormality usually is a deletion of part of the coding DNA and this can be detected by a **Southern blot analysis**. The latter, named after its discoverer, involves cutting DNA with one or more restriction enzymes, running the fragments on an electrophoresis gel, transferring the separated fragments on to a membrane, and probing the membrane with a gene-specific radioactive hybridization probe. This gives a pattern of DNA pieces, visible as bands. Gene abnormalities can show up in a different pattern of bands, because mutations may create a new restriction enzyme site or destroy or delete a pre-existing one. Even though the gel will have vast numbers of different pieces of DNA on it, only those few that hybridize to the specific probe will show up, thus giving a simple pattern of bands.

Even when the gene involved in causing a genetic disease is not precisely localized or isolated, a technique (not described here), known as **restriction fragment length polymorphism (RFLP)** analysis or **genetic linkage analysis**, can be employed to identify which relatives of a person displaying the disease are likely to have the abnormal gene.

A more easily understood use of RFLP analysis is in **DNA fingerprinting**. In this, repetitive human DNA sequences located throughout the genome (page 249) show extreme polymorphisms in restriction sites so that restriction patterns are unique to individuals. This technique is still used in forensic science, but a new technique based on the PCR reaction is replacing it. In this, use is made of the fact that a large number of repeat sequences have been identified in the human genome. These are highly variable from individual to individual in the number of repeats in the sequence. The method is to choose primer sites on either side of the selected repeat sequences and, using PCR, to amplify the latter.

Running the product on an electrophoretic gel gives a band of a size depending on the number of repeats in the amplified segment. By selecting a number of such loci to amplify, a pattern of bands highly characteristic of an individual is obtained.

Further reading

Gene cloning

Jordan, E. and Collins, F. S. (1996). Human genome project: a march of genetic maps. *Nature*, **380**, 111–12.
A short news and views summary of the human genome project, which emphasizes the role of genetic maps in analysing disease genotypes. The article is an excellent guide to future work.

The polymerase chain reaction

Saiki, R. K., Gelfand, D. H., Stoffel, S., Scharf, S. J., Higuchi, R., Horn, G. T., Mullis, K. B., and Erlich, H. A. (1988). Primer-directed enzymatic amplification of DNA with a thermostable DNA polymerase. *Science*, **239**, 487–91.
Detailed account of the technology of the PCR reaction.

Arnheim, N. and Erhlich, H. (1992). Polymerase chain reaction strategy. *Ann. Rev. Biochem.*, **61**, 131–56.
Readable account of all aspects of the polymerase chain reaction.

Gene therapy

Morgan, R. A. and Anderson, W. F. (1993). Human gene therapy. *Ann. Rev. Biochem.*, **62**, 191–217.
Deals with the candidate diseases, gene transfer methods, treatment of adenosine deaminase deficiency and concludes with safety and ethical questions.

Capechi, M. R. (1994). Targeted gene replacement. *Sci. Amer,* **270**(3), 34–41.
Describes gene replacement by homologous recombination and transgenesis techniques.

Anderson, W. F. (1995). Gene therapy. *Sci. Amer.*, **273**(3), 96–8B.
An account of the successful insertion of adenosine deaminase into white cells, and their return to the patient to cure the combined immunodeficiency syndrome. Includes a list of 14 diseases being tested in clinical trials of gene therapy.

Problems for Chapter 24

1 How does a restriction enzyme differ from pancreatic DNase?

2 *E. coli* RI cuts at a hexamer base sequence; such a sequence must occur many times in *E. coli* DNA. Why doesn't the enzyme destroy the cell's own DNA?

3 What is meant by the term 'sticky ends', as applied to DNA molecules?

4 What is a genomic clone? What is a cDNA clone? How do they differ? Refer to eukaryotes in your answer.

5 What are the steps in making a genomic library in lambda bacteriophage?

6 What are the steps in isolating a particular clone from a genomic library in lambda?

7 What is a dideoxynucleoside triphosphate? What is the precise function of these in the Sanger technique of DNA sequencing?

8 Explain in a few sentences the importance and principle of the polymerase chain reaction, specifying the requirements for it to be used.

9 What is a bacterial plasmid expression vector?

10 What is restriction analysis?

Chapter summary

Chapter 25

. . . .

The immune system

The body is constantly under the threat of invasion by pathogenic organisms such as bacteria and viruses. The immune system is the main protection against this. Its absolute necessity is illustrated by the lethal results of an impaired immune system due to genetic defects (for example, adenosine deaminase deficiency—see Fig. 18.6) or to infection by the AIDS virus.

The immune system responds to macromolecules of several types—proteins are the most important class. It does not protect against small foreign molecules such as drugs that enter the body; other chemical protection systems deal with these as was described in Chapter 17. A foreign macromolecule in the body is a reliable warning signal that a defensive response may be needed against an invader. Patients can be allergic to small molecules such as penicillin, the allergy being initiated by an immune response. However, the response is not to penicillin *per se* but rather to penicillin combining with, for example, a protein of the body, and in doing so making that protein 'look' foreign to the immune system.

You may recall from Chapter 4 that one of the major problems of the digestive system is how to digest food, the components of which are essentially identical to those of the body, without autodigestion occurring, that is, without destroying the body itself. An analogous problem is present in immune protection. While, for example, the lipopolysaccharide components of a bacterial cell wall are different from any present in the human body and therefore easily recognizable as foreign, the problem with foreign proteins is very different. The body itself has thousands of proteins differing from foreign proteins only in the detail of amino acid sequences and yet the distinction between 'self' and foreign proteins must be made with accuracy. When the system makes an error in this and attacks one of its own proteins, an **autoimmune disease** may result. As an example, in the disease myasthenia gravis an autoimmune reaction destroys the acetylcholine receptors of muscle (page 393), thus preventing nervous stimulation of contraction. In rheumatic fever, an immune response against a protein produced by certain strains of *Streptococcus* produces antibodies that crossreact with a protein component of the heart, causing damage to the heart valves. Insulin-dependent diabetes is caused by an autoimmune attack on pancreatic cells. The avoidance of the immune system attacking the body's own components is of overriding importance.

There are two protective mechanisms in the immune system

First. there is the production of **antibodies**. These are soluble proteins secreted from immune system cells that specifically combine with the foreign **antigens**. The latter term refers to any molecule that is capable of producing a specific immune response, in this case with *anti*body *gen*eration. The attachment of the antibody to the antigen results in protective events that will be described later. This type of immunity is called **humoral immunity**; in times past, bodily fluids were referred to as humours and, in this mechanism, protection is due to soluble antibodies in body fluids.

The second type of immunity is known as **cell-mediated immunity**. In this, special cells known as **cytotoxic** or **killer cells** recognize abnormal cells in the body. For instance, the abnormality might result from a virus infection in the cell. The killer cell attaches to the virus-infected cell and destroys it. This is an effective way of stopping a virus infection from spreading since cell destruction aborts virus replication inside the cell. The important point is that direct contact is made between the killer cell and its target—hence, the term cell-mediated immunity. Thus the two mechanisms complement each other—to use the virus infection as the example, the humoral system is part of the defence against virus in the blood or in the mucous membrane before it infects a cell, while the cell-mediated system destroys the host cell after the latter has become infected and in doing so aborts virus replication in that cell.

Antibodies are produced by a class of lymphocytes called **plasma cells** (one sort of white cell), which develop when **B cells** are appropriately stimulated. The cytotoxic or killer cells are **T lymphocytes** (T for thymus-derived). Lymphocytes are all produced continuously in bone marrow (or liver in the fetus).

The B cells also mature in the bone marrow but the T cells multiply and have to undergo their primary maturation in the thymus gland which, in humans, is located behind the breast bone. However, there is a very important qualification in the division of labour between B cells and T cells. A B cell, to produce antibody, must first meet an antigen and, secondly, it must be contacted by a special type of T cell *that has met the same antigen*. The T cell thus 'helps' the B cell to do its job and is known as a **helper T cell**. These are in a class separate from that of cytotoxic T cells.

To summarize then, we have B cells that produce antibodies (after activation and maturation into plasma cells), helper T cells that help the B cells to do this, and cytotoxic or killer T cells that are responsible for cell-mediated immunity.

Where is the immune system located in the body?

There is no single organ such as the liver for the immune system. Instead there are vast numbers of separate cells which, *en masse*, would be equivalent to a large organ. They are distributed roughly as follows: 30% in spleen, 20% in lymph nodes, 40% in intestinal and mucosal lymphoid tissue, and 10% in blood and lymph circulation. The main cells are lymphocytes but other white cells known as macrophages are also involved. The B and T cells, as noted above, are lymphocytes. They circulate in the lymph; the body has two circulating fluids, blood and the lymph. In tissues, a 'filtrate' leaks out of the blood and bathes cells to provide nutrients and remove waste products. This is drained into thin-walled lymphatic vessels that return the lymph to the blood via two main ducts. The white cells are able to squeeze through certain capillary walls and so can migrate from blood to lymph. The lymph is not pumped directly but is propelled by the continual movements of the body. Along the route of the lymph returning to the blood, the lymphatic vessels expand at places into **lymph nodes** where high concentrations of B and T cells are found. The nodes are strategically placed in the armpit, groin, tonsils, adenoids, and intestine so that the lymphocytes in them are likely to encounter any foreign antigens in tissues from which the lymph was drained. The nodes are in this sense monitoring for infection in the tissues by acting as a 'filter' and also providing the correct environment for lymphocyte multiplication and differentiation, by which we mean the changes that occur when they meet antigens, as described later. All lymphocytes, as already stated, originate in the bone marrow where **stem cells**, in fact, give rise to all blood cells. Also as stem cells develop, some become committed to becoming B cells, some to T cells, and others to red blood cells, etc. (Fig. 25.1). As stated, the B cells develop in the bone marrow, but the T cells migrate to the thymus gland where they proliferate and undergo their primary maturation. The migration of T cells from bone marrow to thymus and release of matured T cells from the thymus occurs mainly early in development so that removal of the thymus from adult animals

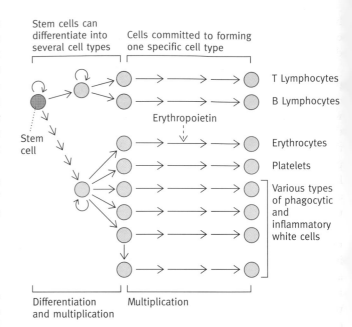

Stem cells can differentiate into several cell types

Cells committed to forming one specific cell type

Stem cell

T Lymphocytes

B Lymphocytes

Erythropoietin

Erythrocytes

Platelets

Various types of phagocytic and inflammatory white cells

Differentiation and multiplication

Multiplication

Fig. 25.1 Simplified diagram of hemopoiesis. At each arrow cell multiplication can occur and this may be specifically controlled by various protein cytokines—the colony-stimulating factors, interleukins, and erythropoietin. For example, the latter stimulates erythrocyte production at the step indicated. Differentiation of stem cells into committed cells is likewise controlled.

has little effect on immune responses. B cell production and release from bone marrow persists in adults. The bone marrow and thymus are called primary lymphoid organs; the lymph nodes secondary lymphoid organs. Lymphocyte *maturation* in the primary lymphoid organs is antigen-*independent*, while all subsequent differentiation (cellular change) of the *released* mature lymphocytes in secondary lymphoid tissue is dependent on the presence of antigen.

In this chapter we shall go into details of the immune process but it may be useful if, first, we give a brief overview of the strategies used by the body in immune protection. Read this overview without worrying about the questions that will arise in your mind about how things are done.

An overview of the strategy involved in immune protection

The problems are to arrange that B cells produce antibodies to foreign antigens but *not* to the body's own components, and that cytotoxic T cells attack only abnormal cells of the body. Vast numbers of immature B and T cells mature in the bone marrow and thymus respectively. A vital principle is that *a given B cell can produce only one specific antibody* (in terms of its antigen-binding capacity) and that *a given T cell responds to only a single antigen*. Vast numbers of different antibodies can be produced so that means that an equivalent number of different B cells have to be formed, each containing a different gene encoding the antibody it produces. (That, of course, raises the big question of how so many different genes can exist; it is explained later.) Similarly, there are vast numbers of T cells with different receptors, each specific for a different antigen.

The immature B cells between them have the potential to produce antibodies to attack just about every macromolecular component in the body. Any B cell whose antibody, if produced, would be against a 'self'-component is best eliminated. Most of this is done during primary maturation in the bone marrow. As they mature there, the B cells are presumably exposed to a large proportion of the components of self-antigens that they are ever likely to meet. (The existence of other mechanisms to suppress autoimmune reactions has been suggested, but this is an area of uncertainty.)

The immature B cell in the bone marrow, produces about 10^5 molecules of its antibody and displays them on its surface (fixed into the membrane) inviting, as it were, recognition by an antigen. It doesn't release this antibody; in this role it is a receptor or recognition 'aerial'. If, *during primary maturation* of the B cell in the bone marrow, an antigen comes along that combines with the displayed antibody, it is assumed (as it were) that it is a self-component that must be tolerated by the immune system. *At this stage of development*, the B cells die (or are otherwise functionally eliminated) on receipt of such a signal. After this selective killing process, the surviving B cells should respond only to foreign antigens. They are now released into circulation. The *released* population of virgin B cells, as they are called, now has a totally different response to meeting their respective antigens. They are activated instead of being killed; this is logical, since the antigen now encountered is likely to be foreign.

The activated B cell does not mature into an antibody-secreting plasma cell until it contacts a mature helper T cell *that has been activated by exposure to the same antigen;* (the latter has to be 'presented' to it by a process described later). The T cells also individually recognize single specific antigens by a different type of receptor; *during their maturation in the thymus* they produce antigen-specific receptors, and exposure to self-components of the body when they are still immature somehow leads to their elimination. The principle is that, during T cell maturation in the thymus (which, you will recall, occurs early in development), they will, like the B cells, meet most of the self-antigens they are likely to encounter. Survivors of the selection process in the thymus are released as inactive helper T cells (or cytotoxic T cell precursors—see below) whose response to meeting specific antigens is, as with B cells, activation. The role of the activated helper T cell is to deliver the signals for a specific activated B cell to multiply and mature into an antibody-secreting plasma cell by mechanisms to be described shortly (Fig. 25.2). This is the meaning of the term **clonal selection theory**. The antigen selects the appropriate cells for multiplication. There is an exception to the above described mechanism. Some components of Gram-negative bacterial cell walls, such as **lipopolysaccharides**, are able to stimulate the

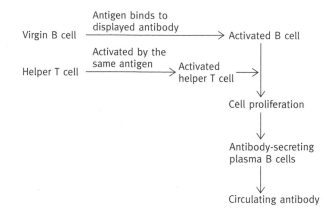

Fig. 25.2 Simplified diagram of the route of activation of B cells to secrete antibody. As will be described later, the mechanism of activation of helper T cells by antigen is not the same as that of B cell activation by the same antigen.

relevant virgin B cell to produce its antibody without needing a helper T cell signal. Contact of this type of antigen with the B cell is sufficient to cause maturation into an antibody-secreting plasma cell.

The plasma cell is the end-stage product of B cell differentiation, a cell specialized for antibody synthesis and secretion. In keeping with the role of the endoplasmic reticulum in protein secretion (page 306) this organelle is very well developed in plasma cells. We shall describe the multiplication controls on B cells and on helper T cells later in this chapter.

Some of the immature cells in the thymus develop into precursor cytotoxic or killer T cells. The same selection procedure occurs; any potential cytotoxic cell whose receptor is such that it would respond to a self-component of the body is eliminated or inactivated. Survivors are released from the thymus into the circulation and can be induced to become fully functional cytotoxic cells by exposure to the appropriate foreign antigen displayed on a host cell. When an activated cytotoxic T cell encounters a host cell carrying a foreign antigen with which its receptor combines, it releases **perforin**, a protein that kills the cell by rendering its membrane leaky and/or it induces apoptosis of the cell. **Apoptosis** is programmed self-destruction of the cell.

With that overview, we will now describe in greater detail the mechanisms of the above processes. We will start with what antibodies are.

The mechanisms of immunity—B cells, helper T cells, and the production of antibodies

Structure of antibodies

Antibodies are **immunoglobulins**—the term 'globulin' is from an early classification system denoting a protein soluble in dilute salt solution. **Ig** is the abbreviation for immunoglobulin. There are several different classes of immunoglobulins, which will be described later. IgG is the one produced usually in largest amounts if exposure to an antigen is prolonged so we'll look at its structure first. IgG is a 'Y'-shaped protein made up of two identical light (L) polypeptide chains and two identical heavy (H) chains held together by disulfide bonds (see page 30) as shown in Fig. 25.3. The ends of the 'Y' arms have variable regions on both the heavy and light chains. It is these variable regions of the two chains that form the antigen-binding site on each of the arms. The binding to the antigen is by noncovalent

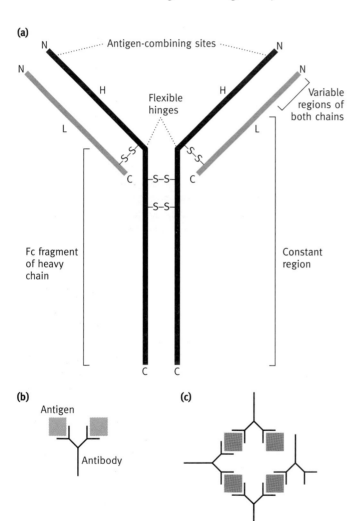

(a)

(b)

Antigen

Antibody

(c)

Fig. 25.3 (a) Diagram of an IgG molecule. IgM and IgA molecules differ in the Fc fraction but are similar in the variable part. H, heavy chain; L, light chain. N and C refer to the amino and carboxyl terminal ends of the polypeptides, respectively. The Fc fragment is the C-terminal half of the two H chains, bound covalently into a single fragment by –S–S–bonds. The name arises from crystallizable (c) Fragment (F)—one of the products of papain hydrolysis of the antibody molecule at the two peptide bonds at the flexible hinges. **(b)** Crosslinking of antigen with single specific epitopes. **(c)** Antigens with multiple antigenic determinants form a crosslinked insoluble cluster with antibody.

bonds. An antibody thus has two identical antigen-binding sites, which means that it can crosslink antigen molecules. At the fork of the 'Y' there are flexible 'hinge' regions that increase the cross-linking ability of the molecule. The other immunoglobulins show this basic but not identical structure although some, such as IgM, are polymeric structures comprised of five subunits, each similar to IgG molecules.

Although antigens need to be large to produce an immune response, a given antibody binds to only a small part of the antigen—in the case of a protein antigen only a few amino acids. The specific part of the antigen that is recognized by an antibody is called an **epitope**. Thus a large protein antigen will usually provoke the production of a number of different antibodies, each combining with a specific epitope and each produced by a different clone of B cells.

The body is potentially exposed to vast numbers of different antigens and it is potentially capable of producing vast numbers of different antibodies that combine with these. The specific antibodies differ in the precise amino acid sequences of their binding sites. This is achieved by the L and H chains at their ends being individually variable in their amino acid sequences in different cells so that vast numbers of different combining sites are possible (one type per cell) and hence there is potentially a correspondingly large number of different antibodies. As indicated above, the principle is that the immune system arranges to randomly produce B cells differing in their genes coding for antibodies. Each newly developed B cell can produce, in terms of antigen specificity, only a single antibody species, but each cell produces a different one, which, at this stage, is not released but molecules of it are fixed into the plasma membrane with the antigen-combining sites on the outside. It puts them in the shop window, on display, as it were. Whatever antigen comes along, one or more of these displayed antibodies happens to bind to it by sheer chance, initiating B cell activation.

What are the functions of antibodies?

An IgG antibody molecule is bivalent—the two identical binding sites can each bind to the same epitope on separate antigen molecules. If the antigen has more than two identical epitopes, a single antibody will form a large antibody–antigen network. Most antibody preparations raised in an animal will contain multiple antibodies recognizing different epitopes on a given antigen so that a network will form between antigens and antibodies. Such a network of molecules activates the **complement system**. The name of the latter indicates that it 'complements' the action of antibodies in lysing bacterial or other invading cells. Complement is the name of a group of proteins in blood, nearly 20 in number. Antibody–antigen complexes on a bacterial cell cause **fixation** (to the cell) of some of these components starting a cascade in which complement proteins assemble on (and around) the antigen–antibody complex, resulting in the bacterial or other cell being killed due to perforation of the cell membrane. Another fate of the antibody-coated bacterium is that the whole complex is engulfed by **phagocytic leucocytes** (for example, macrophages and neutrophils), which kill and digest microorganisms. The phagocytic cells have receptors (**Fc receptors**) on their surface, some of which bind to the constant Fc region of the antibodies (the stem of the Y, see Fig. 25.3(a)) that sticks out from the antibody–antigen complex while others (complement receptors) bind attached complement components. These events allow adhesion between the phagocyte and its target, and trigger engulfment of the particle. During the whole process of the complement reaction, substances that dilate blood vessels and attract phagocytes to the infected site are released—a major factor in inflammation.

What are the different classes of antibodies?

There are five major classes of immunoglobulins, differing from one another in the **constant regions** of their H chains. These differences have nothing to do with the antigen-binding sites but are related to the precise physiological role of the antibody.

When the body responds to an antigen, the antibody produced first is **IgM**. This is a multisubunit, or polymeric, form of antibody with 10 antigen-combining sites. It is particularly efficient at binding viruses and bacteria into a network because of its multiple combining sites and is also efficient in activating complement and promoting phagocytosis. Repeated challenges by the same antigen result in massive production of **IgG**. It also activates complement and is efficiently transferred to a fetus via the placenta. **IgA** is of special importance as a first line of defence. It is transported through epithelial cell membranes, by combination with a special secretory polypeptide, into the mucous layer of the intestine, respiratory tract, etc. and is secreted into milk. IgA is important, for example, in acquired immunity to the bacterium responsible for cholera which attaches, via special adhesions, to the epithelial lining of the intestine at places known as Peyer's patches, and elsewhere. Bacteria coated with specific IgA cannot attach, and so cannot infect. IgA has characteristics appropriate to its role in external mucous membranes. It is not readily destroyed by intestinal proteolytic enzymes and it does not efficiently activate complement. **IgE** sensitizes mast cells and is involved in histamine release and the distressing symptoms of allergies, in the presence of specific antigen or allergen. It can also result, in sensitized individuals, in the synthesis of **platelet-activating factor** (**PAF**) by certain white cells. PAF resembles lecithin except that the central fatty acyl group is replaced by acetyl. PAF has dramatic effects, inflaming air passages and causing platelet aggregation, lowered blood pressure, and other effects. **IgD** functions are unknown.

Although, as stated, a given B cell can only produce a single antibody in terms of its antigen binding site, it can switch from one *class* to another (see above) while preserving its antigen specificity, and this switching allows a B cell clone to increase the range of physiological functions that its antibodies can perform. Class switching is due to DNA recombination events.

How is antibody diversity achieved?

A human cell contains probably somewhere between 50 000 and 100 000 genes to code for all the proteins of the body. To code for, say, a million different antibodies requires a million different genes. Clearly, something special must arrange for the enormous diversity of antibodies.

To understand how this generation of diversity is achieved let us look first at the L chain. It has two sections—a terminal variable region that participates in the antigen-binding site and a constant region that is identical in all Ig molecules. This state of affairs is arrived at by the joining together of sections of DNA encoding the constant portion and one of a large number encoding alternative variable regions. Consider a stem cell in the bone marrow *before* it has become committed to becoming a B cell. The DNA relevant to the L chain consists of several sections. There is a section coding for the constant (C) section; there are four separate sections each of which codes for different short peptide sequences that can be used to join the constant to the variable region, called J (for joining) sections. And, finally, there are about 300 sections (V) each of which codes for alternative variable ends of the L chain. The 300 V sections are all different from one another, as are the four J sections, and they will thus code for different peptide sequences in each case. When the bone marrow stem cell is committed to becoming a B lymphocyte, a rearrangement occurs by recombination events and excision of DNA, resulting in the joining of *one* of the 300 V sections to *one* of the four J sections. This results in a new composite gene in each of the B lymphocytes, as shown in Fig. 25.4. As the figure shows, the immunoglobulin gene may contain more than one J section but, during splicing of the primary RNA transcript, all but one of the J sections are eliminated from the mRNA. *The DNA recombination is a completely random process.* Thus, in the final mRNA coding for an IgG L chain molecule, any one of 300 V sections is joined to any one of the four J sections giving 1200 different combinations. The actual recombination is also imprecise in that the place of cutting and rejoining varies slightly. This increases the number of L chain variants to about 3000.

The H chain gene is assembled in a similar random manner from even more variable sections giving yet larger numbers of H chain gene variants. Since the antigen-binding site is made from the combination of the variable parts of the H and L chains and there are large numbers of different H and L genes, the number of different binding sites coded by the assembled L and H genes is sufficient to account for the immune system's ability to produce a vast number of different antibodies. In summary, each developing B cell assembles, at random, genes coding for one antibody with its specific antigen-binding site.

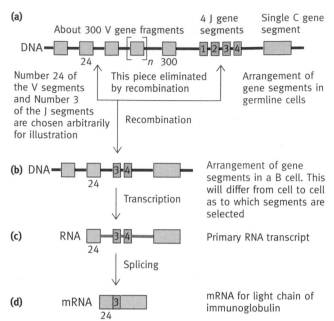

(a)

About 300 V gene fragments

4 J gene segments

Single C gene segment

DNA

24   *n* 300

Number 24 of the V segments and Number 3 of the J segments are chosen arbitrarily for illustration

This piece eliminated by recombination

Arrangement of gene segments in germline cells

Recombination

(b) DNA

24

Transcription

Arrangement of gene segments in a B cell. This will differ from cell to cell as to which segments are selected

(c) RNA

24

Splicing

Primary RNA transcript

(d) mRNA

24

mRNA for light chain of immunoglobulin

Fig. 25.4 The process of rearrangement leading to a functional L chain immunoglobulin gene. **(a)** Arrangement of gene segments in the stem cell where the immunoglobulin genes are not being expressed. **(b)** The randomly chosen V gene (*V24* in this example) is moved next to one of the J genes (*J3* in this example), the intervening DNA being excised. **(c)** Transcription begins at the *V24* gene segment, the J genes (*J3* and *J4*) also being transcribed. **(d)** After RNA splicing to remove transcripts of *J4* and introns, those corresponding to *V24*, *J3*, and *C* now make up the mRNA. Allelic exclusion ensures that only one of the pair of alleles becomes a functional immunoglobulin gene (see text) so that a given cell produces only one immunoglobulin, not two.

With two sets of chromosomes, the generation of diverse immunoglobulin-coding genes would be expected to result in each B cell producing two, not one, antigen-specific combining sites. However, a process known as **allelic exclusion** causes only one of the homologous chromosomes to express the assembled immunoglobulin genes for the light and heavy chains.

There is a further sophistication in the immune system in that, after activation of a B cell, rapid mutation in the V region occurs, resulting in the production of yet further antibody diversity. Many of these mutations may destroy the binding affinity but some will increase it. Antigen will bind preferentially to those cells displaying higher affinity antibodies on its cell membrane. There is a selection of cells producing the most efficient antibody against that particular antigen since antigen binding leads to cell multiplication. In short, after an antigen is encountered there is a rapid process of 'evolution' (known as **affinity maturation**), which refines the antibody effectiveness by increasing its affinity for the antigen.

What is the role of the helper T cells in activating B cells to secrete antibody?

The B cell, which has been activated by antigen binding to its displayed antibody, internalizes the antigen and processes it into small fragments. These pieces are captured by special peptide receptors called MHC proteins and transported to the cell surface, where the processed antigen pieces are displayed in the binding sites of the MHC proteins. The process is outlined in Fig. 25.5. MHC molecules are glycoproteins encoded by the **major histocompatibility complex** (**MHC** for short). The MHC molecules involved in this case, belong to class II (see below) and they are expressed by those cells in the immune system that function as specialized antigen-presenting cells, including B cells. Processed antigens presented in association with the MHC molecules on the surface of B cells can be 'recognized' by helper T cells, with receptors specific for that MHC–antigen complex. Helper T cells do not produce antibodies. However, they do produce antigen-specific surface receptors (**T cell receptors or TCRs**), which are displayed on the external surface of the cell membrane. These receptors are never released. Although quite

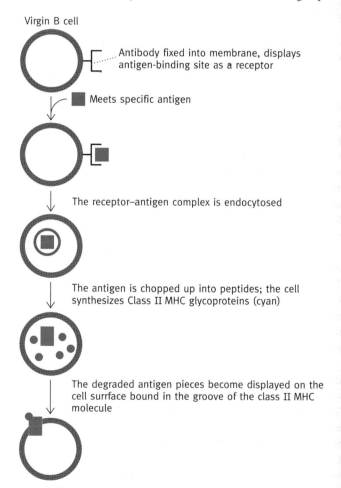

Virgin B cell

Antibody fixed into membrane, displays antigen-binding site as a receptor

Meets specific antigen

The receptor–antigen complex is endocytosed

The antigen is chopped up into peptides; the cell synthesizes Class II MHC glycoproteins (cyan)

The degraded antigen pieces become displayed on the cell surface bound in the groove of the class II MHC molecule

The B cell is now in an activated state; it does not produce antibodies unless it binds to a helper cell that has been activated by an antigen-presenting cell displaying the same antigen. If this happens, the B cell matures into an antibody-secreting plasma cell and also multiplies due to cytokine release by the helper cell. (This is illustrated in Fig. 25.7.)

Fig. 25.5 Conversion of a virgin B cell to an activated state, but not yet to an antibody-producing plasma cell.

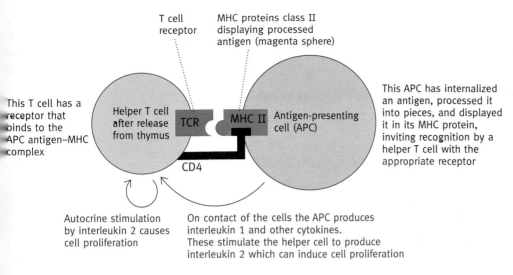

T cell receptor

MHC proteins class II displaying processed antigen (magenta sphere)

This T cell has a receptor that binds to the APC antigen–MHC complex

Helper T cell after release from thymus

TCR

MHC II

Antigen-presenting cell (APC)

This APC has internalized an antigen, processed it into pieces, and displayed it in its MHC protein, inviting recognition by a helper T cell with the appropriate receptor

CD4

Autocrine stimulation by interleukin 2 causes cell proliferation

On contact of the cells the APC produces interleukin 1 and other cytokines. These stimulate the helper cell to produce interleukin 2 which can induce cell proliferation

Fig. 25.6 Activation of a helper T cell by an antigen-presenting cell (APC). The autocrine stimulation of the helper T cell allows cell multiplication to occur after separation of the cells. CD4 is a glycoprotein on the T cell that interacts with the MHC II protein of the APC. This interaction is necessary in addition to the binding of the T cell receptor (TCR) with the APC MHC–antigen complex. The CD4 protein is the receptor by which the AIDS virus (HIV) infects the helper T cell.

different in structure from immunoglobulin, the T cells are able to generate diversity in the TCRs. Again, as with B cells, this process is quite random so that in the thymus there are vast numbers of immature T cells each with a different TCR capable of binding to a specific antigen. As already indicated, at this stage (occurring early in development of the individual animal) ability to bind to a self-antigen leads to elimination of the cell. Survivors of this selection process are released into the circulation.

There is another difference. The virgin B cell surface antibody combines with an antigen molecule directly, but the helper cell TCR does not. For the TCR to recognize an antigen the antigen has to be engulfed by a *special* macrophage (called the **APC** or **antigen-presenting cell**), processed into pieces, and displayed by the macrophage in association with the class II MHC receptor molecules.

The TCR on a T cell recognizes *not* the antigen piece alone displayed by the APC, *not* the MHC molecule alone, but the *combined* class II MHC molecule–antigen fragment assembly. The interaction of the APC and the helper T cell is facilitated by a CD4 protein on the latter which binds to a constant protein of the class II MHC protein of the former (see Fig. 25.6). (The term CD4 means 'cluster of differentiation', a somewhat difficult term to define; it refers to the method used to define specific differentiation markers in cells. The cluster refers to binding of different antibodies to a particular antigen on the cells. 4 refers to a particular site determined experimentally. The CD4 protein is the receptor to which the AIDS virus attaches.) When a helper T cell, released as a mature cell into the circulation, combines with the appropriate MHC–antigen assembly displayed on circulating APC, it becomes activated (Fig. 25.6). If, now, an

encounter occurs between an *activated* helper T cell and a B cell on 'standby' *displaying the same MHC–antigen complex as was originally presented to the T cell by the APC*, the T cell binds to it and releases cytokines. The B cells multiply as a result of

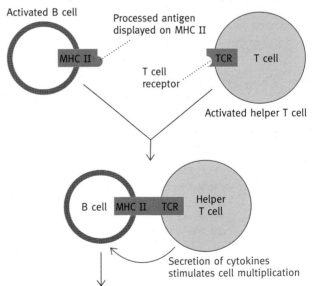

Activated B cell

Processed antigen displayed on MHC II

MHC II

TCR

T cell

T cell receptor

Activated helper T cell

B cell

MHC II

TCR

Helper T cell

Secretion of cytokines stimulates cell multiplication

The B cell, now activated by the helper cell, multiplies and matures into a clone of antibody-secreting B cells known as plasma cells

Fig. 25.7 Diagram of a helper T cell activating a B cell to become a clone of plasma cells secreting antibody. The cytokines are growth factors that stimulate local cells to divide (paracrine action). Upon activation of a B cell to become a plasma cell, differential processing of RNA transcripts eliminates the membrane-anchoring section of the antibody. The cytokines (which include interleukins) secreted by the helper T cells are also believed to activate cytotoxic cells (see Fig. 25.8).

cytokine release (Fig. 25.7) and the resultant plasma cells (secreting B cells) pour out protective antibody. The plasma cells are rich in ER needed for protein secretion (page 306). The T cells also multiply. Figure 25.7 summarizes the process. As shown in this figure, the release of control proteins, called cytokines, is an essential part of the process.

Why is the antibody produced now secreted instead of being fixed in the membrane? The answer lies in differential splicing of introns from the immunoglobulin genes referred to earlier (pages 281–2). At the 3′ end of the immunoglobulin gene is a section coding for a polypeptide sequence that fixes the antibody into the membrane. The onset of secretion is caused by a switch in the splicing mechanism that eliminates this from the mRNA.

Memory cells

When B cells are activated and proliferate, not all of the resultant clone of cells mature into antibody-secreting plasma cells. A proportion become long-lived memory cells that may circulate for years. They are the basis of long-term immunity from a repeat infection. If an appropriate antigen is encountered at a subsequent time, the immunological response is very rapid. As described below, T cells are also involved in immunological memory.

T cells and cell-mediated immunity

As described already, the T cell receptors on helper T cells recognize antigen displayed in association with class II MHC molecules, the latter being present on APC and B cells. The other class of T cells, the cytotoxic or killer cells, has receptors that recognize antigens on class I MHC molecules. Class I MHC molecules are present on most cells of the body. Killer T cells interact only with host cells displaying a foreign antigen on its class I MHC molecules. (While this statement applies generally, more recent work indicates that, under some circumstances, cytotoxic cells may recognize MHC class II proteins. While we do not want to confuse the situation, it is important to realize the very great complexity of the immune system and much remains to be understood.) The killer T cell has a surface glycoprotein CD8 (cf. CD4 on helper T cells) that interacts with the constant protein of MHC I proteins, thus restricting attack of the cytotoxic cells to host cells with MHC I proteins, since the CD8–MHC interaction is essential for effective binding of the cells. The cytotoxic cells recognize, and kill, body cells that have become abnormal, for example, through viral infection. An example of the mechanism for this is as follows. In a virus-infected cell, viral protein (synthesized in the cell) is processed by the cell into peptides and displayed on the cell surface with the class I MHC molecules (Fig. 25.8). An antigen-specific killer T cell 'sees' and binds to the antigen–MHC complex of the infected cell and destroys the latter by introducing membrane perforations (Fig. 25.8), or by inducing apoptosis.

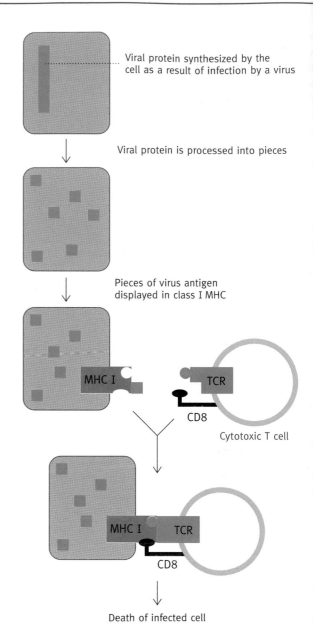

Viral protein synthesized by the cell as a result of infection by a virus

Viral protein is processed into pieces

Pieces of virus antigen displayed in class I MHC

MHC I — TCR
CD8
Cytotoxic T cell

MHC I — TCR
CD8

Death of infected cell

Fig. 25.8 Sequence of events in cell-mediated immune reaction to a foreign antigen synthesized inside cells (for example, as a result of virus infection). The diagram does not show that the cytotoxic cell is also stimulated to divide by the cytokines released by helper T cell–B cell interaction. This results in a clone of cytotoxic cells specific for the particular antigen; a proportion of the clone of cells develops into memory cells. The infected cell may be killed by perforation of its cell membrane due to the release of the protein perforin or death may be due to apoptosis.

The role of cytokines in the immune system

In Chapter 26 we describe how cells signal to one another by means of hormone-like molecules. An important class of these are collectively called **growth factors**—proteins that signal cells to multiply—but other responses can result. As stated in the accounts above, when a helper T cell is activated by an antigen-presenting cell, it secretes cytokines, including interleukins, which stimulate cell multiplication and maturation of both types of cells. An array of cytokines is secreted by helper T cells in contact with B cells and some of these are also involved in killer T cell activation. Some stimulate phagocytes, others cause antibody class switching in B cells, and they stimulate cell division and maturation of B cells into plasma cells. They mobilize the entire protective system. The immunosuppressant antibiotic, cyclosporin, used in organ transplantation, inhibits the production of interleukins by helper T cells.

Why does the human immune system so fiercely reject foreign human cells?

It is somewhat puzzling at first sight why the body should have a mechanism for immune attack on cells transferred from one individual to another—why is there tissue graft rejection? During evolution there was not any known danger of such transfer until tissue grafts and organ transplants were developed by surgeons. The main reasons for such rejection of foreign tissue grafts are the MHC molecules on cells. These molecules are a family of glycoproteins of which class I and class II are structurally different subfamilies. The genes coding for them are all clustered in the major histocompatibility complex. This is highly polymorphic—there are, in the population, many **alleles** (or **variants**) for each gene coding for an MHC protein. It is therefore unlikely that the MHC molecules of one individual are exactly the same as those of the next. Those that are different will therefore be antigenic. The major histocompatibility antigens are the same molecules as the class I and II antigen–peptide receptors discussed above. A foreign MHC molecule may be 'seen' by killer T cells as the body's own MHC molecule, complexed to foreign antigen, and the cell bearing this is therefore attacked by cytotoxic T cells. Again, we remind you that the killer T cell receptor recognizes the *complex* of MHC molecule plus antigen, not either component alone. The evolution of such variation in the MHC molecules of the population might be a protection of the species as a whole from death due to infection. Suppose that, by chance, a virus evolves such that, when its processed antigen is displayed on a host MHC molecule for some reason, it is 'passed' by the cell-mediated immune system as self and normal or is not displayed at all. The virus might then multiply and kill its host. However, the next individual it infects is unlikely to have the same MHC alleles and the chance of the same thing happening again is reduced; thus, while individuals can be killed, it is unlikely that the disease will sweep through the whole population.

Further reading

The immune system

1993. *Sci. Amer.*, **269**(3), 20–108.
A complete issue of *Scientific American* devoted to all aspects of immunity, superbly illustrated.

Hemopoeisis and hemopoietic growth factors

Metcalf, D. (1991). The 1991 Florey Lecture. The colony-stimulating factors: discovery to clinical use. *Phil. Trans. Series B, Roy. Soc., Lond.*, **333**, 147–73.
An in-depth review by the discoverer of these important proteins. While there is more research level material than most students would want, it provides an excellent general review of the field, and a fascinating account of the progression from unidentified factors to their widespread clinical use.

Kelso, A. (1993). The immunophysiology of T cell-derived cytokines. *Today's Life Science*, **5**(4), 30–40.
An excellent summary of cytokines in the immune system.

Elimination of self-specific cells

Collins, M. (1991). Death by a thousand cuts. *Current Biology*, **1**, 140–2.
A concise account of immature T cell-progammed cell death in the thymus.

Monoclonal antibodies

Yelton, D. E. and Scharff, M. D. (1981). Monoclonal antibodies: a powerful new tool in biology and medicine. *Ann. Rev. Biochem.*, **50**, 657–80.
General review of techniques and applications.

Chien, S. and Silverstein, S. C. (1993). Economic impact of applications of monoclonal antibodies to medicine and biology. *FASEB J.*, **7**, 1426–31.
Despite the title, includes a nice summary of how monoclonal antibodies were developed; this is followed by their practical application and monetary value to industries.

Antigen presentation

Engelhard, V. H. (1994). How cells process antigens. *Sci. Amer.*, **271**(2), 44–51.
An account of antigen processing and display of fragments on MHC proteins.

Pathogens and the immune system

Goodenough, U. W. (1991). Deception by pathogens. *American Scientist*, **79**, 344–55.
A fascinating account of how, at the molecular level, bacteria and viruses attempt to escape immunological detection. Summarizes immune mechanisms clearly.

Nowak, M. A. and McMichael, A. J. (1995). How HIV defeats the immune system. *Sci. Amer.*, 273(2), 42–9.
Presents hypothesis that continuous evolution of the human immunodeficiency virus in the body leads to the immune devastation of AIDS—good graphics of the immune system.

Problems for Chapter 25

1 Describe the structure of an IgG molecule.

2 Explain how the body can have genes to code for so many different antibodies.

3 What are the different classes of lymphocyte and their function?

4 What is an antigen-presenting cell?

5 When a host cell displays, for example, a viral antigen on its surface which class of MHC molecule is it displayed on? Which class of MHC molecule does a cytotoxic T cell recognize?

6 What is meant by the term clonal selection theory?

7 What is the principle of avoidance of autoimmunity or how does the immune system become self-tolerant

8 Production of an antibody by a B cell after stimulation (to become a plasma cell) by a helper T cell does not change the antigen-binding site of the antibody, but the antibody has to be released instead of residing in the membrane. How is this achieved?

9 When a B cell (plasma cell) starts to secrete antibody, the latter may have a relatively poor affinity for its antigen but this is rapidly improved. How is this achieved?

10 Which immune systems have the glycoprotein CD4 and which CD8? What are their roles in the immune reaction? What is the relevance of CD4 to the AIDS virus?

Chapter 26

....

Chemical signalling in the body

Note It would be useful in dealing with this subject if you first refreshed your memory of Chapter 12, in which the control of **metabolism** by extrinsic factors is discussed. An extrinsic factor refers to a controlling agent reaching a cell from other cells in the body.

In an animal, the activities of individual cells are controlled so that they are integrated with, and appropriate to, the needs of the organism as a whole. It would not be feasible for an animal to exist without a complex system of cell–cell communication any more than it would be feasible for a large army to exist without a complex system of command to determine the activities of individuals within it. An end result of cellular activities that are uncoordinated with those of the whole animal is seen in cancer cells.

In recent years it has emerged that the extrinsic controls operating on cells are far more complex than was ever imagined. The mechanisms by which some important individual controls operate are now known but the understanding of how all of the regulatory factors add up to a control system that works in the whole body or, indeed, in a whole cell, is far from complete.

Communication between cells is dependent on the release from cells of signalling molecules that migrate to other cells and deliver stimuli to those equipped to receive the signals. The latter are called **target cells**. A target cell for a given signal is one equipped with a receptor to which the signal molecule binds, resulting in a biochemical response within the cell. Variations of this general picture exist in situations in which cell–cell contact delivers the signal. An example is the activation of helper T cells (page 343) by antigen-presenting cells, but there is no real difference of principle here—it is just that the signalling molecule is attached to the signalling cell, rather than free. Gap junctions, or controllable pores between adjacent cells, are another exception, permitting molecules to migrate from one cell to another; they put direct communication channels between the cytoplasm of cells and help to coordinate cellular activities.

In this chapter we will deal with released signalling molecules, under the following headings.
- What sorts of cellular activities are controlled by extrinsic signals?
- What are the signalling molecules involved?
- Which cells release them, how is this controlled, and how do they reach their target cells?
- How are signals detected by target cells?
- How does the cell respond to detected signals?

....

What sorts of cellular activities are controlled by extrinsic signals?

We might, for convenience, divide these into two groups.
- *Controls that do not involve regulation of gene expression.* Examples are: modulation of enzyme activities in fat and carbohydrate metabolism (Chapter 12); activation of voluntary striated muscle contraction by acetylcholine (Chapter 28); opening of ligand-gated pores in nerve cell membranes to trigger a nervous impulse (Fig. 26.24).
- *Controls that do involve modulation of specific gene expression.* The majority of extrinsic controls operate in this way (including some effects of certain signals that also operate by mechanisms in the category above). Since gene expression regulation affects virtually everything in cells, the list of processes controlled is almost all-embracing.

Control of gene expression is predominantly (but not entirely) at the level of initiation of gene transcription in animals and this, in turn, is dependent on the activity of specific transcriptional factors (page 284).

Therefore a major part of this chapter is concerned with mechanisms by which extrinsic control factors exert effects on transcription of specific genes. The majority of extrinsic signalling molecules, in delivering their stimulus, do not enter

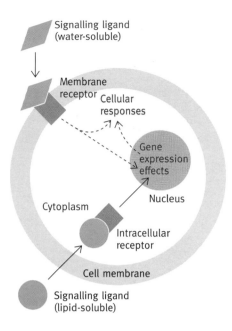

Fig. 26.1 Outline of receptor-mediated signalling. Water-soluble signalling molecules cannot pass through the lipid bilayer. They bind to external domains of receptors. The lipid-soluble signalling molecules such as steroids and thyroxine enter the cell directly and bind to intracellular receptors. Note that some intracellular receptors reside in the nucleus as described later; cytoplasmic steroid receptors move into the nucleus upon ligand binding. In the case of nitric oxide, the intracellular receptor produces cGMP, which can have direct metabolic effects.

the cell but rather combine with external protein domains of membrane-bound receptors. Questions to consider are, therefore: (1) how such signals outside of the cell membrane result in something happening inside the cell membrane (*trans*membrane signal transduction); (2) how the signal generated at the cytoplasmic side of the membrane is transmitted to the site of control. In the case of gene control this means transmission of the signal from membrane to nucleus (Fig. 26.1). By contrast, a small number of lipid-soluble signalling molecules directly traverse the lipid bilayer into the cell where they encounter intracellular receptors involved in gene regulation (Fig. 26.1).

What are the signalling molecules?

In chemical terms these include proteins, large peptides, small peptides, steroids, eicosanoids, catecholamines, thyroxine, and nitric oxide. Examples of the first six groups, respectively, are insulin, glucagon, vasopressin, sex hormones, prostaglandins, and epinephrine. A biological classification is as follows (nitric oxide being almost in a class of its own (see below)):

- endocrine hormones;
- growth factors and cytokines;
- neurotransmitters.

This classification is important in terms of nomenclature and physiology, but all of the signalling molecules bind to cellular receptors and elicit responses so that they are, in principle, all the same, the differences being secondary. With that qualification we will now describe the basis of that classification.

Endocrine hormones

These are the 'classical' signalling molecules, most of which have been known for a long time, probably because they exist in the body in relatively large amounts. Hormones are produced by cells specialized to produce them and assembled into glands that secrete the hormones into the circulation directly, rather than via a duct as occurs with the secretion of pancreatic digestive enzymes into the intestine. For this reason, endocrine glands are also referred to as ductless glands. Hormones reach their target cells by flooding the whole body via the circulation and so reach tissues distant from the secreting gland. A large number of hormones are known and their biological effects well-documented though these are not always understood in mechanistic terms. Table 26.1, given for reference purposes, summarizes them and their effects.

Growth factors

A recent publication has suggested that the name 'developmental regulatory factors' would be a more appropriate term for this class of signalling molecules, for they are much more understandable if regarded as developmental regulators, rather than as factors that always stimulate cell growth.

Growth factors are regulatory proteins secreted by cells that, unlike the endocrine gland cells, are not specialized for this function, but that are typical of whatever tissue they belong to, for example, hepatocytes, lymphocytes, etc. They are not assembled into glands. Growth factors are proteins that attach to external cellular receptors. The first growth factor discovered was PDGF (platelet-derived growth factor) but since then it has been found that many cells produce it. Epidermal growth factor (EGF) is another 'classical' one. A large variety of such factors have been discovered. About 20 are known to be involved in regulation of hemopoietic cell development in the bone marrow. Many of these are known as CSFs (colony-stimulating factors) and interleukins. CSFs are named because they were discovered in experiments in which they stimulated the growth of colonies of certain white cells, cultured in the form of separate colonies on plates. Interleukins are named from white cells affecting other white cells or leukocytes by their secretion. Most growth factors were initially discovered because of experimentally observed mitogenic (growth) effects on cells but their actions are much more complicated than this. In some circumstances, a given growth factor may stimulate or inhibit cell division; in others, it may stimulate or inhibit cell differentiation. They may have quite different effects on different cells or on the same cell under different conditions. This is why 'developmental regula-

Table 26.1 The principal hormones

Secreting organ	Hormone	Target tissue	Function
Hypothalamus	Hormone-releasing factors	Anterior pituitary	Stimulation of circulating hormone secretion
	Somatostatin (also from pancreas)	Anterior pituitary	Inhibits release of somatotropin
Anterior pituitary	Thyroid-stimulating hormone (TSH)	Thyroid	Stimulates T_4 and T_3 release
	Adrenocorticotropic hormone (ACTH)	Adrenal cortex	Stimulates release of adrenocorticosteroids
	Gonadotropins (luteinizing hormone, LH and follicle-stimulating hormone, FSH)	Testis and ovary	Stimulates release of sex hormones and cell development
	Somatotropin (growth hormone)	Liver	Stimulates synthesis of insulin-like growth factors, IGFI and IGFII
	Prolactin	Mammary gland	Required for lactation
Posterior pituitary	Antidiuretic hormone (ADH) or vasopressin	Kidney tubule	Promotes water resorption
	Oxytocin	Smooth muscle	Stimulates uterine contractions
Thyroid	Thyroxine (T_4), triiodothyronine (T_3)	Liver, muscles	Metabolic stimulation
Parathyroid	Parathyroid hormone	Bone, kidney, intestine	Maintains blood Ca^{2+} level, stimulates resorption and dietary Ca^{2+} uptake
	Calcitonin (also from thyroid)	Bone, kidney	Inhibits resorption of Ca^{2+}
Adrenal cortex	Glucocorticosteroids (cortisol)	Many tissues	Promotes gluconeogenesis
	Mineralocorticosteroids (aldosterone)	Kidney, blood	Maintains salt and water balance
Adrenal medulla	Catecholamines (epinephrine, norepinephrine)	Liver, muscles, heart	Mobilize fatty acids and glucose into bloodstream
Gonads	Sex hormones (testosterone from testes estradiol, progesterone from ovaries)	Reproductive organs, secondary sex organs	Promote maturation and function in sex organs
Liver	Somatomedins (insulin-like growth factors: IGFI, IGFII)	Liver, bone	Stimulate growth
Pancreas	Insulin	Liver, muscles	Stimulates gluconeogenesis, lipogenesis, protein synthesis
	Glucagon		Stimulates glycogen breakdown, lipolysis

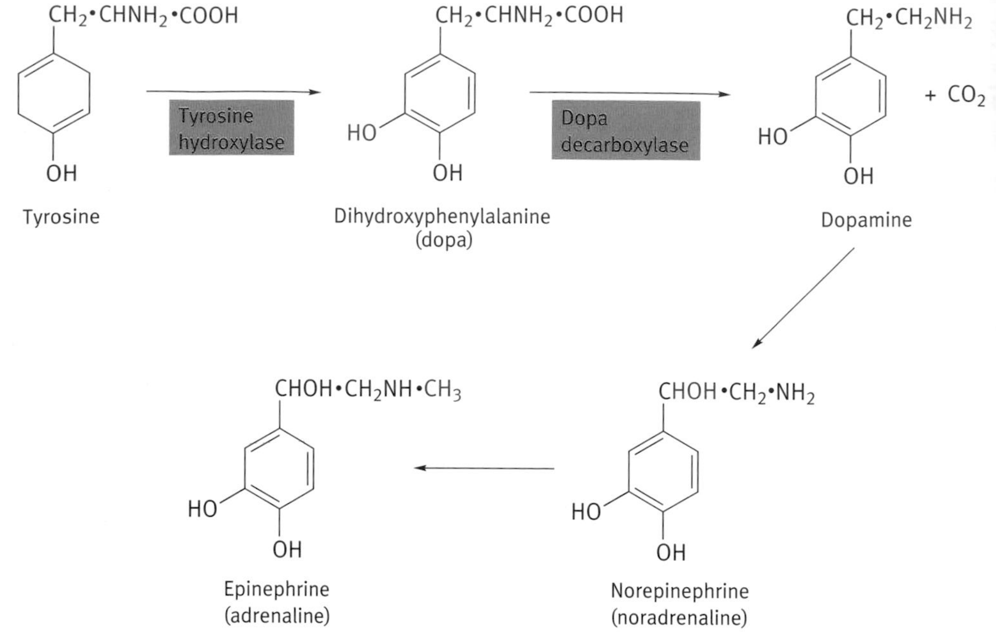

Fig. 26.2 The synthesis and structures of the catecholamines

tory factor' seems a more appropriate description for growth factors but the latter term is widely used so we will use it also. The term **cytokine** is used by immunologists for growth factors involved in the immune response (see page 344). The one thing that stands out like a beacon is that growth factors are of absolutely central importance and, consequently, are the subject of the most intensive research.

Despite all of this biological complexity, as stated, all growth factors combine with external cell receptors and indirectly produce effects on the initiation of specific gene transcription. Since, in this chapter, we are concerned with the signalling routes from receptors to nucleus, all the biological complexity, described above, does not prevent us from discussing this.

How does a growth factor differ from a hormone other than in its production characteristics mentioned above? As stated, hormones are released into the circulation and flood the whole body, and in this way reach their target cells. Most growth factors are **paracrine** in their action—they diffuse only short distances and act only on local cells—while some are **autocrine** (**self-stimulatory**) in action—that is, they act on the cells secreting them (for example, interleukin 2, which stimulates T cell proliferation).

Neurotransmitters

A variety of these are involved in nerve function but we will mention only a few that are relevant to our immediate topic. The sympathetic (involuntary) nervous system, which innervates, for example, fat cell depots, secretes epinephrine and norepinephrine on receipt of a nerve impulse; the motor neurones, which innervate voluntary striated muscle, trigger contraction by the release of acetylcholine from the nerve endings. Thus, a fat cell may be regulated by epinephrine from the adrenal glands via the blood (Chapter 12) or from nerve endings. The difference is that the latter delivery route is faster and the signal is released precisely at the target cell site. All of these work by binding to external cell receptors. The synthesis of catecholamine neurotransmitters is given in Fig. 26.2; you have already seen the relationship of epinephrine to catechol (page 204).

How is the release of signalling molecules controlled?

Control of hormone release

The endocrine system is under a hierarchical control system with part of the brain, the hypothalamus, being at the top of the chain of command, but itself being subject to classic feedback loop controls (Fig. 26.3). The hypothalamus is a small part of the brain, weighing 4 or 5 grams, and represents an ancient, primitive part of the brain, essential to life. It produces the hypothalamic hormones that control release of hormones from the anterior part of the pituitary gland located at the base of the brain. For this reason **hypothalamic hormones** are alternatively called **hormone-releasing factors**.

The anterior pituitary hormones cause target endocrine glands of the body to release *their* hormones—the latter are the 'final' hormones in the chain, which elicit the appropriate

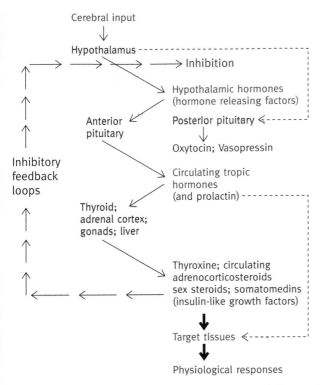

Cerebral input

Hypothalamus

Inhibition

Hypothalamic hormones
(hormone releasing factors)

Anterior
pituitary

Posterior pituitary

Oxytocin; Vasopressin

Inhibitory
feedback
loops

Circulating tropic
hormones
(and prolactin)

Thyroid;
adrenal cortex;
gonads; liver

Thyroxine; circulating
adrenocorticosteroids
sex steroids; somatomedins
(insulin-like growth factors)

Target tissues

Physiological responses

Fig. 26.3 Diagram of hormone release control by the hypothalamus. The list of hormones is not complete. The glands are shown in green. The final tissue-targeting hormones are in blue and tropic hormones in yellow. The hypothalamus hormone-releasing factors are carried directly to the anterior pituitary by blood capillaries in the pituitary stalk.

responses from target tissues in the body rather than affecting other glands. Because the **anterior pituitary hormones** bring forth hormones from other glands they are called **tropic hormones** or **tropins**. These tropic hormones affect the thyroid gland, the adrenal cortex, and the gonads. As already indicated, the release of the anterior pituitary hormones is itself regulated by classic feedback loops in which end-products (the final cell-targeted hormones) inhibit the first step (release of hypothalamic hormones), thus maintaining appropriate levels of circulating hormones. The hypothalamus receives input from the rest of the brain so that a degree of cerebral control on the endocrine system exists. Figure 26.3 summarizes the situation.

Not all endocrine glands are controlled by this hypothalamus-based mechanism. The posterior pituitary is under different control and the release of insulin and glucagon from the pancreas is affected directly by the level of circulating glucose (page 74).

Control of release of growth factors

Relatively little is known of this and, with so many factors, the mechanisms will be numerous. Damage to the epithelial cell layer of blood vessels results in platelet breakage and release of

PDGF, which plays a role in stimulating cell division and repair, but PDGF is produced by many different cells of the body. In the immune system, activation of B and T cells causes the release of growth factors (cytokines) stimulating B and T cell proliferation (page 344). Growth factors can stimulate cells to release other growth factors, illustrating the biological complexity. In short, growth factors are involved in developmental processes as well as in cellular repair processes and protective mechanisms such as the immune response, which involves extensive control of cell proliferation and differentiation.

Control of neurotransmitter release

A nerve impulse passing down the axon of a nerve cell causes vesicles at nerve endings to eject their store of neurotransmitter by exocytosis using a mechanism to be described later.

Removal of signalling molecule

The essence of control is reversibility. Released signal molecules must be removed; otherwise an initial signal would last indefinitely. A classic case is that of acetylcholine destruction by acetylcholine esterase, present in nerve synapses. Removal of the signal is necessary to prepare the synapse for the arrival of the next nerve impulse. This happens by the following reaction.

$$CH_3 - \underset{\underset{CH_3}{|}}{\overset{\overset{CH_3}{|}}{N^+}} - CH_2 - CH_2 - O - CO - CH_3 + H_2O$$

Acetylcholine

Acetylcholine esterase

$$CH_3 - \underset{\underset{CH_3}{|}}{\overset{\overset{CH_3}{|}}{N^+}} - CH_2 - CH_2OH + CH_3 - COO^- + H^+$$

Choline   Acetate

Epinephrine and norepinephrine released at nerve endings are taken up by neighbouring cells, or removed by metabolic destruction.

How are signals detected by target cells?

As stated already, target cells for a given signal have receptors, either on their external surface or inside the cell, to which a signal binds, causing a change in the receptor protein(s) that results in a response. A few hormones (steroids, thyroxine, and nitric oxide) are lipid-soluble and enter the cell directly through the lipid bilayer, and bind to intracellular receptors (Fig. 26.1).

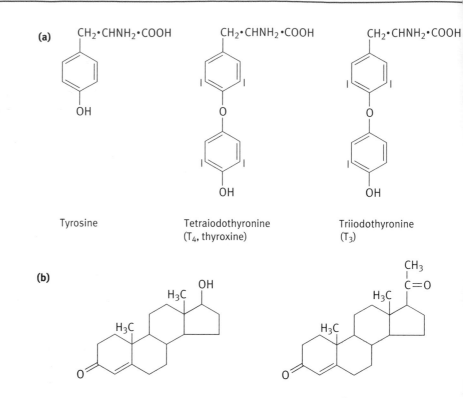

Fig. 26.4 (a) The structures of tyrosine and the thyroid hormones. **(b)** The structures of two steroid hormones.

Have a look at the structures of thyroid hormones and of two steroids in Fig. 26.4. The rest of the signalling molecules are water-soluble and attach to external receptors (Fig. 26.1).

The essence of it all is specificity—the signal molecule (called a **ligand** or **agonist**) has the same exquisitely precise relationship to its receptor as a substrate has to its enzyme. This explains how a hormone flooding the entire body affects only its target cells. If the cell has no specific receptor, it is blind to that signal, just as a house without a receiving dish is blind to satellite television broadcasts.

How does ligand binding result in cellular responses?

Intracellular receptor-mediated responses

Steroid hormones transcriptionally regulate expression of specific genes in target cells at the level of initiation of transcription. The steroids include glucocorticoids, estrogen, and progesterone (see Table 26.1).

These hormones, being lipid-soluble, traverse the lipid bilayer of cells without difficulty and their receptors exist inside the cell rather than on the membrane. A whole family of related steroid/thyroxine receptors cater for their action.

The receptors for estrogen and progesterone are located in the nucleus and the glucocorticoid receptor in the cytoplasm. This localization difference appears to be a secondary one (see below)

for they all have zinc fingers (page 286) as their DNA binding domain and they exist in association with the heat shock proteins (Hsps). Heat shock proteins, you will recall (page 306), are chaperones.

The zinc finger glucocorticoid receptor has been studied intensively and we will describe this as an example here. The receptor itself exists in the cytoplasm, attached to a complex of heat shock proteins that makes unavailable its nuclear targeting label (a peptide sequence on the protein).

When a glucocorticoid hormone binds to a specific site on the receptor, the heat shock protein complex dissociates from it, revealing the targeting label and the receptor is transported into the nucleus. There, it combines as a dimer with DNA at the specific glucocorticoid-responsive element (Fig. 26.5). The latter is in the form of a palindrome, each half combining with the two zinc fingers (Fig. 21.22) of each molecule of the receptor dimer. In the case of those members of the steroid family of receptors that exist in the nucleus, the nuclear targeting signal is not obscured by the Hsp complex so that the whole structure is transported as a unit into the nucleus but, until the Hsp is released by hormone attachment, it does not bind to DNA.

Membrane receptor-mediated responses

In the cell–cell communication story, most of the signalling in the body occurs via membrane receptors. In Fig. 26.6 two types of receptors for membrane-mediated responses are illustrated. In the first, binding of the ligand activates an internal domain of

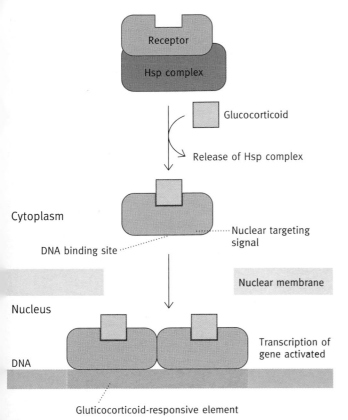

Fig. 26.5 Diagram of gene activation by steroid hormones. The example shown is the glucocorticoid receptor, one of a superfamily of steroid/thyroxine receptors, all of which involve complexes with heat shock proteins (Hsps) and binding to DNA sites by zinc fingers. In the case of those members of the superfamily that reside in the nucleus, the Hsp complex does not obscure the nuclear targeting signal. The locations of the latter and the various binding sites are drawn arbitrarily in the diagram.

the receptor. In the second, binding of the ligand causes dimerization of receptors and this is followed by, or associated with, activation of internal domains. How the signal generated at the cytoplasmic side of the membrane is transmitted to the site of control is a complex topic, so we will give an overview first and then deal with the individual signalling pathways that lead to the cell responses.

Overview of membrane receptor-mediated responses

The predominant theme in this topic is **protein phosphorylation** by protein kinases using ATP. It would be difficult to overstate the importance of protein phosphorylation in cellular control—in animals it is involved in hormone and growth factor control, in the control of cell cycle phases by cyclins and in metabolic controls in smooth muscle contraction controls. Figure 26.7 illustrates the central principle of control; it is

essentially a repeat of Fig. 12.8, illustrating the central role of phosphorylation in metabolic control by extrinsic signals.

It obviously follows from this that most cases of receptor-mediated signalling involve activation of protein kinases. Dephosphorylation of phosphoproteins is catalysed by protein phosphatases and this provides the potential for control at this stage also. Figure 26.8 gives an overview summary of control systems mediated by membrane receptors. The diagram also shows reversal of a signal by destruction of a second messenger.

Modulation of receptor response to signal

The repeated exposure of a cell to a signal often results in a diminished response. In the case of one class of epinephrine receptors, a feedback loop results in phosphorylation of the receptor protein itself and desensitization of the receptor to the ligand (page 171). (To avoid later confusion, in other types of receptors to be described shortly, phosphorylation of receptors is part of the activation process.) Another type of modulation of response is the control of the number of functional receptors in the membrane. In the case of insulin receptor, the insulin-receptor complex is internalized by endocytosis (see page 208). The extent of recycling the receptor back to the membrane controls the displayed number. This can have the effect of

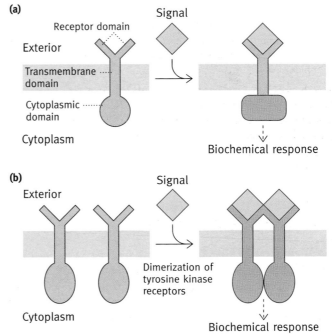

Fig. 26.6 Diagram of transmembrane signal transduction by receptors. In **(a)** binding of the signal to the receptor is postulated to produce an allosteric change in the cytoplasmic domain of the receptor protein molecule. In **(b)** binding of the signal causes dimerization of membrane receptors. This applies to the tyrosine kinase receptors described later (page 361). The bringing together of the two cytoplasmic domains is followed by biochemical response. The nature of the biochemical responses is dealt with later in this chapter.

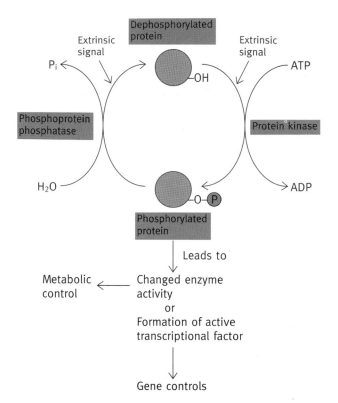

Fig. 26.7 Central principle of control by the majority of extrinsic signals. The phosphorylation of a protein leads to changed properties of that protein, resulting in a cellular response. Note that phosphorylation may have an activating or deactivating effect on different proteins. The extrinsic signals lead to effects on the protein kinases and the protein phosphatases.

diminishing the response of a cell to a given signal, and thus has a dampening effect on the control system; it is known as **downregulation**.

....

Description of individual signalling pathways inside the cell linking receptor activation to cellular response

cAMP-mediated pathways

Since in Chapter 12 (page 170) we have already dealt with the control of metabolism by cAMP-modulation of protein kinase activities, in this section we will deal only with the way in which the level of cAMP inside a cell itself is modulated and how cAMP regulates specific gene expression.

A wide array of hormones (that is, **first messengers**) use cAMP as **second messenger**, including the adrenocorticotropic hormone (ACTH), the antidiuretic hormone (ADH), the gonadotrophins, the thyroid-stimulating hormone (TSH), the parathyroid hormone, glucagon, the catecholamines (epinephrine/norepinephrine), and somatostatin (Table 26.1). The list, which is not exhaustive, is given to illustrate that cAMP can have

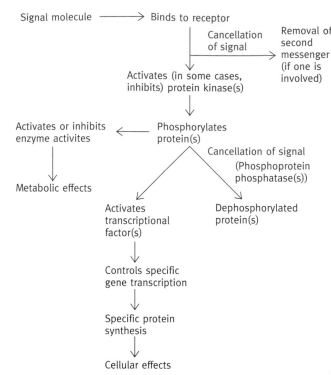

Fig. 26.8 Simplified diagram of main events in membrane receptor-mediated signalling. Note that an arrow may represent several steps in the process.

vastly different effects in different cells. The system works because the response to cAMP in a given cell is appropriate to the signal that that cell is capable of recognizing via its specific receptors. Thus, in cell A, signal X elevates cAMP levels that produce cellular responses appropriate to signal X. Cell B does not have receptors for X but does for signal Y, which also elevates cAMP, but in cell B this causes responses appropriate to signal Y.

The two aspects of these effects then, which we will now discuss, are: (1) how does signal binding to receptor control cAMP levels in the cell; and (2) how does cAMP affect gene transcription?

Control of cAMP levels in cells

cAMP is produced from ATP by an enzyme, adenylate cyclase, that is attached to the inside of the cell membrane. We have met this reaction in Chapter 12, but it is shown again in Fig. 26.9(a).

Let us consider, as a typical example, the way in which adenylate cyclase becomes activated by epinephrine to produce cAMP in liver and skeletal muscle (page 170). The receptor (in this case, called the β_2-adrenergic receptor, after the alternative name for epinephrine, adrenaline) is a protein whose polypeptide chain criss-crosses the membrane seven times (Fig. 26.10). Each of the seven transmembrane segments has about 19 hydrophobic amino acids, which, arranged in α-helix conforma-

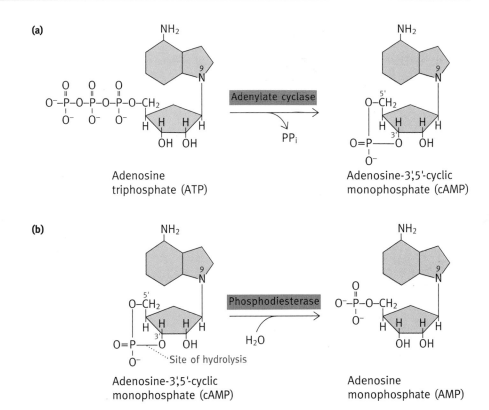

Fig. 26.9 (a) Formation of 3′, 5′-cyclic AMP by adenylate cyclase. **(b)** Hydrolysis of 3′, 5′-cyclic AMP by phosphodiesterase.

tion, are sufficient to span the hydrophobic membrane core. The epinephrine binds to a cleft formed by the transmembrane helices.

Associated with the cytoplasmic face of the receptor protein is a regulatory G (for GTP-binding) protein, which has three subunits, α, β, and γ. Because these three polypeptides are

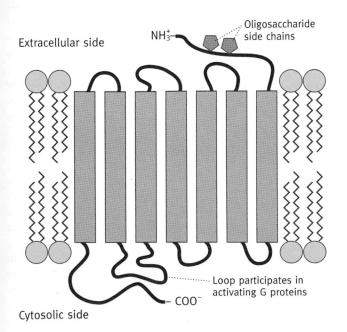

Fig. 26.10 Diagram of the structure of the β_2-adrenergic receptor.

different, it is known as a **heterotrimeric G protein** or, simply, as a **trimeric G protein**. The α subunit has a site on it that can have bound to it *either* GTP or GDP. When the site on the α subunit is occupied by GDP, nothing happens. This is the situation in the absence of hormone, illustrated in Fig. 26.11(a).

When a molecule of epinephrine binds to the receptor, the latter undergoes a conformational change that alters its cytoplasmic domain. This, in turn, affects the G protein bound to it, presumably also by conformational change. The change causes the G protein (called G_s in this system—'s' for stimulatory) to exchange its GDP for GTP. The G protein cannot make this exchange unless it is attached to a receptor to which the hormone is bound. The α subunit–GTP complex detaches, migrates to, and activates an adenylate cyclase molecule that now produces cAMP (Fig. 26.11(b), (c)).

The hormone thus switches on cAMP production using the G protein as an intermediary (Fig. 26.11(c)). Activation of cAMP production by the α-GTP subunit must be limited in time—it must be switched off for otherwise a single hormone stimulus would last indefinitely, long after the response would be appropriate. When the receptor is no longer bound by epinephrine, cAMP production must cease; destruction of cAMP by phosphodiesterase in the cell completes the reversal (Fig. 26.9(b)).

The situation is rather like that of a timed light switch on a staircase where, when you press a button, the light goes on and the button slowly comes back out and switches off the light after a minute or two. You have to keep on pressing the button at

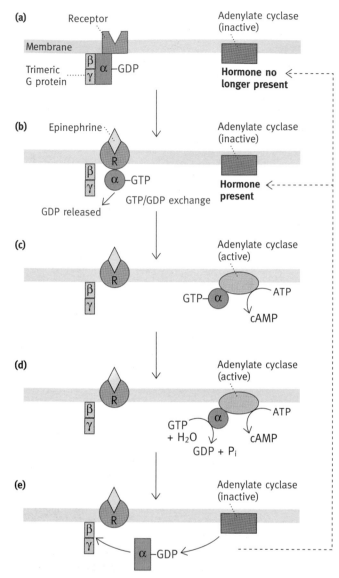

Fig. 26.11 Diagram of the control of adenylate cyclase activity by a hormone such as epinephrine. The geographical locations of the subunits shown are hypothetical, the essentials being that the α-GTP complex activates adenylate cyclase. The steps **(a) → (e)** are referred to in the text.

intervals to keep the light on. The G protein is a timing device to limit the period of activation of adenylate cyclase.

The α subunit of the G protein, in its GTP-bound form (which activates the adenylate cyclase), has enzyme activity—it is a GTPase. It hydrolyses the GTP molecule attached to itself to GDP and P_i (Fig. 26.11(d)). The activity is low so that this hydrolysis occurs only after a delay. As soon as GDP is formed by this hydrolysis, the α subunit reverts to its original state; it detaches from the adenylate cyclase which is now inactivated, the α-GDP joins up to its old partners, the β and γ subunits, and the G protein–GDP complex is reassembled, in contact with receptor (Fig. 26.11(e)). If the latter still has hormone bound,

the whole cycle can start again (that is, back to Fig. 26.11(b)). GTP is accepted in place of GDP and the α subunit goes off to activate a molecule of enzyme. If the receptor no longer has hormone attached (that is, back to Fig. 26.11(a)), the process comes to a halt. Thus, to continue cAMP production, the α subunit runs back and forth from receptor to adenylate cyclase, the duration of it on the adenylate cyclase being that required to hydrolyse its attached GTP molecule.

The system has an amplifying effect, for one hormone-bound receptor can activate one G protein molecule after another and, therefore, one molecule of hormone binding to a receptor can result in numbers of adenylate cyclase molecules being activated and large numbers of cAMP molecules being produced.

The importance of the GTP hydrolysis switch-off is very dramatically seen in cholera. Gut mucosal cells secrete Na^+ into the intestinal lumen and this activity is activated by cAMP. The cholera toxin causes inactivation of the GTPase activity of the G protein α subunit. Thus, once a hormone stimulus activates the adenylate cyclase, it cannot be switched off; it is frozen into the α-GTP state. The prolonged cAMP production results in massive Na^+ loss, accompanied by water molecules, causing debilitating diarrhea and possible death from fluid and electrolyte loss.

In the case given in Fig. 26.11 the α-GTP subunit stimulates adenylate cyclase activity. Another type of receptor for epinephrine (known as the $α_2$-receptor) operates similarly, except that the α-GTP (G_i) subunit is different and it *inhibits* adenylate cyclase activity. The α- and β-adrenergic receptors mediate different responses to catecholamines. As we have seen (page 170), epinephrine acts to mobilize the liver and skeletal muscle for sudden action. This is accomplished via the $β_2$-adrenergic receptor activating adenylate cyclase. By contrast, the $α_2$-adrenergic receptor inhibits adenylate cyclase; the $α_1$-adrenergic receptor is different again—it does not use cAMP as second messenger, but rather causes Ca^{2+} release by a mechanism to be described later (the phosphoinositide pathway, page 359). Thus, one hormone (epinephrine) can exert quite different effects, according to the type of receptor.

How does cAMP exert effects on gene transcription?

In Chapter 12, we described how cAMP activates PKA (protein kinase A). PKA, as well as being involved in enzymic control, is part of an important pathway of gene control. The promoters of several cAMP-inducible genes contain cAMP-responsive elements (CREs). A **CRE binding protein (CREB)**, when phosphorylated by PKA, becomes an active transcriptional factor of the leucine zipper type (page 286). It binds to CREs and promotes specific gene transcription (Fig. 26.12). A family of CREB proteins exist, presumably with varied functions in the cell.

In what we have said in this section, cAMP is a second messenger (page 169) in the transmission of an extrinsic signal to functional elements both in the cytoplasm and the nucleus of

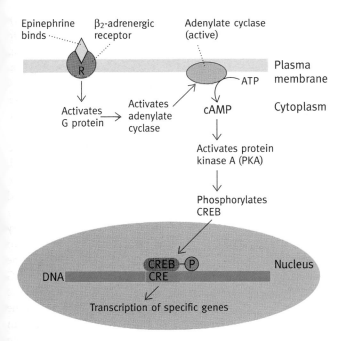

Fig. 26.12 Diagram illustrating the β_2-adrenergic receptor function. Receptor-mediated stimulation of adenylate cyclase and cAMP levels activates the cAMP-dependent protein kinase A (PKA) to modulate receptor activity. Phosphorylation of CREB (cAMP-responsive element binding protein) produces an active transcriptional factor for specific genes.

the cell. Another cyclic nucleotide, cGMP, also has this function in some control systems, and we will now summarize these.

Signal transduction using cyclic GMP as second messenger

A number of signals cause the production of **cGMP** inside the cell, which, in turn, activates protein kinases. cGMP is analogous to cAMP and can be formed by two different guanylate cyclase enzymes from GTP (Fig. 26.13). In one of these, the hormone combines with an external receptor whose inner domain is thereby activated to synthesize cGMP. This is different from the way in which the epinephrine receptor triggers cAMP production, since the inner domain of the receptor *itself* is the guanylate cyclase and is activated directly by attachment of hormone (Fig. 26.14). The raised cGMP level mediates the cell

response by activating specific protein kinases. An example of this type of control is the atrial natriuretic peptide, produced by endothelial cells, which stimulates the kidney to secrete Na^+. The effects of cGMP appear to be more specialized than those of cAMP. They include relaxation of smooth muscle and effects on nerve cells and vision (the latter described below).

There is a second, quite different, guanylate cyclase present in soluble form in the cytosol. This has a heme molecule as its prosthetic group to which binds a cell–cell signalling molecule of surprising simplicity—NO or **nitric oxide**. The heme molecule is, in effect, functioning as an extremely sensitive detector of NO and transduces the signal into activation of its attached enzyme. NO is produced from the arginine guanidino group, by an enzyme called **nitric oxide synthase**, in cells lining parts of the vascular system. The NO diffuses into the smooth muscle of blood vessels, causing cGMP production which, in turn, causes muscle relaxation and vessel dilatation. Production of nitric oxide can also be stimulated by acetylcholine so that it can mediate nervous control. It also is produced in response to a shearing force exerted by blood flow on the endothelial cells lining the vessels, thus resulting in vasodilatation and reduction of shearing force. Trinitroglycerine, a drug long used in the treatment of angina, slowly produces nitric oxide thereby relaxing blood vessels, thus reducing the work load of the heart muscle. Since NO is oxidized to NO_2 and NO_3 in seconds, it is a locally acting hormone; being lipid-soluble it easily escapes from the cells producing it and enters adjacent cells. Nitric oxide is part of a complex regulatory system with multiple physiological effects. Figure 26.14 summarizes the production of cyclic nucleotides as second messengers in cell–cell signalling.

Hormones that work via different second messengers; the phosphatidylinositol cascade

So far, let us remind you, we have been dealing with a diverse group of hormones that use the cyclic nucleotides, cAMP and cGMP, as second messengers. Another large group of hormones and growth factors use different intracellular mechanisms for transducing the signals. The hormones in question bind to specific external cell receptors that also interact with G proteins (that is, GTP-binding proteins), but from there on it is all quite different; cyclic nucleotides are *not* involved. Examples of

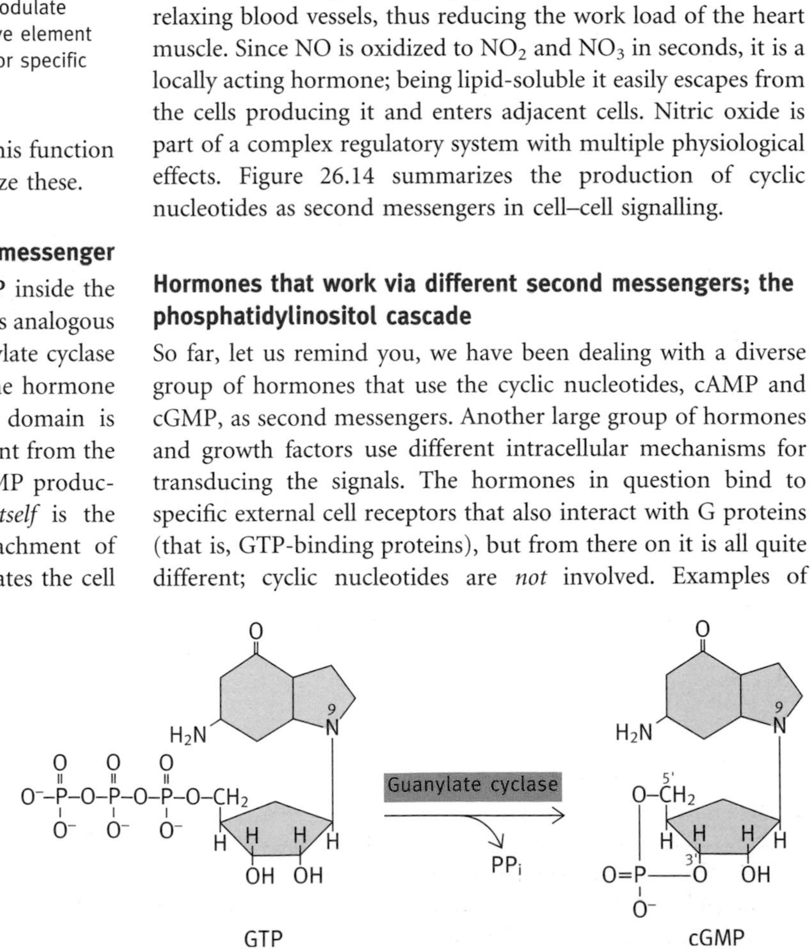

Fig. 26.13 Formation of 3′, 5′-cyclic GMP by guanylate cyclase.

GTP

Guanylate cyclase

PPi

cGMP

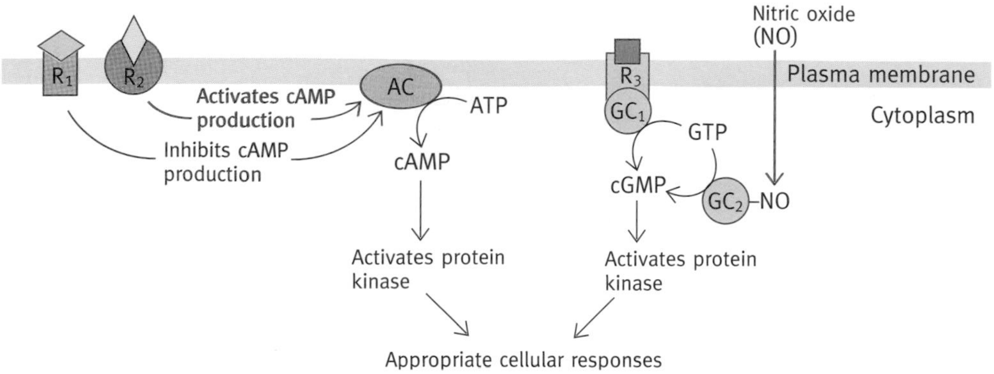

Fig. 26.14 Summary of the action of cyclic nucleotides as second messengers in cell signalling. (Note that the different pathways may occur in different tissues (see text).) R_1, A receptor such as the α_2-adrenergic receptor; R_2, a receptor such as the β_2-adrenergic receptor; R_3, the binding site for the atrial natriuretic peptide which is part of a single polypeptide chain with a membrane-spanning sequence and a guanylate cyclase catalytic site on the cytoplasmic face. (Note that cGMP has a separate function in the visual process, which is described later.) AC, Plasma membrane adenylate cyclase; GC_1, plasma membrane guanylate cyclase; GC_2, cytoplasmic guanylate cyclase, activated by nitric oxide (NO).

hormones that use this mechanism include the thyrotropin-releasing hormone and the gonadotropin-releasing hormone as well as the growth factor, PDGF (page 350).

The start of this story is the phospholipid—**phosphatidylinositol-4,5-bisphosphate (PIP$_2$)**. This is formed in the membrane by phosphorylation of phosphatidylinositol whose structure is given in Chapter 10 (page 42). Once again, as indicated, a G protein links the receptor stimulus to the intracellular signal pathway and acts as a molecular timing device. In this case, binding of the hormone to a receptor causes an associated G protein (this one is called G_p) to exchange its GDP for GTP. The G_p–GTP complex migrates to, and activates, a membrane-bound enzyme, phospholipase C (PLC), which splits PIP$_2$ into **IP$_3$ (inositol trisphosphate)** and **DAG (diacylglycerol)** (Fig. 26.15). The G protein hydrolyses GTP to GDP and P$_i$, thus inactivating itself.

The IP$_3$ causes release of Ca^{2+} from the lumen of the endoplasmic reticulum (ER) where the ion is stored in high concentration relative to that in the cytoplasm. It opens ligand-gated Ca^{2+} channels (see below) in the membrane allowing the ion to escape back into the cytoplasm. Ca^{2+} is continually returned, by a Ca^{2+}/ATPase, to the ER lumen; IP$_3$ is enzymically removed, as also is DAG. Thus, the combination of a hormone, or other agonist, with a receptor associated with the phosphatidylinositol cascade results in increases of intracellular DAG and Ca^{2+} (Fig. 26.16) and reversal of the signal when hormone is no longer attached to the receptor.

DAG is the physiological activator of a protein kinase distinct from that which is activated by cAMP (Ca^{2+} is also required for maximal activation); the DAG-activated one is called **PKC** and occurs on the cytoplasmic face of the cell membrane attached to phosphatidylserine molecules in the membrane bilayer. PKC is a protein kinase of central importance; it can exert multiple cellular effects by phosphorylating different target proteins. As one example, it is involved in phosphorylating certain

transcriptional factors. The importance of PKC in cell division control is illustrated by the tumour-promoting effect of phorbol esters which are analogues of DAG capable of activating PKC. The structures of DAG and a phorbol ester are compared below; it is not necessary to memorize the phorbol ester structure.

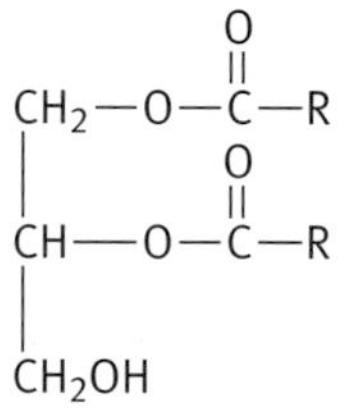

Diacylglycerol (DAG)
(R = Fatty acid chain)

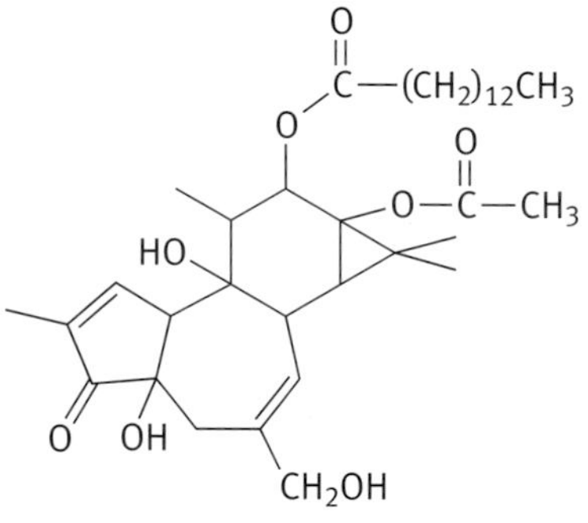

A phorbol ester

It may seem incongruous that DAG has the same effect in activating PKC as does an agent which is a promoter of tumour formation. The difference, presumably, is that DAG is rapidly destroyed and activates PKC only when this is required, while phorbol esters are longer-lived and deliver a correspondingly prolonged inappropriate signal.

In Fig. 26.16 it is seen that stimulation of a receptor of the type described in this section produces two second messengers—DAG and Ca^{2+}, both of which are needed for maximal activation of PKC. However, quite apart from this, Ca^{2+} is an important second messenger on its own.

What are the other second messenger roles of calcium?

A general second messenger role of the ion involves combination with a widely distributed protein called **calmodulin**. This

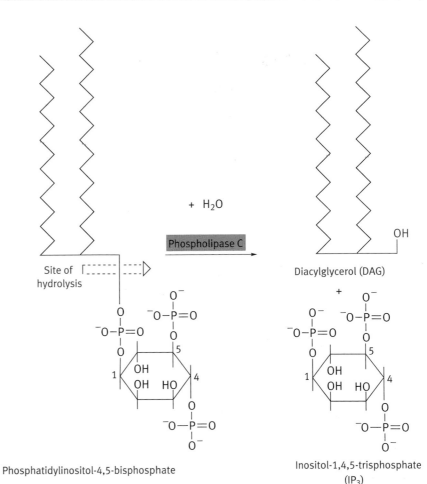

Fig. 26.15 Hydrolysis of phosphatidylinositol-4,5-bisphosphate (PIP$_2$) to diacylglycerol (DAG) and inositol trisphosphate (IP$_3$).

has four sites for binding Ca^{2+} with high affinity, the binding causing a conformational change in the protein. Calmodulin is found in association with the enzymes it controls; on the binding of the Ca^{2+}, the induced conformational change alters the activity of its associated enzyme. The important point is that a number of calmodulin–Ca^{2+}-activated protein kinases exist and thus Ca^{2+} can, via this route, exert multiple cellular effects.

The target proteins of these kinases include glycogen phosphorylase and glycogen synthetase (page 79). Since one type of epinephrine receptor (the α_1) activates the phosphoinositide pathway and hence causes Ca^{2+} increase in the cytoplasm of cells, epinephrine can affect glycogen metabolism via this mechanism as well as via cAMP. Other target proteins of calmodulin–Ca^{2+} kinases are myosin light chains (pages 394–5).

Tyrosine kinase-associated receptors

Although we are still dealing with membrane receptors in this section, we now come to an area quite different from that which has been described and one of immense importance.

Again, phosphorylation is a predominant theme (as on page 356), with two important differences. One is the amino acid

group on the target protein that is phosphorylated. The second is that the first target protein is the receptor itself, that is, a **self-phosphorylation**.

So far, all the protein kinases described phosphorylate serine or threonine groups in their target proteins. Activation of the group of receptors we will now discuss, in response to signal binding, depends on phosphorylation of the —OH group of specific tyrosine residues of proteins (Fig. 26.17(a)). In itself this might seem to be a trivial point, but it so happens that **tyrosine kinase-associated receptors** are of a different type from those described, and activate quite different intracellular signalling pathways. Their importance is underlined by the fact that most growth factors and certain hormones such as insulin have receptors of this type.

We will use the effect of the epidermal growth factor (EGF) to illustrate this type of signalling system. EGF receptors in the cell membrane are caused to dimerize when the ligand binds (Fig. 26.17(b)). This results in multiple phosphorylation of the tyrosine residues of the cytoplasmic domains of the *receptors*. The cytoplasmic domains *are* tyrosine kinases, and probably mutually phosphorylate each other on dimerization. The phosphorylation results in the activation of a signalling pathway

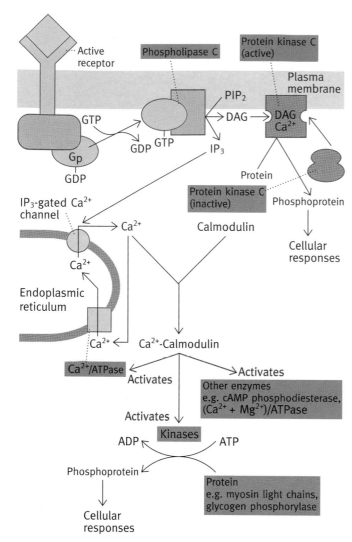

Fig. 26.16 The phosphatidylinositol cascade: interactions of DAG, IP_3, and Ca^{2+} as second messengers. Note that the G_p protein binds to phospholipase C (PLC) only in the GTP-complexed form. On hydrolysis to GDP the process reverses.

from membrane receptor to nucleus terminating in the activation of transcriptional factors (Fig. 26.17(b)). The question is how this pathway works; we will now describe this.

The Ras pathway—the widespread signalling pathway from membrane to gene in eukaryotes

A protein known as **Ras** exists in all eukaryote cells. It is a small, monomeric GTP-binding protein with GTPase activity (yet another one). It turns out to be part of a major signalling pathway from growth factor-responsive tyrosine kinase-associated receptors to activation of gene transcriptional factors. *There are no small molecular weight second messengers in this pathway*—all of the components are proteins.

It was known that the protein Ras and other components of the intracellular signalling pathway are important in cell

regulation since mutations resulting in abnormal forms of the proteins are oncogenic—they are associated with transformation to the cancerous state and uncontrolled cell division.

Figure 26.18 illustrates the system. In the cytoplasm there is a **growth factor receptor-binding protein (GRB)** associated with a protein called SOS (see below for what this means). The GRB associates with the phosphorylated receptor (*but not to the unphosphorylated form*) by means of a domain of the GRB, called SH2, which is found in a family of cytoplasmic proteins that are believed to have this phosphorylated-receptor binding role. The name SH2 derives from the fact that the proteins have a domain homologous with region 2 of the Src (pronounced 'sark') protein coded for by the oncogene of the Rous sarcoma virus (see page 320). Hence Src homology region 2 or SH2 for short.

Combination of the GRB–SOS with the activated receptor causes the Ras protein to exchange its GDP for GTP (Fig. 26.19). This is precisely analogous to the activation of the G protein in cAMP production (Fig. 26.11). Ras has low GTPase activity and a molecular clock function.

The next component of the pathway is the Raf protein, a protein kinase of the serine–threonine type, requiring activation by Ras-GTP. This is the start of a **serine/threonine protein kinase cascade** ending in the phosphorylation of transcriptional factor proteins and their activation to promote attachment to their responsive elements and specific gene transcription.

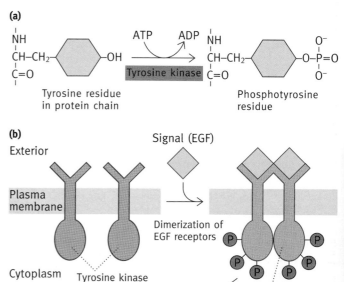

Fig. 26.17 (a) Tyrosine phosphorylation by tyrosine kinase.
(b) Diagram illustrating the overall effect of EGF on a target cell.

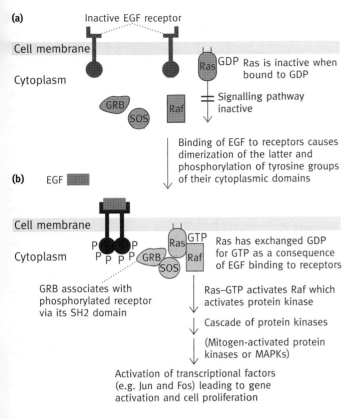

Fig. 26.18 Simplified diagram of the activation of the Ras pathway by EGF. The red colour indicates inactivity; green, the activated state. **(a)** Inactive, without EGF. **(b)** Activated by EGF. GRB, Growth factor receptor-binding protein; SOS, son of sevenless, a protein involved in the pathway named after a sevenless mutation in *Drosophila*. Raf, Ras, Jun, and Fos are named from their oncogenes (see text). Note that the receptor is phosphorylated on tyrosine groups but the Raf kinase and subsequent kinases of the cascade are of the serine/threonine-specific type. Note also that the mechanism of Raf kinase activation by Ras-GTP is not currently understood and may not be direct as indicated here.

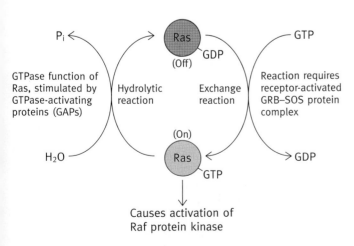

Fig. 26.19 Diagram illustrating activation of the Ras protein.

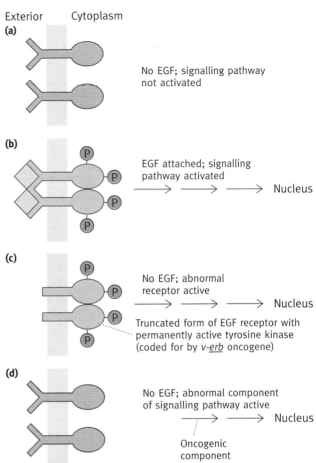

Fig. 26.20 Oncogene products as components of signalling pathways from the membrane to the nucleus—the EGF receptor is used as an example. **(a)** Normal receptor—not activated. **(b)** Normal receptor—activated. **(c)** Permanently active truncated EGF receptor. **(d)** Normal receptor—pathway component permanently active.

The Ras pathway and oncogenes

In Chapter 23 (page 320) we explained that cancer-producing genes carried by retroviruses are mutated forms of genes (protooncogenes) present in normal cells. The proteins, Ras and Raf, are protooncogene products; the corresponding oncogenes, when introduced by retroviruses into cells, are cancer-producing. It appears that the oncogenic counterparts of Ras, Raf (and other components of the Ras pathway in Fig. 26.18) are in a permanently 'activated' state without a signal being present on the receptor. Therefore, the intracellular signalling pathway behaves as if the signal is present on the receptor permanently. Such an oncogene is v-*erbB*, which codes for an abnormal EGF receptor. This lacks the external domain but the cytoplasmic domain is permanently phosphorylated (Fig. 26.20). Thus, any one of these oncogenic proteins, if present, makes the Ras pathway behave in an uncontrolled fashion. Oncogenes are not solely acquired from retroviral infection—normal protooncogenes can become oncogenes as a result of mutations.

Conversion of protooncogenes to oncogenes can occur by several mechanisms other than by single-base mutations. Chromosomal rearrangements can result in translocation of genes involved in control of cell differentiation and cell division. This can cause oncogene formation in several ways.

1. A protooncogene may be placed under the control of a different regulatory element causing overexpression of a particular protein. Burkitt's lymphoma, for example, is due to the c-*myc* gene being put under the control of an immunoglobulin gene promoter.

2. Translocation may cause gene fusion and result in the production of an oncogenic hybrid protein.

3. Gene amplification may cause overproduction of a control protein.

Cancer is the result of a cell's acquiring multiple genetic abnormalities. In addition to the conversion of protooncogenes to oncogenes, cancer production often involves the loss of a functional tumour suppressor gene. About half of all human cancers are associated with loss of the p53 tumour suppressor gene. The p53 gene product has complex effects, including the stimulation of DNA repair mechanisms, but, most crucially, it somehow controls the transition from the G_1 to the S phase of the cell cycle and thus is involved in the prevention of uncontrolled cell division.

A note on terminology

As stated earlier, oncogenes are named after the retroviruses carrying them (and printed in italics). Thus, *ras* was found in the *Rat Sarcoma* virus. The oncogene is identified by the letter 'v' (for virus) and the protooncogene by 'c', for cytoplasmic. Thus we have v-*ras* and c-*ras*, etc. The derived proteins are written as Ras, Raf, etc. (not printed in italics).

The **SOS protein** is named after a receptor mutant in fruit flies called '*sevenless*'; a protein required for this receptor function was called Son of sevenless—hence the name SOS; it was subsequently found to be present in eukaryotes in general. It is not to be confused with the SOS response of *E. coli* (page 322).

Why is the Ras pathway so long?

It is tempting to speculate that the protein kinase cascade is an amplifying mechanism, analogous to those in glycogen breakdown (page 173) and blood clotting (page 214).

In addition, and most importantly, it is possible that the Ras pathway interconnects at various points with other pathways. Evidence is mounting that extensive 'crosstalk' occurs. Thus PKC, activated by the phosphoinositide pathway, can phosphorylate and activate Raf. The various protein kinases in the Ras pathway may have other target proteins, thus activating alternative pathways. In addition, the activated tyrosine kinase type of receptor may associate with SH2-proteins of other signalling pathways so that one receptor might activate several pathways (Fig. 26.21). To use an electrical analogy, the Ras and

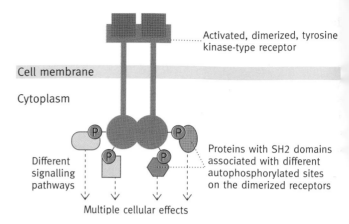

Fig. 26.21 Hypothetical diagram of how multiple proteins with SH2 domains of target proteins might function to link multiple pathways to an activated (phosphorylated) receptor.

other signalling pathways are electronic circuits on a very complex board with crosslinking connections, but we cannot yet see how the whole functions as an integrated control system. The salient feature of the signalling pathways is that they involve protein–protein interactions that probably form networks as complex as metabolic pathways.

In stark contrast to the Ras pathway is another receptor-mediated intracellular signalling pathway of relative simplicity in which tyrosine kinase activity is involved. We will now describe this.

A direct intracellular signalling pathway from receptor to nucleus

Signalling by interferons

As long ago as 1957 it was reported that vertebrate cells, infected with a virus, secrete proteins called **interferons** that protect cells from the same or other unrelated viruses (called interference). Ten or 20 different interferons are known. The antiviral activity of the interferons is not fully understood, but it is known that there are **interferon-stimulated response elements (ISREs)** on genes, that is, interferon attachment to a cell receptor causes specific gene responses.

γ-Interferon has a different type of receptor from that described for EGF. It has two cytoplasmic tyrosine kinase molecules (JAK kinases) associated with it. Ligand binding causes dimerization of receptors; this activates the associated kinases, and, by a complex series of phosphorylations, tyrosine residues of the receptor dimer are phosphorylated. This, in turn, causes two proteins in the cytosol to bind to the phosphorylated receptor. The proteins have the same SH2 domains as that of the GRB protein of the Ras pathway (page 362); as already stated, the SH2 domain is now recognized as the structure found in proteins that bind to phosphorylated receptors of the tyrosine type. The protein kinases, still bound to the receptor, phosphorylate the two SH2 proteins, which now move into he nucleus where they assemble into an active transcriptional actor and promote transcription of ISRE elements (Fig. 26.22).

How general is this direct pathway?

A major question is whether this system is a widespread one or elatively uncommon and specific to a few signals. If widespread, t might provide the mechanism by which unlimited numbers of ignals could have specific effects on cells. It is known that a second interferon and several cytokines involved in control of

blood cell production use this type of mechanism. It is possible to envisage that different phosphorylated receptors could bind different cytosolic SH2 proteins that could then be converted to different specific transcriptional factors and provide the basis for the transduction of a wide variety of signals into specific cellular controls.

A suitable concluding remark might be that from a recent *Nature* review on this topic that stated, 'in the next few years there will be hundreds and then thousands of papers describing how diverse proteins funnel their signals into intracellular signalling pathways.'

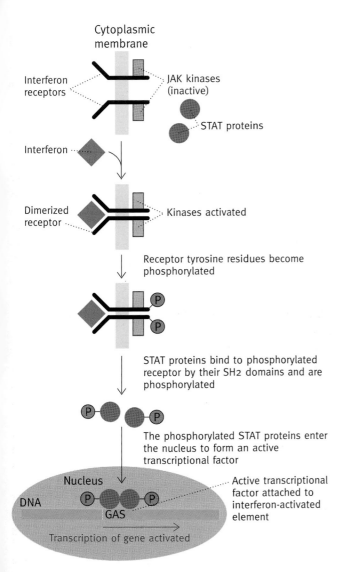

Fig. 26.22 Diagram of the signalling pathway by which γ-interferon activates specific gene transcription. STAT, Signal transducer and activator of transcription; GAS, γ-interferon-activated sequence element. Note that, unlike the EGF receptor (Fig. 26.18), which is a tyrosine kinase, in this pathway the dimerized receptor activates cytoplasmic tyrosine kinases; these are known as Janus kinases (JAK kinases) after the Roman god with two faces. (JAK kinases have two kinase sites.) α-Interferon and several cytokines have analogous signalling pathways but may differ somewhat in detail.

Extrinsic signals that control ligand-gated ion channels

We come now to a quite different type of control in which the immediate result of **ligand binding** to a membrane receptor is to open a **gated channel** in the membrane (Fig. 26.23). Neuronal signalling will be used to illustrate this.

How does acetylcholine binding to a membrane receptor result in a nerve impulse?

As mentioned earlier, the motor neuron, which signals voluntary striated muscle to contract, receives an **acetylcholine signal** from the presynaptic neuron. Let us start at the point at which acetylcholine is released at a synapse and triggers the nerve impulse to flow in the next neuron, which causes the muscle to contract (see Fig. 26.24).

Neurons, like other animal cells, have a high K^+ level inside and a low level outside, while the Na^+ level is high outside and low inside. This is the result of the Na^+/K^+ ATPase ion pump in the membrane (page 50).

In the resting stage (where nothing is happening in terms of nerve impulses) the cell membrane is more permeable to K^+ than to Na^+. K^+, high in concentration inside, therefore tends to leak out but, since the membrane is impermeable to anions, the latter do not leak out. This creates a positive charge outside and a negative charge inside; the membrane is said to be polarized. The K^+ leakage is self-limiting for the charge difference created

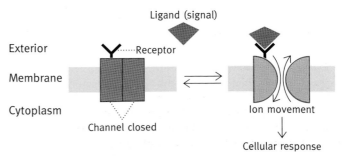

Fig. 26.23 Diagram of a ligand-gated channel. Channels specific for Ca^{2+}, Na^+, and K^+ ions are common.

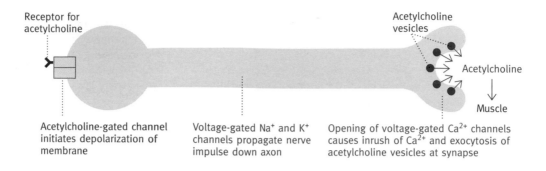

Receptor for acetylcholine

Acetylcholine vesicles

Acetylcholine

Muscle

Acetylcholine-gated channel initiates depolarization of membrane

Voltage-gated Na⁺ and K⁺ channels propagate nerve impulse down axon

Opening of voltage-gated Ca²⁺ channels causes inrush of Ca²⁺ and exocytosis of acetylcholine vesicles at synapse

Fig. 26.24 Simplified diagram of the chain of events by which acetylcholine stimulation of a motor neurone results in liberation of acetylcholine at the neuromuscular junction to trigger striated muscle contraction.

opposes further leakage. An equilibrium is achieved at which the **resting cell potential** across the membrane is about -70 mV (that is more negative inside).

The acetylcholine receptor is a gated Na^+/K^+ channel. It transitorily opens when acetylcholine binds to it and allows Na^+ to move in and K^+ out (Fig. 26.24). However, because of its higher concentration, Na^+ moves in faster than K^+ moves out and reduces the negative charge inside; the membrane is now said to be **depolarized**. This affects only the membrane in the immediate vicinity of the acetylcholine receptors in the synaptic region. It is the initial stimulus.

In effect, the nerve impulse moves along the axon by a repeat of the membrane depolarization caused by acetylcholine, but different types of Na^+ and K^+ channels are involved, called **voltage-gated channels**, which open when the resting cell potential falls (from about -70 to -40 mV). The voltage change occurs in a narrowly localized zone of the membrane and triggers the opening only of the adjacent voltage-gated channels. The duration of opening is very brief and, after closing, there is a recovery period before the channels can respond again to a signal. The depolarization can only go forward. The Na^+/K^+ pump restores the situation after the wave has passed. (The mechanism of nerve conduction beyond this general summary is regarded as beyond the scope of this book.) In this way, a wave of membrane depolarization passes down the axon to the neuronal endings. In the membrane of the latter, are voltage-gated Ca^{2+} channels that open when the depolarization reaches them, permitting extracellular Ca^{2+} to enter the cell. This results in vesicles containing stored acetylcholine fusing with the cell membrane to release their contents by **exocytosis** (Fig. 26.24). A $Ca^{2+}/ATPase$ ejects Ca^{2+} to restore the resting situation. The released acetylcholine binds to muscle cell receptors, triggering the events described in Chapter 28.

This Ca^{2+}-induced exocytosis of vesicles has a more general biological role than simply that involved in neurotransmitter release. Ca^{2+}-stimulated exocytosis of vesicle content occurs in the release of digestive enzymes from the pancreas; only here the Ca^{2+} release is via the phosphoinositide pathway (page 359).

Also fertilization of an ovum causes the Ca^{2+}-induced exocytosis of vesicles that forms a barrier to further sperm penetration.

Another important area in which ligand-gated pore control is involved is in vision, described below.

Vision: a process dependent on ligand-gated pore control

The basic problem of vision is to convert the stimulus of light photons into chemical changes that result in impulses in the optic nerve carrying signals to the brain.

The retina of vertebrates has two types of cells for light detection, namely **rods** for black and white and dim light vision and **cones** for colour and bright light vision. The rod cell has

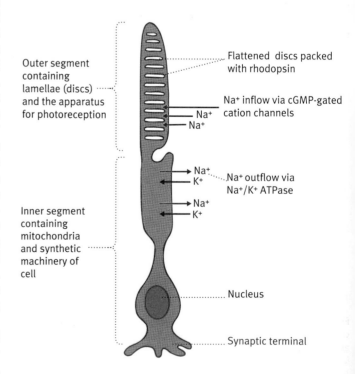

Outer segment containing lamellae (discs) and the apparatus for photoreception

Flattened discs packed with rhodopsin

Na⁺ inflow via cGMP-gated cation channels

Na⁺

Na⁺

Na⁺

K⁺

Na⁺ outflow via Na⁺/K⁺ ATPase

Na⁺

K⁺

Inner segment containing mitochondria and synthetic machinery of cell

Nucleus

Synaptic terminal

Fig. 26.25 Structure of a rod cell.

three sections. The middle part has the usual complement of organelles. One end of the cell makes a synapse with a bipolar cell that connects with the optic nerve. At the other end is a cylindrical rod-shaped section in which there is a stack of membranous discs (as many as 2000) embedded in the cytoplasm (Fig. 26.25). These contain the light detection machinery.

Transduction of the light signal

The process of light detection is complex in detail; we will here omit the detail and deal with the essential principles only (Fig. 26.26).

In the dark, the cone cells have a relatively high level of cGMP synthesized by a guanylate cyclase (Fig. 26.14). In the cell membrane there are ligand-gated cation channels that are kept *open* by cGMP.

The constant inrush of Na^+ in the dark means that the cell membrane has a lower potential across it—the equilibrium between the inrush and pumping out of Na^+ establishes a potential of −30 mV (quite different from the situation in a neuron where, at rest, Na^+ channels are closed and a membrane potential of −70 mV exists).

The light receptor in the rod discs is **rhodopsin**, a complex of the protein opsin and 11-*cis*-retinal. Light causes activation of rhodopsin due to a conformational change in the visual pigment to become all-*trans*-retinal. Have a look at the structures in Fig. 26.27 and the relationship of the retinal molecule to vitamin A and β-carotene, also shown there.

When this happens, a series of events occurs, involving a trimeric G protein (and GTP hydrolysis in the usual 'clock'

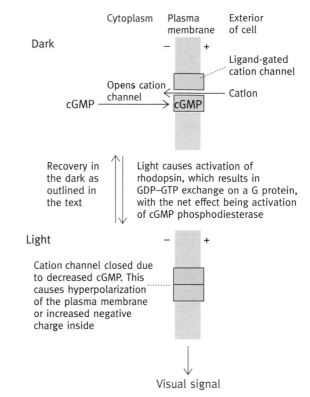

Fig. 26.26 Simplified diagram of the visual process. The action of light on rhodopsin results in the activation of cGMP phosphodiesterase. Decreased cGMP results in the closure of a cation channel, hyperpolarization of the cell membrane, and an optic nerve impulse.

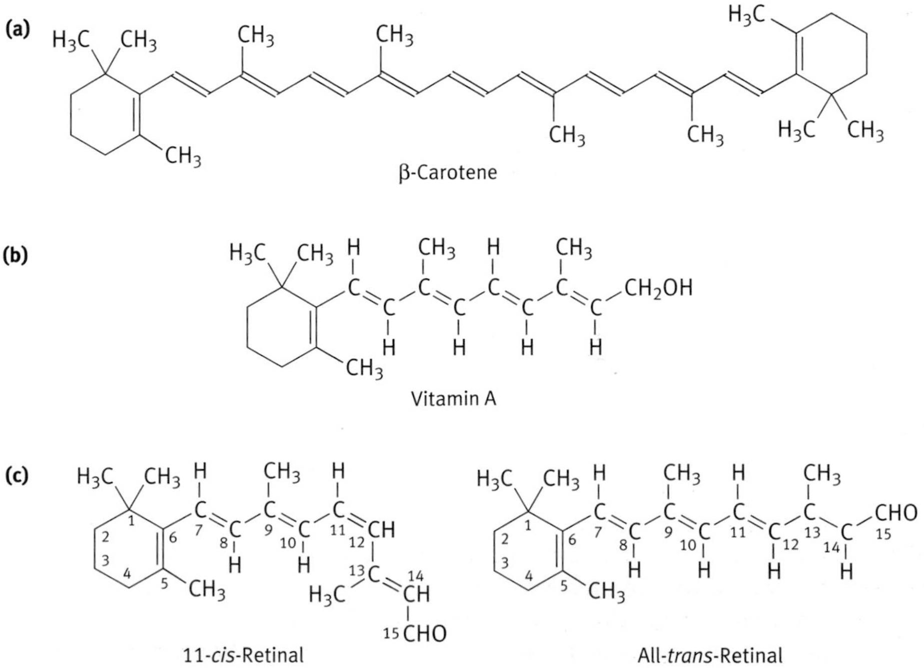

(a)

β-Carotene

(b)

Vitamin A

(c)

11-*cis*-Retinal

All-*trans*-Retinal

Fig. 26.27 (a) The structure of β-carotene, which is the principal accessory photosynthetic pigment in plants and a precursor of vitamin A in animals. **(b)** The structure of vitamin A. **(c)** The structures of 11-*cis*-retinal and all-*trans*-retinal.

role). The net effect is the activation of an enzyme that destroys cGMP by hydrolysis. This enzyme (**cGMP phosphodiesterase**) has a similar action to that of the cAMP phosphodiesterase shown in Fig. 26.5(b). Lowering cGMP results in the closure of the Na^+ channels and a hyperpolarization of the membrane occurs. This is collected as a visual signal into the optic nerve.

The recovery of the cell after illumination is complex. It involves: (1) inactivation of the activated rhodopsin; (2) lowering of the Ca^{2+} level in the cell, which stimulates cGMP synthesis; and (3) recycling of the rhodopsin by a complex pathway. The net effect is that, in light, cGMP levels fall, Na^+ channels close, and hyperpolarization of the cell membrane leads to visual signal in the optic nerve. After illumination, cGMP levels are restored, Na^+ channels open, and we are ready for the next photon.

Further reading

Signal transduction

1992. *Trends Biochem. Sci.*, **17**, 367–443.
A complete issue devoted to all aspects of cell signalling including membrane receptors, second messenger generation and destruction, protein kinases and phosphatases, and activation of transcription factors.

G proteins

Linder, M. E. and Gilman, A. G. (1992). G proteins. *Sci. Amer.*, **267**(1), 36–43.
The G proteins on the internal surface of the cell's outer membrane coordinate cellular responses to a multitude of signals. An excellent review.

Receptors

Lincoln, T. M. and Cornwell, T. L. (1993). Intracellular cGMP receptor proteins. *FASEB J.*, **7**, 328–38.
Describes the diverse biological roles of cGMP as a second messenger.

Pawson, T. (1994). Look at a tyrosine kinase. *Nature*, **372**, 726–7.
A news and views item on the three dimensional structure of the insulin receptor.

Heldin, C.-H. (1995). Dimerisation of cell surface receptors in signal transduction. *Cell*, **80**, 213–23.
Dimerisation or oligomerisation of receptors is the initial response to many growth factors and cytokines. This article reviews the important field.

Nitric oxide

Snyder, S. H. and Bredt, D. S. (1992). Biological roles of nitric oxide. *Sci. Amer.*, **266**(5), 28–35.
Excellently illustrated article on the roles and mechanism of action of nitric oxide.

Bredt, D. S. and Snyder, S. H. (1994). Nitric oxide: a physiologic messenger molecule. *Ann. Rev. Biochem.*, **63**, 175–95.
Readable general review, but particularly valuable for the survey of the several roles of nitric oxide in the whole body.

Signalling pathways

Feg, L. A. (1993). The many roads that lead to Ras. *Science*, **260**, 767–9.
Reviews different signalling pathways converging on the Ras protein.

Nishizuka, Y. (1994). Protein kinase C and lipid signalling for sustained cellular responses. *FASEB J.*, **9**, 484–96.
Excellent summary of the signalling pathways using inositol phospholipid hydrolysis, and the degradation of various other membrane lipid constituents.

Karin, M. and Hunter, T. (1995). Transcriptional control by protein phosphorylation: signal transmission from the cell surface to the nucleus. *Current Biology*, **5**, 747–57.
Well illustrated comparison of the Ras pathway involving translocation of activated protein kinases from the cytoplasm to the nucleus and the JAK kinase pathway involving translocation of activated transcription factors into the nucleus.

Quilliam, L. A., Khosravi-Far, R., Huff, S. Y., and Der, C. J. (1995). Guanine nucleotide exchange factors: activators of the Ras superfamily of proteins. *BioEssays*, **17**, 395–404.
A fairly advanced, but readable review of the control of Ras activity.

Burgering, B. M. T. and Bos, J. L. (1995). Regulation of Ras-mediated signalling: more than one way to skin a cat. *Trends Biochem. Sci.*, **22**, 18–22.
Extensively reviews interaction of the Ras pathway with other signalling routes.

Ihle, J. N. (1996). STATs and MAPKs: obligate or opportunistic partners in signalling. *BioEssays*, **18**, 95–8.
A speculative discussion on the possible cooperation of the Ras and STAT signalling pathways.

Hemopoietic growth factors

Metcalf, D. (1991). The 1991 Florey Lecture. The colony-stimulating factors: discovery to clinical use. *Phil. Trans. Series B, Roy. Soc., Lond.*, **333**, 147–73.
An in depth review by the discoverer of these important proteins. While there is more research level material than most

students would want, it provides an excellent general review of the field, and a fascinating account of the progression from unidentified factors to their widespread clinical use.

Interferons

Johnson, H. M., Bazer, F. W., Szente, B. E., and Jarpe, M. A. (1994). How interferons fight disease. *Sci. Amer.*, **270**(5), 40–9.
Describes structure of interferons and the signalling pathways by which they influence cells.

Ihle, J. N., Witthuhn, B. A., Quelle, F. W., Yamamoto, K., Thierfelder, W. E., Kreider, B., and Silvennoinen, O. (1994). Signalling by the cytokine receptor superfamily: JAKs and STATs. *Trends Biochem. Sci.*, **19**, 222–7.
Comprehensive review on the signalling pathway from cell membrane to nucleus for interferons, interleukins, colony stimulating factors, erythropoietin, and growth hormone.

Calcium homeostasis

Bootman, M. D. and Berridge, M. J. (1995). The elemental principles of calcium signalling. *Cell*, **83**, 675–8.
Minireview on the sources of Ca^{2+} for generating signals: Ca^{2+} release from intracellular stores and Ca^{2+} entry across the plasma membrane.

Monteith, G. R. and Roufogalis, B. D. (1995). The plasma membrane calcium pump—a physiological perspective on its regulation. *Cell Calcium*, **18**, 459–70.
The role of the plasma membrane Ca^{2+}/Mg^{2+}-dependent ATPase in cellular signalling and in intracellular calcium homeostasis.

Oncogenes

Cavanee, W. K. and White, R. L. (1995). The genetic basis of cancer. *Sci. Amer.*, **272**(3), 50–7.
Describes the accumulation of genetic defects that can cause normal cells to become cancerous.

The cell cycle and cancer

Elledge, R. M. and Lee, W.-H. (1995). Life and death by p53. *BioEssays*, **17**, 923–30.
Summarizes the many functions of the p53 gene product, including its role in cancer suppression.

Nigg, E. A. (1995). Cyclin-dependent protein kinases: key regulators of the eukaryotic cell cycle. *BioEssays*, **17**, 471–80.
Describes the role of phosphorylation by cyclin-dependent protein kinases in the regulation of the cell cycle.

Ion channels

Neher, E. and Sakmann, B. (1992). The patch clamp technique. *Sci. Amer.*, **266**(3), 44–51.
Enables studies to be made on ligand-gated and voltage-gated ion channels.

Changeux, J.-P. (1993). Chemical signalling in the brain. *Sci. Amer.*, **269**(5), 30–7.
The acetylcholine receptor and how it works.

Catterall, W. A. (1995). Structure and function of voltage-gated ion channels. *Ann. Rev. Biochem.*, **64**, 493–531.
Covers sodium, calcium, and potassium channels.

Problems for Chapter 26

1 What are the classes of extrinsic signalling molecules between cells?

2 By means of a simple diagram, compare the action on cells of lipid-soluble and lipid-insoluble extrinsic signalling molecules.

3 How is cAMP production controlled? In particular, describe the role of GTP in the process. What is the relevance of the latter to cholera?

4 How does cAMP exert its effect as second messenger?

5 How does nitric oxide exert its controlling effect?

6 cAMP and cGMP are not the only second messengers. Describe another system.

7 Important gene-activating control pathways exist that do not involve small molecular weight second messengers. Describe such a pathway that is of widespread importance.

8 If a protein is found to have an SH2 domain, what is its likely role?

9 Explain how abnormal components of cellular signalling pathways may promote cancer.

10 Interferon controls gene activity by a pathway quite different from the Ras pathway. Describe this.

11 Make a general comparison of the activation mechanism of three types of receptors: (a) the epinephrine receptor activating adenylate cyclase; (b) the EGF receptor of the Ras pathway; and (c) the interferon receptor.

12 What is a voltage-gated Ca^{2+} channel? Give an example.

13 What is a ligand-gated channel? Give an example of one in the visual process.

Chapter summary

Previous page: Scanning electron micrograph of human red blood cells (erythrocytes). The cells, packed with hemoglobin, transport oxygen from lungs to tissue, and are involved in carbon dioxide transport back to the lungs. The biconcave shape maximizes the surface area available for gas exchange across the membrane. The nucleus disappears as the red blood cell matures.

Photograph: CNRI/Science Photo Library

Chapter 27

The red blood cell and the role of hemoglobin

The centrally important molecule in the delivery of oxygen from the lungs to the tissues is hemoglobin, which is not just a protein molecule passively capable of combining with oxygen but rather is a sophisticated molecular machine. The red blood cells need to pick up a maximum load of oxygen in the lungs, which implies a high oxygen affinity, and surrender it in the tissue capillaries, which implies a relatively low oxygen affinity. Conversely, it must pick up CO_2 in the tissues and surrender it in the lungs. Because hemoglobin has been studied so intensively, it is the best understood protein and a great deal is understood as to how physiological needs are satisfied.

In this chapter we will discuss several topics all involved with the gas transport system. First, the red cell and its production will be dealt with. Hemoglobin is a protein (globin) combined with heme; the synthesis of heme and its control will therefore be discussed, which, in turn, leads us to iron biochemistry. The control of synthesis of globin will then be discussed. Finally, we will turn to the major topic of how hemoglobin functions.

The red blood cell

The human body produces about 160 million red cells or **erythrocytes** per minute; they circulate for about 110 days and are then destroyed. The mammalian red cell is a flattened biconcave disc lacking a nucleus (Fig. 27.1). The shape is maintained by the submembranous cytoskeleton (see page 52), giving the cell a larger surface area than would a spherical shape. This, together with the smaller diffusion distances involved, facilitates gaseous exchange in and out of the cell. (The total surface area of a human's red cells is about 4000 m^2.)

In the fetus, the liver and spleen produce red cells but in the adult this occurs in the red marrow of flat bones in which **hemopoietic stem cells** continually multiply and produce progenitors of all types of blood cells, as is illustrated in Fig. 25.1. (A stem cell is one that continually reproduces itself, its progeny being capable of differentiating.) Individual cells become irreversibly committed to differentiate into one or other blood cell types (erythrocytes, different white cells, and blood platelets). The production of the various blood cell types is regulated. Different regulatory proteins have been isolated that have the effect of causing multiplication of specific progenitor cells. The protein **erythropoietin**, produced by the kidney medulla, stimulates erythrocyte production, and, since the rate of erythropoietin production is inversely controlled by oxygen tension in the tissue, automatic regulation of red blood cell levels occurs—in oxygen deficiency more hormone and therefore more red cells are produced.

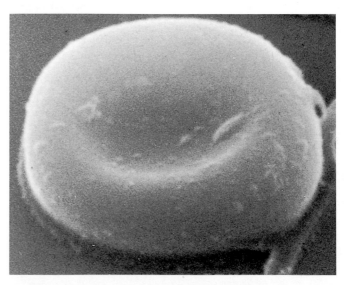

Fig. 27.1 Scanning electron micrograph of an erythrocyte. Kindly provided by Professor W. G. Breed, Department of Anatomy, University of Adelaide.

The early cell stages of the erythrocyte are nucleated, but later, in mammals, the nucleus is lost to form the immature red cell known as the **reticulocyte**, which is released into circulation. Reticulocytes contain mRNA and protein synthesis occurs until the cell finally matures into the erythrocyte. When cells are stained, the ribosomes and mRNA result in a reticulum of dark patches—hence the name reticulocyte. The mature cell has no mitochondria and relies on glycolysis for energy production, producing lactate; it has the pentose phosphate pathway to supply NADPH (see page 181).

Heme and its synthesis

The full structure of heme, the complete details of which probably few biochemists carry in their heads, is given in Fig. 27.3; you do not need to learn this detailed structure. It is a ferrous iron complex with protoporphyrin (Fig. 27.2). The latter is a tetrapyrrole, the four substituted pyrroles being linked by methene ($=$CH$-$) bridges such that a **conjugated double bond system** exists (that is, you can go right round the molecule via alternating single and double bonds). This gives protoporphyrin and heme their deep red colour.

In heme, the four pyrrole N atoms are bound to Fe^{2+} as shown in Fig. 27.4, leaving two more of the six ligand positions of the Fe^{2+} available for other purposes.

Synthesis of heme

Red blood cells have the vast majority of the body's heme content, so this topic fits well here though heme synthesis occurs in all aerobic cells to supply prosthetic groups for cytochromes and other proteins.

The synthesis of heme appears to be a formidable task but the essentials are surprisingly simple, requiring in animals only two starting reactants, glycine and succinyl-CoA. You have met the latter in the citric acid cycle (page 118). An enzyme,

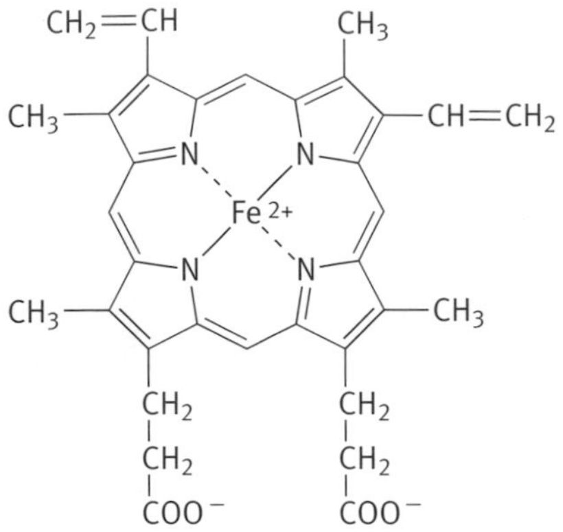

Fig. 27.3 The structure of heme. This form is found in hemoglobin; other hemes differing in their side chains are found in cytochromes. Note that, at one side of the molecule, there are two hydrophilic propionate groups ($-CH_2CH_2COO^-$) while the remaining side groups are all hydrophobic. In myoglobin, heme sits in a cleft of the molecule with the hydrophilic side pointing out towards water and the hydrophobic groups buried into the nonpolar interior of the protein.

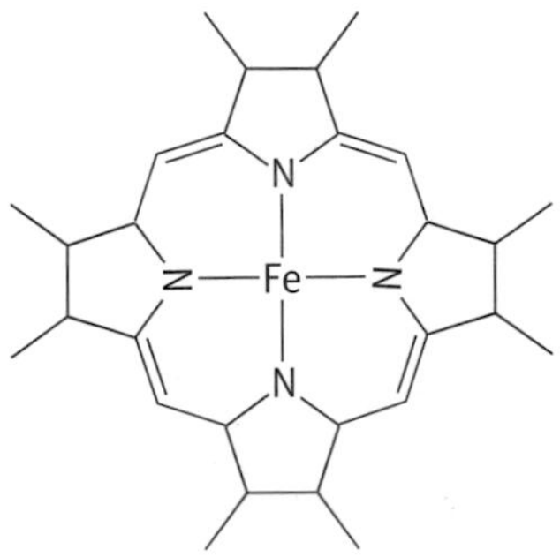

aminolevulinate synthase, or ALA-S, carries out the reaction shown in Fig. 27.5. 5-Aminolevulinic acid (ALA) is the precursor solely committed to porphyrin synthesis. Two molecules of ALA are used to form the pyrrole, **porphobilinogen (PBG)**, the two molecules being dehydrated by ALA dehydratase (Fig. 27.6). The remainder of the heme biosynthetic pathway consists of linking four PBG molecules together, modifying the side groups, and chelating an atom of ferrous iron to form heme. The intermediate tetrapyrroles between PBG and heme are the colourless uro- and coproporphyrinogens (methylene bridges) and the red protoporphyrin (methene bridges). The pathway is given in Fig. 27.7; have a quick look at this.

The heme synthesis pathway has a curious feature. The first reaction, synthesis of ALA, occurs inside mitochondria, after which the ALA moves out to the cytoplasm, but the final three steps also occur in the mitochondria. Why this should be so is not clear.

Fig. 27.2 Outline of the heme molecule. (The full structure is given in Fig. 27.3.)

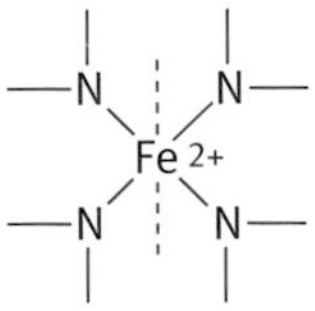

Fig. 27.4 Binding ability of Fe^{2+} in heme. The iron atom in heme can bond to six ligands in total: four bonds to the pyrrole nitrogen atoms as shown and the other two above and below the plane of the page.

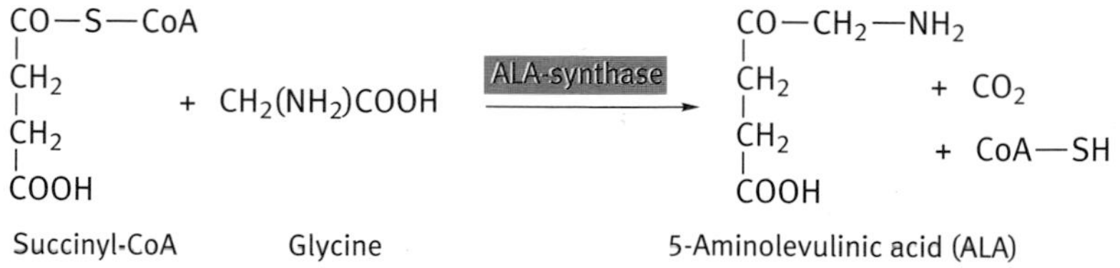

Fig. 27.5 The first step in heme synthesis, catalysed by 5-aminolevulinate synthase (ALA-synthase). For clarity of illustration, nonionized structures are given.

Regulation of heme biosynthesis and iron supply to the erythrocyte

The control of heme biosynthesis is of unusual interest and is intimately tied up with regulation of iron uptake and storage in the cell.

The situation is as follows.

- ALA synthase activity is the rate-limiting step in heme synthesis.
- ALA synthase synthesis and therefore ALA synthase levels are controlled by Fe levels.
- Fe levels are controlled by transferrin receptor levels.
- Fe levels feedback-control the synthesis of transferrin receptor protein.

The rate-limiting step is, as mentioned, the one first committed to heme production—ALA synthesis. The synthesis of the enzyme ALA synthase itself is regulated in erythrocytes at the translational level. mRNA for the enzyme has, at its $5'$ untranslated end (that is, the translational initiation end), a hairpin loop called an **iron-responsive element** or **IRE**. At low iron levels, an IRE binding protein attaches to the IRE and prevents translation (Fig. 27.8). At higher cellular levels of iron, the latter complexes with an iron–sulfur cluster (page 122) which is attached to the protein, causing release of the protein, and translation occurs, thus coordinating the rate of heme

synthesis with the iron supply. Heme synthesis starts in the nucleated proerythroblast, well before the reticulocyte stage, so that mRNA translation occurs in these cells and continues into the reticulocyte stage.

The question now follows as to what controls the cellular iron level. Iron is transported in the blood plasma as a complex with the protein, **transferrin**, produced in the liver. This is taken up by receptor-mediated endocytosis (page 210). In *non-erythroid cells*, uptake is known to be regulated by control of the synthesis of the receptor protein by cellular iron, which regulates the stability of the transferrin receptor-protein mRNA. The mRNA for the receptor protein has iron-responsive elements at the $3'$ end of the molecule. In the presence of low cellular levels of iron, the IRE-binding protein binds to the receptor-protein mRNA, which apparently protects the mRNA from degradation. In conditions of high iron levels, the binding protein detaches and removes the protection. Thus, iron causes downregulation or reduction in the number of transferrin receptors which, in turn, reduces iron intake. At low iron levels the reverse applies. It is not established whether this applies to erythroid cells, but it is known that red blood cells have transferrin receptors that increase in number as the cell develops.

Iron is stored in red cells as **ferritin**—a complex of the protein, apoferritin, and inorganic iron. (The liver is also an important site for ferritin storage.) It is interesting that apoferritin synthesis is regulated by a mechanism essentially the same as that for ALA synthase—by iron causing detachment of an IRE-binding protein for apoferritin mRNA, permitting synthesis of the protein.

As a slight diversion from blood cells (because this is a convenient place for it), all cells of the body, so far as we know, synthesize heme by the same route; it is needed for cytochromes of the electron transport chain (page 121) and liver cytochrome P450 (page 216) and other enzymes. Red cells have the greatest need, for the mature cell is packed with hemoglobin; liver is next because of its large cytochrome P450 content (page 217). The ALA synthase of erythroid cells is different from that of the 'housekeeping' version (as that in other cells is called). The reaction catalysed is the same, but the control of the ALA-S reaction is different. In liver, synthesis of ALA-S is controlled by heme, which destabilizes its mRNA and also inhibits transport of the enzyme into mitochondria, its functional site. Its mRNA,

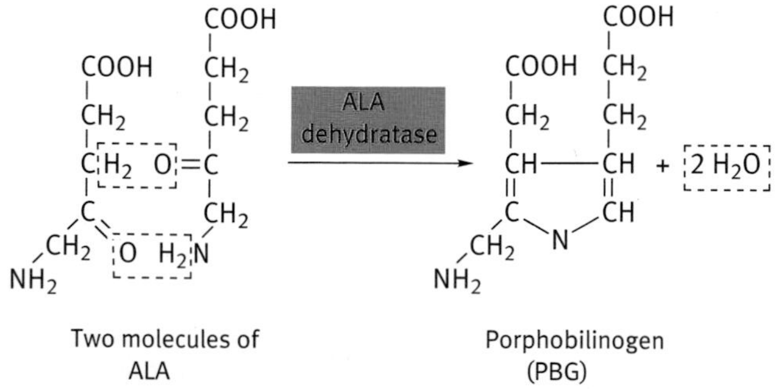

Fig. 27.6 Synthesis of porphobilinogen (PBG)—the ALA-dehydratase step. Nonionized structures are given for simplicity. PBG is a monopyrrole; heme is a tetrapyrrole. The conversion of PBG to heme is shown in Fig. 27.7.

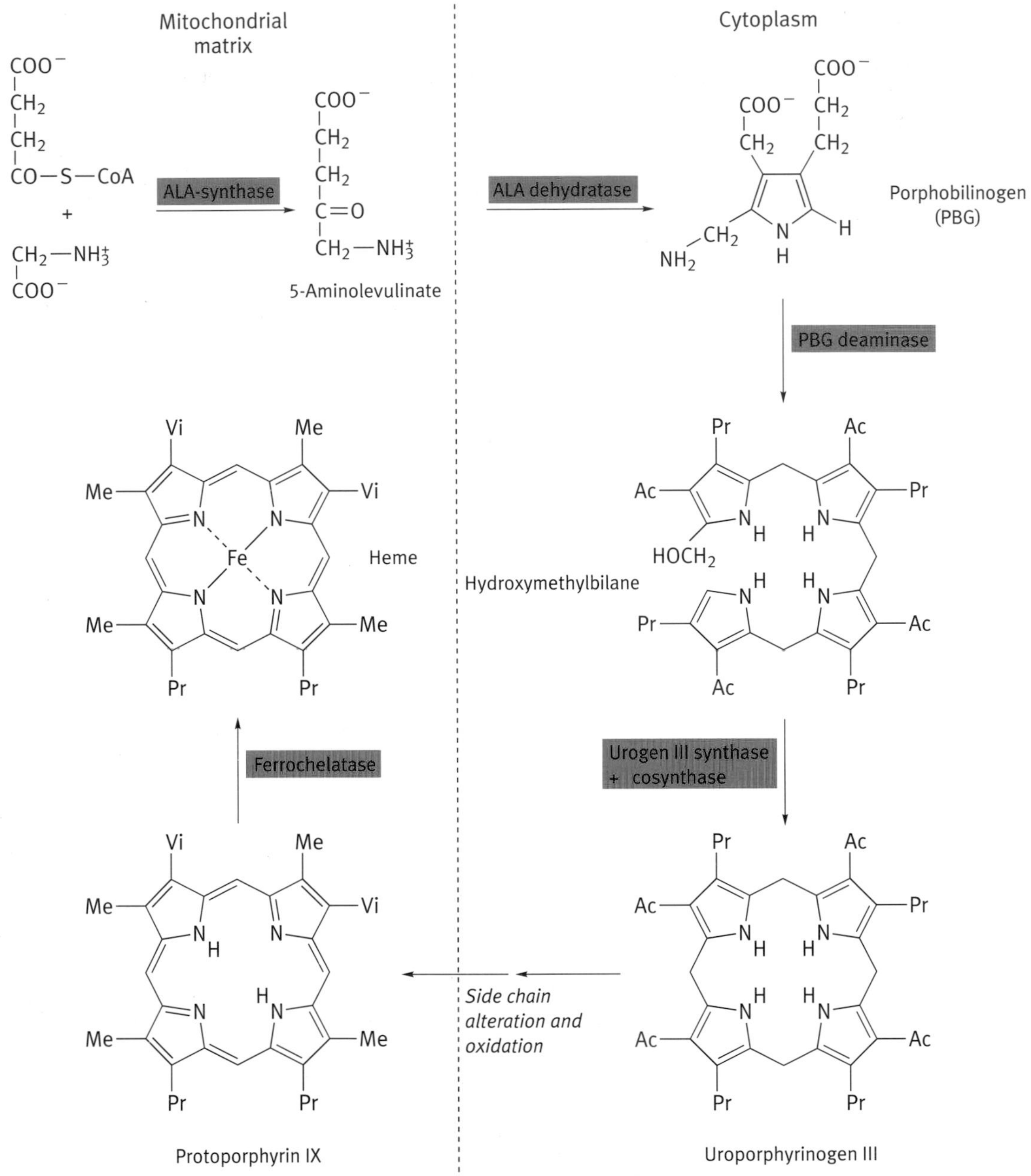

Fig. 27.7 An abbreviated pathway of heme biosynthesis, given for reference purposes.

however, has no iron-responsive elements, as has the erythroid version. Overproduction of hepatic ALA synthase and high blood levels of ALA in acute intermittent porphyria ('mad king' disease) patients correlate with the onset of neurological symptoms in that disease, though the molecular basis of the symptoms is not known. Such patients have a partial block in the heme biosynthetic pathway. Chlorophyll is also a tetra-pyrrole but, interestingly, ALA is synthesized in plants by a quite different reaction.

After that diversion, we'll now return to the red blood cell.

Destruction of heme

Red blood cells are destroyed mainly by **reticuloendothelial cells** of spleen, lymph nodes, bone marrow, and liver. Removal of sialic acid groups from the red cell membrane glycoproteins is a signal that the cell is aged and ready for destruction. The degraded carbohydrate attaches to cell receptors, which leads to endocytosis of the erythrocyte. The enzyme, **heme oxygenase** opens up the tetrapyrrole ring, releasing the iron for re-use and forming biliverdin, a linear tetrapyrrole (Fig. 27.9). Biliverdin is reduced to **bilirubin**. This is water-insoluble but is transported

Situation A: Fe concentration in the cell is high

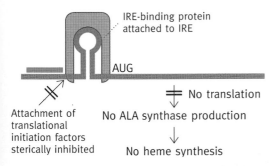

Situation B: Fe concentration in the cell is low

Fig. 27.8 Diagram of heme synthesis control by Fe levels in the red blood cell. The regulatory mechanism presumably depends on the ALA synthase having a short half-life in the cell so that, once synthesis of the enzyme stops, heme synthesis stops shortly afterwards. The enzyme is known to be unstable in reticulocyte lysates *in vitro* and presumably is so in red blood cells. The question of how Fe levels in the cell are controlled is dealt with separately. IRE, Iron-responsive element, a stem loop structure in the mRNA. When the IRE-binding protein attaches to the IRE, it probably sterically inhibits the attachment of the translational initiation proteins. Binding of Fe to the IRE-binding protein abolishes its affinity for the IRE.

in the blood, attached to serum albumin, to the liver, where it is rendered much more polar by the addition of two glucuronate groups (see page 216) and then excreted, in the bile, into the gut; modification and partial re-absorption of some bile compounds leads to the yellow colour of urine. (Bilirubin has been postulated to function as an antioxidant—see page 218.)

How is globin synthesis regulated?

Hemoglobin consists of heme attached to the protein globin. The two components, **globin** and **heme**, need to be produced in the correct relative amounts if excess of one or the other is to be avoided. A coordinating regulatory link exists. It has been demonstrated in experiments with reticulocyte lysates that, in

the absence of heme, a protein kinase phosphorylates one of the factors (eIF2) involved in initiation of protein synthesis (page 303). This has the effect of halting the initiation of translation in the red blood cell and thus prevents globin synthesis. In the presence of heme, the kinase is inactivated and a phosphatase restores the normal activity of the initiation factor. Thus globin synthesis can proceed only when heme is present. This mechanism is found specifically in erythroid cells.

Transport of gases in the blood

The solubility of oxygen in simple aqueous solution is grossly insufficient for tissues to be supplied with oxygen in this manner. A carrier is therefore needed in the blood. Likewise CO_2 cannot be carried from the tissues to the lungs in simple solution at a sufficient rate.

What is the chemical rationale for a heme–protein complex as the oxygen carrier?

Inorganic Fe^{2+}, but not Fe^{3+} ions, are capable of combining with oxygen. However, Fe^{2+} ions are spontaneously, and rapidly, oxidized to Fe^{3+} ions so that inorganic iron, on its own, is no good as an oxygen carrier. The Fe^{2+} in heme can also bind with oxygen, but again this rapidly oxidizes to the Fe^{3+} form of heme, known as **hematin**, so that *free* heme is no good for oxygen-carrying purposes either. For heme to oxidize, two molecules must associate on either side of an oxygen molecule. The binding of heme in a crevice of protein prevents this association so that the Fe^{2+} in the heme of hemoglobin is much more resistant to oxidation than is free heme. It is important to be clear that, while heme in the electron transport cytochromes (page 121) oscillates between the Fe^{3+} and the Fe^{2+} states, hemoglobin functions only in the Fe^{2+} state. Binding of oxygen does not affect this. The iron atom in hemoglobin is bonded to the four pyrrole nitrogens; the fifth and sixth coordination positions are directly above and below the plane of the heme structure; one of these is bonded to a histidine residue of the protein, the other is available for oxygen binding (Fig. 27.13).

Structure of myoglobin and its oxygen binding role

Myoglobin is the red pigment of muscles. A preliminary discussion of this helps the understanding of the more complex hemoglobin.

Myoglobin has the relatively simple job of providing a reservoir of oxygen in the muscle cell. It takes up oxygen from the blood and then gives it up to be used by the mitochondria inside the cells for energy generation. It consists of a single heme molecule bound to a single protein molecule. The protein is roughly spherical in shape and is made up of a series of α helices (Fig. 27.10(a), (b)) with the heme molecule situated in a crevice between two of these.

Figure 27.11 shows the percentage saturation of myoglobin at

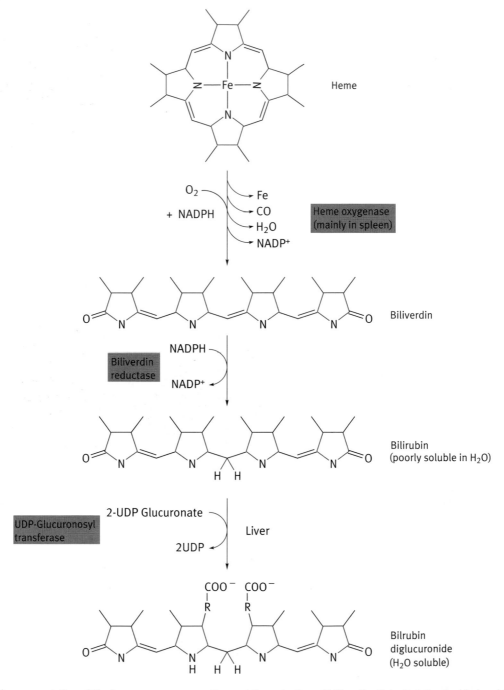

Fig. 27.9 Diagrammatic representation of the heme oxygenase reaction and the reduction of biliverdin. Note that the double bonds of the ring structures are omitted for simplicity. The reaction converts one methene group to carbon monoxide, thus opening up the ring structure. Biliverdin is reduced to bilirubin, conjugated with two glucuronic acid molecules, and excreted. There appears to be some uncertainty as to the stoichiometry of the heme oxygenase reaction; three O_2 molecules are believed to be involved. By analogy with cytochrome P450, it might be assumed that three molecules of NADPH are required to convert three oxygen atoms to H_2O.

increasing oxygen tensions; the result is a hyperbolic curve—the same, in fact, as that observed with an enzyme–substrate concentration curve, described earlier for the 'classical' type of enzyme (see page 162). The affinity of myoglobin for oxygen is high so that it readily extracts oxygen from oxyhemoglobin in the blood (see Fig. 27.11 for the relative affinities of the two carriers). In muscle cells, as the oxygen level falls due to electron transport and its conversion to H_2O, myoglobin gives up its oxygen. It is thus functionally a very simple carrier.

The three-dimensional structure of myoglobin was the first to

(a)

(b)

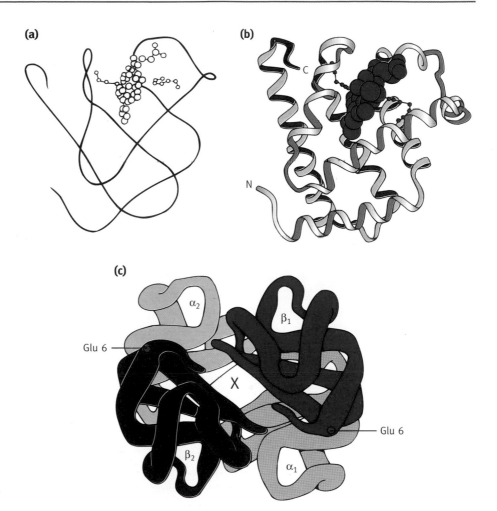

(c)

Fig. 27.10 Models of myoglobin and hemoglobin. **(a)** Simple computer-generated schematic diagram showing the folding of the polypeptide in myoglobin and the positioning of heme in the molecule. **(b)** A more elaborate diagram of the same thing. **(c)** Model of the subunits and their arrangement in hemoglobin showing the central cavity into which 2:3-bisphosphoglycerate fits in the deoxygenated state (marked X). In sickle cell anemia the two glutamic acid residues at position 6 in the β side chains are mutated to valines, creating hydrophobic patches on the molecule.

be determined for a protein and thus is a landmark in biochemistry.

Structure of hemoglobin

Hemoglobin is a tetramer of protein subunits (Fig. 27.10(c) each of whose folded polypeptide structures closely resembles that of myoglobin and each of which has a similarly located heme molecule capable of binding oxygen. In normal adult hemoglobin, which is the type we will discuss here, the two α subunits are identical, as also are the two β subunits.

Binding of oxygen to hemoglobin

A molecule of hemoglobin binds four molecules of oxygen, one per subunit. When an oxygen-saturation curve is plotted, instead of the hyperbolic curve seen with myoglobin, there is a sigmoid curve that is well to the right of that of myoglobin (Fig. 27.11). Higher oxygen concentrations are required to 50% saturate hemoglobin that in the case for myoglobin, indicating as already stated that the latter has a higher affinity for oxygen.

Hemoglobin needs to pick up as much oxygen as it can in the lungs but surrender it in the tissue capillaries. The sigmoidal oxygenation curve means that it is most steep (that is, surrenders the most oxygen) at oxygen pressures encountered in the capillaries (see Fig. 27.11) but, nonetheless, it is still capable of becoming virtually saturated at the higher oxygen pressures encountered in the lungs.

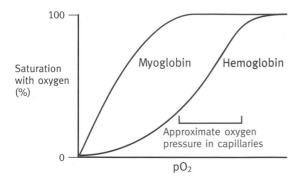

Fig. 27.11 Oxygen–hemoglobin saturation curve. The higher affinity for oxygen of myoglobin as compared to hemoglobin means that myoglobin in the muscles readily accepts oxygen from the blood.

How is the sigmoidal oxygen saturation curve achieved?

You have met sigmoidal kinetics already when metabolic control was discussed (page 163). It was mentioned there that allosteric enzymes are usually multisubunit proteins that undergo a conformational change on binding of substrate molecules. It is this change of structure that results in changes in the affinity of the enzyme for further attachments of its substrate. The same applies in oxygen binding to hemoglobin.

Although the heme molecules in hemoglobin are distant from one another, the initial binding of oxygen to one subunit facilitates the binding of further molecules of oxygen to the other subunits. This is known as a **homotropic positive cooperative effect** (homotropic, because only oxygen is involved). It is this that causes the sigmoidal curve. The effect is that the initial affinity of deoxyhemoglobin for oxygen is 200 times less than that involved in the final addition of oxygen. At the end of the curve, oxygenation becomes limited by availability of binding sites.

Mechanism of the allosteric change in hemoglobin

When oxygen molecules bind to a hemoglobin molecule, as stated, it makes it easier for the remaining sites to accept oxygen. According to the concerted model (see page 164), hemoglobin exists in two conformational states: one is the 'tense' or T state with *low* oxygen affinity and the other is the 'relaxed' or 'R' state with *high* oxygen affinity, the two being in free equilibrium and the T state predominating in the absence of oxygen. As oxygen binds, it increases the probability of all four subunits of a given molecule being in the R, *high*-affinity, state—it swings the $T \leftrightarrow R$ equilibrium to the right,

$$T \leftrightarrow R \xrightarrow{O_2} RO_2.$$

Since, in a study of the oxygenation of hemoglobin, large numbers of individual molecules are involved, as more and more oxygen binds, more molecules of hemoglobin are in the relaxed state and, therefore, the observed oxygen affinity in the

solution increases. This is analogous to the situation with many allosteric enzymes, described on page 163.

By means of X-ray diffraction studies, the conformation of the hemoglobin tetramer has been determined in the oxygenated and deoxygenated states. As already mentioned, hemoglobin has four subunits, two α and two β. Although the α subunits are identical to one another, and the β subunits are also identical to one another, in the tetramer they are distinguishable and are named α_1, α_2, β_1 and β_2. The two α subunits differ slightly in their contacts with the β subunits and vice versa for the β_1 and β_2, something that is evident from the X-ray diffraction studies.

The most useful way to view hemoglobin is to say that the α_1 and β_1 subunits are firmly associated as a dimer and, similarly, that the α_2 and β_2 form a dimer. It is the interaction between the two dimers in the tetramer that undergoes rearrangement in the $T \leftrightarrow R$ conversion. Figure 27.12 illustrates the relative rotation between the dimers' conversion from the T to R state caused by binding of oxygen.

What causes this allosteric change to the entire hemoglobin molecule when one or more oxygen molecules bind? The movement is initiated by the fact that, on binding of oxygen to heme, the iron atom moves slightly and, in doing so, brings about the $T \leftrightarrow R$ transformation. Although heme, as usually written, appears as a planar molecule, in the deoxygenated state, the Fe atom lies above the plane of the molecule because it is too big to fit into the tetrapyrrole (Fig. 27.13(a)). In the deoxygenated hemoglobin, the tetrapyrrole itself is not quite flat (Fig. 27.13(a)). The Fe^{2+} atom itself is bonded to a histidine residue in one of the α helices of which the subunit is composed—in fact, the α helix designated F, the histidine group being identified as F8 (Fig. 27.13(a)).

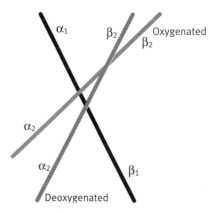

Fig. 27.12 Diagrammatic representation of the relative subunit positioning in deoxygenated and oxygenated hemoglobin molecules, the axes of which are represented by straight lines. The α_1, β_1 pair and the α_2, β_2 pair should be regarded as single dimer units. On oxygenation, these dimers rotate and slide, relative to each other, by about 15°. The black and blue lines represent the relative positions of the dimers in the T state. In the oxygenated (R) state the red line shows the rotation of the α_2/β_2 dimer relative to the α_1/β_1 dimer, which is represented as fixed. The change alters the $\alpha_1\beta_2/\alpha_2\beta_1$ contacts between the dimers. A more realistic model of hemoglobin is shown in Fig. 27.10 (c).

(a) F$_8$ α helix of globin subunit

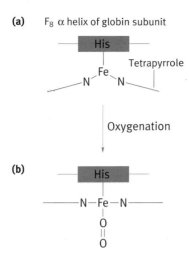

Fig. 27.13 Diagrammatic representation of the changes in the heme of hemoglobin upon oxygenation. **(a)** Diagram of the heme molecule in deoxyhemoglobin with the tetrapyrrole structure strained into a slightly domed shape. **(b)** Attachment of heme in oxyhemoglobin.

On binding of oxygen, the iron atom of heme becomes effectively smaller in diameter, due to electronic changes and moves into the plane of the tetrapyrrole, thus flattening the molecule. The protein rearranges itself as a result of the movement of the Fe atom (Fig. 27.13(b)). This tiny movement is amplified because of a lever-like effect of the arrangement of the polypeptide chain of the protein, so that a bigger movement occurs elsewhere in the molecule.

This 'elsewhere' is the point where the α unit of one dimer interacts with the β unit of the other dimer (α_1–β_2/α_2–β_1 interfaces). At each of these locations there is a group of weak bonds between amino acid residues of the two protein subunits involved. In the T state there is one set, in the R state another set of bonds, holding the dimers together. When it moves into the new position in the R state, an alternative set of weak bonds is formed because of close association of different groups on the subunits. The movement initiated by the Fe atom thus causes the two dimers to rearrange from one set of weak bonds to the other set, causing the relative rotation shown in Fig. 27.12.

It should be noted that the structures of the T and R states of hemoglobin are fully proven by X-ray studies on crystals of hemoglobin and of oxyhemoglobin, respectively. However, it is not possible to obtain crystals of intermediate, partially oxygenated hemoglobin and therefore the structure(s) of the latter are not known. The intermediate stages (if any) by which the T and R forms interconvert are therefore not established. On page 164 we described two hypotheses (the concerted and sequential models) both of which could account for the observed changes.

The essential role of 2:3-bisphosphoglycerate (BPG) in hemoglobin function

You are familiar with the glycolytic intermediate 1:3-bisphosphoglycerate (Fig. 8.6) but we have not so far mentioned that **2:3-bisphosphoglycerate (BPG)** is also synthesized in the cell.

1:3-Bisphosphoglycerate   2:3-Bisphosphoglycerate

It plays an important physiological role in oxygen transport by lowering the affinity of hemoglobin for oxygen and thus increases the unloading of oxygen to the tissues; it moves the dissociation curve of oxyhemoglobin to the right. The explanation is both interesting and simple. The hemoglobin tetramer, looked at from the appropriate viewpoint, has a cavity running through the molecule (see Fig. 27.10(c)). Projecting into this cavity are amino acid side chains with positive charges. The BPG molecule has four negative charges at the pH of the blood and has just the correct size and the configuration to fit into the cavity and make ionic bonds with the positive charges on the protein. This helps to hold hemoglobin in its deoxygenated position; in effect it crosslinks the β units in that position. In the deoxygenated (T) state, the hemoglobin can accommodate a molecule of BPG. However, on oxygenation, because of the conformational change in the protein, the cavity of the R state becomes smaller and is unable to accommodate BPG. If we consider oxyhemoglobin in the capillaries, the ability of BPG to strongly bind to, and stabilize the deoxygenated state favours unloading of the oxygen. In effect, the process is

(a) Hb—oxygen ↔ Hb + oxygen;
    (relaxed)   (tense)
(b) Hb + BPG ↔ Hb—BPG.
    (tense)   (tense)

Reaction (b) with BPG will tend to pull the equilibrium of reaction (a) over and favour oxygen release.

If blood is stripped of all of its BPG, the hemoglobin remains virtually saturated with oxygen even at oxygen concentrations below that encountered in the tissue capillaries. It therefore would be incapable, in that state, of delivering oxygen to the tissues efficiently. The effect of BPG on oxygen binding by hemoglobin is illustrated in Fig. 27.14.

The higher the concentration of BPG, the more the deoxygenated form is favoured. This constitutes a regulatory

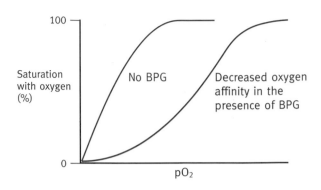

Fig. 27.14 Oxygen saturation curves for hemoglobin, illustrating the effect of 2:3-bisphosphoglycerate (BPG).

system—if oxygen tension in the tissues is low, synthesis of more BPG in the red blood cells (which occurs as a result of regulatory mechanisms we will not go into) favours increased unloading of oxygen. Acclimitization at high altitudes involves the establishment of higher BPG levels in red blood cells. It is to be noted that, while BPG causes greater delivery of oxygen to the tissues, the decreased oxygen affinity has little effect on the degree of oxygenation in the lungs. The normal molar concentration of BPG in the blood is roughly equivalent to that of tetrameric hemoglobin.

There is yet another refinement of the BPG-based regulatory system that illustrates how small changes to proteins can have major physiological effects. For a mother to deliver oxygen to a fetus, it is necessary for the fetal hemoglobin to extract oxygen from the maternal oxyhemoglobin across the placenta. This requires the fetal hemoglobin to have a higher oxygen affinity than that of the maternal carrier. This is achieved by a fetal hemoglobin subunit (called γ) replacing the adult β chains, each of which lacks one of the positive charges of the β subunit. The missing charges are those that, in adult hemoglobin, line the cavity into which BPG fits; therefore fetal hemoglobin has two fewer ionic groups to bind BPG, the latter therefore being held less tightly. BPG is therefore less efficient in lowering the oxygen affinity, giving fetal hemoglobin a higher O_2 affinity than that of maternal hemoglobin. Thus, the latter readily transfers its oxygen load to the fetus.

Effect of pH on oxygen binding to hemoglobin

Hemoglobin in the deoxygenated state has a higher binding affinity for protons than has oxyhemoglobin. Put in another way, the R form is a stronger acid than is the T (deoxygenated) form, resulting in dissociation of protons from the molecule when oxygen binds, a phenomenon known as the **Bohr effect**,

$$(1) \quad Hb + 4O_2 \leftrightarrow Hb(O_2)_4 + (H^+)_n.$$

(where n is somewhere around 2; the number depends on a complex set of parameters.)

The protons are released from, for example, histidine side

groups of the protein. The release is caused by the $T \rightarrow R$ conformational change affecting the ionization (pK_a) of such groups (see page 17) for dissociation of histidine).

Role of pH changes in oxygen and CO₂ transport

The Bohr effect, described above, has important physiological repercussions. In the tissues, CO_2 is produced and must be transported to the lungs. It enters the red blood cell where the enzyme **carbonic anhydrase** converts it to H_2CO_3 which dissociates into the bicarbonate ion and a proton,

$$(2) \quad CO_2 + H_2O \leftrightarrow H_2CO_3 \leftrightarrow H^+ + HCO_3^-.$$

The latter will drive the equilibrium shown in equation (1) above, to the left, causing the HbO_2 to unload its oxygen, the effect thus being in harmony with physiological needs.

The HCO_3^- passively moves out via the anion channel (page 50) down the concentration gradient into the serum. The HCO_3^- movement is not accompanied by H^+ movement for there is no channel allowing passage of this across the membrane of the red cell. To electrically balance exit of HCO_3^-, Cl^- moves into the cell via the same anion channel. The dual movement is known as the **chloride shift** (Fig. 27.15). The HCO_3^- travels in solution in

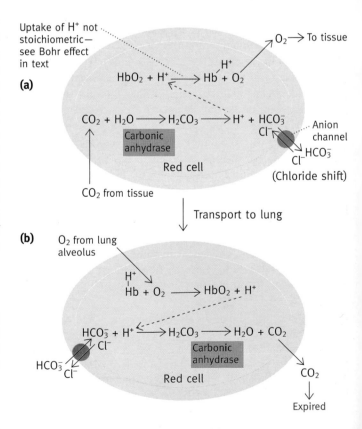

Fig. 27.15 Transport of CO_2 in the blood. **(a)** Reactions in the tissue capillaries. **(b)** Reactions in the lungs. The diagrams omit transport of CO_2 as carbamino groups of hemoglobin.

the serum of the venous blood back to the lungs. Here, the proton concentration changes help to achieve the physiologically desirable results too. On oxygenation, release of protons from hemoglobin does two main things. It produces H_2CO_3 from HCO_3^- by a simple equilibrium effect,

$$HCO_3^- + H^+ \leftrightarrow H_2CO_3,$$

that permits carbonic anhydrase to form CO_2 to be expired because H_2CO_3 (not HCO_3^-) is the substrate of the enzyme

$$H_2CO_3 \leftrightarrow H_2O + CO_2.$$

The destruction of HCO_3^- in the red blood cell causes HCO_3^- in the serum to enter the cell, down the concentration gradient, and Cl^- exits, that is, a reverse chloride shift occurs in the lungs, resulting in CO_2 expiration.

A small amount of CO_2 is transported in simple solution in the blood, but by far the greatest amount (about 75%) is transported as HCO_3^-. About 10–15% of the CO_2 is carried by the hemoglobin itself. CO_2 chemically and spontaneously reacts with uncharged $-NH_2$ groups of the globin to form carbamino groups

$$RNH_2 + CO_2 \leftrightarrow RNHCOOH \leftrightarrow RNHCOO^- + H^+.$$

The RNH_2 groups available are mainly the terminal amino groups—lysyl and arginine side groups have too high a pK_a (page 26) to be significantly uncharged.

pH buffering in the blood

From the above, it is clear that there are major changes in the hydrogen ion concentration of blood associated with CO_2 and oxygen transport.

When acid is produced in the tissues, the pH of red blood cells cannot be allowed to fall by very much. The buffering power of HCO_3^-, phosphates, and hemoglobin itself is important in maintaining a physiological pH. The Bohr effect, described above, also buffers. When oxygen is unloaded, hemoglobin takes up protons (equation 1). This carries roughly half of the H^+ ions generated by CO_2 in the tissues, and this again helps to prevent a drop in the pH of the red blood cell to unphysiological levels.

Sickle cell anemia

This disease is usually described in textbooks because it illustrates how a single amino acid change in a protein can have a profound effect, and gave rise to the concept of a molecular disease. In the normal β chains of the human hemoglobin tetramer, amino acid number 6 is glutamic acid whose side group is negatively charged and highly hydrophilic (Fig. 27.10(c)). In the hemoglobin of sickle cell anemia patients, this is replaced by the hydrophobic valine residue. The codons in mRNA for glutamic acid are GAA and GAG; mutation of the central A base to a U is all that is required for this substitution since GUA and GUG are both valine codons. The abnormal hydrophobic patch on the globin subunit caused by valine, by chance, causes a deoxygenated hemoglobin molecule to bind to a particular hydrophobic pocket on another molecule and so on, resulting in the formation of a long multistrand rigid rod. Oxygenated hemoglobin, because of its different conformation, does not permit this. The long deoxyhemoglobin rods distort the normal biconcave disc into sickle-shaped cells that tend to block capillaries and also to break up, causing anemia. If both chromosomes are affected, the disease may be lethal, especially at low oxygen tensions such as occur at high altitude causing unusually high deoxygenation of the hemoglobin.

The disease is prevalent only in geographical areas with a malignant form of malaria (ignoring the effects of immigration) where the high incidence can be explained only by positive selection of the disease genome. The sickled red blood cell is unfavourable for the development of the malarial parasite and thus protects against the parasite. Since malarial infection of normal persons has a higher death rate than does the heterozygous sickle cell state, the latter increases survival of persons with the genetic trait.

Further reading

Heme and its synthesis

Straka, J. G., Rank, J. M., and Bloomer, R. (1990). Porphyria and porphyrin metabolism. *Ann. Rev. Med.*, **41**, 457–69.
The review deals with the defects in the enzymes of heme biosynthesis which lead to porphyrias, and the therapeutic uses of porphyrins.

Chen, J.-J. and London, I. M. (1995). Regulation of protein synthesis by heme-regulated eIF-2α kinase. *Trends Biochem. Sci.*, **20**, 105–8.

A comprehensive review of the control of hemoglobin synthesis by this mechanism.

Hemoglobin and myohemoglobin

Branden, C. and Tooze, J. (1991). *Introduction to protein structure.* Garland Publishing.
A well illustrated account of all aspects of protein structure, including hemoglobin and myoglobin; very readable.

Problems for Chapter 27

1   Describe the first two steps in heme biosynthesis in animals.

2   What is the medical relevance of ALA synthase control in liver?

3   Explain how the red blood cell ALA synthetase activity is coordinated with the availability of iron.

4   Compare the oxygen dissociation curves of myoglobin and hemoglobin.

5   What is the likely mechanism of the sigmoid oxygen-binding curve of hemoglobin?

6   Binding of an oxygen molecule to hemoglobin causes a conformational change in the protein. Describe the mechanism of this.

7   Explain how the fetus is able to oxygenate its hemoglobin from the maternal carrier.

8   Explain the significance of the chloride shift in red blood cells.

Part 6

Mechanical work by cells

Previous page: Electron micrograph of a section of mammalian skeletal muscle. The photograph shows five myofibrils which run the length of the muscle cell. Each is surrounded by the sarcoplasmic reticulum (seen here as green and bubble-like). Each myofibril is divided into sarcomeres, the contractile units—five complete ones are shown in the central vertical column. Making up each sarcomere can be seen myosin thick filaments (orange) and actin thin filaments (blue), giving the striated appearance. The structure corresponds to that shown in Fig. 28.2.

Photograph: Biology Media/Science Photo Library

Chapter 28

. . . .

Muscle contraction

Muscles are highly specialized organs in which devices for mechanical movement, present in all cells, whether muscle or nonmuscle, have been developed to the highest and most easily studied state. It might seem logical to deal with the 'molecular motors' present in all cells first but it is probably easier to understand muscle contraction and then move on to movement in nonmuscle cells. This is the arrangement adopted here—we will deal with muscle contraction in this chapter and then with the more generalized movement systems in nonmuscle cells in the following chapter.

A reminder of conformational changes in proteins

The most astonishing thing about muscle contraction is that it occurs at all. The only basis for biological processes are molecules and, for contraction, individual molecules must move in a purposeful fashion. What type of molecular activity is available for this? The only type we know of is conformational change in proteins, minute changes in shape of individual protein molecules that occur as a result of different ligands binding to the surface of proteins.

For movement to occur, whatever exerts force must have something to exert it against. A 'molecular motor' exerting force by conformational change must always have a partner structure to react against—if the motor is fixed, the partner molecule will move; if the latter is fixed, the motor molecule will move. To anticipate the mechanism of muscle contraction, the motor is myosin, which is fixed, and the structure it acts against, and moves, is the actin filament. But more of this later.

. . . .

Types of muscle cells and their energy supply

The two main classes of muscles are **smooth** and **striated**. Smooth muscle is found in the intestine and blood vessels, which are typically under involuntary nervous control and frequently under control by several hormones. It contracts slowly and can maintain the contraction for extended periods. It

is called smooth because it does not have the same striated appearance under the microscope as does the second class of muscle—striated muscle. This is found in skeletal muscle, which is under voluntary nerve control. Striated muscle contracts rapidly. Heart muscle is striated but not identical in structure to skeletal muscle and is under involuntary control.

If we turn now to voluntary striated skeletal muscle there is further specialization of biochemical interest. There are fast twitching fibres and slow twitch fibres. In animals such as lobster and fish, the white striated muscle, consisting of fast twitching fibres with poor blood supply, permits very fast 'escape' reactions dependent on extremely rapid stimulation of ATP production, resulting from the breakdown of glycogen reserves by glycolysis (the mechanism of the almost instant control of this is given earlier—see page 172). Because of the low yield of ATP from glycolysis and lactic acid accumulation (page 98) the white muscle is quickly exhausted. Slow twitch muscle is slower to contract. It is richly supplied with blood and has a high mitochondrial content and a reserve of oxygen bound to myoglobin (page 377). It generates ATP mainly by oxidative phosphorylation. This is much more efficient in terms of ATP yield than is glycolysis but, in emergency, oxygen supply becomes limiting and it takes time for blood vessels to expand and the heart rate to speed up. It is therefore slower in response than fast twitch muscle but can function much longer without exhaustion. In man, voluntary striated muscles contain mixtures of both types of fibres, the proportions depending on the precise role of the muscles in the body. Human back muscle, which has to maintain tension for the body posture for extended periods, is mainly slow twitch muscle. The ocular muscles that move the eyeball are good examples of the fast twitch type. In fish, it is the thin stripes of red muscle that cause the contractions involved in ordinary swimming.

In all muscles, the reserve of ATP on which contraction depends is remarkably low—only enough for a very brief period of intensive contraction. This explains a role for a reserve of phosphoryl groups in the form of **creatine phosphate**, whose hydrolysis is associated with a $\Delta G^{0'}$ value of -43.0 kJ mol^{-1}

(cf. -30.5 kJ mol^{-1} for the hydrolysis of ATP to ADP and P$_i$). When ATP is converted to ADP and P$_i$ by contraction, the **creatine kinase** reaction regenerates ATP using the phosphoryl group of creatine phosphate.

Contraction event: $ATP + H_2O \rightarrow ADP + P_i$.

ATP regeneration event: $ADP + creatine - ⓅP \rightarrow$

$ATP + creatine$.

During muscle recovery, ATP is regenerated by oxidative metabolism and the pool of creatine phosphate is regenerated by the creatine kinase reaction, which is readily reversible in the presence of high levels of ATP and low levels of ADP.

Creatine phosphate has the structure shown below; the compound is of the 'high-energy' type and the phosphoryl group can be reversibly transferred to ADP to form ATP. Because of the reactivity of the phosphoryl group, there is a spontaneous (noncatalysed) formation of creatinine; this has no function and is excreted in the urine at a daily rate essentially proportional to muscle mass.

Creatine phosphate

Creatinine

Structure of skeletal striated muscle

A muscle is composed of multinucleate cells called **myofibres** (Fig. 28.1), which may be very long. The cell membrane (the **sarcolemma**) is excitable (see page 365)—it has nerve endings associated with it, at the neuromuscular junctions, to deliver the nervous signal that triggers contraction. The plasma membrane of the cell surrounds the cytoplasm, multiple nucleii, and many mitochondria. Embedded in this cytoplasm and running lengthwise through it are many long **myofibrils**, each surrounded by a membraneous sac, the **sarcoplasmic reticulum**.

Structure of the myofibril

The myofibril is the structure that does the contracting and, as stated, there are many running the length of the muscle cell. Each myofibril is divided into small segments bounded by **Z discs** (Z for *zwischen* or division). Each small segment is called a **sarcomere** (Fig. 28.2). On contraction, the Z discs are pulled closer together, thus shortening the individual sarcomeres and hence the myofibril. This shortens the entire myofibre, a process that is seen as muscle contraction.

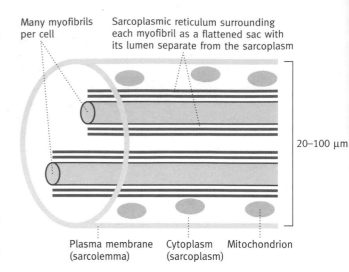

Fig. 28.1 Diagram of a myofibre (muscle fibre) or muscle cell from striated muscle. Bundles of myofibres make up a muscle.

How does the sarcomere shorten?

The Z discs are robust discs of protein at each end of the sarcomere to which are attached 'rods' pointing to the centre of the sarcomere. These rods are thin filaments made up mainly of the protein actin, and are firmly attached by one of their ends to

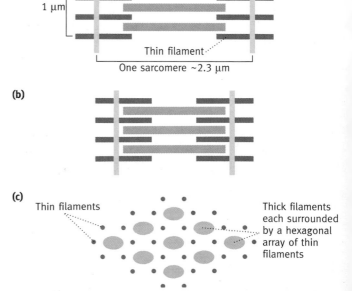

Fig. 28.2 Arrangement of thick and thin filaments in a sarcomere. **(a)** Relaxed. The striated appearance of myofibril sections in electron microscopy (page 387) is caused by the amount of protein the beam traverses; **(b)** contracted; **(c)** arrangement in cross-section. The diagram shows only a few filaments but each sarcomere has a large number, all in lateral register.

a Z disc. They are essentially inert and serve as 'rachets' on which force can be exerted to drag the Z discs together. In vertebrate muscle, the thin filaments are in hexagonal array on the two discs, and each sarcomere has a number of these arrays. Inside each hexagonal 'cage' formed by the six thin filaments, is a thick filament. The thick filament has finger-like projections that do the actual work of contraction. By a ratchet-like mechanism they 'claw' the thin filaments towards the centre and in doing so pull the Z discs closer together (Fig. 28.2).

That, then, is the overall picture of muscle contraction. To understand how contraction happens we must look at the molecular structures involved.

Structure and action of thick and thin filaments

Thin filaments

Actin is a globular protein molecule (called **G actin**; G for globular) (Fig. 28.3(a)(b)). Each actin molecule has two dissimilar globular parts connected by a narrow waist, giving the molecule a polarity. Under cellular conditions it readily polymerizes into long fibres, always in a definite head to tail polarity. A thin filament consists of two such fibres wound around each other with a long pitch (and the same polarity), the ends being referred to as (+) and (−).

Thick filaments

We now come to a most remarkable molecule—**myosin**, the protein of the thick filament. Each molecule has a straight rod section made up of two polypeptides in coiled coil configuration (Fig. 28.4(a))—that is, two α helices (page 28) coiled around each other in helical form. The structure is such that each α helix has regularly spaced hydrophobic residues that form hydrophobic attachments to its partner molecule, the two forming a rigid rod structure. Each chain terminates in a globular head; two other small polypeptide chains, called light myosin chains, also occur in each head.

A thick filament is made up of several thousand myosin molecules arranged in a bipolar fashion as shown in Fig. 28.4(b). The thick filaments are held in a central position relative to the thin filaments by other proteins involved in sarcomere structure. One is **titin**, so named for its huge (titanic) molecular weight.

(a)

G actin molecules
The arrows indicate the polarity of the molecules

(b)

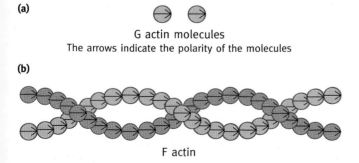

F actin

Fig. 28.3 (a) diagram of G (globular) actin. **(b)** Diagram of F (fibrous) actin made from the polymerization of G actin.

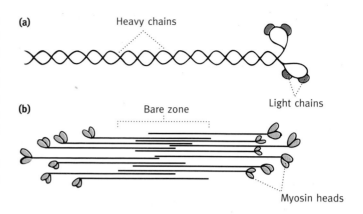

Fig. 28.4 (a) Diagram of a myosin molecule consisting of a dimer structure of two heavy chains, each terminating in a myosin head. The latter has two dissimilar light chains attached to it. **(b)** A thick filament made of myosin molecules arranged in a bipolar fashion.

This central positioning is at an optimal distance to allow the myosin heads to contact the actin filaments. The thin filaments are anchored to the Z discs at their (+) ends so that the myosin heads, at both ends of the thick filament, are oriented the same way, relative to the thin filament polarity (Fig. 28.5).

How does the myosin head convert the energy of ATP hydrolysis into mechanical force on the actin filament?

The **myosin head** is an enzyme capable of hydrolysing ATP to ADP and P_i. It is generally accepted that the myosin head undergoes conformational changes, driven by ATP hydrolysis, which is likely to cause confusion unless explained. You are, by now, used to ATP breakdown being used to perform work and it would be natural to assume that the work of muscle contraction occurs when myosin hydrolyses ATP to ADP and P_i. In the ATP-driven processes that you have learned about until now, the hydrolysis has been indirect—that is, first there is a transfer of a phosphoryl group, or an AMP group, to a reactant molecule followed by a second step in which P_i or AMP is liberated—in other words, coupled reactions are involved (page 12). In muscle contraction, mechanical work is driven by the direct hydrolysis of ATP (without covalent intermediates), but the actual 'power stroke' in contraction does not occur on ATP hydrolysis as you might expect. It is when the ADP leaves the protein that the main liberation of free energy occurs so that the 'power step' in muscle contraction correlates, not with ATP hydrolysis, but with ADP release from the myosin head. The conversion of protein-bound ATP to protein-bound ADP plus P_i involves little free energy change. The conversion of *free* ATP, in solution, to *free* ADP and P_i, in solution, has the expected, large, negative $\Delta G^{0'}$ value so that the overall energetic concept of ATP hydrolysis driving contraction is unaltered.

The myosin head has two states, a **primed configuration** ready to exert a force and a **resting configuration** after it has exerted a force. The head attaches to the actin filament in the

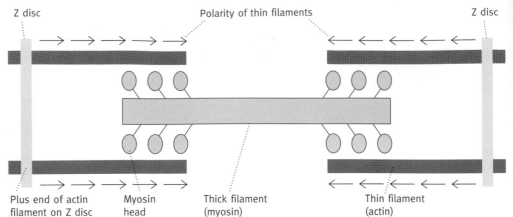

Fig. 28.5 Diagram showing the arrangement of thick and thin filaments in a sarcomere. The arrows in the latter are to show the polarity of the actin filaments. In a contraction event, in effect, the myosin heads track along the actin filaments towards their (+) ends and, in doing so, claw the two discs together.

primed state and then performs a **power stroke** on the actin (Fig. 28.6) as it moves to the resting state. Let us now look more closely at the steps involved in this cycle starting at the point at which a myosin head has just exerted its power stroke. The head is attached to actin in the resting configuration. The following sequence of events occurs as depicted in Fig. 28.7.

1. ATP attaches to the myosin head and causes the latter to detach from actin.

2. ATP is hydrolysed to ADP and P_i. P_i is released and the myosin head enters into the primed configuration attached to the actin.

3. The myosin head enters into the power stroke associated with ADP release. This drives the conformational change of the myosin head, which swings on its hinge region.

4. The myosin head in the unprimed configuration is locked on to the actin until a molecule of ATP arrives to release it—that is, back to step 1.

Vast numbers of myosin molecules are involved in a muscle

contraction. If no ATP is available, all of the myosin heads become locked into crosslinkages to actin (since step 1 cannot occur) resulting in the rigid structure of muscles in the rigor mortis state whose onset coincides with ATP exhaustion. That the myosin head itself is capable of carrying out the whole process is shown by the *in vitro* experiment in which latex beads, coated with isolated myosin heads, literally walk along immobilized actin fibres in the presence of ATP and Ca^{2+}—a most dramatic experiment.

How is contraction in voluntary striated muscle controlled?

In skeletal muscles, contraction is initiated by a nerve impulse causing Ca^{2+} ions to be liberated into the myofibril from the sarcoplasmic reticulum and contraction is triggered. The Ca^{2+} ions are rapidly removed from the myofibril (see below), so that, unless more Ca^{2+} is liberated by a continuing nerve impulse, contraction ceases.

How does Ca^{2+} trigger contraction?

Thin filaments are primarily made of actin but they have associated with them additional protein molecules. One of these is **tropomyosin**, consisting of two subunits. This is a fairly short, elongated molecule that lies along the helical groove between the two actin fibres of a thin filament. Each tropomyosin molecule has on it seven actin attachment sites, each one binding to an actin monomer within the groove. The tropomyosin molecule thus is about seven actin monomers long, and the successive molecules overlap one another to form a continuous thread right along the thin filament. A thin filament has two grooves, with a tropomyosin thread in each of them (Fig. 28.8).

Each tropomyosin molecule has attached to it, at one end, an additional complex of three more globular proteins called

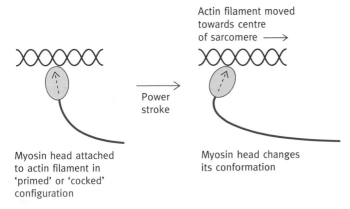

Fig. 28.6 Simplified diagram of the principle of contraction. All details of ATP involvement are omitted.

Actin filament moved towards centre of sarcomere ⟶

Power stroke

Myosin head attached to actin filament in 'primed' or 'cocked' configuration

Myosin head changes its conformation

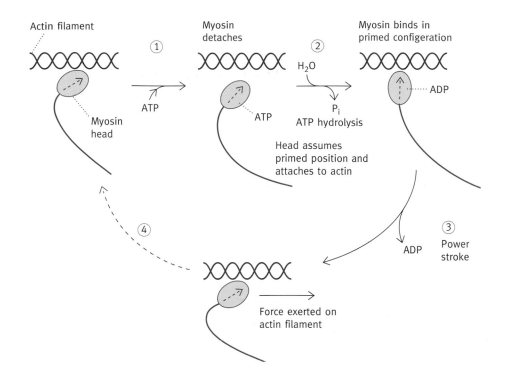

Fig. 28.7 Simplified diagram of contractile events. The numbers refer to the steps in the text. We start with the myosin head having just performed its power stroke on the actin thin filament. The important point to note is that the power stroke is associated with ADP release from the myosin head.

troponin. This is all leading up to the fact that the **troponin–tropomyosin complex** is sensitive to Ca^{2+}, which causes the tropomyosin thread to move slightly, relative to the thin filament. This is the event that permits the myosin head to attach to the actin filament. It was believed that the tropomyosin prevents this attachment to the actin by a simple blocking effect, but other evidence suggests that Ca^{2+} does not affect the binding of myosin to the thin filament. What is known is that, unless Ca^{2+} is present, the **myosin–actin power cycle** does not occur.

In short, Ca^{2+} combines with troponin, causing a conformational change in tropomyosin that somehow activates the myosin–actin power cycle resulting in ATP hydrolysis and contraction. Withdrawl of Ca^{2+} reverses the whole sequence of events and shuts down the contraction event. Which brings us to the mechanism of the control of Ca^{2+} release and withdrawal.

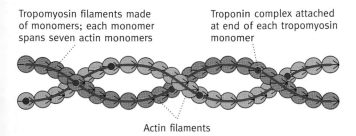

Tropomyosin filaments made of monomers; each monomer spans seven actin monomers

Troponin complex attached at end of each tropomyosin monomer

Actin filaments

Fig. 28.8 The relationship of the actin and tropomyosin molecules to each other. Each tropomyosin molecule binds to seven actin monomers forming a continuous filament in each of the actin filament grooves. Each tropomyosin molecule also has a troponin complex bound at one end.

Release and uptake of Ca^{2+} in the muscle

Each myofibril within the muscle cell, is surrounded by a membranous sac—the **sarcoplasmic reticulum**. The membrane is highly specialized in two regards. First, it is rich in a Ca^{2+}/ATPase (see page 52) that pumps Ca^{2+} from the cytosol surrounding the myofibril (sarcoplasm) into the lumen of the reticulum, using ATP hydrolysis energy to drive the process. This continually depletes the myofibril of Ca^{2+} and prevents muscle contraction. Secondly, the sarcoplasmic reticulum membrane has a protein, the **Ca^{2+} channel** or **Ca^{2+} gate**, which is a passive Ca^{2+}-transporting channel; it is normally closed, but, on receipt of a nerve impulse to the muscle cell, it opens and releases Ca^{2+} from the lumen of the reticulum into the myofibril, causing contraction.

For a long muscle to contract, all of the sarcomeres in a myofibril and all the myofibres in the muscle need to respond to a motor nerve impulse essentially simultaneously, for otherwise the contraction will be uncoordinated and relatively ineffective. The nerve impulse causes liberation of acetylcholine at the neuromuscular junction; this causes a local depolarization of the plasma membrane, that rapidly propagates throughout the plasma membrane (see page 366). The depolarization causes the voltage-gated Ca^{2+} channels of the sarcoplasmic reticulum to open and release the ion on to the myofibril. In order to ensure that the signal reaches all of the sarcoplasmic reticulum within a cell very rapidly, the plasma membrane is invaginated into transverse (T) tubules that enter the cell at Z discs and make direct contact with the sarcoplasmic reticulum membrane, thus permitting the electrical signal to reach, virtually simulta-

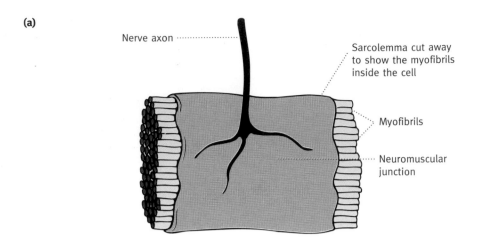

(a)

Nerve axon

Sarcolemma cut away
to show the myofibrils
inside the cell

Myofibrils

Neuromuscular
junction

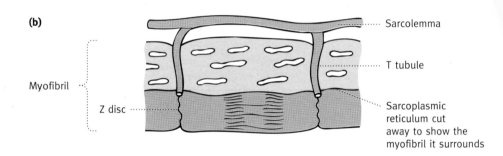

(b)

Sarcolemma

Myofibril

Z disc

T tubule

Sarcoplasmic
reticulum cut
away to show the
myofibril it surrounds

Fig. 28.9 (a) Plasma membrane (sarcolemma) of the myofibre, showing the neuromuscular junction. **(b)** Diagram of the transverse (T) tubules that carry the plasma membrane depolarization signal to the sarcoplasmic reticulum (SR). It is postulated that the plasma membrane depolarization is directly transmitted to the SR Ca^{2+} channels causing rapid Ca^{2+} release from the SR.

neously, all of the contractile units controlled by that nerve impulse (Fig. 28.9).

How does smooth muscle differ in structure and control from striated muscle?

Smooth muscle is found in the walls of blood vessels, in the intestine, and in the urinary and reproductive tracts. The long spindle-shaped cells have a single nucleus and associate together to form a muscle in patterns appropriate to their function, such as an annular arrangement in blood vessels and a criss-cross network in the bladder.

The basic principles of contraction are the same as in striated muscle in that myosin molecules exert force on actin filaments, using ATP hydrolysis as the source of energy, and employing a similar 'power cycle'. However, the contractile components are not so highly organized. There are no myofibrils and no repeating sarcomers, the latter being the reason for the absence of a striated appearance under the microscope. Instead of the sarcomere structure, actin filaments run the length of the whole cell (which is small, compared with a striated muscle cell) and are anchored at one end into the cell membrane.

Control of smooth muscle contraction

Although Ca^{2+} is responsible for causing contraction, the control mechanism is quite different from that in striated muscle. A smooth muscle contracts much more slowly than a striated muscle, typically about 50 times more slowly, taking about 5 seconds. There is no requirement in smooth muscles for contraction to be almost instantaneous throughout the structure as there is in striated muscle; the contraction of individual cells can spread at a more leisurely pace. In smooth muscle cells, a neurological impulse from the autonomic nervous system causes Ca^{2+} gates in the cell membrane to open and allow an inrush of the ion into the cells from the extracellular medium. There is no sarcoplasmic reticulum. The relatively slow diffusion of Ca^{2+} throughout the cell can be tolerated because of the small distances involved and the slow response requirements. Special junctions between cells allow a neurological signal to spread throughout the muscle.

How does Ca^{2+} control smooth muscle contraction?

In the head of each myosin molecule, as already described, are two small polypeptides known as **myosin light chains**; this is in addition to the heavy myosin chains. In smooth muscle, one of these light chains (the p-light chain) inhibits the binding of the myosin head to the actin fibre, and thus prevents contraction. Ca^{2+} activates a myosin kinase that, with ATP,

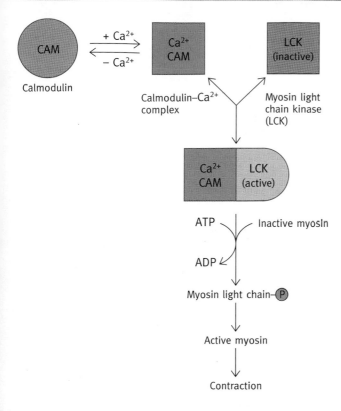

Fig. 28.10 Diagram of the mechanism of activation of smooth muscle contraction by Ca^{2+}.

phosphorylates the p-light chain and abolishes its inhibitory effect, thus triggering contraction.

The Ca^{2+} does not directly activate the kinase; instead it combines with a protein, calmodulin (see page 360), which induces a conformational change in the latter such that it combines with the inactive kinase and activates it. When the Ca^{2+} levels falls, the process reverses and a phosphatase dephosphorylates the myosin light chain causing muscle relaxation. The scheme is summarized in Fig. 28.10.

As well as neurological control of smooth muscle contractions, several hormones exert control; prostaglandin (page 146) is one of these; norepinephrine causes contraction of certain blood vessel muscles.

Further reading

Bray, D. (1992). *Cell movements.* Garland Publishing.
A comprehensive and authoritative, but readable, account of all aspects of movement in cells. Good simple diagrams and also illustrated with electron micrographs.

Muscle contraction

Cooke, R. (1995). The actomyosin engine. *FASEB J.*, 9, 636–42.
Beautifully illustrated structure of the myosin head and of the actin filament, plus discussion of how a mechanical force is generated in muscle contraction.

Smooth muscle contraction

Small, J. V. (1995). Structure-function relationships in smooth muscle: the missing links. *BioEssays*, 17, 785–92.
Discusses the structural organization of the smooth muscle cell and control of contraction.

Problems for Chapter 28

1   What are the biochemical differences between fast twitch muscle fibres and slow twitch ones? What are their biological roles?

2   Illustrate the strucutre of: (a) a myofibril with the sarcoplasmic reticulum and (b) a sarcomere.

3   Explain how the myosin head converts the energy of ATP hydrolysis into mechanical energy.

4   How is contraction of a voluntary striated muscle sarcomere controlled?

5   How is smooth muscle contraction controlled?

Chapter 29

···

The role of the cytoskeleton in the shape determination of cells and in mechanical work in cells

It is easy to form the incorrect mental image that mechanical work is not important in nonmuscle cells. It is also easy to envisage an animal cell as a bag of protoplasm without an internal scaffolding. This is also incorrect, for animal cells have a complex internal scaffolding—the cytoskeleton—that is involved in the movement of cells, the movement of structures within them, and in conferring shape.

It may not be self-evident why these activities are needed. First, shape: an animal cell without a rigid cell wall would be an amorphous blob unless something conferred shape on it. We have already described (page 52) how the erythrocyte maintains its flattened biconcave disc shape by means of a submembranous protein scaffolding. Most, or all, animal cells have distinct shapes and, for this, internal structures are important.

Many cells are constantly changing their shape. This means that the cytoskeleton, which determines shape, must, in most situations, be capable of rapid disassembly and reassembly. This is not true for permanent internal structures such as that of the red cell, or of the intestinal microvilli (Fig. 4.1) but it is true for animal cells in general. Plant cells and bacterial cells with rigid cell walls are in a different category.

Secondly, movement of animals cells: the most obvious examples are those of the macrophages and other white cells of the body that migrate through tissues via ameboid-like action. However, motility is a much more widespread property of cells; many migrate over surfaces by a crawling action and this is important in embryonic development and in normal wound healing. A different type of active contraction is that involved in cell division. When a cell divides, a constriction develops around the equatorial region which progresses to the point of cell separation. Yet another form of movement is due to cilia, whose beating causes movement of surface mucous layers, for example,

in the respiratory passages, or of the flagellae that sperm use for swimming.

Thirdly, there is internal transport. A eukaryote cell is a very large object when its dimensions are considered in relation to the low rate of diffusion of chemicals in solution. A bacterial cell, being so much smaller, can get away with diffusion of solutes but many eukaryote cells require an elaborate internal transport system. Vesicles from the Golgi apparatus (page 307) must move to the cell surface or alternative destinations and may need active transport. The movement of mRNA molecules from the nucleus to the cytoplasm also may not rely simply on diffusion. The problem is most easily seen in exaggerated cases. A nerve axon may be half a metre or more in length: synthesis of proteins and vesicles occurs in the cell body and some of these have to be transported to the tip of the axon. Diffusion could never do this. The giant algal cell *Nitella* is so large that solute diffusion would not support the life of the cell; to cope with this, the cell keeps its cytoplasm continually streaming so as to achieve constant mixing. Another example is that, at cell division, chromosomes move apart into daughter cells.

The message of all this is that cells have a most elaborate internal organization devoted to shape, cell movement, and internal transport. It is less obvious than the contraction of muscles and particularly so because: (1) much of it is not visible in cells without appropriate staining; and (2) apart from a few permanent structures, the whole system is dynamic and capable of continuous modification; structures are assembled and disassembled at disconcerting speeds.

A brief statement of some essential points may help before we get to the details.

- Actin fibres, essentially the same as in muscles (page 391), are present in almost all cells; they are involved in the shape determination of animal cells and in their ameboid and crawling movements.

- In addition to their role in shape and movement, actin fibres act as transport tracks—almost ratchet railways—along which cellular constituents are actively moved.
- Myosin-like molecules exist in nonmuscle cells. One type (conventional myosin as in muscle) exerts force on adjacent actin fibres and is responsible for ameboid and crawling cell movement. Other types of myosins (minimyosins) run along actin transport tracks pulling a load. ATP supplies the energy.
- A totally distinct transport system exists in the form of microtubules (to be described). These also act as transport tracks along which run molecular motors pulling a load. Microtubule–specific 'motors' using ATP also act on adjacent microtubules causing a sliding motion and thus perform mechanical work much as in the actin–myosin system.
- A third type of cytoskeleton component is the intermediate filament (intermediate in size between actin filaments and microtubules). One type (the lamins) forms a network as part of the nuclear envelope but the cytoplasmic intermediate filaments are not essential for the survival of cells in culture. The roles of intermediate filaments are less clear than those of the other filaments but seem to be especially important in conferring toughness in certain cells such as those of the epidermis; hair cells are largely composed of the keratin intermediate filaments. The intermediate filaments are in a class quite distinct in function from actin filaments and microtubules.

After this overview we will now describe the systems more fully.

The role of actin and myosin in nonmuscle cells

Actin is an abundant protein in most eukaryote cells and, in many, it is the most abundant single protein. There are slightly different actins in different types of cells but it is a highly conserved protein and all form actin filaments, as described for muscle cells (see page 391). Myosin also appears to be an almost universal constituent of eukaryote cells but is present in much smaller amounts. The imbalance between actin and myosin in nonmuscle cells correlates with the fact that actin has a purely structural role in addition to that in contraction.

Structural roles of actin and its involvement in cell movement

A particularly clear example of a structural role for actin is seen in the microvilli of intestinal brush border cells (page 62). These are the finger-like projections that enormously increase the absorptive area of the gut lining (Fig. 29.1). Apart from this specialized role, actin filaments pervade most animal cells. They

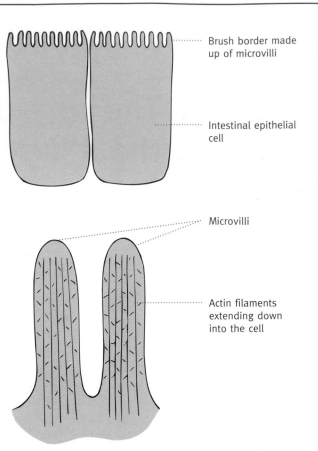

Brush border made up of microvilli

Intestinal epithelial cell

Microvilli

Actin filaments extending down into the cell

Fig. 29.1 Diagram of actin filaments in microvilli. Note that the filaments have a complex crosslinking and anchoring arrangement with other proteins to form a robust cytoskeletal structure.

are particularly dense near the cytoplasmic membrane where bundles of them anchored into the membrane form stress fibres, involved in maintaining cell shape (Fig. 29.2(a)). The arrangement of actin filaments, their anchoring and crosslinking into networks, depends on an array of other proteins that associate with the actin. The actin filaments in microvilli are permanent structures but, in general, actin filaments are disassembled and reassembled. When a phagocyte is triggered to engulf a particle, actin filaments are seen to extend into the pseudopodia (these are the foot-like extensions the cell puts out in the direction of movement).

As implied, actin filaments are involved in the ameboid and crawling actions of cells. This is the result of the action of myosin II (the 'conventional' muscle-like myosin molecule). About 16 of these molecules assemble, as required, into small bipolar filaments analogous to striated muscle thick filaments. In an action very similar to that described for sarcomeres (page 390), these exert a contracting force on adjacent actin filaments and exert a pull on the cell membrane to which the filaments are anchored (see Fig. 29.3). Control of contraction is associated with the phosphorylation of myosin light chains as in smooth muscle (page 394).

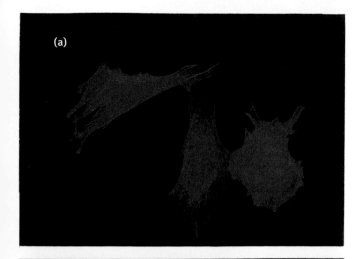

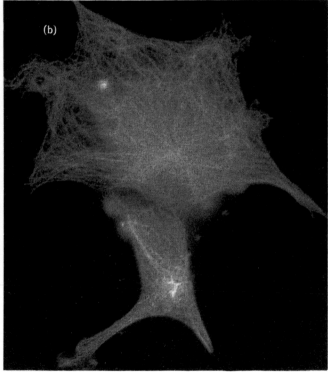

Fig. 29.2 (a) Actin filaments in mouse myoblasts visualized with an actin antibody. The filaments pervade the cell and bundles of them (called stress fibres) are attached at focal points to the cell membrane, often at points at which the membrane makes contact with a solid surface. The fibres are rapidly disassembled and assembled. Photograph kindly provided by Dr. P. Gunning, Children's Medical Research Institute, Sydney. **(b)** Microtubules in cytoplasm of a cell radiating from the microtubule organizing centre (MTOC) or centrosome (see text).

A rather precise and defined contractile role of the actin–myosin system is in **cytokinesis**—the constriction of a dividing cell into two daughter cells (see page 402). During **anaphase** an annulus of actin fibres assembles at the equatorial plane where constriction occurs. These actin filaments are postulated to be anchored at the plasma membrane. Overlapping of these filaments (with opposite polarities) presents the opportunity for cytoplasmic myosins to exert a contractional force. The set up is disbanded as the contraction progresses. The action results in cell division. The system is an excellent example of the temporary nature of the contractile arrangements that are set up in nonmuscle cells.

The role of actin and myosin in intracellular transport of materials

In our brief overview, we mentioned that actin filaments can act as transport tracks along which ATP-driven molecular motors can move cellular components. In Chapter 28, it was mentioned that, if, experimentally, detached myosin heads (muscle type) are attached to latex beads and supplied with $ATP + Ca^{2+}$, they will move along immobilized actin filaments. The 'tail' of the muscle-type myosin molecule is designed so that myosin can arrange itself in bipolar (thick) filaments that can cause adjacent actin filaments to slide over one another and thus cause a contraction to occur (page 390).

Another family of myosin molecules exists with essentially the same myosin heads but, instead of the long rod-like tail of muscle myosin, the tail is small—hence, the name **minimyosins**. Minimyosin molecules cannot form bipolar bundles but they run along the actin filament, at the expense of ATP. The tails of the minimyosin are designed to attach to other structures such as a vesicle membrane—if such an attachment occurs, the minimyosin will pull the vesicle along the actin filament 'track'. In the giant alga *Nitella*, minimyosin motors pull the endoplasmic reticulum along stationary actin filaments that circuit the cell. The minimyosin tails attach to the ER that pervades the cytoplasm and, by pulling this along, cause the protoplasm to continually circulate (or stream) in its entirety.

A variety of unconventional (that is, nonmuscle type) myosins have been identified. All have the globular actin binding heads(s) with different tails that appear to bind to specific membranes and have been found in yeast and vertebrates, including brain and other tissues.

Microtubules, cell movement, and intracellular transport

What are microtubules?

Microtubules are made by the polymerization of tubulin protein subunits. The latter are dimers of α and β tubulin molecules, the dimers being referred to as **tubulin**. Tubulin polymerizes to form a hollow tube bounded by 13 longitudinal rows of subunits. The polymerization occurs in a head to tail fashion so that a microtubule has a definite polarity with the ends referred to as plus $(+)$ and minus $(-)$ ends. Microtubules are basically unstable in the cell. They can assemble and disassemble very rapidly. The instability is associated with unprotected ends—where these occur, the tubules undergo catastrophic collapse by

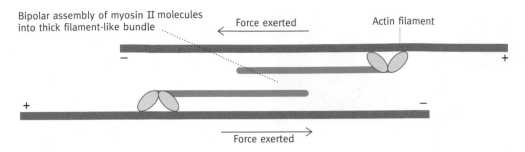

Fig. 29.3 Diagram of the pulling action of myosin II on actin filaments. Note that ATP hydrolysis supplies the energy. The achievement of useful work by this system will depend on appropriate anchoring of the actin filaments to the cell membrane so as to change cell shape. See page 391 for an explanation of the (+) and (−) polarities of the actin filaments Each myosin molecule has a region capable of interacting with other molecules and thus allowing self-assembly of bipolar filaments. The diagram is a cross-sectional representation—the bipolar assemblies are cylindrical. In the arrangement of actin fibres and myosin bundle shown, a contraction results from the relative sliding of the actin filaments.

depolymerizing into free tubulin subunits. In the cell the (−) ends are associated with the **microtubule organizing centre (MTOC)**, a structure near the nucleus and containing (in animal cells) a pair of small bodies, the **centrioles**, made of fused microtubules. The cell is pervaded by microtubules radiating out from the MTOC (Figs. 29.2(b) and 29.4).

What protects the positive (+) ends of microtubules?

As stated, microtubules rapidly collapse unless their ends are protected. The MTOC protects the (−) end. Microtubules grow (and collapse) in the cell from the (+) ends. It is thought that, when the microtubule reaches an appropriate target, target proteins 'cap' and protect the microtubule. The arrangement means that microtubules can grow out of the MTOC in completely random directions—those that make contact with an appropriate component of the cell will, in this mode, be capped and stabilized while the remainder will collapse. Obviously what constitute suitable target proteins and how these are arranged is a separate major question. There is still the problem of what

protects a growing microtubule, *before* it reaches a target protein, from catastrophic collapse. This is an area of intensive research but a current model is that tubulin subunits have GTP attached to them; a GTP-tubulin subunit protects the (+) end. The GTP is hydrolysed to GDP and P_i after the addition of a subunit to a tubule, but not instantly. Thus newly added subunits will be in the GTP form for a brief period, thus temporarily 'capping' the microtubule, which protects the end from collapse. The GTP hydrolysis in this model is acting as a clock (as in other systems—page 357); provided the addition of new subunits occurs before the last one added loses its GTP, the microtubule is protected. If not and if the GTP on the exposed end subunit has been hydrolysed, then collapse, in this model, ensues. In cilia and flagella where permanent microtubules occur (see below), covalent modification of the protein occurs after assembly. This then is the basic outline of microtubules. We should now turn to their functions.

Functions of microtubules

As with actin fibres, microtubules are involved in cellular morphology, cell movement, and intracellular transport.

Molecular motors involved in microtubule-associated movement

Two types of motors have been identified, kinesin and dynein, both of which are ATP-dependent. Kinesin travels along a microtubule in the (−) → (+) direction, and dynein in the opposite direction.

The motors have the general design of a pair of globular heads reminiscent of those of myosin (Fig. 29.5(a)) with light chains at the opposite and of the molecule.

There are families of kinesin and dynein molecules specialized for different tissue functions. The heads that perform the actual movement along the microtubule are probably a constant motif while the tail structures are modified to attach to specific structures (Fig. 29.5(b)). As an example, vesicles are pulled along microtubules by molecular motors (see below). Kinesin and dynein molecules (supplied with ATP) will walk along

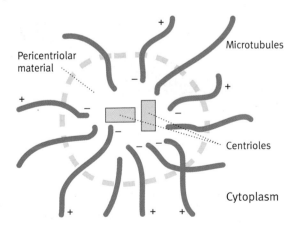

Fig. 29.4 Diagram of microtubules radiating out from the MTOC. The (+) and (−) indicate the polarity of the microtubules. The centrioles are a pair of tube-like structures made of fused microtubules; the radiating microtubules originate in the material surrounding the centrioles. (There is no precise boundary to this.)

(a)

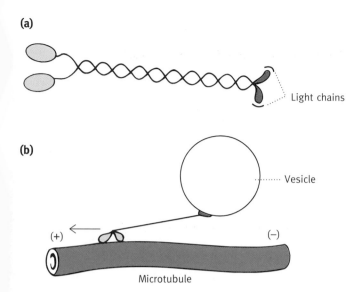

Light chains

(b)

Vesicle

(+)   (−)

Microtubule

Fig. 29.5 (a) Diagram of the structure of a kinesin molecule.
(b) Diagram illustrating how a kinesin molecule can transport vesicles along a microtubule. Also see Fig. 29.6 for an electron micrograph of vesicles being transported within a cell.

microtubule fibres immobilized on a solid, just as myosin heads will walk along immobilized actin fibres.

The role of microtubules in cell movement

This is confined to cilia and flagellae. Cells lining the respiratory passages have large numbers of cilia whose beating motion sweeps along mucus and its entrapped foreign particles. Sperm propel themselves with flagellae. Cilia and flagellae are basically similar. Microtubules run down the length of the organelle, originating in a basal body closely resembling a centriole; attached to the microtubules are **dynein molecules**—these are the 'motors'. Using ATP energy, they 'walk' along microtubules (towards the (−) end) causing the latter to slide relative to each other and resulting in a wave motion in the cilium or flagellum.

The role of microtubules in vesicle transport inside the cell

Microtubules have an intracellular transport role. Vesicles produced by the Golgi apparatus (see page 307) are targeted on different destinations and active, guided transport may be involved. How the detailed programming of movements and vesicles is achieved is unknown, but microtubules and associated ATP-driven motors appear to be involved. The clearest cases of this can be seen, as expected, in cells where the need is most exaggerated. As mentioned in the introduction, the axon of a nerve cell can be enormously long so that the axon tip is far removed from the main body of the cell where synthesis of vesicles and proteins occurs. To transport these, microtubules extend down the axon to the tip, with the (+) ends oriented

towards the tip. A molecular motor, **kinesin**, is attached to membrane vesicles at one end and moves along the microtubule in the (+) direction (Fig. 29.5(b)). Another motor, **cytoplasmic dynein**, transports in the reverse direction from the tip to cell body. The movement of pigment vesicles (Fig. 29.6) is a clearly demonstrated case of vesicle transport along microtubules.

The role of microtubules in mitosis

The stages of cell cycle mitosis are shown in Fig. 29.7(a).

In **prophase**, the **centrosome** (equals the MTOC in nondividing cells) is duplicated and the two migrate to the opposite poles of the cell; microtubules begin to radiate out from them, to become the **mitotic spindle**. The DNA condenses to form the compact chromosomes seen in the light microscope. At **prometaphase** the nuclear membrane disappears and the microtubules of the spindle enter the nuclear area and attach to the centromeres of the chromatids. (The latter are identical sister chromosomes, formed by the DNA replication at interphase, and associated at their central regions into the familiar 'X'-shaped structures (Fig. 19.9). When separated, each chromatid will be a chromosome.) The chromatids at **metaphase** become arranged on a central equatorial plate ready for **anaphase**. In the latter, the chromatids separate and move to the poles (**anaphase A**) while at the same time the poles move apart (**anaphase B**). In **telophase**, re-establishment of a nuclear membrane occurs. This is followed by cell division, the cleavage or cytokinesis being a function of the actin–myosin system. The centrosome in each daughter cell then becomes the site of formation of the microtubule array of the interphase cell.

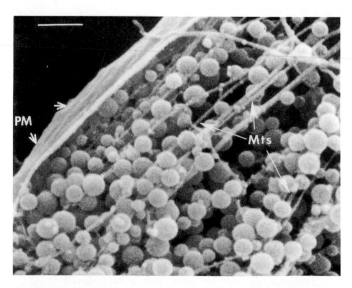

PM

Mts

Fig. 29.6 Vesicle transport by microtubules. Scanning electron micrograph of pigment-containing vesicles being transported along microtubules in a chromatophore of a squirrel fish. The change of colour of the cell is effected by the movement of the vesicles to and from the cell centre. Mts, microtubules; PM, plasma membrane. Bar scale = 0.5 μm

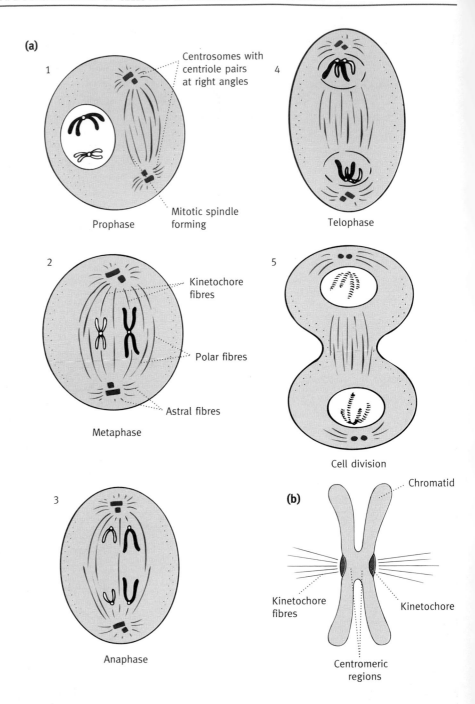

Fig. 29.7 (a) A simplified diagram of the phases in mitosis. The polar fibres are shown in red; these are believed to be responsible for the poles moving apart. The fibres in black are kinetochore fibres responsible for chromosome migration to the poles. All of the fibres are microtubules. **(b)** Diagram of kinetochore fibre attachment.

The mechanism of chromosome migration (anaphase A)

It is not certain how the chromosomes are caused to move to the poles as a result of their attachment to **kinetochore fibres** (Fig. 29.7(b)) though clearly a molecular motor, such as dynein, is a possibility. During the process, the kinetochore fibres shorten. Microtubules do not contract and it is believed that the chromosome moves along the fibre, the microtubule being depolymerized as it goes.

The mechanism of pole separation (anaphase B)

This is the function of the **polar fibres** (Fig. 29.7(a)) which do not attach to chromosomes but overlap in the equatorial area. As the poles move apart they elongate to maintain the overlap. The fact of overlap makes it tempting to speculate that molecular motors operate between them, causing the poles to move apart. The **astral fibres** radiating out in all directions from the poles could conceivably be associated with motors pulling the poles apart. There is evidence that, whatever the mechanism, ATP is needed.

Intermediate filaments

The actin network, described above, has microfilaments 6 nm in diameter; the microtubule filaments are 20 nm in diameter. The third network in eukaryote cells is made of filaments averaging 10 nm in diameter—hence their name of **intermediate filaments (IFs)**.

There are several types of IF composed of a diverse group of proteins that, nonetheless, have close homology in their structures. Typically there is a core filament about 350 amino acid residues in length with ends varying in the different types of IF. This diversity of proteins contrasts with the single protein subunits in actin and microtubule filaments.

Different types of IF occur in different eukaryote cells and are expressed at specific stages of development and differentiation, suggesting that they play important roles. Despite this, there is no known general functional role for IF networks—all that is available in this respect are some specific cases where likely roles may be envisaged. As mentioned earlier, these include keratin IFs in epidermal cells, which confer toughness (page 34); a number of lamin proteins form a network associated with the inner surface of the inner nuclear membrane and are associated with nuclear pores; neurofilaments exist in nerve cells to give necessary mechanical support to the long axons; desmin filaments are located in the Z discs of sarcomeres. Disconcertingly, however, cultured cells lacking IFs (due to mutation or injection of specific antibodies) appear to grow and divide quite happily.

Further reading

Bray, D. (1992). *Cell movements.* Garland Publishing.
A comprehensive and authoritative, but readable, account of all aspects of movement in cells. Good simple diagrams and also illustrated with electron micrographs.

Lewin, B. (1994). *Genes V.* Oxford University Press.

Actin in nonmuscle cells

Stossel, T. P. (1994). The machinery of cell crawling. *Sci. Amer.*, **271**(3), 40–7.
The regulated assembly and disassembly of actin filaments is the mechanism for cell movement; it involves calcium, phospholipid and actin-binding protein.

Microtubules in cell movement

Byard, E. H. and Lange, B. M. H. (1991). Tubulin and microtubules. *Essays in Biochemistry*, **26**, 13–25.
Excellent general review.

Goldstein, L. S. B. (1993). With apologies to Scheherazade: tails of 1001 kinesin motors. *Ann. Review Genetics*, **27**, 319–51.

An in depth review of these minimotors with useful overviews.

Walker, R. A. and Sheetz, M. P. (1993). Cytoplasmic microtubule-associated motors. *Ann. Rev. Biochem.*, **62**, 429–51.
Detailed review of the minimotors and their roles.

Wallee, R. B. and Sheetz, M. P. (1996). Targeting of motor proteins. *Science*, **271**, 1539–44.
Discusses the movements of kinesins and dyneins on microtubules.

Microtubules in vesicle transport

McNiven, M. A. and Ward, J. B. (1988). Calcium regulation of pigment transport in vitro. *J. Cell Biol.*, **106**, 111–25.
A research paper but readable.

Microtubules in mitosis

Glover, D. M., Gonzalez, C., and Raff, J. W. (1993). The centrosome. *Sci. Amer.*, **268**(6), 32–9.
Very useful review of this relatively little known structure.

Problems for Chapter 29

1   Actin is found in nonmuscle cells. What are its roles there?

2   What are microtubules?

3   Microtubules with unprotected ends undergo catastrophic collapse. What protects the ends?

4   What are kinesin and dynein? Illustrate by a diagram.

5   At cell division, chromosomes on the metaphase equatorial plate move apart. Microtubules are attached to the kinetochores and shorten as the chromosomes move apart. Does this mean that microtubules contract? Explain your answer.

6   What are intermediate filaments?

7   What are the functions of intermediate filaments?

Chapter summary

Answers to problems

Chapter 1

1   The amount of ATP that can be synthesized using 5000 kJ of free energy is 5000/55 which equals 90.91 mol. The weight of disodium ATP produced is thus 551×91 g daily, or 50 141g which is equal to 72% of the man's body weight. The reason why this is possible is that ATP is being continuously recycled to $ADP + P_i$ and back again to ATP.

2   The $\Delta G^{0'}$ value refers to standard conditions where ATP, ADP, and P_i are present at 1.0 M concentrations. In the cell the concentrations will be very much lower, the actual ΔG value for ATP synthesis will be different from the $\Delta G^{0'}$ value according to the relationship

$$\Delta G = \Delta G^{0'} + RT\, 2.303 \log_{10} \frac{[ADP][P_i]}{[ATP]}.$$

3   ATP and ADP are high-energy phosphoric anhydride compounds, whereas AMP is a low-energy phosphate ester. The factors that make hydrolysis of the former more strongly exergonic include the following.

Release of phosphate relieves the strain caused by the electrostatic repulsion between the negatively charged phosphate groups. The released phosphate ions fly apart. A factor also contributing to the exergonic nature of the hydrolyses is the resonance stabilization of the phosphate ion which exceeds that of the phosphoryl group in ATP. Hydrolysis of AMP causes little increase in resonance stabilization.

4   Very little of signifcance because the plot will be arbitrary depending on the time period selected for the measurement of activity. This is because, at higher temperatures, destruction of the enzyme is likely to occur and, at any given temperature, the amount of destruction will be proportional to the time the enzyme is exposed to that temperature. If information is sought on the heat stability of the enzyme it is much better to expose it for a standard time

at the different temperatures and, after cooling, to measure the activity in each sample at the normal incubation temperature.

5   (a) The nonpolar molecule cannot form hydrogen bonds with water molecules so that those of the latter surrounding the benzene molecule are forced into a higher-order arrangement in which they can still hydrogen bond with each other (for the total bonding does not change). This increased order lowers the entropy and increases the energy of the system; insertion of the benzene molecules into the water is therefore opposed and they are forced into a situation with minimal benzene/water interface as spherical globules and then as a separate layer. The effect is known as hydrophobic force.

(b) The polar groups on glucose can form hydrogen bonds with water.

(c) The Na^+ and Cl^- ions become hydrated and the free energy decreases as a result of this exceeds that of the ionic attraction between them. The separation also has a large negative entropy value.

6   The enzyme AMP kinase transfers a phosphoryl group from ATP to AMP by the reaction

ATP + AMP → 2ADP.

Hydrolysis is not involved so that there is no significant $\Delta G^{0'}$ in the reaction.

7   In the cell, PP_i will be hydrolysed to $2P_i$, whereas with a completely pure enzyme there will be no inorganic pyrophosphatase present. In the former case, the $\Delta G^{0'}$ of the total reaction will be $-32.2 - 33.4 + 10$ kJ mol^{-1} $= -55.6$ kJ mol^{-1}. For the completely pure enzyme, the $\Delta G^{0'}$ will be -22.2 kJ mol^{-1}.

8   Ionic bonds, hydrogen bonds, and van der Waals forces with average energies of 20, 12–29, and 4–8 kJ mol^{-1}, respectively. The energies of activation for formation of

weak bonds are very low and so can occur without catalysis. Weak bonds in large number can confer definite structures on molecules, but nonetheless can be broken easily, resulting in flexibility of structures. The facts that (a) numbers of weak bonds are required for molecular association and (b) they are shortrange forces are the basis of biological specificity.

9 (a)

$$
\begin{array}{cccc}
\underset{\text{pH0}}{\text{HO}-\overset{\displaystyle\overset{O}{\|}}{\underset{\displaystyle\underset{OH}{|}}{P}}-\text{OH}} &
\underset{\text{pH4}}{\text{HO}-\overset{\displaystyle\overset{O}{\|}}{\underset{\displaystyle\underset{OH}{|}}{P}}-\text{O}^-} &
\underset{\text{pH9}}{\text{HO}-\overset{\displaystyle\overset{O}{\|}}{\underset{\displaystyle\underset{O^-}{|}}{P}}-\text{O}^-} &
\underset{\text{pH14}}{{}^-\text{O}-\overset{\displaystyle\overset{O}{\|}}{\underset{\displaystyle\underset{O^-}{|}}{P}}-\text{O}^-}
\end{array}
$$

(b)

$$
pH = pK_a + \log\frac{[\text{Salt}]}{[\text{Acid}]}
$$

The relevant pK_a is 7.2.

$$
\begin{aligned}
pH &= 7.2 + \log\frac{[\text{Na}_2\text{HPO}_4]}{[\text{NaH}_2\text{PO}_4]} \\
&= 7.2 + \log 1 \\
&= 7.2.
\end{aligned}
$$

(c) Histidine with a pK_a value of 6.5.

10 The $\Delta G^{0'}$ value of a reaction determines whether a reaction may proceed, but says nothing about the rate at which it does proceed (if at all). The latter is determined by the energy of activation of the reaction and the rate at which the transition state is formed.

11 This can be due to several factors.

(a) The active site binds the transition state much more firmly than it does the substrate and, in doing so, lowers the activation energy.

(b) It positions molecules in favourable orientations.

(c) It can exert general acid–base catalysis on a reaction.

(d) It may position a metal group which facilitates the reaction.

Chapter 2

1 It is a polypeptide chain composed of large numbers of amino acids linked together by peptide (−CO−NH−) bonds. The structure of such a chain is given on page 24.

2 The polypeptide chain of a protein is folded into a specific, usually compact shape, the precise folding being dependent on noncovalent bonds and covalent disulfide bonds between the amino acid side chains. Heat disrupts the former bonds, causing the polypeptide chains to unravel and become entangled into an insoluble mass, devoid of biological activity. This is known as protein denaturation.

3 Examples of all are given on pages 25 and 26.

4 (a) Around 4; (b) around 10.5–12.5; (c) around 6.0.

5 Aspartic and glutamic acids, lysine, arginine, and histidine. The carboxyl and amino groups of all other amino acids are bound in peptide linkage (excluding the two end ones).

6 Primary, secondary, tertiary, and quaternary. Illustrated in Fig. 2.2

7 The structures are the α helix, the β-pleated sheet, and the random coil or connecting loop region. The first two are structures that satisfy the hydrogen bonding potential of the polypeptide backbone. The structures are given in Figs. 2.3 and 2.4. The random coil has no consistent structure. It usually is on the exterior of proteins where hydrogen bonding can be satisfied by water.

8 The structure of collagen fibres is given in Fig. 2.8.

9 It lies in the four-way desmosine group shown in Fig. 2.9.

Chapter 3

1 Two hydrophobic tails rather than a single one.

2 No.

3 Only by exocytosis, endocytosis, or by special transport mechanisms (described in Chapter 22).

4 It acts as a fluidity buffer by preventing the polar lipids from associating too closely.

5

$$
\begin{array}{l}
\text{CH}_2-\text{O}\cdot\text{CO}-\text{R} \\
| \\
\text{CH}-\text{O}\cdot\text{CO}-\text{R} \\
| \qquad\quad \overset{\displaystyle O}{\underset{\displaystyle \|}{}} \\
\text{CH}_2-\text{O}-\overset{\displaystyle\overset{O}{\|}}{\underset{\displaystyle\underset{O^-}{|}}{P}}-\text{O}^-
\end{array}
$$

6 Lecithin (choline); cephalin (ethanolamine); phosphatidyl-serine (serine).

7 As on page 43.

8 The choline is replaced by galactose and an oligosaccharide, respectively.

9 As on page 44.

10 They put a kink into the tail which prevents close packing of the fatty acyl tails of lipid bilayers and thus make the membrane more fluid.

11 To do so would require stripping the molecules of H_2O molecules, an energetically unfavourable process.

12 It involves a membrane protein permitting solutes to traverse the membrane in either direction, according to the direction of the concentration gradient. No energy input is involved. The anion transport of red blood cells is an example; it allows Cl^- and HCO_3^- ions to pass through in either direction.

13 C_1 = the concentration outside and C_2 that inside.

$$\Delta G = RT\,2.303 \log_{10} \frac{[C_2]}{[C_1]}$$
$$= (8.315\,\text{J mol}^{-1})(298)2.303 \log_{10} \frac{10}{1}$$
$$= 5706\,\text{J mol}^{-1}$$

14 This is described in Fig. 3.10.

Chapter 4

1 Pepsin as pepsinogen; chymotrypsin as chymotrypsinogen; trypsin as trypsinogen; elastase as proelastase; carboxypepidase as procarboxypepidase.

2 The proteolytic enzymes are potentially dangerous in that they could attack proteins lining the ducts. There are no components that amylase might attack. Mucins coating the gut cell lining of the intestine protect cells against proteolytic attack.

3 In the stomach the acid pH causes a conformational change in pepsinogen that activates molecules to self-cleave the extra peptide in pepsinogen that inactivates the enzyme. Once some pepsin is formed, it activates more pepsinogen so that an autocatalytic activation cascade occurs. In the small intestine, enteropeptidase, an enzyme produced by gut cells, activates trypsinogen to trypsin. This, in turn, activates the other zymogens.

4 Pancreatitis, or inflammation of the pancreas, due to proteolytic damage of pancreatic cells.

5 Because the gut cells can absorb only monomers (plus monoacylglycerols). Fat, protein, polysaccharides, and di-saccharides cannot be absorbed.

6 See Fig. 4.4

7 Milk contains lactose, which must be hydrolysed to glucose and galactose by the enzyme lactose. After infancy, many individuals lose the capacity to produce this. The lactose is not absorbed and is fermented in the large intestine. It

attracts water into the gut by its osmotic effect, producing a diarrhea.

8 TAG or triacylglycerol.

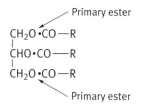
Primary ester

Primary ester

9 The fat is emulsified by intestinal movement and by the monocylglycerol and free fatty acids produced by initial digestion together with bile salts, so that the lipase can attack the emulsified substrate. The products of digestion (free fatty acids and monoacylglycerol) are carried to the intestinal cells as disc-like mixed micelles with bile salts that have a high carrying capacity for such products. Probably at the cell surface the micelle breaks down.

10 The free fatty acids and monoacylglycerol are resynthesized into TAG. The TAG cannot traverse cell membranes but it is incorporated into lipoprotein particles, called chylomicrons. These have a shell of stabilizing phospholipid, some proteins, and a centre of TAG and cholesterol and its ester. The chylomicrons are released by exocytosis into the lymphatics and eventually discharge into the blood as a milky emulsion via the thoracic duct.

Chapter 5

1 The osmotic pressure of glucose precludes this. The osmotic pressure of a solution is related to the number of particles of solute in it. By polymerizing thousands of glucose molecules into a single glycogen molecule, the osmotic pressure is correspondingly reduced.

2 Glycogen stores in the liver are sufficient to last through 24 h starvation only, but there may be sufficient TAG for weeks. TAG is a more concentrated energy source—it is more highly reduced and is not hydrated. If the energy in fat were present as glycogen, the body would have to be much larger.

3 Yes—glucose can be converted to acetyl-CoA.
No—acetyl-CoA cannot be converted into pyruvate.
No—there are no special amino acid storage proteins (excluding milk) in the body.

4 (a) It is the glucostat of the body. It stores glucose as glycogen in times of plenty; it releases glucose in fasting to keep a constant blood sugar level which the brain needs.

(b) During prolonged starvation it synthesizes glucose for the same reason; it converts fats to ketone bodies which the brain (and other tissues) can use for part of its energy supplies and so conserve glucose.

(c) It synthesizes fat and exports it to other tissues.

5 No.

6 They store fat in times of plenty; they release fatty acids in fasting.

7 They are completely dependent on glucose, which they convert to lactate. Since mature red blood cells lack mitochondria, they are unable to produce ATP by any metabolic route except glycolysis.

8 When blood glucose is high, insulin is released. This is the signal for tissues to store food. As the blood glucose levels fall, insulin levels fall and glucagon levels rise. The latter is the signal for the liver to release glucose and for adipose cells to release free fatty acids for the tissues to utilize. The brain is not insulin-dependent for its uptake of glucose, so this can proceed even in starvation, when insulin levels are extremely low.

In addition, epinephrine can override all controls, causing massive turnout of glucose and fatty acids to cope with an emergency requiring the 'fight or flight' reaction.

Chapter 6

1 Glucose-1-phosphate $+$ UTP $\rightarrow$ UDP-glucose $+$ PP$_i$.
PP$_i$ $+$ H$_2$O $\rightarrow$ 2P$_i$.
UDPG $+$ glycogen$_{(n)}$ $\rightarrow$ UDP $+$ glycogen$_{(n+1)}$.

The hydrolysis of PP$_i$ makes the overall reaction strongly exergonic.

2 The reaction named never occurs in the cell because the required substrate inorganic pyrophosphate is immediately destroyed. It is the reverse reaction that occurs in the cell but, for systematic nomenclature reasons, the enzyme is named for the forward reaction.

3 Only the liver (and the less quantitatively important kidney) does so. These alone possess the enzyme glucose-6-phosphatase.

4 Glucose $+$ ATP $\rightarrow$ Glucose-6-phosphate $+$ ADP.

Glucokinase has a much higher K_m than hexokinase. During starvation the liver turns out glucose into the blood, primarily to supply the brain (and red blood cells). The first reaction in the uptake of glucose into brain and liver is phosphorylation of glucose. The lower affinity of glucokinase for glucose, as compared with hexokinase, means that the liver does not compete with brain for blood sugar. It efficiently takes up glucose only when blood sugar levels are high. Since it is not inhibited by glucose-6-phosphate as is

hexokinase, it can take up glucose and synthesize glycogen even when cellular levels of glucose-6-phosphate are high.

5 Many glycolipids and glycoproteins contain galactose. Removal of galactose from the diet does not prevent synthesis of these because UDP-glucose can be epimerized to UDP-galactose.

6 In the capillaries, lipoprotein lipase splits free fatty acids from TAG, and these immediately enter adjacent cells.

7 VLDL are lipoproteins, resembling chylomicrons, that carry cholesterol and TAG from the liver to peripheral tissues.

8 Cholesterol moves outwards from the liver to peripheral tissues in VLDL. The VLDL become depleted of TAG and pass through lipoprotein stages known as IDL and finally LDL. The latter are taken up by peripheral tissues. However, some of the IDL and LDL are re-taken up by the liver. The IDL and LDL receive peripheral tissue cholersterol via HDL and this constitutes a reversal flow of cholesterol from tissues to liver.

9 Conversion to bile salts in the liver.

10 The enzyme LCAT transfer fatty acyl groups from lecithin to cholesterol.

11 TAG is *not* released; a hormone-sensitive lipase, stimulated by glucagon (or epinephrine in emergency situations), hydrolyses off free fatty acids that are carried in the blood to the tissues attached loosely to serum albumin.

Chapter 7

1 Glycolysis, the citric acid cycle, and the electron transport chain; cytoplasm, mitochondrial matrix, and the inner mitochondrial membrane, respectively.

2 Structures on page 96.

$$AH_2 + NAD^+ \leftrightarrow A + NADH + H^+$$
$$B + NADH + H^+ \leftrightarrow BH_2 + NAD^+$$

3 FAD is another hydrogen carrier; it is reduced to FADH$_2$. It is not a coenzyme but a prosthetic group attached to enzymes.

4 In aerobic glycolysis, glucose is broken down to pyruvate. The NADH produced is reoxidized by mitochondria. In anaerobic glycolysis, the rate of glycolysis exceeds the capacity of mitochondria to reoxidize NADH. This can occur, for example, in the 'fight or flight' reaction. Since the supply of NAD$^+$ is limited, if NADH were not reoxidized sufficiently rapidly to NAD$^+$, glycolysis and its ATP

production would cease. An emergency mechanism reoxidizes NADH by reducing pyruvate to lactate.

$$CH_3COCOO^- + NADH + H^+ \rightarrow CH_3CHOHCOO^- + NAD^+$$

Pyruvate   Lactate   Lactate
dehydrogenase

5   Structures on page 99.
-31 as compared with -20 kJ mol^{-1} for the carboxylic ester—that is, the thiol ester is a high-energy compound.

6   Pyruvate $+$ CoASH $+$ NAD$^+$ $\rightarrow$ Acetyl-S-CoA $+$ NADH$^+$ $+ H^+ + CO_2$ $\Delta G^{0'} = -33.5$ kJ mol^{-1}.

7   The acetyl group is fed into the citric acid cycle.

8   They are reoxidized by the electron transport chain producing H_2O and generating ATP in the process.

9   The Nernst equation relates $\Delta G^{0'}$ and $\Delta E_0'$ values as follows.
$\Delta G^{0'} = -nF \, \Delta E_0'$ (F is the Faraday constant $= 96.5$ kJ V^{-1} mol^{-1})
and $\Delta E_0'$ is the difference between the redox potentials of the electron donor and acceptor.
In the example given $\Delta E_0'$ equals -1.035 V $(-0.219 - 0.816$ V).

Therefore,
$\Delta G^{0'} = -2$ (96.5 kJ V^{-1} mol^{-1}) (-1.035 V)
$= -193 \, (-1.035) = 194.06$ kJ mol^{-1}.

10   Breakdown of fatty acids to acetyl-CoA.

11   (a) Yes. Glucose is converted to pyruvate and pyruvate dehydrogenase converts this to acetyl-CoA. The latter can be used to synthesize fat.

(b) No. Fatty acids are broken down to acetyl-CoA. To synthesize glucose, pyruvate is needed. The pyruvate dehydrogenase reaction is irreversible. There can, in animals, be no net conversion of acetyl-CoA (and therefore of fatty acids) to glucose.

(c) Yes, but not by the reverse of the pyruvate dehydrogenase reaction (see legend to Fig. 7.11).

Chapter 8

1   The splitting of the C_6 molecule into two C_3 molecules occurs by the aldol split catalysed by aldolase. This requires the structure

```
      R
      |
      C=O
      |
R'—C—R''
      |
  H—C—OH
      |
      R
```

The conversion of glucose-6-phosphate to fructose-6-phosphate produces the aldol structure capable of splitting in this way. In straight chain formulae,

```
CHO                CHOH
|                  |
CHOH               CO
|                  |
CHOH               CHOH
|                  |
CHOH               CHOH
|                  |
CHOH               CHOH
|                  |
CH2OPO3^2-         CH2OPO3^2-
```

Glucose-6-phosphate   Fructose-6-phosphate

2   It occurs when a 'high-energy phosphoryl group' transferable to ADP is generated in covalent attachment to a substrate. Oxidation of glyceraldehyde-3-phosphate by the mechanism given on page 110 is an example.

3   Pyruvate kinase works in the reverse reaction but the nomenclature of kinases always derives from the reaction using ATP. The reaction is irreversible because the product of the reaction, enolpyruvate, spontaneously isomerizes into the keto-form, a reaction with a large negative $\Delta G^{0'}$.

4   3 and 2, respectively. Phosphorylsis of glycogen using P$_i$ generates glucose-1-phosphate, convertible to glucose-6-phosphate. To produce the latter from glucose, a molecule of ATP is consumed.

5   Cytoplasmic NADH itself cannot enter the mitochondrion; instead the electrons must be transported in by one of two alternative shuttles. The malate–aspartate shuttle (Fig. 8.9) reduces mitochondrial NAD$^+$, but the glycerophosphate shuttle reduces FAD attached to glycerophosphate dehydrogenase of the inner mitochondrial membrane. The redox potential of the former is more negative than that of the latter and the two enter the electron transport chain at different respiratory complexes. The yield of ATP is lower from the glycerophosphate shuttle.

6   In the breakdown of succinyl-CoA, a molecule of GTP is formed from GDP and P$_i$. This effectively releases one molecule of H_2O (which doesn't appear as such), but supplies the elements of it to the cycle, thus balancing the equation.

7   Oxidation forms a β-keto acid that readily decarboxylates.

8   The reaction is catalysed by pyruvate carboxylase.

$$
\begin{array}{c}
COO^- \\
| \\
C{=}O \\
| \\
CH_3
\end{array} + HCO_3^- + ATP
$$

$$
\begin{array}{c}
O \\
\| \\
C{-}COO^- \\
| \\
H_2C{-}COO^-
\end{array} + ADP + P_i + H^+
$$

9   Biotin, a B group vitamin. Using ATP, it forms a reactive carboxybiotin that can donate a carboxy group to substrates. The reaction is

10   This is shown in Fig. 8.19.

11   They are both mobile electron carriers. Ubiquinone connects respiratory complexes I and II with III and cytochrome c connects complexes III and IV. The former exists in the hydrophobic lipid bilayer; the latter in the aqueous phase on the outside of the inner mitochondrial membrane.

12   To pump protons from the mitochondrial matrix to the outside of the inner mitochondrial membrane and so generate a proton gradient and a charge gradient that can be used to drive ATP synthesis.

13   In the eukaryote, NADH generated in the cytoplasm has to have its electrons transferred to the electron transport pathway. If the glycerol-3-phosphate shuttle is employed for this we lose one ATP generated per NADH that has to be so handled. Since there are two NADH molecules generated per glucose glycolysed, this leads to a potential loss of two ATP

molecules. In *E. coli* there is no such problem. In addition, in *E. coli* there is no expenditure of the energy used in eukaryotes to exchange ATP and ADP across the mitochondrial membrane.

Chapter 9

1   (a) From FFA carried by the serum albumin in the blood; these originate from the fat cells.

(b) From chylomicrons—lipoprotein lipase releases FFA and TAG.

(c) From VLDL produced by the liver—released in the same way as (b).

2   The brain, red blood cells (the latter have no mitochondria). See Chapter 5.

3   (a) Activation to convert them to acyl-CoA derivates.

(b) On the outer mitochondrial membrane.

(c) In the mitochondrial matrix.

(d) As carnitine derivatives, as shown in Fig. 9.1

4   The reactions in Fig. 9.2 are analogous to the succinate $\rightarrow$ fumarate $\rightarrow$ malate $\rightarrow$ oxaloacetate reactions in the citric acid cycle, both in the reactions and in the electron acceptors involved.

5   One molecule of palmitic acid produces eight of acetyl-CoA and generates seven $FADH_2$ and seven NADH molecules. NADH oxidation is estimated to produce 2.5, and $FADH_2$ 1.5 molecules of ATP per molecule of NADH and $FADH_2$. If you add to this the yield of ATP from the oxidation of eight molecules of acetyl-CoA (ten per molecule), the total is 108 molecules of ATP (counting the GTP from the citric acid cycle as ATP). From this must be subtracted two used in the activation reaction = a net yield of 106 molecules.

6   After two rounds of β-oxidation, the *cis*-Δ^3-enoyl-CoA is isomerized to become the *trans*- Δ^2-enoyl-CoA (see page 9.5 for reactions).

7   In situations of rapid fat release from adipose cells such as occurs in starvation or diabetes, the liver converts acetyl-CoA to ketone bodies that are released into the blood. These are preferentially utilized by muscle, thus conserving glucose; most importantly, the brain can obtain about half its energy needs from ketone bodies.

8   Acetoacetate synthesis occurs in the mitochondrial matrix and cholesterol synthesis in the cytoplasmic compartment, the process occuring on the ER membrane.

Chapter 10

1   Fatty acids are synthesized two carbon atoms at a time, but the donor of these is a three-carbon unit, malonyl-CoA. Acetyl-CoA is converted to the latter by an ATP-dependent carboxylation. The subsequent decarboxylation results in a large negative $\Delta G^{0'}$ value. In other words, the point of the carboxylation and decarboxylation is to make the process of adding two carbon atom units to the growing fatty acid chain irreversible.

2   This is given in Fig. 10.1.

3   In eukaryotes, all of the enzyme reactions are organized into a single protein molecule with the enzymic functions catalysed by separate domains. The functional unit is a dimer with the two molecules cooperating as a single entity. In *E. coli* the different activities are catalysed by separate enzymes. The advantage of the eukaryote situation is that the intermediates are transferred from one active centre to the next. In *E. coli* the products must diffuse to the next enzyme so that the process is slower.

4   See page 141 for structure. NAD^+ is used in catabolic reactions—it accepts electrons for oxidation and energy generation. $NADP^+$ is involved in the reverse—in reductive syntheses. The existence of the two is a form of metabolic compartmentation that facilitates independent regulation of the processes.

5   The main sites in the body are liver and adipose cells and the mammary gland during lactation.

6   The acetyl-CoA in the mitochondrion is converted to citrate; the latter is transported into the cytosol where citrate lyase cleaves citrate to acetyl-CoA and oxaloacetate. This is an ATP-requiring reaction that ensures complete cleavage.

$$Citrate + ATP + CoA—SH + H_2 \rightarrow$$
$$acetyl\text{-}CoA + oxaloacetate + ADP + P_i$$

7   The oxaloacetate is reduced to malate by malate dehydrogenase, an NADH-requiring reaction. The malate is oxidized and decarboxylated to pyruvate by the malic enzyme, an $NADP^+$ requiring reaction. This scheme effectively switches reducing equivalents from NADH to NADPH. The pyruvate generated returns to the mitochondrion, as illustrated in Fig. 10.4

   This generates only one NADPH per malonyl-CoA produced, while the reduction steps in fatty acid synthesis require two. The rest is generated by the glucose-6-phosphate dehydrogenase system, described in Chapter 13.

8   The scheme is shown in Fig. 10.5.

9   In the synthesis of glycerol-based phospholipids there are two routes. In one, phosphatidic acid is joined to an alcohol such as ethanolamine (see Fig. 10.6). For this the alcohol is activated; in phospholipid synthesis, the activated molecule is always a CDP-alcohol formed as described on page 10.10. For some glycerophospholipid syntheses, the diacylglycerol component is activated (see Fig. 10.6). Again, this is by formation of the CDP-diaclyglycerol complex. The situation is reminiscent of the use of UDP-glucose whenever an activated glucose moiety is wanted.

10 (a) Eicosanoids have 20 carbon atoms (*eikosi* = 20). They include prostaglandins, thromboxanes, and leukotrienes.

    (b) All are related to and synthesized from polyunsaturated fatty acids.

    (c) Prostaglandins cause pain, inflammation, and fever. Thromboxanes affect platelet aggregation. Leukotrienes case smooth muscle contraction and are a factor in asthma, by constricting airways.

    (d) Aspirin inhibits cyclooxygenase, an enzyme involved in their synthesis, and thus it can suppress pain and fever, and also inhibit blood clotting.

11 Mevalonic acid is the first metabolite committed solely to cholesterol synthesis. Structural analogues of mevalonic acid have been found to inhibit HMG-CoA reductase, which is the enzyme responsible for mevalonate production. The drugs act in the body as transition-state analogues of HMG-CoA reductase, which they inhibit.

Chapter 11

1   The brain cannot use fatty acids; it must have glucose. So must red blood cells, which have no mitochondria and can generate energy only from glycolysis.

2   The substrate of pyruvate kinase is the enol form of pyruvate, but the keto–enol equilibrium is overwhelmingly to the keto form and hence the enzyme has no substrate. The solution lies in a metabolic route in which two high-energy phosphate groups are expended.

$$Pyruvate + ATP + HCO_3^- \rightarrow Oxaloacetate + ADP + P_i$$
$$(Pyruvate\ carboxylase)$$
$$Oxaloacetate + GTP + H_2O \rightarrow PEP + GDP + CO_2$$
$$(PEP\text{-}CK)$$

3   Fructose-1:6-bisphosphate, producing fructose-6-phosphate and glucose-6-phosphatase.

4   No. Only the liver and kidney produce free glucose.

5   In normal nutritional situations (non-starvation) strenuous muscular activity can generate lactate by anaerobic glycolysis. This travels in the blood to the liver where it is converted back to glucose. Release of glucose into the blood and its uptake by muscle completes the cycle. See Fig. 11.4.

6   Glycerol kinase is required to convert glycerol to glucose by the route shown in Fig. 11.5. Glycerol release occurs in starvation where a prime concern is to produce blood glucose. Since only the liver can do this it makes sense for the glycerol to travel to the liver rather than being metabolized by adipose cells that cannot release blood glucose.

7   Via the glyoxalate cycle shown in Fig. 11.6. The principle is that the two decarboxylating reactions of the citric acid cycle are bypassed.

....

Chapter 12

1   One, by allosteric control; two, by covalent modification of the enzyme, the chief mechanism for this being phosphorylation.

2

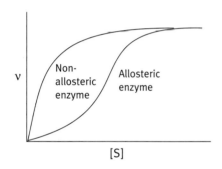

3   Allosteric effectors (with a few exceptions in which v_{max} is changed) usually work by changing the affinity of an enzyme for its substrate. This requires that the substrate concentrations for such enzymes are sub-saturating, which is usually the case. A positive allosteric effector moves the sigmoid substrate–velocity curve to the left and a negative effector to the right (Fig. 12.5). A sigmoid relationship amplifies the effect of such changes on the reaction velocity and so increases the sensitivity of control. This can be seen in Fig. 12.4.

4   They would have no effect since the effectors change the affinity of the enzyme for its substrate. At saturating concentrations this cannot be used to increase the rate of the reaction.

5   The concerted and sequential mechanisms as illustrated in Figs 12.7 and 12.8.

6   It is that an allosteric effector need have no structural relationship to the substrate of the enzyme. This means that completely distinct metabolic systems can interact in a regulatory fashion.

7   Intrinsic regulation is usually allosteric and can apply to a single cell. It keeps the metabolic pathways in balance. However, it cannot determine the overall direction of metabolism of a cell—whether, for example, it will store glycogen and fat or release these. These are determined by extrinsic controls—hormones, etc.—which direct the activities of cells to be in harmony with the physiological needs of the body.

8   See Fig. 12.9. AMP activates glycogen phosphorylase and phosphofructokinase while ATP inhibits the latter. The salient feature is that high ATP/ADP ratios stop glycolysis while a lower ATP level (resulting in increased AMP) speeds it up. High citrate levels also logically slow the feeding of metabolite via glycolysis into the citric acid cycle. High acetyl-CoA levels may be an indication that oxaloacetate is low—hence activation of pyruvate carboxylase to perform its anaplerotic reaction. At the same time high acetyl-CoA levels indicate that the supply of pyruvate by glycolysis is adequate and inhibition of glycolysis at the PEP step is therefore logical.

9   Direct allosteric controls, controls on a kinase that phosphorylates and inactivates PDH, and controls on protein phosphatase that reverses the phosphorylation (Fig. 12.10).

10 As in Fig. 12.11.

11 High blood glucose levels stimulate insulin release; low glucose levels stimulate glucagon release.

12 A hormone such as epinephrine is a first messenger; in combining with a cell receptor it causes an increase in a second molecule that exerts metabolic effects. This is a second messenger. For the two hormones mentioned, this is cAMP. It allosterically activates a protein kinase PKA whose activity has diverse metabolic effects.

13 By mobilizing glucose transporters into a functional position in the cell membrane (Fig. 12.7).

14 It causes a cascade of activation, starting with PKA activation as shown in the scheme on page 173.

15 The differences are summarized in Fig. 12.22.

16 Different cells have receptors for different hormones. Thus, cell A has a receptor for hormone X; cAMP in cell A has effects appropriate to hormone X. Cell B does not have a receptor for hormone X but does for hormone Y. cAMP in cell B elicits responses appropriate to hormone Y but not those appropriate to hormone X.

17 Fructose 2:6-bisphosphate. cAMP increases its level. See Fig. 12.25.

18 Gluconeogenesis is switched on by glucagon whose second messenger is cAMP. The latter activates a kinase that phosphorylates and inactivates pyruvate kinase. In muscle, epinephrine produces cAMP as second messenger. The aim

of epinephrine is to maximize glycolysis so that inactivation of pyruvate kinase here would be inappropriate.

19 cAMP activates a hormone-sensitive lipase that hydrolyses TAG.

Chapter 13

1 It supplies ribose-5-phosphate for nucleotide synthesis; it can supply NADPH for fat synthesis; it provides for the metabolism of pentose sugars.

2 Glucose-6-phosphate is converted to ribose-5-phosphate $+ CO_2$ and $NADP^+$ is reduced. The reactions are given in Fig. 13.1.

3 Transaldolase and transketolase are the main ones whose reactions are given in Fig. 13.2. (Other enzymes of the glycolytic pathway may also participate.)

4 Part of the ribose-5-phosphate is converted to xylulose-5-phosphate, a ketose sugar (since transaldolase and transketolase must have a ketose sugar as donor). The following manipulations now occur.

1. $2 C_5 \rightarrow C_3 + C_7$ (transketolase)
2. $C_7 + C_3 \rightarrow C_4 + C_6$ (transaldolase)
3. $C_5 + C_4 \rightarrow C_3 + C_6$ (transketolase)

The $2C_5$ in reaction 1 are ribose-5-phosphate and xylulose-5-phosphate; the final C_3 compound is glyceraldehyde-3-phosphate, convertible to glucose-6-phosphate with loss of P_i. The net effect is that six molecules of ribose-5-phosphate are converted to five molecules of glucose-6-phosphate $+ P_i$. Thus the cell can produce NADPH with no net increase in ribose-5-phosphate.

5 The oxidative part of the pathway converts glucose-6-phosphate to ribose-5-phosphate and CO_2. If you consider that six molecules of the former, and the six molecules of the latter are recycled to 5 glucose-6-phosphate, then, on a balance sheet, six molecules of glucose-6-phosphate have produced six molecules of CO_2 which, on paper, looks like the oxidation of a glucose molecule. It isn't; one molecule of glucose-6-phosphate is not converted to six molecules of CO_2.

6 NADPH is required to reduce glutathione, a molecule necessary for the protection of the red blood cell. Patients lacking the enzyme are sensitive to the anti-malarial drug, pamaquine, resulting in a hemolytic anemia.

Chapter 14

1 Light reactions involve the splitting of water using light energy, and the reduction of $NADP^+$ to NADPH. Dark reactions mean the utilization of that NADPH to reduce

CO_2 and water to carbohydrate. The term 'dark' implies that light is not essential, not that it occurs only in the dark. In fact, dark reactions will occur maximally in bright sunlight.

2 Photosystems contain many chlorophyll molecules. When excited by a photon, one of the electrons in the latter is excited to a higher energy level. Resonance energy transfer permits this excitation to jump from one molecule to another until it becomes trapped in special reaction centre molecules. The excitation is insufficient for resonance energy transfer but sufficient for an electron to enter the electron transport chain shown in Fig. 14.4.

3 (a) Electrons passing along the carriers of Photosystem II result in ATP synthesis by the chemiosmotic mechanism as they traverse the cytochrome *bf* complex (see Fig. 14.6).

(b) When all of the $NADP^+$ is reduced, electrons passing along the carriers of photosystem I are diverted to the cytochrome *bf* complex and so generate more ATP (see Fig. 14.7).

4 It is the chlorophyll $P680^+$, that is, the reaction centre pigment of PSII that has been excited by resonance energy transfer from antenna chlorophyll molecules to donate an electron to pheophytin, the first component of the PSII electron transport chain. $P680^+$ is an electron short and has a strong tendency to accept an electron, that is, it is a strong oxidizing agent.

5 Thylakoids are formed by invagination of the inner chloroplast membrane (cf. inner mitochondrial membrane), which explains the apparent opposite orientation of proton pumping.

6 The enzyme Rubisco (ribulose-1:5-bisphosphate carboxylase) splits ribulose-1:5-bisphosphate into two molecules of 3-phosphoglycerate, a molecule of CO_2 being fixed in the process (See reaction on page 192.)

7 As in Fig. 14.10.

8 The sequence is given in Fig. 14.9.

9 If CO_2 is fixed first into 3-phosphoglycerate by Rubisco, the plant is known as a C_3 plant. These are subject to full competition between oxygen and CO_2 in the Rubisco reaction. However, in high temperature—high sunlight areas, in C_4 plants CO_2 is first fixed into oxaloacetate by pyruvate-P_i dikinase and PEP carboxylase (see Fig. 14.11). This is reduced to malate which is transported into the underlying bundle sheath cell where the Calvin cycle occurs. Decarboxylation of the malate occurs so that the ratio of CO_2/O_2 in the cell is greatly increased—the concentration of CO_2 in bundle sheath cells may be raised 10–60-fold by this means. The whole scheme is given in Fig. 14.11. Variations in detail of the scheme given occur in different C_4 plants but all are devoted to the same basic strategy.

10 In animals this cannot happen directly—pyruvate kinase cannot form PEP from pyruvate. However, the enzyme in plants is pyruvate-P_i dikinase in which two phosphoryl groups of ATP are used, making the reaction thermo-dynamically feasible.

Chapter 15

1 Oxidation results in the formation of a Schiff base, hydrolysable by water.

$$\underset{\displaystyle |}{\overset{\displaystyle |}{CHNH_2}} \xrightarrow{-2H} \underset{\displaystyle |}{\overset{\displaystyle |}{C}}{=}NH \xrightarrow{H_2O} \underset{\displaystyle |}{\overset{\displaystyle |}{C}}{=}O + NH_3$$

2 Glutamic acid.

3 Transdeamination is the most prevalent mechanism. The amino group of many amino acids are transferred to α-ketoglutarate forming glutamate. The latter is deaminated by glutamate dehydrogenase. As an example,

  1. Alanine + α-ketoglutarate → pyruvate + glutamate;
  2. Glutamate + NAD^+ + H_2O →
  α-ketoglutarate + NADH + NH_4^+
  Net reaction: Alanine + NAD^+ + H_2O →
  pyruvate + NADH + NH_4^+.

4 Pyridoxal phosphate (Fig. 15.2). The mechanism of trans-amination is given in Fig. 15.3.

5 By removal of H_2O and H_2S, respectively, as illustrated in Fig. 15.4

6 A glucogenic amino acid is one that, after deamination, can give rise to pyruvate (or phosphoenolpyruvate). This may be indirect—any acid of the citric acid cycle is glucogenic. A ketogenic amino acid is one that cannot give rise to the above but rather gives rise to acetyl-CoA. Only leucine and lysine are purely ketogenic but some, such as phenylalanine, are mixed. Ketogenic amino acids produce ketone bodies only in circumstances appropriate to this such as starvation. Otherwise the acetyl-CoA is oxidised normally.

7 Phenylalanine is not normally transaminated; it is converted to tyrosine and then metabolized (see Fig. 15.5). If the phenylalanine conversion to tyrosine is defective, phenyla-lanine does transaminate, producing phenylpyruvate, which causes irreparable brain damage to babies and early death.

8 To supply the reducing equivalent for the formation of H_2O from one atom of the oxygen molecule used in the hydroxylation reaction.

9 By formation of 5-adenosylmethionine ('SAM'); this has a sulfonium ion structure conferring a strong leaving tendency on the methyl group. The formation of SAM is illustrated in Fig. 15.6

10 This is shown in Fig. 15.8.

11 Both situations call for high rates of deamination of amino acids (in starvation muscle proteins break down to allow glucose synthesis). The amino nitrogen must be converted to urea.

12 (a) Ammonia is converted to glutamine, which is trans-ported to the liver and hydrolysed.

Glutamate + ammonia $\xrightarrow[\text{ATP} \quad \text{ADP} + P_i]{}$ Glutamine

Glutamine + H_2O $\longrightarrow$ Glutamate + ammonia .

  (b) Amino nitrogen is transported from muscle as alanine. The alanine cycle is shown in Fig. 15.9.

Chapter 16

1 To destroy unwanted molecules and structures imported into the cell by endocytosis and to destroy components of the cell destined for destruction.

2 They are vesicles produced by the Golgi sacs containing an array of hydrolytic enzymes. The enzymes are acid hydrolases with an optimum pH of 4.5–5.0. Proton pumps in the membrane maintain this pH inside the vesicle.

3 The target object is inside either an endocytotic vesicle or an autophagosome. The primary lysosome fuses with these forming a single vesicle, a secondary lysosome, in which digestion occurs.

4 A large variety of genetically determined lysosomal storage disorders exist in which the absence of a specific hydrolytic enzyme causes the lysosomes to become overloaded with material normally disposed of.

5 In Pompe's disease, a fatal genetically determined one, there is a lack of a lysosomal α-1:4-glycosidase that normally degrades glycogen. The lysosomes become overloaded with glycogen. However, it is not clear why disposal of glycogen by this means should be needed. It is not part of the usual accounts of glycogen metabolism.

6 These are membrane-bounded vesicles in the cytosol that contain oxidases. These enzymes oxidize a variety of substrates, using oxygen, and generate H_2O_2.

  $R'H_2 + O_2 \rightarrow R' + H_2O_2$.

The peroxide is used to oxidize other substrates.

$$RH_2 + H_2O_2 \rightarrow R + 2H_2O$$

Substrates include those not metabolized elsewhere, for example, very long chain fatty acids are shortened. Oxidation of the cholesterol side chain to form bile salts is believed to occur here. The essential role of peroxisomes is underlined by a fatal genetic disease in which some tissues lack peroxisomes.

Chapter 17

1   The initial stimulus for blood clotting is, in quantitative terms, very small. To obtain a sufficiently rapid response amplification is needed as cascades are the biological method of amplification. The final enzyme activation is that of prothrombin to thrombin, an active proteolytic enzyme.

2   Fibrinogen monomer proteins are prevented from spontaneous polymerization by negatively charged fibrinopeptides (Fig. 17.2) that cause mutual repulsion. Removal of these by thrombin permits the association of fibrin monomers, illustrated in Fig. 17.3.

3   Covalent crosslinks are formed between monomers in the polymer by an enzymic transamidation process between glutamine and lysine side chains,

$$-CONH_2 + H_3N^+ - \rightarrow -CO-NH- + NH_4^+.$$

4   In the conversion of prothrombin to thrombin, a glutamic acid side chain is carboxylated; the carboxyglutamate binds Ca^{2+}. Vitamin K is a cofactor in the carboxylation reaction.

Other conversions in the proteolytic cascade involve this conversion.

5   Is is involved in the conversion of hydrophobic xenobiotics to soluble ones. A typical reaction is

$$AH + O_2 + NADPH + H^+ \rightarrow AOH + H_2O + NAD^+.$$

6   To reduce one oxygen atom to H_2. It is known as a mixed function oxygenase.

7   The glucuronyl group is donated to an $-OH$ on the foreign molecule, rendering it much more polar (see Fig. 17.5).

8   It involves an ATP-driven membrane transport system that removes a variety of compounds from cells. It transports steroids in steroid-secreting adrenal cortical cells, but all transported compounds (including anticancer drugs used in chemotherapy) are lipid-soluble amphipathic compounds, indicating a more general role.

9   Neutrophils secrete elastase in the mucous lining of lung tissue that destroys elastin, the elastic structural material of lungs. This, unchecked, converts the minute alveoli into much larger structures with a greatly reduced surface area for gas exchange. α_1-Antitrypsin in the blood diffuses into the lung and keeps elastase in check and thus prevents damage. Smoking has two effects: (a) it inactivates α_1-antitrypsin by converting a crucial methionine side chain to a sulfoxide ($S \rightarrow S=O$); (2) by irritating the lungs it attracts neutrophils resulting in more elastase being liberated.

10   It is an oxygen molecule that has acquired an extra electron.

$$O_2 + e^- \rightarrow O_2^-$$

Superoxide is extremely reactive, causing damage.

11   Most eukaryote cells have superoxide dismutase and catalase

$$2O_2^- + 2H^+ \rightarrow H_2O_2 + O_2 \text{ (dismutase)};$$

$$2H_2O_2 \rightarrow 2H_2O + O_2 \text{ (catalase)}.$$

Secondly, there are antioxidants such as ascorbic acid and vitamin E. Superoxide is dangerous because it attacks a molecule and generates another free radical, causing a self-perpetuating chain of reactions. Antioxidants are quenching agents. When attacked by superoxide, the free radical produced by these is insufficiently reactive to perpetuate the chain of reactions.

Chapter 18

1   PRPP or 5′-phosphoribosyl-1-pyrophosphate is the universal agent. Its formation from ribose-5-phosphate and ATP and its mode of action are given on page 225.

2   Tetrahydrofolate (FH_4).

Formyl-FH_4 =

3 Serine. Serine hydroxymethylase transfers $-CH_2$ OH to FH_4 leaving glycine and forming N^5,N^{10}-methylene FH_4. This is oxidized to N^5,N^{10}-methylene FH_4 by an $NADP^+$ requiring reaction. Hydrolysis of this produces formyl-FH_4. The reactions are shown on page 228.

4 It ribotidizes guanine and hypoxanthine in the purine salvage pathway.

5 HGPRT is missing so that purine salvage cannot occur. However, the brain has the direct pathway and it may be that overproduction of purine nucleotides *de novo* occurs due to increased levels of PRPP (because it is not used by salvage). In such patients, excess uric acid is formed but prevention of this by allopurinol does not relieve the neurological symptoms. Nor do gout patients suffer the latter, so these symptoms are unexplained.

6 It is converted to alloxanthine which inhibits xanthine oxidase. The conversion is due to xanthine oxidase itself—referred to as suicide inhibition.

7 As in Fig. 18.8.

8 No. Thymidylate synthetase transfers a C_1 group from methylene FH_4 to dUMP and at the same time reduces it to a methyl group. The H atoms for this are taken from FH_4 so that FH_2 is a product as shown on page 226.

9 It inhibits reduction of FH_2, generated by thymidylate synthase, and so prevents dTMP production essential for cell multiplication. FH_4 is essential for the thymidylate synthase reaction.

10 Vitamin B_{12} is required for the methylation of homocystene to methionine, the methyl donor being methyl FH_4. Lack of B_{12} causes FH_4 to be 'trapped' as the methyl compound and unavailable for the other folate-dependent reactions.

Chapter 19

1 This is shown on page 241.

2 Genetic material needs to be as chemically stable as possible. DNA is more stable than RNA. This is because, as shown on page 241, the $2'$-OH of RNA can make a nucleophilic attack on the phosphodiester bond making RNA less chemically stable than DNA.

3 The phosphodiester bond is longer than the thickness of a base; a straight chain structure of DNA would leave hydrophobic faces of the bases exposed to water. Because of hydrophobic forces the bases are collapsed together by sloping the phosphodiester bond as shown in Fig. 19.3(a).

4 B DNA; right-handed; 10 base pairs.

5 One chain runs $5' \rightarrow 3'$ in one direction and the other $5' \rightarrow 3'$ in the other direction. Thus in a linear DNA

molecule each end has a $5'$ end of one chain and a $3'$ end of another chain.

6 A $5' \rightarrow 3'$ direction means that you are moving from a terminal $5'$-OH group to a $3'$-OH group of the polynucleotide chain.

7 5′ CATAGCCG 3′
3′ GTATCGGC 5′

Watson–Crick base pairing explains the complementary sequence. By convention a single sequence is written with the $5'$ end to the left.

8 (a) The DNA of a eukaryote chromosome is wound around an octamer of histone proteins, two turns per nucleosome occupying 146 base pairs. Successive nucleosomes are connected by linker DNA.

(b) When chromatin is digested with DNase, the 146 base-pair sections of DNA are protected; this observation led to the discovery of nucleosomes (see Fig. 19.7).

9 Repetitive DNA; an Alu sequence is a few hundred bases long but repeated almost exactly hundreds of thousands of times; scattered throughout the human chromosome.

Chapter 20

1 A section of DNA whose replication is initiated at a single origin of replication.

2 Ahead of the replicative fork positive supercoils occur.

3 In *E. coli*, gyrase, a topoisomerase II, introduces negative supercoils. In eukaryotes, a topoisomerase I relaxes positive supercoils.

4 As shown in Figs 20.6 and 20.7.

5 In winding DNA around a nucleosome, a local negative supercoil is introduced, but, since no bonds are broken, there cannot be a net negative supercoiling. The local negative supercoil is compensated for by a local positive supercoiling elsewhere. This is relaxed by topoisomerase I, leaving a net negative supercoiling in the DNA.

6 dATP, dGTP, dCTP, and dTTP.

7 Cytosine readily deaminates to uracil; if uracil were a normal constituent of DNA it would be impossible to recognize and correct the mutation since A–U pairing is the same as A–T pairing.

8 (a) No. A primer is required. DNA polymerase cannot initiate new chains.

(b) Synthesis proceeds in the $5' \rightarrow 3'$ direction; by this is meant that the new chain is elongated in the $5' \rightarrow 3'$ direction, new nucleotides being added to the free $3'$OH of the preceding nucleotide. It does not refer to the template strand, which runs antiparallel.

9   Hydrolysis of inorganic pyrophosphate; base pairing of the nucleotides.

10   It means that once the enzyme is attached to the template DNA it goes on synthesizing the new chain for very long distances without detaching; this is achieved by a 'ring clamp' or annulus of proteins that assembles behind the polymerase, the DNA chains sliding through it (see Fig. 20.14).

11   Polymerase I is the enzyme that processes Okazaki fragments into a continuous lagging strand. Its activities are summarized in Fig. 20.16 and described on page 259.

12   Correct base pairing of the incoming nucleotide triphosphate with the template nucleotide is the primary essential. The enzyme also has a proofreading activity. It has a $3' \rightarrow 5'$ exonuclease activity that removes the last incorporated nucleotide if this is improperly base paired.

13   The methyl-directed mismatch repair system is described in Fig. 20.19. There is evidence that proteins similar to *E. coli* proteins exist in humans and that, where these are deficient, there is an increased risk of cancer.

14   Thymine dimers are formed when DNA is subjected to UV irradiation. Two adjacent thymine bases become covalently linked. Repair can either be direct by a light-dependent repair system that disrupts the bonds and restores the separate bases; or it can be subject to excision repair (Fig. 20.20).

15   As shown in Fig. 20.22, removal of the $3'$ Okazaki fragment primer leaves an unreplicated section.

16   The answer lies in telomeric DNA synthesized as described in Fig. 20.23.

Chapter 21

1   RNA polymerase uses ATP, CTP, GTP, and UTP as substrates (cf. dATP, dCTP, dGTP, and dTTP). Most importantly, RNA polymerase can initiate new chains; DNA polymerase requires a primer. RNA synthesis never includes proofreading.

2   This is illustrated in Fig. 21.4. In more detail the promoter has a -10 (Pribnow) box and a -35 box.

3   DNA polymerase attaches nonspecifically to the DNA, but, when joined by a sigma factor molecule, it binds firmly to a promoter. The Pribnow box and the -35 box position orientate the polymerase correctly. The enzyme synthesizes a few phosphodiester bonds and then the sigma protein flies off and the polymerase progresses down the gene transcribing mRNA.

4   One is by the G–C stem loop structure shown in Fig. 21.7. In this, internal $G{\equiv}C$ base pairing in the mRNA prevents

bonding of the mRNA to the template DNA strand. Following this, a string of uracil nucleotides further weakens the attachment (only two hydrogen bonds in a $U{=}A$ pair) facilitating detachment. The second method depends on the Rho factor, a helicase that unwinds the mRNA–DNA hybrid. At the termination site, the polymerase pauses (possibly due to a $G{\equiv}C$ rich region in the DNA and the Rho factor catches up and detaches the mRNA.

5   The precise base sequence of the -10 and -35 boxes, their distance apart, and the bases in the $+1$ to $+10$ region.

6   This is covered in Fig. 21.10.

7   The primary transcript has introns that must be spliced out; the $5'$ end is capped; the mechanism of termination is unknown; the mRNA is, with few exceptions, polyadenylated at the $3'$ end.

8   The mechanism is shown in Fig. 21.13. It is based on a consensus sequence that determines where splicing occurs and on a transesterification reaction involving little change in free energy. Split genes may have facilitated evolution by promoting exon shuffling. Differential splicing can also result in a single gene giving rise to different proteins.

9   The eukaryote polymerase III does not attach to the DNA directly but rather attaches to a protein complex that assembles on the DNA (Fig. 21.19). In addition, any number of transcriptional factors may be associated with the complex (Fig. 21.20).

10   Helix–turn–helix proteins, leucine zipper proteins, zinc fingers proteins, and homeodomain proteins.

Chapter 22

1   If only 20 codons were used, there would be 44 not coding for any amino acid. Any mutation in a gene coding region would then be highly likely to inactivate the gene by prematurely introducing a stop codon. Instead, by using 61 codons, a base-change would either cause no change or would substitute a different amino acid. Because of the arrangement of the genetic code, many of these substitutions would be conservative and cause minimal change to the change to the protein structure.

2   The answer lies in the wobble mechanism described on page 295.

3   The convention is that RNA is shown with the $5'$ end to the left. When mRNA is shown in this manner, a tRNA anticodon is base-paired in an antiparallel fashion. The tRNA molecule is therefore shown with the $5'$ end to the right.

4   In the case of some amino acids, at the stage of attaching the amino acid to tRNA. In all cases at the stage of elongation; a

pause in GTP hydrolysis on the EF-Tu gives time for unpaired aminoacyl tRNAs to leave the ribosome.

5 In general GTP hydrolysis to GDP and P_i is believed to result in conformational changes in the relevant proteins. GTP is involved in assemblage of the *E. coli* initiation complex. It is involved in the delivery of aminoacyl-tRNAs to the ribosome by EF-Tu. It is involved in the translocation step.

6 Figure 22.11 is necessary for this explanation. The straddling has the advantage that in moving from one site to the other the tRNA is never completely detached. It also means that the peptidyl group does not have to physically move relative to the ribosome. The two possible mechanisms are shown in the figure. Model II would explain why ribosomes have two subunits.

7 Figure 22.13 explains this. The scanning mechanism of selecting the start AUG means that there can be only one start site. As illustrated in Fig. 22.9, prokaryote initiation permits multiple start sites.

8 They bind to nascent polypeptides and prevent premature, improper folding associations. Appropriate release of chaperones facilitates correct folding though the process is far from fully understood. Other roles are summarized on page 306.

9 The prion diseases described on page 305.

10 This is given in Fig. 22.16.

Chapter 23

1 By exploiting the normal receptor-mediated endocytosis mechanism as illustrated in Fig. 23.1.

2 The latter must carry into the cell an RNA replicase; the RNA of the (+) virus is an mRNA which codes for an RNA replicase.

3 The vaccinia replicates in the cytoplasm; the host polymerase is in the nucleus.

4 The retroviral RNA is copied into double-stranded DNA which integrates into the host chromosome as shown in Fig. 23.5.

5 The mechanism is explained in Fig. 23.3.

6 Antibody protection against influenza is directed at the hemagglutinin protein. This is continually mutated but, because there are many epitopes on the molecule, the immune protection loss is partial and gradual. This antigenic drift leaves people susceptible to infection but residual protection means that it is mild. If by a recombination between different strains a totally new hemagglutinin is present (antigenic shift), a lethal pandemic can result.

7 The virus has, on its surface, proteins that specifically bind to neuraminic acid (sialic acid). The red blood cell glycophorin terminates in this acid. On mixing, the multiple binding sites on the virus crosslink the red blood cells.

8 The virus binds to its host cell by a glycoprotein terminating in neuraminic acid. Why should the virus have an enzyme for destroying the very receptor which it needs to gain entry to the cell? The answer may be (a) to facilitate release of new virtus particles that might adhere to the cell glycoproteins; or (b) to liquefy mucins that have neuraminic acid in their structure and so facilitate the virus reaching the cell surface. Mucin liquefaction might also help spread the virus through sneezing.

9 It may enter the lytic or lysogenic phase as illustrated in Fig. 23.7.

10 (a) Small, infectious, naked RNA molecules without any protein or other type of coat.

(b) No. There are no sufficiently long coding stretches of bases.

Chapter 24

1 Pancreatic DNase randomly cuts DNA. A restriction enzyme cuts only at specific short sequences of bases.

2 The adenine bases in all of the relevant hexamer sequences of *E. coli* strain R DNA are methylated so that the enzyme does not recognize it. Invading foreign DNA will not be so protected.

3 It refers to ends resulting from a staggered cut made by many restriction enzymes.

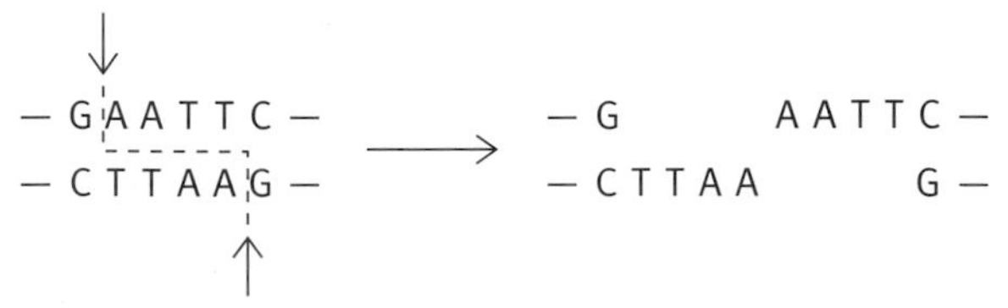

```
      ↓
— G A A T T C —           — G          A A T T C —
— C T T A A G —   ⟶       — C T T A A          G —
          ↑
```

The overhanging ends will automatically base pair and 'stick' together.

4 A genomic clone refers to a cloned section of DNA identical to the sequence of DNA in a chromosome. A cDNA clone means complementary DNA. It refers to cloned DNA identical to the mRNA for a gene. The cDNA lacks introns; genomic clones have them.

5 These are outlined in Fig. 24.1.

6 These are outlined in Fig. 24.2.

7 The nucleotide lacks a 3′-OH and therefore, when added by DNA polymerase to a growing DNA chain, terminates that

CHAPTER 26 **419**

chain. The application of this to sequencing is described on page 330.

8 A specific section of DNA on a chromosome can be amplified logarithmically. It involves copying the section of DNA and copying the copies *ad infinitum*. Its importance is that a minute amount of DNA, too small for any studies, can be amplified at will. The essential, apart from enzymes and substrates, is that you must have primers for copying the section in both directions and this means knowing the base sequences at either end of the piece to be amplified.

9 It is an engineered plasmid with a convenient insertion site for, say, a cDNA clone for a specific protein, but which also contains appropriate bacterial DNA transcriptional signals (a promoter) and translational signals. The DNA is transcribed and the mRNA translated in a bacterial cell. It can be used to produce large quantities of specific proteins.

10 It is a technique for detecting gene abnormalities. If DNA is cut by a pattern of restriction enzymes and the resultant pieces separated on an electrophoretic gel, a pattern of bands can be visualized by hybridization methods. Mutations may produce or remove restriction sites so that the pattern may be altered.

Chapter 25

1 As in Fig. 25.3.

2 The generation of the diversity of immunoglobulin genes is described in Fig. 25.4.

3 B cells produce antibodies (after activation and maturation into plasma cells). Helper T cells are (in most cases) required for B cells to do this. Cytotoxic or killer T cells bind to host cells displaying a foreign antigen and kill them by perforating their membrane or inducing apoptosis.

4 It is a type of phagocyte that engulfs a foreign antigen, processes it into pieces, and displays it in combination with its MHC class II molecules. If an inactive helper T cell combines with the antigen–MHC complex it is activated (see Fig. 25.4).

5 In both cases MHC class I.

6 Each B cell produces a different antibody. For any given antigen there will be very few B cells specific for it, but, when the antigen binds to the displayed antibody, the B cell proliferates into a clone. Thus the antigen automatically selects which B cells are to proliferate.

7 During *primary* maturation of B cells in the bone marrow or of T cells in the thymus, if an antigen binds, the cell is eliminated, the principle being that such antigens will be 'self'. After release from the bone marrow or thymus, the binding to an antigen activates the cells.

8 Differential splicing of introns; at the 3′ end of the immunoglobulin gene is an exon that codes for an anchoring polypeptide sequence. At the onset of secretion a switch in the splicing eliminates this.

9 The onset of secretion is accompanied by rapid somatic mutation in the cells, which modifies the variable site. Since binding of the antigen causes cell proliferation this constitutes a selection mechanism for those cells producing a 'better' antibody.

10 CD4 is present on helper T cells. The protein binds to a constant protein component of MHC class II molecules and thus confines interactions to B cells, helper T cells, and antigen-presenting cells, which are all of this class. Cytotoxic T cells have CD8 which binds to MHC class I molecules; this restricts the interaction of the killer T cells to host cells. The CD4 protein is the receptor by which the AIDS virus infects helper T cells.

Chapter 26

1 Endocrine hormones, growth factors and neurotransmitters.

2 This is summarized in Fig. 26.1.

3 The production is described in Fig. 26.11. GTP hydrolysis has a timing function. In cholera the GTPase is inactivated so that cAMP production remains activated once stimulated.

4 It allosterically activates protein kinase A (PKA). This can have metabolic effects depending on phosphorylation of key enzymes but, in addition, there are genes with cAMP-responsive elements (CREs). Inactive transcriptional factors can be activated by phosphorylation by PKA. Therefore cAMP can have extensive gene control effects.

5 It activates a cytoplasmic guanylate cyclase that produces cGMP. The latter is a second messenger in some systems.

6 The phosphatidylinositol cascade involves the receptor-mediated activation of membrane-bound phospholipase C. This releases inositol trisphosphate (IP_3) and diacylglycerol (DAG). The former increases cytoplasmic Ca^{2+}; the latter activates a protein kinase PKC. Thus, IP_3 and DAG are second messengers. The scheme is shown in Fig. 26.16.

7 The Ras pathway has receptors of the tyrosine kinase type. It is summarized in Fig. 26.18.

8 To associate with a phosphorylated tyrosine kinase type of receptor (see Fig. 26.21) and activate cellular control pathways.

9 If a receptor or a component of, for example, the Ras pathway is abnormally permanently active, the resulting overactivation of gene transcriptional factors may contribute to uncontrolled cell division (see Fig. 26.22).

10 As shown in Fig. 26.22.

11 (a) Involves an allosteric change of the cytoplasmic domain on binding of the hormone.

(b) Involves receptor dimerization and activation of self-phosphorylation of tyrosine residues by an intrinsic kinase.

(c) Involves receptor dimerization and association with a separate tyrosine kinase that phosphorylates the receptor (see Figs 26.6, 26.17, and 26.22).

12 It is a channel in, for example, the membrane of a nerve synapse that liberates acetylcholine, for example, to trigger muscle contraction. The channel opens when a wave of neuronal membrane depolarization reaches it. The inrush of Ca^{2+} causes liberation of acetylcholine (Fig. 26.24).

13 It is a channel that opens when a specific ligand binds to it. In vision, the cation channels of the rod cell membrane are opened by cGMP. Light causes the reduction of cGMP concentration, causing closure of the channels. The resultant hyperpolarization of the rod cell membrane is translated into an optic nerve impulse.

Chapter 27

1 5-Aminolevulinate synthase (ALA) carries out the reaction shown in Fig. 27.4 and ALA dehydratase produces the pyrrole, porphobilinogen, by the reaction in Fig. 27.6.

2 Increased activity in liver is associated with acute intermittent porphyria (the 'mad king' disease) though the connection between ALA production and the neurological symptoms is not understood.

3 The mechanism is shown in Fig. 27.8.

4 As shown in Fig. 27.11, myoglobin has the higher affinity and the curve is hyperbolic as compared with the sigmoid curve of hemoglobin.

5 According to the concerted model, the hemoglobin exists in a low-affinity state and a high-affinity state. Binding of oxygen displaces the equilibrium between the two to increase the amount in the high-affinity state (see page 164).

6 On binding of oxygen to the heme iron, Fe moves into the plane of the porphyrin ring. This requires the slightly domed tetrapyrrole to flatten. The Fe is attached to the F8 histidine of the protein and the molecule rearranges itself. This allows the tetrapyrrole to flatten. The movement affects subunit interactions causing the tetramer to change its conformation from the T to the R state (see Fig. 27.13).

7 Adult hemoglobin in the deoxygenated state has a cavity that binds 2:3-bisphosphoglycerate (BPG) by positive charges on the protein side chains. Binding of BPG is possible only in the deoxygenated state and therefore binding increases unloading of oxygen—it reduces affinity of the hemoglobin for oxygen. Fetal hemoglobin has a γ subunit instead of the adult β one. It lacks one of the BPG charge-binding groups. BPG therefore binds to fetal hemoglobin less tightly, thus raising its affinity for oxygen above that of the maternal hemoglobin.

8 It is necessary for the major transport of CO_2 as HCO_3^- as illustrated in Fig. 27.15.

Chapter 28

1 Fast twitch fibres rely mainly on glycolysis for immediate contraction since this can be rapidly activated. However, they quickly exhaust. The white muscle of fish exemplify this—this muscle is for escape reactions. Fast twitch fibres exist in humans in the eye muscles. Slow twitch fibres obtain their ATP from oxidative metabolism and can function for much longer without exhaustion but the energy supply cannot be increased as rapidly as in slow fibres. In man, muscles have both types. Back muscles, which maintain body posture for long periods, are mainly of the slow twitch type.

2 These are illustrated in Figs 28.1 and 28.2.

3 The answer is summarized in Fig. 28.7.

4 Acetylcholine stimulation of the muscle receptor leads to liberation of Ca^{2+} from the sarcoplasmic reticulum into the myofibril. On the actin filaments are tropomyosin molecules and, in turn, these have a Ca^{2+}-sensitive troponin complex attached to them. When Ca^{2+} binds, a conformational change occurs; it was believed that the tropomyosin blocked the attachment of myosin heads to actin. This is now questioned, A Ca^{2+}/ATPase returns the Ca^{2+} to the sarcoplasmic reticulum and terminates the contraction.

5 In smooth muscle myosin heads there is a p-light chain that inhibits the binding of the myosin head to the actin fibre, preventing contraction. Neurological stimulation opens channels allowing Ca^{2+} to enter the cell. The Ca^{2+} combines with calmodulin regulatory protein, which activates a myosin light chain kinase. Phosphorylation of the light chain abolishes its inhibitory effect and contraction occurs.

Chapter 29

1 Contraction occurs in such cells. Actin filaments anchored in the cell membrane provide the means for small bundles of myosin molecules to exert a contractile force. A second role is for actin filaments to form a transport track along which special minimyosin molecules move. The latter have a

myosin head but the rod-like structure is replaced by a short tail to which vesicles may be attached.

2   Hollow tubes formed by the polymerization of tubulin protein subunits. They have a definite polarity with $(+)$ and $(-)$ ends.

3   The microtubule organizing centre protects the $(-)$ end. The $(+)$ ends of *growing* tubules are protected by a tubulin–GTP cap. The GTP is slowly hydrolysed, removing the protection. Unless a new tubulin–GTP molecule is added before this, the microtubule collapses. When the microtubule reaches a 'target' structure it is then protected.

4   They are molecular motors that move along microtubule tracks and can pull a load with them. Kinesin and dynein travel in opposite directions (in terms of microtubule polarity) on the microtubule track.

5   No, they cannot contract. It is believed that the shortening is due to depolymerization but precisely what causes chromosome movement is not certain.

6   Filaments 10 nm in diameter which are intermediate in this respect between microtubule filaments (20 nm) and actin filaments (6 nm).

7   In specialized cases, they form the structural basis of hair. They may be associated with conferring toughness on structures such as neurofilaments and the Z discs of sarcomeres. However, mutant cultured cells survive without them.

Figure acknowledgements

Fig. 1.8. Courtesy of R. Rogers.

Fig. 2.6 (a)–(d). Redrawn from figs 2.10a, 2.15b, 4.1a, and 4.4, respectively, in Branden, C. and Tooze, J. (1991). *Introduction to protein structure.* Garland Publishing, New York. Figs 2.10a and 4.4 by permission of Garland Publishing. Figs 2.15b and 4.4 by permission of J. Richardson.

Fig. 2.7. Redrawn from fig. 2 in Jentoft, N. (1990). *Trends Biochem. Sci.* **15**, 293.

Fig. 3.15. Kindly provided by Professor W. G. Breed, Department of Anatomy, University of Adelaide.

Fig. 6.13. Kindly provided by Dr Trudy Forte, Lawrence Berkeley Laboratory, University of California.

Fig. 9.4. Reproduced from fig. 25.7 in *Biochemistry* by Stryer. (Fourth edn) Copyright (c) 1995 by Lubert Stryer. Used with permission of W. H. Freeman and Company.

Fig. 12.17. Modified from fig. 7 in Karnieli *et al.* (1981). *J. Biol. Chem.* **256**, 4772.

Fig. 19.3 (a), (d). Adapted from figs 2.4a, and 3.1, respectively, in Calladine, C. R. and Drew, H. R. (1994). *Understanding DNA.* Academic Press.

Fig. 19.4. Reproduced from fig. 2.1 in Ptashne, M. (1987). *A genetic switch.* Cell Press and Blackwell Scientific Publications, Oxford. Copyright by Cell Press.

Fig. 19.5. Kindly provided by Sanjay Kumar.

Fig. 19.8. Fig. 28.19 in Lewin, B. (1994). *Genes V.* Oxford University Press, Oxford. (The electron micrograph was provided to Lewin by Barbara Hamkalo.)

Fig. 19.22. Kindly provided by Sanjay Kumar.

Fig. 20.6. Fig. 2.24 in Singer, M. and Berg, P. (1991). *Genes and genomes.* University Science Books, Sausalito, California.

Fig. 20.7. Fig. 2.26a in Singer, M. and Berg, P. (1991). *Genes and genomes.* University Science Books, Sausalito, California.

Fig. 20.14. Fig. 3a from Krishna, T. S. R., Kong, X-P., Gary, S., Burgers, P. M. and Kuriyan, J. (1994). *Cell* **79**, 1233. Copyright by Cell Press. Photograph kindly provided by Dr. John Kuriyan.

Fig. 20.15. Adapted from fig. S11-12 in *DNA Replication,* Supplement by Kornberg, A. Copyright (c) 1982 W. H. Freeman and Company. Used with permission.

Fig. 20.17. Fig. 3 from Echols, H. and Goodman, M. F. (1991). *Ann. Rev. Biochem.* **60**, 490; Reproduced with copyright permission of Annual Reviews Inc. Original from Kennard, O. (1987). *Nucleic Acids and Molecular Biology* **1**, 25.

Fig. 20.24. After fig. 34.1 from Lewin, B. (1994). *Genes V.* Oxford University Press, Oxford.

Fig. 21.16. Reproduced from fig. 29.10 in Lewin, B. (1994). *Genes V.* Oxford University Press, Oxford.

Fig. 21.20 (b). Fig. 3a in Ohlendorf, D. H., Anderson, U. F. and Matthews, B. W. (1983). *J. Molec. Evolution* **19**, 109. Reproduced with copyright permission of Springer-Verlag.

Fig. 21.21 (b). Adapted from Fig 3b in Ellenberger, T. E. *et al.* (1992). *Cell* **71**, 1223. Copyright by Cell Press.

Fig. 21.23. Adapted from fig. 4 in Ross, J. (1995). *Microbiol. Rev.* **59**, 423.

Fig. 21.24. Fig. 2 from Ross, J. (1995). *Microbiol. Rev.* **59**, 423.

Fig. 22.2. Fig. 2 from Holbrook, S. R., Sussman, J. L., Warrant, R. W. and Kim, S. H. (1978). *J. Mol. Biol.,* **123**, 631. Reproduced with copyright permission of Springer-Verlag.

Fig. 22.5. Fig. 9.5 from Lewin, B. (1994). *Genes V.* Oxford University Press, Oxford.

Fig. 22.11. Adapted from fig. 4 of Noller, H. F. (1991). *Ann. Rev. Biochem.* **60**, 191; modified with copyright permission of Annual Reviews, Inc.

Fig. 22.12. From fig. 1 of Noller, H. F. (1991). *Ann. Rev. Biochem.* **60**, 193. Reproduced with copyright permission of Annual Reviews, Inc.

Fig. 23.2. Fig. 1.1 from Ptashne, M. (1986). *A genetic switch* (2nd edn). Cell Press and Blackwell Scientific Publications, Oxford. Copyright by Cell Press.

Fig. 23.8. Courtesy of Professor R. H. Symons, Department of Plant Science, Waite Institute, University of Adelaide.

Fig. 24.6. Photograph courtesy of Dr Chris Hahn.

Fig. 25.1. After Metcalf, D. (1991). *Phil. Trans. Roy. Soc., Lond.* **B333**, 147. Reproduced (modified) by permission of the author and the Royal Society.

Fig. 26.10. After Dohlman, H. G., Caron, M. G., and Lefkowitz, R. J., (1987). *Biochem.* **26**, 2660. Reprinted with copyright permission of the American Chemical Society.

Fig. 27.1. Kindly provided by Professor W. Breed, Department of Anatomy, University of Adelaide.

Fig. 27.10. Parts (a)–(c) are, respectively, from fig. 2.9 b, fig. 2.10a, and fig. 3.7 of Branden, C. and Tooze, J. (1991). *Introduction to protein structure.* Garland Publishing, New York. Fig. 2.9b by permission of A. Lesk. Figs. 2.10a and 3.7 by permission of Garland Publishing.

Fig. 29.2 (a) Courtesy of Dr P. Gunning, Children's Medical Research Institute, Sydney. (b) Fig. 2.13 from Lewin, B. (1994). *Genes V.* Oxford University Press, Oxford. (Photograph provided to Lewin by Frank Solomon).

Fig. 29.6. Fig. 1e from McNiven, M. A., and Ward, J. B. (1988). *J. Cell Biol.* **106**, 111. Reproduced with copyright permission of the Rockefeller University Press.

Index

References to structural formulae are given in **boldface** type